. . . à une époque où des savants éminents, doués de grands talents géométriques, s'efforçaient de ne jamais dévoiler les idées directes et simples qui les avaient guidés et de faire dépendre leurs résultats élégants d'une théorie générale abstraite qui, souvent, ne s'appliquait que dans les cas particuliers en question. La géométrie devenait une étude des équations algébriques, différentielles ou aux dérivées partielles: elle perdait ainsi tout le charme qu'elle doit au fait d'être un art, et presque un art plastique.

Henri Lebesgue
Leçons sur les Constructions Géométriques,
Paris, 1950

GEOMETRY

A COMPREHENSIVE COURSE

Dan Pedoe

Professor Emeritus, Mathematics
University of Minnesota

DOVER PUBLICATIONS, INC.
NEW YORK

To
William, Nadine,
Susan, Simon,
Oliver and Ian

Coypright © 1970 by Cambridge University Press.
All rights reserved under Pan American and International Copyright Conventions.

Published in Canada by General Publishing Company, Ltd., 30 Lesmill Road, Don Mills, Toronto, Ontario.
Published in the United Kingdom by Constable and Company, Ltd.

This Dover edition, first published in 1988, is an unabridged, corrected republication of the work first published by the Cambridge University Press, Cambridge, England, 1970, under the title *A Course of Geometry for Colleges and Universities*.

Manufactured in the United States of America
Dover Publications, Inc., 31 East 2nd Street, Mineola, N.Y. 11501

Library of Congress Cataloging-in-Publication Data

Pedoe, Daniel.
   [Course of geometry for colleges and universities]
   Geometry, a comprehensive course / by Dan Pedoe.
      p.   cm.
  Reprint. Originally published: A course of geometry for colleges and universities. London : Cambridge University Press, 1970.
   Bibliography: p.
   Includes index.
   ISBN 0-486-65812-0 (pbk.)
   1. Geometry. I. Title.
QA445.P43 1988   88-18691
516—dc19   CIP

CONTENTS

Chapter III: Coaxal Systems of Circles

Chapter IV: The Representation of Circles by Points in Space of Three Dimensions

Chapter V: Mappings of the Euclidean Plane

Chapter VI: Mappings of the Inversive Plane

Chapter VII: The Projective Plane and Projective Space

Chapter VIII: The Projective Geometry of n Dimensions

Chapter IX: The Projective Generation of Conics and Quadrics

Chapter X: Prelude to Algebraic Geometry

X   **CONTENTS**

PREFACE

This book is based on a course given for the past few years at the University of Minnesota to junior and senior students, to first-year graduate students and to a Year Academic Institute of College teachers of geometry who had returned to the University for a year to learn more geometry. The main purpose of the course was to increase geometrical, and therefore mathematical understanding, and to help students to enjoy geometry. This is also the purpose of my book.

I believe, and in this I am supported by many reputable mathematicians, that a reasonable knowledge and appreciation of the scope and methods of elementary geometry is desirable for those who wish to study and to relish more abstract and advanced notions.

By 'elementary geometry' I mean geometry as it is usually taught up to, but not including, a rigorous theory of algebraic curves. I include projective geometry of two and three dimensions, with some geometry of n dimensions in my definition of elementary geometry. The full scope of my definition can be initially realized by a look at the Table of Contents.

In this book I discuss the algebraic methods available for a study of elementary geometry. I have tried not to be tedious, and on the whole only those theorems of elementary geometry are studied which are of significance for further study. Whenever possible, pointers to this significance are indicated, in the text or in the Exercises, of which there are more than 500. The algebraic approach, although it is the main thread of my discourse, is never over-stressed. If a theorem appears to be more attractive with a geometric proof, this is also given. In any case, geometric insights are stressed all through.

The student reading this book is expected to have completed a course in linear algebra, of the kind indicated in my '*Geometric Introduction to Linear Algebra*' (Chelsea, New York, 1978). That is, some acquaintance with the ideas of a vector space is assumed, including some theorems on the solution of linear homogeneous equations, and also some matrix theory. But in other algebraic matters the book is self-contained, and there is a Preliminary Chapter 0 with a summary of results the student will need.

The chapters of this book need not necessarily be taken in strict sequence. There is ample material for a year's course, or selections can be made which can be covered in one, two or three quarters. The second half of the book, which deals with the projective space of 2, 3 and n dimensions can be used independently of the first part.

Chapter I, which need not be taken first, shows what kinds of theorem in real

Euclidean and affine geometry can be proved by using vectors. The student who has only come across vectors in a few sections of a book on the calculus may be surprised at the scope of the theorems which are amenable to the algebra of vectors. Later in the chapter the exterior algebra of Grassmann is introduced, and developed a little, and the power of the methods then available is demonstrated.

Chapters II and III contain those parts of the theory of circles which are of importance for later developments in geometry. A fairly detailed study of coaxal systems is made. After all, these are the simplest linear systems of curves, after linear systems of lines, and in Chapter IV coaxal systems of circles are mapped onto lines in space of three dimensions, and we therefore need to know something about the systems which have such a simple map. Inversion is also studied, not only for its theoretical importance, but because it never fails to fascinate the student. The idea that geometry is an attractive study is not overlooked in this book.

Chapter IV discusses the representation of circles by points in E_3, real Euclidean space of three dimensions, and here the interplay between theorems in space of different dimensions (two and three in this particular case) first becomes clear. The ability to regard theorems from different points of view is probably the hall-mark of a geometer, and in this chapter a student may receive his first training in this desirable art. At the end of the chapter an "algebra of circles" is introduced. This is a further application of the ideas of Grassmann, and although the use made of the algebra is not new, the explanation is new and, it is hoped, no longer almost metaphysical. As in the preceding chapters, there are numbers of Exercises, not completely trivial, on which the student can test his understanding of the text. Nobody has, as yet, discovered a better way of learning mathematics.

Chapter V deals with mappings of the Euclidean plane, especially isometries. These mappings have already appeared in earlier chapters, but are now considered from the point of view of the algebra of complex numbers. This algebra has many advantages, and some disadvantages, but it is important for the student to become familiar with the field of complex numbers. Wherever proofs using complex numbers are not completely satisfactory, aesthetically speaking, an alternative proof using Euclidean geometry is indicated. Some group theory is used in this chapter. Elementary mappings of the Euclidean plane have, of course, been an important part of elementary geometry for many years. Recently they have become important in the foundations of geometry.

Chapter VI continues the discussion of mappings by considering the Moebius transformations of the inversive plane, and ends with a detailed discussion of the Poincaré model of a hyperbolic non-Euclidean geometry. There is not too much development here, since once the student is convinced that Euclid's parallel axiom can be dispensed with, the rest is not necessary for our purpose, and can be found in other texts. The Appendix lists the Propositions of Euclid, as a useful reference.

Chapter VII introduces projective geometry, the treatment being algebraic. Examples are given in two and three dimensions, some important constructs appearing.

Chapter VIII considers the projective geometry of n dimensions, but only essential topics are discussed. The main theorems of projective geometry are derived, and applied to geometric proofs of theorems treated algebraically in earlier chapters. At the end of the chapter the geometrical interpretation of the Jordan normal form for a projective transformation of S_n is discussed. The algebraic derivation of the normal form must be considered to require more algebra than the readers of this book are expected to have, and so stress is laid on the geometrical meaning of the form when it is presented.

Chapter IX uses the material of the preceding chapters to develop the projective theory of conics and quadrics. This is a powerful and attractive theory, and there is constant interplay between theorems in two and three dimensions. Stereographic projection is one of the methods developed to show this interplay. Once the main theorems are established, the proofs and the thinking are geometrical.

Chapter X, entitled 'Prelude to Algebraic Geometry', ventures to the threshold of the theory of algebraic curves, Bezout's Theorem on the number of points of intersection of two plane algebraic curves of given orders. Twisted cubic curves on a quadric are discussed, the curve of intersection of two quadrics is investigated, the theorem on eight associated points is proved, and applications made to theorems encountered much earlier, proved by different methods. The chapter ends with a discussion of oriented circles, and proves some attractive theorems which hold only for oriented circles.

The Appendix lists the Propositions of Euclid and gives a brief account of the Grassmann-Pluecker coordinates of lines in projective space of three dimensions. These coordinates occur in the text, but are not used in the development. There is also a pulling together of various threads of the discourse, with an indication of developments which are beyond the scope of this book.

Bibliography and References list some of the books which have influenced me during the time spent in composing the lectures on which this book is based, and details are also given of books and articles which the student may wish to consult for further information. (Bold-face numerals refer to the Bibliography, p. 442.)

I wish to thank Prof. Erwin Engeler for drawing my attention to Henri Lebesgue's comments, (printed as a Frontispiece to this volume), on those he regarded as mathematical vandals. Readers of Felix Klein's 'Hoehere Geometrie' (9) will come across some tart comments there on the curious attitude of some mathematicians towards geometry. Without trying to explain why the teaching of geometry has slumped so badly in the United States during the past thirty years, and far more than in any other country, it is heartening to notice that there are signs of a revival, and I hope that my book may help in this.

I also wish to thank Mr. Wolfgang Nauck, who helped me with some of the drawings, and Prof. W. Wunderlich, who used the resources of his Institut fuer Geometrie in the Technische Hochschule of Vienna to supply other drawings.

Many of the Sections into which this book is divided contain at most one Theorem. If this is the case, the reference to the Theorem is given by the number of the Section containing the Theorem. If the Section contains more than one Theorem, the reference is to Thm p, § q.

D.P.

School of Mathematics,   May, 1970
University of Minnesota,
Minneapolis, Minn. 55455.

O

PRELIMINARY NOTIONS

In this chapter we introduce some of the notions which the reader will need when he begins his study of the main part of this text. We go into rather more detail in some cases than in others, our aim being to proceed fairly informally at first, and to introduce more precise notions as we proceed, and where necessary. Some ideas will therefore be discussed more than once, but if the reader is already familiar with them, he will turn the pages until he comes across something new. The idea of a group, for example, is needed at the beginning of our work, but we do not go into more detail until Chapter V, when we need to know more about groups. Of course, we could assume that the reader is already familiar with all the algebraic concepts which we wish to use, or give references to another book, but we prefer to make this book fairly self-contained, and have inserted sections on algebraic topics where teaching experience has shown that the student needs to be reminded of these topics. This manner of presentation may be less elegant, but we hope it will make this book more useful than a completely stream-lined presentation might be.

0.1 The Euclidean plane

This is the familiar plane of ordinary geometry, and we shall investigate it algebraically, using the real number system. The properties of the real numbers which we use are (*a*) that they form a field, which means that the ordinary processes of arithmetic are possible with real numbers, and (*b*) that they are ordered.

The points of our Euclidean plane are given by ordered pairs of real numbers (x_1, x_2), these being coordinates with respect to a pair of orthogonal cartesian axes OX_1, OX_2 (Fig. 0.1) The distance between a pair of points $P = (x_1, x_2)$, $Q = (y_1, y_2)$ is given by the formula

$$[d(P, Q)]^2 = (x_1 - y_1)^2 + (x_2 - y_2)^2.$$

The lines of our plane are sets of points which satisfy equations of the form

$$a_1 X_1 + a_2 X_2 + a_3 = 0.$$

We shall use upper-case letters P, Q, A, B, . . . for points and lower-case letters l, m, n, . . . for lines.

[1]

We can set up a coordinate system on any given line m, so that if the point A has coordinate x_A, the point B has coordinate x_B, then

$$|x_B - x_A| = d(A, B)$$

for all points A, B on the line.

Definitions. A point B is *between* A and C, where $A \neq C$ if

$$d(A, B) + d(B, C) = d(A, C).$$

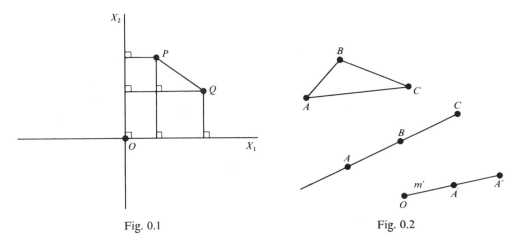

Fig. 0.1 Fig. 0.2

Note that this definition implies that B lies on the line AC (Fig. 0.2). The points A and C together with all points B between A and B form the *segment AC*. The *half-line m'* with *endpoint O* is defined by two points O, A in line m ($A \neq O$) as the set of all points A' of m such that O is not between A and A'. A half-line is sometimes called a *ray*.

If A, B, C are three distinct points the three segments AB, BC and CA are said to form a *triangle ABC* with *sides AB*, *BC* and *CA* and *vertices A, B* and C. If A, B and C are on the same line the triangle is said to be *degenerate*. A non-degenerate triangle is a *proper* triangle.

If ABC is a proper triangle, the sum of the lengths of any two sides is always greater than the length of the third side, and so we have

$$d(A, B) + d(B, C) > d(A, C), \qquad d(B, C) + d(C, A) > d(B, A),$$
$$d(C, A) + d(A, B) > d(C, B).$$

This inequality is often referred to as *the triangle inequality*. In fact, we can say that *if A, B, C are any three points,*

$$d(A, B) + d(B, C) \geq d(A, C),$$

with equality if and only if the point B lies on the line AC between A and C, that is in the segment AC.

An *angle* is formed by three ordered points A, O, B, where $A \neq O, B \neq O$. The point O is called the *vertex* of the angle. The half-lines $m = OA$, $n = OB$, through any point O can be put into $1:1$ correspondence with the real numbers so that if a_m corresponds to m, a_n to n, then the *measure* of AOB, which we call $\sphericalangle AOB$, is $a_n - a_m$ (mod 2π).

This can be done, for example, by drawing a circle, center O, and putting the half-lines through O into $1:1$ correspondence with the points in which half-lines through O intersect the circle (Fig. 0.3).

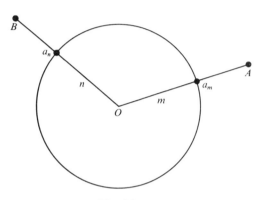

Fig. 0.3

We note that this definition of angle distinguishes between $\sphericalangle AOB$ and $\sphericalangle BOA$, and that $\sphericalangle AOB = - \sphericalangle BOA$. We call such angles *sensed* or *oriented* angles. We can also call them *signed* angles.

We know that in Euclidean geometry, for any triangle ABC,

$$\sphericalangle BCA + \sphericalangle CAB + \sphericalangle ABC = \pi(\text{mod } 2\pi).$$

Similar triangles occur frequently in our work. Triangles ABC, $A'B'C'$ are said to be *similar* if for some constant $k > 0$,

$$d(A', B') = kd(A, B), \quad d(B', C') = kd(B, C), \quad d(C', A') = kd(C, A),$$

and

$$\sphericalangle B'C'A' = +\sphericalangle BCA, \quad \sphericalangle C'A'B' = +\sphericalangle CAB, \quad \sphericalangle A'B'C' = +\sphericalangle ABC,$$

when the triangles are said to be *directly* similar; or corresponding sides are proportional, but corresponding angles are the *negatives* of each other, when the triangles are said to be *indirectly* similar.

If in two triangles ABC and $A'B'C'$ we have $d(A', B') = kd(A, B)$, $d(A', C') = kd(A, C)$ and also $\sphericalangle B'A'C' = \pm\sphericalangle BAC$, then the triangles are either directly or indirectly similar.

If, in the above definition, $k = 1$, the triangles are *congruent*.

The theorem that, if in two triangles ABC, $A'B'C'$ we are given $d(A, B) = d(A', B')$,

$d(A, C) = d(A', C')$ and $\sphericalangle BAC = \pm \sphericalangle B'A'C'$ the triangles are congruent, is often referred to as *the side-angle-side theorem*, or SAS-theorem, for short.

A point O on a given line m is the vertex of two half-lines. We may assign positive coordinates to the points of one half-line, taking O as origin, and negative coordinates to points of the other half-line. If A, B are any two points on m, we define the *sensed*, or *directed* segment AB as

$$\overline{AB} = x_B - x_A.$$

Then $\overline{BA} = -\overline{AB}$, and for any three points A, B, C on m we have

$$\overline{AB} + \overline{BC} + \overline{CA} = 0.$$

If A, B, P are distinct points on m (*collinear* points), we define *the ratio in which P divides the segment AB* to be the ratio $\overline{AP}/\overline{PB}$. This ratio is independent of the position of the origin of coordinates O on the line, and independent also of which of the two half-lines vertex O is chosen to have positive coordinates. If P lies between A and B the division is said to be *internal*; otherwise the division is said to be *external*. We note that for *external* division the ratio is *negative*. If we denote the ratio $\overline{AP}/\overline{PB}$ by r, we note that if P lies on the prolongation of $\overline{BA}$, then $-1 < r < 0$; if P lies between A and B then $0 < r < \infty$. If P lies on the prolongation of $\overline{AB}$ then $-\infty < r < -1$. If A and B are distinct and P coincides with A, we set $\overline{AP}/\overline{PB} = 0$. If A and B are distinct and P coincides with B the ratio $\overline{AP}/\overline{PB}$ is undefined, and we indicate this by writing $\overline{AP}/\overline{PB} = \infty$.

We note that for any given r, where $-\infty < r < +\infty$, the point P is uniquely defined, so that there is a $1:1$ correspondence between the positions of P on the line and the real numbers. As P moves along the line *in either direction*, the position-ratio r tends towards the value -1. This suggests that the line has *a unique point at infinity*.

Of course *points at infinity* have no place in Euclidean geometry, but it is very useful to extend our ideas of points so as to embrace this concept.

In Euclidean geometry of the plane a unique line can be drawn to contain two given distinct points, and two distinct lines either intersect in a unique point, or they do not intersect at all, in which case they are said to be parallel. Through a given point which does not lie on a given line there is a unique line parallel to the given line.

We introduce a new kind of point, which we call an *ideal* point. This is defined to be *the set of lines parallel to a given line*. If we denote such a point by $\mathscr{P}$, we say that the point $\mathscr{P}$ lies on a given line l if and only if l is a member of the set of parallel lines which define $\mathscr{P}$. If now Q is an ordinary point of the plane, there is a unique line which contains Q and $\mathscr{P}$, since there is a unique parallel through Q to the lines defining $\mathscr{P}$. Two lines now always intersect, either in an ordinary point, or in an ideal point. We say that the set of ideal points lie on an ideal line ω, and we sometimes refer to this as *the line at infinity* in the plane we are considering. We see that any line l intersects ω in a unique ideal point $\mathscr{P}$, and so it is reasonable to call the set of ideal points an ideal *line*. These ideas are discussed algebraically in §80.1.

Finally, it is sometimes convenient to assign a *sense*, or *orientation* to the areas of triangles lying in the Euclidean plane. A triangle ABC will be considered as *positive* or *negative* according as the tracing of the perimeter from A to B to C to A is counterclockwise or clockwise. Such a signed triangular area is called a *sensed* or *signed* area, and will be denoted by $\triangle \overline{ABC}$, in contrast to the unsigned area $\triangle ABC$.

We note that the area of triangle ABC is given by the determinant

$$\frac{1}{2} \begin{vmatrix} x_1, & x_2, & 1 \\ y_1, & y_2, & 1 \\ z_1, & z_2, & 1 \end{vmatrix},$$

where $A = (x_1, x_2)$, $B = (y_1, y_2)$ and $C = (z_1, z_2)$, and this determinant is positive if the sense of rotation from the axis OX_1 to the axis OX_2 is counterclockwise, that is in the same sense as the tracing of the perimeter of triangle ABC. This determinant is a function of the vertices of a triangle which enables us to compare the orientation of different triangles. It will appear again when we consider geometries as the set of properties unchanged by linear transformations of the plane.

Circles will figure largely in this book, and although we shall prove many of the properties we wish to use, we shall assume the fundamental property that if A, B, C, D are four points in the Euclidean plane, a necessary and sufficient condition that they are *concyclic* (that is, that they lie on a circle) is the equation

$$\sphericalangle ABC = \sphericalangle ADC.$$

These are directed angles, and the relation is to hold mod π.

0.2   The affine plane

We consider a plane in which the points are given by ordered pairs of real numbers (x_1, x_2), and the line joining the points $P = (x_1, x_2)$ and $Q = (y_1, y_2)$ is defined as the set of points R where

$$R = \left(\frac{k_2 x_1 + k_1 y_1}{k_1 + k_2}, \quad \frac{k_2 x_2 + k_1 y_2}{k_1 + k_2} \right)$$

and k_1, k_2 vary over all real values, but $k_1 + k_2 \neq 0$. If we eliminate the ratio $k_1 : k_2$, we find that the coordinates (X_1, X_2) of R satisfy a linear equation of the form

$$a_1 X_1 + a_2 X_2 + a_3 = 0.$$

If we take the values $(0, 1)$ and $(1, 0)$ respectively of (k_1, k_2) we see that the set of points R contains both P and Q, and the line can be described as *the line PQ*. Without introducing the idea of distance between a pair of points we can say that the point R is *between*

P and Q if the ratio $k_1:k_2$ is positive, and we can also define *the ratio of the signed lengths*

$$\overline{PR}:\overline{RQ} = k_1:k_2.$$

We do not compare lengths between points in general, that is we have no general formula for the distance between two points. But besides being able to compare signed lengths between pairs of points on the same line, we can also compare them on *parallel* lines.

The two lines $a_1X_1 + a_2X_2 + a_3 = 0$, $a_1X_1 + a_2X_2 + b_3 = 0$, where $a_3 \neq b_3$, have no point of intersection, and are said to be *parallel*. We can show that through a given point P which does not lie on a given line l there is a unique line parallel to l. Suppose that the points, P, Q lie on a line l and the points P', Q' lie on a parallel line l'. If the line PP' is parallel to the line QQ', we say that the signed lengths $\overline{PQ}$ and $\overline{P'Q'}$ are equal. If PP' intersects QQ' at the point V, we find that $\overline{VP}:\overline{VP'} = \overline{VQ}:\overline{VQ'}$ and we define $\overline{PQ}:\overline{P'Q'}$ to equal either of these ratios.

We do not define angle-measure in this geometry, which is called *affine* geometry, but we shall see in §§ 1.1–6.1, where we use vectors to prove theorems, that many theorems true in the Euclidean plane are also true in the affine plane.

We can introduce *ideal* points into the affine plane exactly as we did for the Euclidean plane, and talk of *the line at infinity*. The affine plane will appear again (§80.2), considered from another point of view.

Finally, we can *compare* areas in the affine plane, using the same formula for the area of a triangle ABC that is used in Euclidean geometry (§0.1). This involves the possibility of comparing the orientation of two triangles ABC, $A'B'C'$, the orientation being the same if the ratio of the corresponding determinants is positive, and opposite if the ratio of the two determinant functions is negative.

0.3 Notation

It is, of course, desirable to have a consistent notation, but in geometry this is not always possible. The important thing is that the reader should always understand the notation used at any given moment! We shall refer to the unique line which contains the points A and B as *the line* AB, but we shall usually shorten $d(A, B)$, which is the distance between the points A and B in the Euclidean plane to $|AB|$. We already have the notation $\overline{AB}$ for the *signed* length $|AB|$ in the Euclidean plane, and we can compare parallel signed segments in the affine plane. If we refer to the angles of a triangle ABC in the Euclidean plane, which are $\sphericalangle BCA$, $\sphericalangle CAB$ and $\sphericalangle ABC$, we shall sometimes, for brevity, refer to them as the angles C, A and B respectively, or $\hat{C}$, $\hat{A}$ and $\hat{B}$.

We shall denote points by upper-case letters, P, Q, R and so on, and lines by lower-case letters, l, m, n and so on. If AB and CD are two lines which intersect at a point V, we shall write $V = AB \cap CD$. In three-dimensional space, Euclidean, affine or projective, we shall denote the plane through the points A, B and C by ABC, or, if it makes reading easier, by (ABC). We shall also do this in space of n dimensions. If we wish to refer to the plane defined by a point P and a line l which does not contain P, we shall write $[Pl]$. In all cases it is hoped that the text will amplify the notation, so that there is no misunderstanding.

The situation with regard to coordinate axes is made troublesome by the fact that for centuries mathematicians have talked of Ox and Oy-axes, and Ox, Oy and Oz-axes in three dimensions. We shall use both this notation and a more modern one, talking about the OX_1 and OX_2-axes, or the OX_1, OX_2 and OX_3-axes. We have tried to be consistent in using $(X_0, X_1, \ldots, X_n)$ to denote a variable point, and $(x_0, x_1, \ldots, x_n)$ to denote a given point in projective space S_n, but purists will probably be able to detect lapses.

0.4   Complex numbers

Associated with every point (a, b) of the Euclidean plane, where the axes are the usual orthogonal cartesian axes, there is a complex number $a + ib$, where we may regard i as an algebraic symbol which obeys the usual laws of algebra, but in addition the law $i^2 = -1$.

This is one approach to complex numbers. We may also introduce complex numbers as *ordered pairs* (a, b) of real numbers, and say that two complex numbers (a, b) and (c, d) are equal if and only if $a = c$ and $b = d$. This is sometimes called: 'equating the real and imaginary parts of the two sides of an equation in complex numbers'. We define addition thus:

$$(a, b) + (c, d) = (a + c, b + d),$$

and scalar multiplication $k(a, b)$, where k is real, thus:

$$k(a, b) = (ka, kb).$$

We define the multiplication $(a, b) . (c, d)$ of two complex numbers as follows:

$$(a, b) . (c, d) = (ac - bd, ad + bc).$$

With these rules, we soon see that the usual laws of algebra, the associative law, the distributive law and the commutative law, hold. There is an obvious zero complex number, the number $(0, 0)$, and we have

$$(a, b) + (0, 0) = (a, b), \quad \text{and} \quad (a, b) . (0, 0) = (0, 0),$$

so that we can define $-(a, b)$ to be $(-a, -b)$, and

$$(a, b) - (a, b) = (0, 0).$$

If (a, b) is not equal to $(0, 0)$, we can find *an inverse*, that is a number (c, d) such that $(a, b) . (c, d) = (1, 0)$. We choose $(1, 0)$ as our unity because it is clear that $(a, b) . (1, 0)$

$= (a, b)$. The inverse of (a, b) may be calculated as $(a/(a^2 + b^2), -b/(a^2 + b^2))$, and it is immediately verified that this is indeed the inverse.

If we consider the special complex numbers $(a, 0)$, we see that since

$$(a, 0) + (c, 0) = (a + c, 0), \quad \text{and} \quad (a, 0) \, . \, (c, 0) = (ac, 0),$$

the numbers $(a, 0)$ are isomorphic to the ordinary real numbers a. We may therefore identify the numbers $(a, 0)$ with the corresponding numbers a, and consider the complex numbers (a, b) as an *extension* field of the real numbers. If we call the special complex number $(0, 1)$ by the name i (engineers use j), and note that $(0, 1) \, . \, (0, 1) = (-1, 0)$, which we have agreed to call -1, we have $i^2 = -1$. We also have

$$(a, b) = a(1, 0) + b(0, 1) = a + bi,$$

by our conventions and procedures, and we have returned to the point of view that the complex numbers are an extension of the real numbers, of the form $a + ib$, where a and b are real, and i obeys the usual algebraic laws, with the additional fact that $i^2 = -1$.

0.5   The Argand diagram

The contemplation of a complex number as an ordered pair (a, b) makes the representation of a complex number immediate and non-metaphysical. We represent the complex number (a, b) by the point with cartesian coordinates (a, b), usually referred to orthogonal axes. This representation is known as *the Argand diagram* (Fig. 0.4). For historical

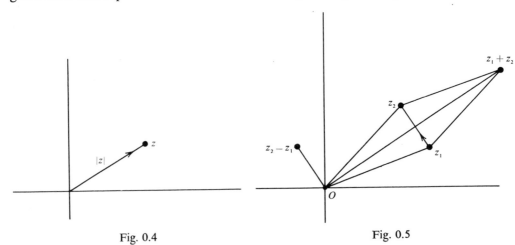

Fig. 0.4   Fig. 0.5

reasons the number b is called the *imaginary* part of the complex number, and the OX_2-axis in the Argand diagram, is called *the imaginary axis*, the OX_1-axis being called *the real axis*. This does confuse some students, especially if their first introduction to complex numbers is via certain technical books. One is usually told that there is no number whose square is equal to -1, and subsequently asked to imagine a number i whose square is equal to -1. 'Such numbers are imaginary'.

We shall often use a single symbol, such as z, to represent a complex number, and we may write $z = a + ib$, or $z = (a, b)$. If we represent the complex number not only by the point-pair (a, b) in the Argand diagram, but also by an arrow from the origin to the point (a, b), we are considering complex numbers as vectors (Fig. 0.5). We shall examine vectors in Chapter I.

Since $(a, b) + (c, d) = (a + c, b + d)$, we see that the parallelogram law for the

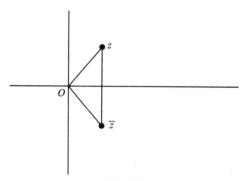

Fig. 0.6

addition of complex numbers z_1, z_2 holds as it does for vectors, and in particular we deduce the important result that the join of the complex number $z_1 = (a, b)$ to $z_2 = (c, d)$ is a vector which represents the complex number

$$z_2 - z_1 = (c - a, d - b).$$

0.6　The conjugate of a complex number

If $z = a + ib$, we define the *conjugate* of z, written $\bar{z}$, as

$$\bar{z} = a - ib.$$

Geometrically, $\bar{z}$ is the reflexion of z in the real axis. It is clear that

$$\bar{\bar{z}} = a + ib = z,$$

corresponding to the fact that geometrical reflexion in a line is an operation of period two (Fig. 0.6).

It is also clear that

$$\overline{(z_1 + z_2)} = \bar{z}_1 + \bar{z}_2,$$

and it is quickly verified that

$$\overline{(z_1 . z_2)} = (\bar{z}_1) . (\bar{z}_2).$$

We note that

$$z . \bar{z} = (a + ib)(a - ib) = a^2 - i^2b^2 = a^2 + b^2,$$

which gives the square of the distance of the point (a, b) from the origin. We define the *modulus* of z, written $|z|$, to be the length $(a^2 + b^2)^{1/2}$, so that

$$z . \bar{z} = |z|^2.$$

Since

$$|z_1 . z_2|^2 = (z_1 . z_2)(\overline{z_1 . z_2}) = (z_1 . z_2)(\bar{z}_1)(\bar{z}_2) = (z_1 . \bar{z}_1)(z_2 . \bar{z}_2) = |z_1|^2|z_2|^2,$$

we have the fundamental relation

$$|z_1 . z_2| = |z_1||z_2|.$$

We note that $|z| > 0$ if $z \neq 0$, where the second 0 now stands for the complex number $(0, 0)$. The introduction of the conjugate of a complex number enables us to find the inverse of a complex number $z \neq 0$ with speed. We have

$$z . \bar{z} = |z|^2 \neq 0,$$

and therefore

$$z^{-1} = \bar{z}/|z|^2 = a/(a^2 + b^2) - ib/(a^2 + b^2).$$

0.7   The triangle inequality

In Fig. 0.5 the diagonal of the parallelogram represents $z_1 + z_2$, and since one side of a triangle is always less than or equal to the sum of the lengths of the other two sides, we have

$$|z_1 + z_2| \leqslant |z_1| + |z_2|.$$

This is the basic inequality connecting the moduli of complex numbers. It may also be written in the form

$$|z_2| \geqslant |z_1 + z_2| - |z_1|,$$

or

$$|z_1 + z_2 - z_1| \geqslant |z_1 + z_2| - |z_1|$$

If we rename our complex numbers, and write $Z_1 = z_1 + z_2, Z_2 = z_1$, we have

$$|Z_1 - Z_2| \geqslant |Z_1| - |Z_2|.$$

Since $|Z_1 - Z_2| = |Z_2 - Z_1|$, we also have

$$|Z_1 - Z_2| \geqslant |Z_2| - |Z_1|,$$

and all this information may be put together in the inequality

$$|Z_1 - Z_2| \geqslant ||Z_1| - |Z_2||.$$

It should always be remembered that if $z_1 = (a, b)$ and $z_2 = (c, d)$, the *distance* between the points (a, b) and (c, d) is $|z_2 - z_1|$.

Exercises

0.1   Show, by induction, that

$$\overline{z_1 + z_2 + \ldots + z_n} = \bar{z}_1 + \bar{z}_2 + \ldots + \bar{z}_n,$$

and that

$$\overline{z_1 . z_2 . \ldots . z_n} = \bar{z}_1 . \bar{z}_2 . \ldots . \bar{z}_n.$$

0.2   Show that

$$|z_1 + z_2 + \ldots + z_n| \leqslant |z_1| + |z_2| + \ldots + |z_n|.$$

0.3   Prove that $|z_1 + z_2|^2 + |z_1 - z_2|^2 = 2(|z_1|^2 + |z_2|^2)$, and give a geometric interpretation of this formula.

0.4   Verify the identity below which holds for any four complex numbers z_1, z_2, z_3 and z_4:

$$(z_1 - z_4) . (z_2 - z_3) + (z_2 - z_4) . (z_3 - z_1) + (z_3 - z_4) . (z_1 - z_2) = 0.$$

Deduce that if A, B, C and D are any four points in a plane,

$$|AD||BC| \leqslant |BD||CA| + |CD||AB|.$$

(Compare the stronger result of § 24.1.)

0.5   Using the theorem $|z_1 . z_2| = |z_1| . |z_2|$, express the product

$$(a_1{}^2 + b_1{}^2)(a_2{}^2 + b_2{}^2)$$

of the sum of two squares as the sum of two squares. Prove that the product

$$(a_1{}^2 + b_1{}^2)(a_2{}^2 + b_2{}^2) \ldots (a_n{}^2 + b_n{}^2)$$

may also be written as the sum of two squares.

0.6   Show that the equation of a circle, center z_0 and radius r, may be written as $|z - z_0| = r$, or as $z\bar{z} - z\bar{z}_0 - \bar{z}z_0 + |z_0|^2 - r^2 = 0$.

0.8   De Moivre's theorem

The Argand diagram representation of complex numbers leads to an important representation in terms of polar coordinates (Fig. 0.7). Let $z = a + ib$, and let the modulus of z be written as r. Then if the radius vector from the origin to the point (a, b) makes an angle ϕ with the positive direction of the X_1-axis, we have

$$a = r \cos \phi, \qquad b = r \sin \phi,$$

so that

$$z = a + ib = r(\cos \phi + i \sin \phi).$$

If we are given a and b we may determine r without ambiguity as $r = (a^2 + b^2)^{1/2}$, but there is obvious ambiguity in the determination of ϕ. If we restrict ϕ to the values given by the inequalities $-\pi < \phi \leqslant \pi$, then ϕ is determined without ambiguity, and this value of ϕ is then called a *principal value*. We write this value of ϕ as am z (*amplitude of z*). It is also sometimes written as arg z (*argument* of z). We note that r is the *modulus* of z.

If we multiply together the two complex numbers

$$z_1 = r_1(\cos \varphi_1 + i \sin \varphi_1), \qquad z_2 = r_2(\cos \varphi_2 + i \sin \varphi_2),$$

we find that

$$z_1 . z_2 = r_1 r_2(\cos (\varphi_1 + \varphi_2) + i \sin (\varphi_1 + \varphi_2)).$$

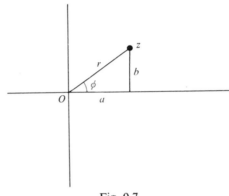

Fig. 0.7

Hence we obtain once more the theorem:

$$|z_1 . z_2| = |z_1| . |z_2|,$$

but we *cannot* say that

$$am(z_1 . z_2) = am(z_1) + am(z_2),$$

because of the definition of principal value. For a simple counterexample, take both z_1 and z_2 equal to -1. As a complex number,

$$-1 = 1(\cos \pi + i \sin \pi),$$

and the amplitude is π. Since $(-1)^2 = 1$, which has amplitude zero, and $2\pi \neq 0$, a simple additive relation between principal values of amplitudes does not hold. But we can say that the amplitude of the product of two complex numbers is equal to the sum of the amplitudes of the two numbers, *or differs from it by* $\pm 2\pi$.

If $z_i = r_i(\cos \varphi_i + i \sin \varphi_i)$, we find by induction that

$$z_1 . z_2 \ldots . z_n = r_1 r_2 \ldots . r_n(\cos (A) + i \sin (A)),$$

where $A = \varphi_1 + \varphi_2 + \ldots + \varphi_n$.

If we take all the z_i to be equal to $r(\cos \varphi + i \sin \varphi)$, we have

$$[r(\cos \varphi + i \sin \varphi)]^n = r^n(\cos n\varphi + i \sin n\varphi),$$

or

$$(\cos \varphi + i \sin \varphi)^n = \cos n\varphi + i \sin n\varphi,$$

and this is De Moivre's Theorem.

This theorem also holds for negative integers, since

$$\frac{1}{r(\cos \varphi + i \sin \varphi)} = \frac{1}{r}(\cos \varphi - i \sin \varphi) = r^{-1}(\cos (-\varphi) + i \sin (-\varphi)),$$

and

$$(\cos \varphi + i \sin \varphi)^{-n} = (\cos (-\varphi) + i \sin (-\varphi))^n = \cos(-n\varphi) + i \sin(-n\varphi).$$

We have also shown that if $z = r(\cos \varphi + i \sin \varphi)$, the modulus of z^{-1} is r^{-1}, and its amplitude is $-\varphi$, unless $\varphi = \pi$, when the amplitude of z^{-1} is still π.

More generally, if $z_1 = r_1(\cos \varphi_1 + i \sin \varphi_1)$, $z_2 = r_2(\cos \varphi_2 + i \sin \varphi_2)$, $z_1/z_2 = r_1 r_2^{-1}(\cos (\varphi_1 - \varphi_2) + i \sin(\varphi_1 - \varphi_2))$, and the modulus of z_1/z_2 is r_1/r_2, and the amplitude is either equal to the difference of the amplitudes of z_1 and z_2, or differs from it by $\pm 2\pi$.

Exercises

0.7 Express the following numbers in the form $A + iB$, where A and B are real numbers:

$$(1 + i)^2, (1 - i)^2, \quad (1 + i)^2/(1 - i)^2, \quad (1 - i)^2/(1 + i)^2.$$

Determine the modulus and amplitude of each of these complex numbers.

0.8 The complex numbers a, b and c are represented by the points P, Q and R in the Argand diagram, and the acute angle between the lines PR and QR is equal to θ. Show that am $(c - a)/(c - b)$ may equal θ, $\pi - \theta$, $-\theta$ or $-\pi + \theta$, according to the relative positions of the points P, Q and R in the plane.

0.9 If $(z - a)/(z - b) = r(\cos \varphi + i \sin \varphi)$, show that the curves $r = $ constant, and $\varphi = $ constant, for fixed complex numbers a and b and varying z, are circles of orthogonal coaxal systems, the one which has the points a and b as limiting points, and the other of circles which pass through these two points. (See Exercise 27.1.)

0.10 Show that the equation $z^3 = 1$ may be solved by writing $1 = \cos 2k\pi + i \sin 2k\pi$, and observing that $\cos (2k\pi/3) + i \sin (2k\pi/3)$ satisfies the equation. Show that the three distinct values obtained for z are the vertices of an equilateral triangle on the circle $|z| = 1$ in the Argand diagram.

0.11 Show that if the roots of $z^3 = 1$ are 1, ω, ω' then $\omega' = \omega^2$, and $1 + \omega + \omega^2 = 0$.

0.12 Show that $x^3 + y^3 + z^3 - 3xyz \equiv (x + y + z)(x^2 + y^2 + z^2 - xy - yz - zx)$
$$\equiv (x + y + z)(x + \omega y + \omega^2 z)(x + \omega^2 y + \omega z).$$

0.13 Solve the equation $z^4 = 1$ by using De Moivre's theorem, and plot the solutions on the Argand diagram. Do the same for $z^5 = 1$. Can you state a general theorem about the geometrical configuration formed by the solutions of $z^n = 1$ when plotted on the Argand diagram?

0.9 Real and imaginary in geometry

The geometric representation of complex numbers by means of points in a Euclidean plane, the complex number $a + ib$ being mapped onto the point (a, b), leads to a number of embarrassing features in our terminology, different matters being conventionally described by the same word. The numbers a, b, . . ., are all real numbers, and these are also viewed as complex numbers which all lie on the real axis of the Argand representation. The purely imaginary numbers are associated with the points of another axis in this representation. But if we think of the points of a Euclidean plane, we think of the points (a, b), where a and b are real, as real points. We shall wish to consider *real* curves, such as lines, circles, conics, etc., and these are curves defined by equations with real coefficients. A real line $a_1 X_1 + a_2 X_2 + a_3 = 0$ always contains real points, such as the point $(-a_3/a_1, 0)$, if $a_1 \neq 0$, but it also contains the 'imaginary' point $(-(ia_2 + a_3)/a_1, i)$. A real circle, $X_1^2 + X_2^2 + 1 = 0$ has a real center at $(0, 0)$, but contains no real points. What shall we call such a curve? Felix Klein suggested the adjective 'nullteilig', which means 'having no real part', since he wished to retain the adjective 'imaginary' for curves given by equations whose coefficients are complex numbers. We shall use the term 'virtual circle' for a circle, given by an equation with real coefficients, which contains no real points. The term 'ideal circle' has also been used. We note finally that a real circle can also contain both real and imaginary points. Thus the circle $X_1^2 + (X_2 - 2)^2 - 1 = 0$ is real, and intersects the real line $X_2 = 0$ in the points $(\pm i\sqrt{3}, 0)$, which are imaginary points!

We shall see later that there are great advantages in extending the real field of co-ordinates to the complex field, and in considering points whose coordinates are possibly complex numbers.

The so-called 'fundamental theorem of algebra' is very relevant to this discussion. This says that any polynomial of degree n, with complex coefficients, factorizes over the complex field into n linear factors; that is:

$$z^n + p_1 z^{n-1} + \ldots + p_{n-1} z + p_n \equiv (z - t_1)(z - t_2) \ldots (z - t_n),$$

where the t_i are complex numbers. This means that an equation with complex coefficients can always be solved, with solutions in the field of complex numbers. The equation $z^2 + 1 = 0$ has no real solutions, but does have the two complex solutions $z = \pm i$. Because of this fundamental theorem of algebra we know, for instance, that a line *always* intersects a given circle in two points, which may coincide. The points may not be real points, but if they are not, they are points with complex coordinates, which we may call *complex* or *imaginary* points.

0.10 Equivalence relations

In a set S let there be defined a relation, which we write $\sim$, between any two elements of the set. This relation may be true or false. Suppose that this relation $\sim$ satisfies the following conditions:

(i) The relation of reflexivity, that is $x \sim x$ for all x in the set S,
(ii) The relation of symmetry, that is $x \sim y$ implies $y \sim x$,
(iii) The relation of transitivity: if $x \sim y$ and $y \sim z$, then $x \sim z$.

Such a relation is called *an equivalence relation*. It is the natural extension of the relation of equality.

Suppose that we choose an element x_0 in S, and form a subset of all the elements of S which satisfy the relation $x \sim x_0$. Such a subset is called *an equivalence class*. Since $x_0 \sim x_0$, this equivalence class contains x_0. If $y_0 \sim x_0$, so that $x_0 \sim y_0$, and we consider the elements of S which satisfy the relation $x \sim y_0$, the relation of transitivity shows that all these elements satisfy the relation $x \sim x_0$, and so lie in the equivalence class of elements satisfying $x \sim x_0$. If we now take an element z_0, if there is one, *not* satisfying $z_0 \sim x_0$, and form the equivalence class of elements z of S satisfying $z \sim z_0$, this class has no elements in common with the first equivalence class, by the transitivity relation. We can continue thus, and eventually find that the equivalence relation $\sim$ has divided S up into mutually disjoint subsets, and every element of S lies in one, and only one such subset.

Conversely, if a set S is covered by mutually disjoint subsets, an equivalence relation can be set up between the elements of S, by simply saying that $x \sim y$ if and only if x and y lie in the same subset.

There are many examples of equivalence relations, some of which will be given in the Exercises which follow.

Exercises

0.14 Show that in the set of integers '$x - y$ is divisible by m', where m is a fixed integer, defines an equivalence relation. If $m = 2$, show that the relation divides the set of integers into subsets of odd and even integers.

0.15 Show that in the set of integers "x and y are both even" is not an equivalence relation.

0.16 Criticize the following proof: if $x \sim y$, then $y \sim x$, by the symmetry relation, and if $x \sim y$ and $y \sim x$ then $x \sim x$, by the transitivity relation. Does this mean we can dispense with the relation of reflexivity in our definition of equivalence relations? Consider the preceding Exercise as an illustration.

0.11   Mappings or transformations

Let S and T denote two sets. We use the symbol $\in$ to stand for the phrase 'is a member of', so that $s \in S$ means 's is a member of S'. The sets S and T may be the same set. If to every $s \in S$ there is assigned a unique $t \in T$, we say that the set S is *mapped* or *transformed* into T, or we say that *a function is defined on S*, with *values* in T.

If the mapping, transformation, or function is denoted by the symbol α, we shall use the notation

$$\alpha : S \rightarrow T, \text{ for the sets,}$$
$$\alpha : s \rightarrow t, \text{ for the points.}$$

We shall also write $s\alpha = t$. Some authors use $\alpha s = t$, and others $s^{\alpha} = t$, which is more difficult to print, but has some advantages.

Since the sets S and T need not consist of numbers or points, it is clear that the definition of function given above is a generalization of the older definition.

The element $s\alpha$ is called the *image* of s *under the mapping* α, and since this term is taken from the science of optics, we shall call s the *object* of $t = s\alpha$. (Some authors use the term *preimage*). The set S is called the *domain* of α, and the set of images, which is the set of all $s\alpha$ $(s \in S)$ is called the *range* of α, and denoted by $S\alpha$.

If the equation $s_1\alpha = s_2\alpha$ implies that $s_1 = s_2$, which means that two distinct elements of S are never mapped onto the same element of $S\alpha$, the mapping is said to be $1:1$, or one-to-one, in words.

If $S\alpha = T$, so that all the elements of T are images under α, the mapping α is said to be *surjective*, and we say that α maps S *onto* T, instead of *into* T. Of course, a mapping which is only known to be *into*, at first, can turn out to be *onto*. A $1:1$ mapping which is onto is said to be *bijective*.

In what follows we shall consider bijective mappings of a set onto itself. Every point of the set is mapped onto a unique point of the set. Every point of the set is image-point of a unique point in the set. It will be noticed that we have dropped the term 'element' of a set and use the word 'point'. In most cases our elements will be points, but they need not be.

0.12 Products of mappings

Let α, β be bijective mappings of a set onto itself. We shall write $\alpha\beta$ for the transformation which results if we carry out α first, and then β:

$$s(\alpha\beta) = (s\alpha)\beta.$$

We call the mapping $\alpha\beta$ the *product* of the mappings of α and β. We shall prove that $\alpha\beta$ is a mapping of the set. It is sometimes called a *composition* of the two mappings.

There is a special bijective mapping of a set onto itself, the *identity* mapping, which we denote by ι. This maps every point of the set onto itself

$$\iota: \quad s\iota \to s.$$

It is clear that for any bijective mapping of the set onto itself,

$$\alpha\iota = \iota\alpha = \alpha.$$

The mapping α has an *inverse* α', which is such that

$$\alpha\alpha' = \alpha'\alpha = \iota.$$

For if $s\alpha = t$, let α' be the mapping which maps t onto s. Since t arises from a unique s, s is uniquely defined, given t. Since also the mapping α is onto, we may choose any element of the set S to be t. The mapping α' is therefore defined, and it is clearly one-to-one and onto. We see immediately, since

$$s(\alpha\alpha') = (s\alpha)\alpha' = t\alpha' = s,$$

that $\alpha\alpha' = \iota$. Again, we have

$$t(\alpha'\alpha) = (t\alpha')\alpha = s\alpha = t,$$

so that $\alpha'\alpha = \iota$. We denote α' by α^{-1}.

The set of mappings we are considering obeys the *associative* law: in other words, if α, β, γ are three such mappings, then

$$\alpha(\beta\gamma) = (\alpha\beta)\gamma.$$

To prove this result, we note that $s[\alpha(\beta\gamma)]$ means $(s\alpha)[(\beta\gamma)]$, and $(\beta\gamma)$ means $\beta(\gamma)$, so that $(s\alpha)[(\beta\gamma)]$ means $(s\alpha)(\beta)(\gamma)$, which can be written $s(\alpha)(\beta)(\gamma)$. In the same way we show that $s(\alpha\beta)[\gamma]$ means $s(\alpha)[(\beta)]\gamma$, which is the same as $s(\alpha)(\beta)(\gamma)$.

If we now show that the product of two mappings α, β is not only one-to-one but also onto, we shall have shown that the one-to-one onto mappings of a set form a closed set under composition, and together with the existence of an identity ι, an inverse for any mapping of the one-to-one onto kind we have been considering, and the fulfilment of the associative law, we shall have established that one-to-one onto mappings of a set *form a group* under composition.

Since $(\alpha\beta)(\beta^{-1}\alpha^{-1}) = \alpha(\beta\beta^{-1})\alpha^{-1} = \alpha\iota\alpha^{-1} = \alpha\alpha^{-1} = \iota,$

$$\beta^{-1}\alpha^{-1} = (\alpha\beta)^{-1}.$$

If $$s_1(\alpha\beta) = s_2(\alpha\beta),$$

multiplication on the right of each side of this equation by $(\alpha\beta)^{-1}$ shows that $s_1 = s_2$. Hence the map $\alpha\beta$ is one-to-one. If t is any element of S, and $t(\alpha\beta)^{-1} = s$, then

$$s(\alpha\beta) = t,$$

and so the map $\alpha\beta$ is also an onto map.

We shall be dealing with groups later, and we give the usual group axioms here.

Group Axioms. There is a set of elements p, q, r, . . . and a binary operation which, applied to p and q, is written pq. The set is closed under this single-valued binary operation. We also have the following conditions satisfied:

The Associative Law. $p(qr) = (pq)r$, for all p, q and r in the set.

The Identity Law. $p\iota = \iota p = p$ for all p and for some ι in the set.

The Inverse Law. For any p in the set there exists a p' in the set such that $pp' = p'p = \iota$. (We denote p' by p^{-1}.)

If the group axioms are satisfied for a set, we can show that ι, the identity element, is unique, and we can also show that the inverse of a given element is unique.

For if ι' is another identity element, $\iota' = \iota'\iota = \iota$, by the properties of ι and of ι'.

If p^* is another inverse for p, then

$$p^* = \iota p^* = p'pp^* = p'\iota = p'.$$

Since the inverse is unique, and $pp^{-1} = \iota$, we must have $(p^{-1})^{-1} = p$.

If the group operation is commutative, so that $pq = qp$ for all p, q in the group, the group is said to be *Abelian*, and the group operation applied to p and q is usually written $p + q$, instead of pq.

The set of Exercises which follow give some idea of the general nature of the theory of groups. In the particular case we are examining, mappings of a set onto itself, the associative law is always satisfied, and it is therefore satisfied for the more restrictive bijective mappings. Hence, when we consider bijective mappings of a set onto itself, the group axioms will be satisfied if the composition of two mappings lies in the set of mappings, if the identity mapping of the set onto itself lies in the set of mappings, and if the inverse of any mapping lies in the set of mappings, the inverse being defined thus: if α is the mapping, and $s\alpha = t$, then $s = t\alpha^{-1}$. In tabular form, the axioms to be satisfied are:

I. If α, β lie in the set of mappings, so does $\alpha\beta$.

II. The identity mapping ι given by $s\iota = s$ lies in the set.

III. If α lies in the set, so does α^{-1}, where α^{-1} is defined thus:

$$\text{If } s\alpha = t, \text{ then } s = t\alpha^{-1}.$$

Exercises

0.17 We have associated the complex number $z = a + ib$ with the point (a, b) of the Euclidean plane. Are the following mappings of points of this plane onto? Are they $1:1$? Give reasons for your answers.

$$\text{(i) } Z = z^2, \quad \text{(ii) } Z = x^2 + iy^2, \quad \text{(iii) } Z = |z|.$$

0.18 Show that the set of two elements $\{1, -1\}$ form a group, the group multiplication being ordinary multiplication. What is the inverse of the element -1?

0.19 Which of the following sets are groups?
(a) The rational numbers under addition.
(b) The rational numbers under multiplication.
(c) The rational numbers, with zero omitted, under multiplication.
(d) The set of integers under subtraction.
(e) The set of complex numbers z with $|z| = 1$, under multiplication.

0.20 Let $n > 1$ be an integer. Show that the n roots of the equation $z^n = 1$ form a group under multiplication. (De Moivre's Theorem is not necessary.)

0.21 Over the set of all rationals distinct from -1, let an operation $*$ be defined by $a * b = a + b + ab$ (usual addition and multiplication).

(i) Show that $a * b$ is in the set, and that under the operation $*$ the set forms a group.
(ii) Do the integers $\neq -1$ form a subgroup?

0.13   Linear algebra

We shall usually recapitulate the notions of linear algebra with which we assume the reader is familiar, but we state here that this book assumes some slight acquaintance with vector spaces, the solution of sets of homogeneous equations, the ideas connected with the rank of a matrix, and matrix notation in general. These ideas are covered in the author's book *A Geometric Introduction to Linear Algebra* (Chelsea, New York, 1978).

I

VECTORS

1.1 Vectors in the Euclidean (and affine) plane

The concept of a vector as a *directed segment* is a familiar one, and it can be placed on a firm mathematical foundation. Much of what follows will apply not only to vectors in the real Euclidean plane, but also to vectors in the real affine plane (§0.2). This is indicated by choosing a system of cartesian axes which are not necessarily orthogonal

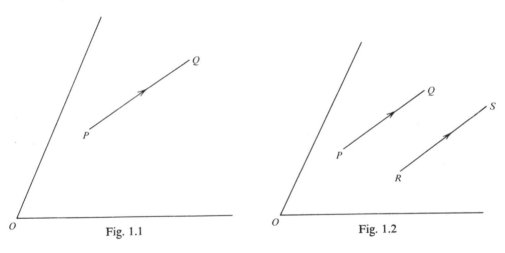

Fig. 1.1   Fig. 1.2

(Fig. 1.1). Let $P = (x_1, x_2)$ and $Q = (y_1, y_2)$ be two points in the plane. Then the *vector* $\overrightarrow{PQ}$ is defined to be *the ordered pair of real numbers:*

$$\overrightarrow{PQ} = (y_1 - x_1, y_2 - x_2).$$

Two vectors $\overrightarrow{PQ}$ and $\overrightarrow{RS}$ are thought of as being equal, in ordinary language, when PQ is parallel to RS, when the length of $PQ =$ the length of RS, and when the direction of motion from P to Q is the same as that from R to S. All this is expressed, in our language, thus: if $R = (z_1, z_2)$ and $S = (t_1, t_2)$ then $\overrightarrow{PQ} = \overrightarrow{RS}$ if and only if *the ordered pair of real numbers* $(y_1 - x_1, y_2 - x_2)$ *is the same as the ordered pair of real numbers* $(t_1 - z_1, t_2 - z_2)$ (See Fig. 1.2).

This definition of equality between vectors sets up an *equivalence relation* between

[20]

ordered pairs of real numbers (§0.10, and Exercise 1.1). In particular, in the equivalence class which contains a given vector there is always one vector with its first number pair at the origin, (0, 0). We take this as *the vector representative of the class.*

Such a vector joins the origin to a point $P = (a_1, a_2)$, say, and we call this vector *the position-vector* of the point P (Fig. 1.3). The remaining vectors of the equivalence

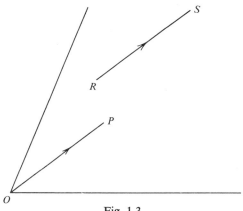

Fig. 1.3

class represented by this vector are called *free* vectors. The representative of the class is called a *bound* vector.

1.2   Addition of bound vectors

Our operations on vectors will be confined to bound vectors. Since the origin is kept fixed, a bound vector is defined by a pair of real numbers, which define the end point of the vector. Thus if $P = (x_1, x_2)$, the vector $\overrightarrow{OP}$ is defined by the ordered number-pair (x_1, x_2), and we may use the symbol P, by itself, to denote the position-vector $\overrightarrow{OP}$. We therefore write

$$P = (x_1, x_2),$$

and regard P as *a vector.* If, similarly, $Q = (y_1, y_2)$, we define *the sum* $P + Q$ of the two bound vectors P, Q by the formula

$$P + Q = (x_1 + y_1, x_2 + y_2).$$

We form a *scalar multiple* of the vector P by the real number r thus:

$$rP = (rx_1, rx_2).$$

The geometric interpretation of these rules is important. If OP and OQ are regarded as two adjacent sides of a parallelogram, the end point of the diagonal through O has coordinates $(x_1 + y_1, x_2 + y_2)$, and therefore *vector addition obeys the parallelogram law*, that is the vector $P + Q$ is given by the diagonal of the parallelogram through O which has OP and OQ as adjacent sides (Fig. 1.4).

The vector rP is a dilatation of the vector P from the origin. If r is positive, rP is in the direction of OP. If r is negative, rP is in the opposite direction. We deduce from this

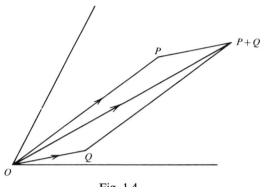

Fig. 1.4

that *if two free vectors, represented by the bound vectors P and Q, are parallel, then for some real number r we have $P = rQ$* (See Fig. 1.5).

The zero vector is the bound vector $(0, 0)$, and if $P = (x_1, x_2)$, so that $-P = (-x_1, -x_2)$, we have $P + (-P) = 0$, where now 0 stands for the zero vector. The

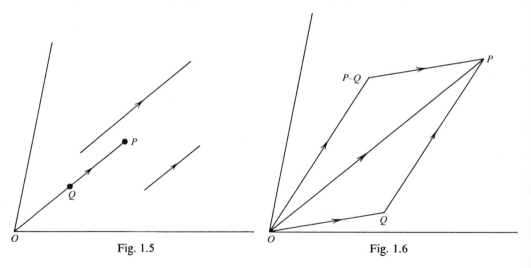

Fig. 1.5   Fig. 1.6

difference $P - Q$ of the vectors P and Q must be the vector $(x_1 - y_1, x_2 - y_2)$. This is a bound vector, *representative of the free vector $\overrightarrow{QP}$, which starts at Q and finishes at P* (Fig. 1.6).

1.3   Vector spaces

Bound vectors, with the rules for addition and scalar multiplication by a real number given above, form a vector space over the reals. The idea of a vector space is developed in most courses on linear algebra, but we give the laws which have to be satisfied below.

A set of vectors V, with an operation of addition, and an operation of multiplication by a real number form a vector space over the reals if:

The vectors $P, Q, R, \ldots \in V$ form an Abelian group under the operation of addition. That is,

If $P, Q \in V$, then $P + Q \in V$. If $P, Q, R \in V$, then

$$(P + Q) + R = P + (Q + R).$$

There exists a zero vector in V, denoted by 0, such that

$$0 + P = P + 0 = P, \text{ for all } P \in V.$$

For every $P \in V$, there exists an inverse, which we call $-P$, also in V, such that

$$P + (-P) = (-P) + P = 0.$$

Also, we always have

$$P + Q = Q + P, \text{ for all } P, Q \in V.$$

Finally, scalar multiplication by the real numbers $r, s, \ldots$ obeys the laws:

$$rP \in V, \text{ for all } P \in V, (rs)P = r(sP), \text{ and}$$
$$(r + s)P = rP + sP,$$
$$r(P + Q) = rP + rQ,$$
$$1P = P.$$

1.4   Linear dependence

It is a simple algebraic verification that our bound vectors form a vector space over the field of real numbers. Another notion which we need from the theory of vector spaces is that of *linear dependence*.

The vectors $P_1, P_2, \ldots, P_n$ in a vector space V are said to be *linearly dependent* if real numbers $k_1, k_2, \ldots, k_n$, *not all zero*, can be found so that

$$k_1 P_1 + k_2 P_2 + \ldots + k_n P_n = 0.$$

A set of vectors which are not dependent are said to be linearly *independent*.

In the case of the vector space formed by bound vectors in the Euclidean or affine plane, let us see what can be said about *two* dependent vectors P_1 and P_2. We assume that neither is the zero vector. Then if

$$k_1 P_1 + k_2 P_2 = 0,$$

and we do not have $k_1 = k_2 = 0$, neither k_1 nor $k_2 = 0$, since $k_1 = 0$, for instance, involves either $k_2 = 0$ or $P_2 = 0$. We may therefore write.

$$P_1 = k_3 P_2,$$

where $k_3 = -k_2 k_1^{-1}$. Hence the two vectors lie along the same line through the origin. Conversely, if two vectors lie along the same line through the origin they are linearly dependent vectors.

We can easily find two independent vectors among the bound vectors in the Euclidean

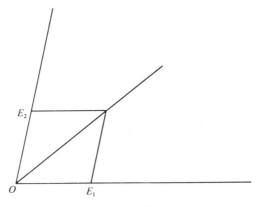

Fig. 1.7

or affine plane. They are $E_1 = (1, 0)$ and $E_2 = (0, 1)$. Moreover, any vector $P = (x_1, x_2)$ can be expressed linearly in terms of E_1 and E_2, since

$$P = (x_1, x_2) = x_1(1, 0) + x_2(0, 1) = x_1 E_1 + x_2 E_2.$$

The vectors E_1, E_2 are said to form a *basis* for our vector space, and it can be shown that *any* two independent vectors can serve as a basis for our space (Fig. 1.7). We say that the *dimension* of the vector space we are studying is two. (See Exercise 1.5.)

Exercises

1.1 Verify that the algebraic conditions given in Section 1.1 for the equality $\overrightarrow{PQ} = \overrightarrow{RS}$ satisfy the conditions for an equivalence relation, namely that they are reflexive, symmetric and transitive.

1.2 'In the equivalence class which contains a given vector there is always one vector with its first number pair at the origin, $(0, 0)$.' To what axiom in Euclidean geometry is this statement equivalent? Formulate rules for the addition of free vectors, and for the scalar multiplication of a free vector by a real number.

1.3 Investigate geometrically the associative law for the addition of vectors:

$$P + (Q + R) = (P + Q) + R.$$

Investigate also the distributive law:
$$r(P + Q) = rP + rQ.$$

1.4   If P, Q are vectors, what is the geometrical significance of the vector $(P + Q)/2$? What theorem in Euclidean geometry are you using to obtain your result?

1.5   If $U_1 = (a_1, a_2)$ and $U_2 = (b_1, b_2)$, and U_1, U_2 are independent vectors, show algebraically that numbers q_1, q_2 can always be found so that

$$P = (x_1, x_2) = q_1 U_1 + q_2 U_2.$$

2.1   Notation

We shall be using vectors to prove theorems in the Euclidean plane. Until we introduce *inner products* in §7.1, all our results will also be true in an affine plane. We shall use upper-case letters P, Q, . . . to denote both position-vectors of points and the points

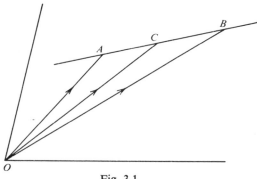

Fig. 3.1

themselves. If we wish to refer to the line joining the end-points of the vectors P, Q, we shall refer to *the line PQ*. The *signed segment* which joins P and Q will be denoted by $\overline{PQ}$, and the *length* of the segment PQ by $|PQ|$. If an expression $P - Q$ appears, it is a bound vector, and given geometrically by the directed segment which joins Q to P, that is by $\overrightarrow{QP}$.

    We make immediate use of this freedom in notation in the formulation and proof of our next theorem.

3.1   Points on a line

Theorem I   *If C is any point on the line determined by two distinct points A and B, then we may always write*

$$C = (1 - t)A + tB,$$

where the ratio of the real numbers $t : 1 - t$ is equal to the ratio of the signed lengths $\overline{AC} : \overline{CB}$ (see Fig. 3.1).

Proof. Let $\overline{AC} = t\overline{AB}$. This equation between signed segments can be translated into an equation between bound vectors, namely

$$C - A = t(B - A),$$

which gives us immediately

$$C = (1 - t)A + tB.$$

Since we also have, for signed segments,

$$\overline{AB} = \overline{AC} + \overline{CB},$$

from $\overline{AC} = t\overline{AB}$ we deduce that $\overline{AC} = t(\overline{AC} + \overline{CB})$, so that

$$(1 - t)\overline{AC} = t\overline{CB}, \text{ and } t:1 - t = \overline{AC}:\overline{CB}.$$

We now look at Thm I in a slightly different way:

Theorem II *If A, B and C are collinear points, then real numbers x, y, z, not all zero, can be found such that*

$$x + y + z = 0, \text{ and } xA + yB + zC = 0.$$

Proof. If $A = B = C$, the theorem is trivial, since $0A = 0$ for any vector A. We assume that A and B are distinct. Then by Thm I, we may write $C = (1 - t)A + tB$, for some value of t, and writing this in the form

$$(1 - t)A + tB + (-1)C = 0,$$

and choosing $x = 1 - t$, $y = t$, $z = -1$, $x + y + z = 0$, and not all of x, y, z are zero.

The converse of this theorem is important, and we state this as:

Theorem III *If A, B and C are given points, and real numbers x, y and z, not all zero, can be found such that $x + y + z = 0$ and $xA + yB + zC = 0$, then the points A, B and C are collinear.*

Proof. By renaming the points we may assume that $x \neq 0$. Then we may write

$$A = -yx^{-1}B - zx^{-1}C.$$

Since $x + y + z = 0$, $-yx^{-1} - zx^{-1} = 1$, so that if we put $-zx^{-1} = t$, then $-yx^{-1} = 1 - t$. We now have

$$A = (1 - t)B + tC.$$

If B and C are not distinct, then of course A, B and C are collinear. If B and C are distinct, then A lies on the line BC.

As an important Corollary to this Theorem we have:

Theorem IV   *If A, B and C are three given points which are* not *collinear, and we can find three real numbers x, y and z such that $x + y + z = 0$ and $xA + yB + zC = 0$, then we must have $x = y = z = 0$.*

Proof. In the proof of Thm III the assumption that any one of x, y, z is *not* zero leads to the proof that A, B and C are collinear. Hence $x = y = z = 0$.

We shall make extensive use of these theorems.

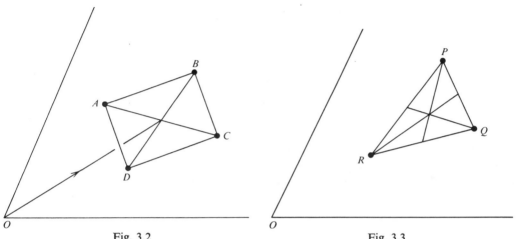

Fig. 3.2   Fig. 3.3

3.2   First applications

If we put $t = 1/2$ in the statement of Thm I, §3.1, we find that the midpoint of the segment AB is the point $(A + B)/2$. We apply this result to prove the theorem that *the diagonals of a parallelogram bisect each other.*

If A, B, C and D are the vertices of a parallelogram (Fig. 3.2), we know that $\overrightarrow{AB} = \overrightarrow{DC}$. In terms of position-vectors this is

$$B - A = C - D.$$

Therefore

$$B + D = C + A,$$

and

$$(B + D)/2 = (C + A)/2,$$

which means that the midpoint of the segment BD is the same point as the midpoint of the segment CA. In other words, the diagonals bisect each other.

3.3 The centroid of a triangle

If we put $t = 2/3$ in the formula $C = (1 - t)A + tB$, we find that the point which divides a segment AB internally in the ratio $2:1$ is the point $(A + 2B)/3$. We use this result to prove that *the medians of a triangle* (the lines joining the vertices to the midpoints of opposite sides) *are concurrent.*

Let the vertices be P, Q and R (Fig. 3.3). The midpoint of QR is $(Q + R)/2$, and the point which divides the segment joining this midpoint to the point P in the ratio $1:2$ is the point

$$\frac{2(Q + R)/2 + P}{3} = \frac{P + Q + R}{3}.$$

The symmetry of this result establishes the theorem. That is, if we carry out similar procedures with the midpoint of RP and the vertex Q, and finally the midpoint of PQ and the vertex R, we obtain the point $(P + Q + R)/3$ in each case. This means that the three medians of the triangle PQR all pass through the point $(P + Q + R)/3$. This point is called *the centroid* of the triangle.

The method used above is a typical one when vectors are used to prove geometrical theorems. To show that three given lines all pass through the same point, we may have to find the point which lies on each line. In §4.3 the Theorem of Ceva will provide sufficient conditions for the concurrency of three given lines.

Exercises

3.1 In the formula $C = (1 - t)A + tB$, trace the position of C on the line AB as t varies through all possible real values.

3.2 Show that Thm I, §3.1, can also be formulated thus:
If C is any point on the line determined by the two distinct points A and B then we may write $C = kA + k'B$, where $\overline{AC}:\overline{CB} = k':k$, and $k + k' = 1$. Where is the point C if $k + k' \neq 1$?

3.3 If D, E are the midpoints of the sides BC, CA of a triangle ABC prove, using vectors, that $\overline{DE}$ is parallel to BA, and that $\overline{DE}:\overline{BA} = 1:2$.

4.1 Theorem of Menelaus

Theorem *A line cuts the sides BC, CA and AB of a triangle ABC in the points L, M and N respectively. If $L = xB + x'C$, $M = yC + y'A$, and $N = zA + z'B$, where $x + x' = y + y' = z + z' = 1$, then $xyz = -x'y'z'$.* (See Fig. 4.1.)

Remark. The statement of this theorem is in terms of the ratios in which the collinear points L, M and N divide the respective sides of triangle ABC. Another way of formulating the theorem is in terms of the ratios of signed segments:

$$(\overline{BL}:\overline{LC})(\overline{CM}:\overline{MA})(\overline{AN}:\overline{NB}) = -1,$$

and the fact that the product of the ratios is a negative number is a consequence of the fact that in real Euclidean or affine geometry a straight line which cuts two sides of a triangle internally must cut the third side externally.

Proof. Since L, M and N are collinear, Thm II, §3.1 asserts that real numbers p, q and r exist, not all zero, such that $p + q + r = 0$, and $pL + qM + rN = 0$. That is

$$p(xB + x'C) + q(yC + y'A) + r(zA + z'B) = 0,$$

or

$$(px + rz')B + (qy + px')C + (rz + qy')A = 0.$$

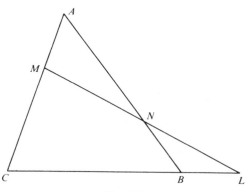

Fig. 4.1

Noting that the sum of the coefficients of A, B and C is

$$(px + rz') + (qy + px') + (rz + qy') = p(x + x') + q(y + y') + r(z + z')$$
$$= p + q + r = 0,$$

we may assert, by Thm 4, §3.1, since A, B and C are not collinear, that

$$px + rz' = qy + px' = rz + qy' = 0.$$

Hence

$$(px)(qy)(rz) = -(rz')(px')(qy'),$$

and if none of p, q, r is zero, we deduce that

$$xyz = -x'y'z'.$$

Since not all of p, q, r may be zero, not more than one may be zero (since $p + q + r = 0$). Suppose that $p = 0$. Then $q + r = 0$, and $q = -r \neq 0$, so that from $qM + rN = 0$, we find that $M = N$. But we are tacitly assuming that L, M and N are distinct points, or else the theorem is trivial, since it is then the statement $0 = 0$.

4.2   Barycentric coordinates

Theorem *If A, B, C are three non-collinear points, any vector P may be expressed in terms of the vectors A, B, C thus:*

$$P = xA + yB + zC, \text{ where } x + y + z = 1.$$

For a given point P the numbers x, y and z are uniquely determined.

Proof. If P is at the point A (Fig. 4.2), we may take $y = z = 0$ and $x = 1$. If P lies on AB we may take $z = 0$, and if P lies on AC we may take $y = 0$, and we are back at Thm I, §3.1.

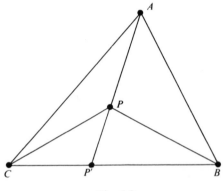

Fig. 4.2

Hence we may assume that $P \neq A$, and if $AP \cap BC = P'$ that P' is distinct from B and from C. We may write $P' = kB + k'C$, where $k + k' = 1$, and $P = lA + l'P'$, where $l + l' = 1$. Then

$$P = lA + l'(kB + k'C) = lA + l'kB + l'k'C,$$

and

$$l + l'k + l'k' = l + l'(k + k') = l + l' = 1.$$

If also $P = x'A + y'B + z'C$, where $x' + y' + z' = 1$, then

$$(x - x')A + (y - y')B + (z - z')C = 0,$$

where $(x - x') + (y - y') + (z - z') = 1 - 1 = 0$, so that by Thm IV, §3.1 we must have $x - x' = y - y' = z - z' = 0$.

Remark. By means of this theorem we may assign a uniquely determined ordered triad of coordinates (x, y, z) to any point P in the plane of a given triangle ABC. These are *normalized* coordinates, in the sense that $x + y + z = 1$. They are sometimes called *areal* coordinates, since it is easy to show that

$$x = \text{Area } \overline{BCP}/\text{Area } \overline{ABC}, \quad y = \text{Area } \overline{CAP}/\text{Area } \overline{ABC}, \quad z = \text{Area } \overline{ABP}/\text{Area } \overline{ABC}.$$

The term *barycentric* coordinates is also used. When we discuss projective geometry (Chapter VII), we shall again refer a point in a plane to the vertices of a given triangle in the plane. We note finally that the proof above shows that $x > 0$, $y > 0$ and $z > 0$ define the *interior* of the triangle ABC.

4.3   Theorem of Ceva

Theorem   *The lines AP, BP, and CP cut the opposite sides of triangle ABC in L, M and N respectively. If $L = xB + x'C$, $M = yC + y'A$, and $N = zA + z'B$, where $x + x' = y + y' = z + z' = 1$, then $xyz = x'y'z'$.* (See Fig. 4.3.)

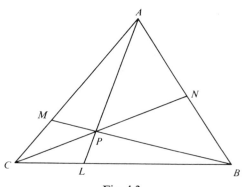

Fig. 4.3

Remark. Once again this is a theorem on ratios, and may be written

$$(\overline{BL}:\overline{LC})(\overline{CM}:\overline{MA})(\overline{AN}:\overline{NB}) = 1.$$

The theorems of Ceva and Menelaus naturally go together, since the one gives the condition for lines through the vertices of a triangle to be concurrent, and the other gives the condition for points on the sides of a triangle to be collinear.

Proof. By Thm 4.2 we may write $P = k_1A + k_2B + k_3C$, where $k_1 + k_2 + k_3 = 1$. This may also be written

$$P = k_1A + (k_2 + k_3)X,$$

where

$$X = (k_2B + k_3C)/(k_2 + k_3).$$

The point X lies on AP, from the first formulation, and it also lies on BC from the second. We therefore have $X = L = xB + x'C$. Hence

$$x:x' = k_2:k_3.$$

Similarly we find that $y:y' = k_3:k_1$, and $z:z' = k_1:k_2$, so that

$$(x:x')(y:y')(z:z') = 1.$$

4.4 Example

The points D, E and F are the respective midpoints of the sides BC, CA and AB of a triangle ABC. If P is any point in the plane of the triangle, and if PD, PE and PF are divided at L, M and N in the same ratio $x:y$, show that AL, BM and CN are concurrent.

We may assume that $x + y = 1$. Then $L = yP + x(B + C)/2$. Hence

$$(xA)/2 + L = yP + x(A + B + C)/2.$$

On division by the correct real number, each side of this equation represents a point. We divide by $x/2 + 1 = x/2 + x + y = y + 3x/2$. If we write the right-hand side of the equation in the form

$$yP + (3x/2)(A + B + C)/3,$$

this is a linear combination of the points P and $(A + B + C)/3$, and to obtain a point of the join we divide by $y + 3x/2$. On division by $1 + x/2$ the left-hand side represents a point on AL, and the symmetry of the right-hand side shows that the same point lies on BM and on CN.

A figure will show that the triangles ABC, DEF and LMN have their corresponding sides parallel, and another proof of the theorem that AL, BM and CN are concurrent is possible (see §5.1).

Exercises

4.1 Prove that the conditions given in the theorem of Menelaus and Ceva (§§ 4.1 and 4.3) are also *sufficient* conditions, for collinearity in the first case, and concurrency in the second.

4.2 If D, E and F be the midpoints of the sides BC, CA and AB of triangle ABC, and AP, BP, CP meet the opposite sides in L, M and N, P being any point in the plane, prove that the joins of D, E and F to the midpoints of AL, BM and CN are concurrent.

4.3 The side AB of a triangle ABC is divided internally at P_{AB} in the ratio $k_1:k_2$, externally at P'_{AB} in the same ratio, the side BC is divided internally at P_{BC} in the ratio $k_2:k_3$, and externally at P'_{BC} in the same ratio, and the side CA is divided internally at P_{CA} in the ratio $k_3:k_1$ and externally at P'_{CA} in the same ratio.

Show that the lines AP_{BC}, BP_{CA} and CP_{AB} are concurrent, that the line $P_{CA}P_{AB}$ contains the point P'_{BC}, and that finally the points P'_{AB}, P'_{BC} and P'_{CA} are collinear. (This theorem has geometrical significance when there are circles with respective radii k_1, k_2 and k_3 and centers at A, B and C respectively. See Exercise 13.2.)

4.4 If the point P lies in the plane of triangle ABC, but is distinct from the vertices of the triangle, and the parallelograms $PBLC$, $PCMA$ and $PANB$ are completed, prove that the segments AL, BM and CN bisect each other.

4.5 Points P, Q and R lie on the sides BC, CA and AB of triangle ABC and are such that

$$\overline{BP}:\overline{PC} = \overline{CQ}:\overline{QA} = \overline{AR}:\overline{RB}.$$

Prove that the centroids of triangles PQR and ABC are the same point. If the parallelograms $QABL$ and $RACM$ be completed, prove that RL and QM are each parallel to the median of the triangle ABC through A.

5.1 The parallel case of the Desargues theorem

Theorem *If triangles ABC, $A'B'C'$ have their corresponding sides parallel, then the joins of corresponding vertices of the two triangles are either concurrent or parallel.*

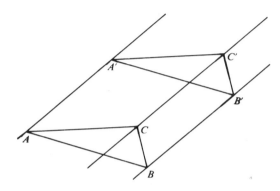

Fig. 5.1

Proof. Since BC is parallel to $B'C'$ (Fig. 5.1), we may write $B - C = k_1(B' - C')$, and for similar reasons $C - A = k_2(C' - A')$, and $A - B = k_3(A' - B')$. Since $B - C + C - A + A - B = 0$, we obtain the equation

$$(k_2 - k_3)A' + (k_3 - k_1)B' + (k_1 - k_2)C' = 0.$$

Since A', B' and C' are *not* collinear, and the sum of the coefficients of A', B' and C' is zero, Thm 4, §3.1 shows that each coefficient must be zero, that is

$$k_2 - k_3 = k_3 - k_1 = k_1 - k_2 = 0.$$

Hence $k_1 = k_2 = k_3 = k$, say, and the first set of equations give us

$$A - kA' = B - kB' = C - kC'.$$

If $k \neq 1$, division by the number $1 - k$ yields a point which is on AA', on BB' and on CC', so that these three lines are concurrent at this point. On the other hand, if $k = 1$, we have

$$A - A' = B - B' = C - C',$$

which shows that the lines AA', BB' and CC' are parallel.

Remark. When AA', BB' and CC' are not parallel, their point of intersection is called the *center of perspective* of the two triangles. More generally, if two triangles ABC, $A'B'C'$ are such that the joins of corresponding vertices AA', BB' and CC' are concurrent at a point V, we say that the triangles are *in perspective* from *the center of perspective* V, and the theorem of Desargues, which we now prove, holds for such a pair of triangles.

5.2　The Desargues theorem

Theorem　*If ABC and $A'B'C'$ are triangles with distinct vertices, and AA', BB' and CC' intersect in a point V, then the intersection Q_1, Q_2 and Q_3 of corresponding sides BC, $B'C'$, of CA, $C'A'$ and of AB, $A'B'$ are collinear, if these intersections exist.*

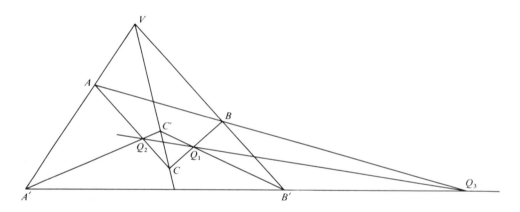

Fig. 5.2

Proof. Since V lies on AA', on BB' and on CC' (Fig. 5.2), we have (Thm I, §3.1)

$$V = k_1A + (1 - k_1)A' = k_2B + (1 - k_2)B' = k_3C + (1 - k_3)C'.$$

Hence

$$k_2B - k_3C = (1 - k_3)C' - (1 - k_2)B'.$$

If the intersection Q_1 of BC and $B'C'$ exists, each of these expressions is equal to $(k_2 - k_3)Q_1$. Hence we have

$$k_2B - k_3C = (k_2 - k_3)Q_1,$$

and similarly

$$k_3C - k_1A = (k_3 - k_1)Q_2,$$

and

$$k_1A - k_2B = (k_1 - k_2)Q_3.$$

Adding these equations, we obtain the equation

$$(k_2 - k_3)Q_1 + (k_3 - k_1)Q_2 + (k_1 - k_2)Q_3 = 0.$$

There are two possibilities. The points Q_1, Q_2 and Q_3 are collinear, or they are not collinear. If they are not, since the sum of the coefficients of Q_1, Q_2 and Q_3 in the equation above is zero, application of Thm 4, §3.1 enables us to say that $k_1 = k_2 = k_3 = k$, say. In this case, we have $k(B - C) = (1 - k)(C' - B')$, with two more similar equations, and we deduce that BC is parallel to $B'C'$, CA is parallel to $C'A'$, and AB is parallel to $A'B'$.

But if this is *not* the case, so that the points Q_1, Q_2 and Q_3 exist, then they are also collinear.

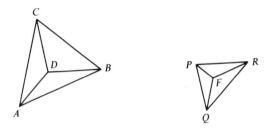

Fig. 6.1

Exercises

5.1   In the proof of Thm 5.2, how can we continue if Q_1 does not exist, but Q_2 and Q_3 exist?

5.2   Without using vectors, can you prove that the Desargues Theorem implies the converse theorem, that is: if ABC, $A'B'C'$ are triangles with distinct vertices, and the intersections Q_1, Q_2 and Q_3 of corresponding sides are collinear, then the joins AA', BB' and CC' of corresponding vertices are concurrent?

5.3   The triangles $A_1B_1C_1$, $A_2B_2C_2$ and $A_3B_3C_3$ have their corresponding sides parallel. Each pair of triangles has a center of perspective. Prove that the three centers of perspective are collinear.

The next two Exercises refer to the Euclidean plane

5.4   If $ABCD$ are any four distinct points, and P,Q are points on AD,BC such that $\overline{AP}:\overline{PD} = |AB|:|DC| = \overline{BQ}:\overline{QC}$, show that $Q - P$ can be written as the sum of vectors of equal length respectively parallel to AB and CD, and deduce that PQ is equally inclined to AB and DC.

5.5   Deduce from the previous Exercise that the internal bisector of an angle of a triangle divides the opposite sides in the ratio of the sides containing the angle.

6.1   Reciprocal figures

Theorem   *If ABC, PQR are two coplanar triangles, and there is a point D such that the sides QR, RP and PQ of triangle PQR are respectively parallel to DA, DB and DC, then there is also a point F such that the sides BC, CA and AB of triangle ABC are respectively parallel to FP, FQ and FR (see Fig. 6.1).*

Remark. Two such triangles are called *reciprocal* triangles. This theorem first arose in graphical statics. Books have been written on the subject of reciprocal figures. Clerk Maxwell gave an ingenious proof of the above theorem, in a paper on the subject. This will appear later, in connection with circles (§28.4).

Proof. By Thm 4.2 we may write

$$D = xA + yB + zC, \text{ where } x + y + z = 1.$$

From the given conditions we also have

$$Q - R = k_1(A - D), \ R - P = k_2(B - D), \ P - Q = k_3(C - D).$$

Hence, by addition,

$$0 = k_1A + k_2B + k_3C - (k_1 + k_2 + k_3)D$$
$$= k_1A + k_2B + k_3C - (k_1 + k_2 + k_3)(xA + yB + zC).$$

The sum of the coefficients of A, B, C in this equation is

$$k_1 + k_2 + k_3 - (k_1 + k_2 + k_3)(x + y + z) = 0,$$

so that, by Thm 4, §3.1, since A, B and C are not collinear points, each coefficient is zero. Hence

$$k_1 = x(k_1 + k_2 + k_3), \ k_2 = y(k_1 + k_2 + k_3), \ k_3 = z(k_1 + k_2 + k_3).$$

We now introduce the point

$$F = xP + yQ + zR, \text{ where we still have } x + y + z = 1,$$

and show that this is the point F described in the theorem.
We have

$$P - F = P - (1 - y - z)P - yQ - zR = y(P - Q) + z(P - R),$$

so that

$$(k_1 + k_2 + k_3)(P - F) = k_2(P - Q) + k_3(P - R)$$
$$= k_2k_3(C - D) - k_2k_3(B - D) = k_2k_3(C - B).$$

Hence PF is parallel to BC, and similarly we prove that QF is parallel to CA, and RF to AB.

Remark. It will be observed that in Fig. 6.1 the lines in one figure which pass through a point are parallel to the lines in the other figure which form the sides of a triangle. This is characteristic of reciprocal figures, in the sense that this is how they are defined, but, starting with a given figure, it may be impossible to find a figure reciprocal to it.

Exercises

6.1 What new theorem do we obtain if in Fig. 6.1, where we are now in the Euclidean plane, we rotate triangle ABC through a right angle about the point D, keeping triangle PQR fixed?

6.2 Draw a figure which does not have a reciprocal in the sense described above.

7.1 Inner products in Euclidean space of two dimensions

We shall now consider theorems which are no longer valid in both the affine and Euclidean planes, but only in the Euclidean plane. We therefore start again, with orthogonal coordinate axes (Fig. 7.1). Let $A = (a_1, a_2)$ and $B = (b_1, b_2)$ be two vectors

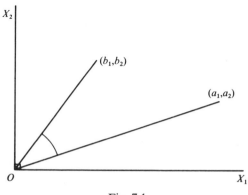

Fig. 7.1

in the Euclidean plane. We define a product $A \cdot B$ for these two vectors, sometimes called dot product, sometimes scalar product and sometimes inner product:

$$A \cdot B = a_1 b_1 + a_2 b_2.$$

This is a real number, and we note the following properties, which are immediately proved:

$$A \cdot B = B \cdot A,$$

$$(kA \cdot B) = (A \cdot kB) = k(A \cdot B),$$

and

$$(A + B) \cdot C = A \cdot C + B \cdot C, \text{ (distributive law)},$$

where $C = (c_1, c_2)$ is any vector.

If $B = A$, then $A \cdot A = a_1{}^2 + a_2{}^2 = $ (length of A)^2.

We write $|A|$ (*modulus* of A) for the length of A, and therefore have the equation

$$A^2 = A \cdot A = |A|^2.$$

If we consider the vector $C = A - B = (a_1 - b_1, a_2 - b_2)$, then

$$C^2 = C \cdot C = (A - B) \cdot (A - B) = (a_1 - b_1)^2 + (a_2 - b_2)^2.$$

We have already remarked that the position-vector $A - B$ is equal to the vector $\overrightarrow{BA}$, where A and B are the respective endpoints of the vectors A and B. We therefore have

$$(A - B)^2 = (\text{distance } AB)^2.$$

This is a formula we shall use often. We now give the geometrical interpretation of the inner product of two vectors A and B.

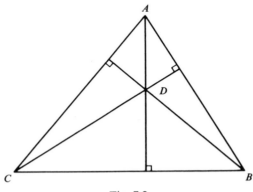

Fig. 7.2

If A makes an angle θ_1 with OX_1, then $a_1 = |A| \cos \theta_1$, and $a_2 = |A| \sin \theta_1$. Similarly, if B makes an angle θ_2 with OX_1, we have $b_1 = |B| \cos \theta_2$, $b_2 = |B| \sin \theta_2$. Hence

$$A \cdot B = |A||B| (\cos \theta_1 \cos \theta_2 + \sin \theta_1 \sin \theta_2)$$

$$= |A||B| \cos (\theta_1 - \theta_2).$$

In particular, if the vectors A and B are *perpendicular*,

$$A \cdot B = 0.$$

This is also a *sufficient* condition for perpendicularity if neither A nor B is the zero vector $(0, 0)$.

7.2 Example

In a triangle ABC (Fig. 7.2) let D be the intersection of the perpendiculars from A onto BC and from B onto CA. Then

$$(D - B) \cdot (C - A) = 0 = (D - A) \cdot (B - C).$$

On adding these two equations and using the distributive law,

$$0 = (D - A) \cdot (B - C) + (D - B) \cdot (C - A)$$

$$= D \cdot (B - A) - A \cdot (B - C) - B \cdot (C - A)$$
$$= D \cdot (B - A) + C \cdot (A - B) = (D - C) \cdot (A - B),$$

and *we deduce that CD is perpendicular to AB.*

The point D is called the *orthocenter* of triangle ABC. We see that *each of the four points A, B, C and D is the orthocenter of the triangle formed by the other three.* Such a set of four points is called an *orthocentric tetrad.*

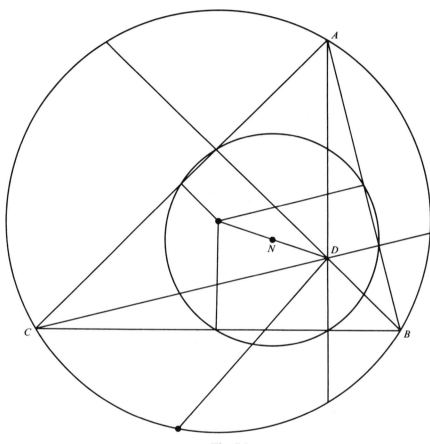

Fig. 7.3

7.3   The nine-point circle

We have already indicated that we shall write $P \cdot P = P^2$ for the inner product of a vector P with itself. Suppose that A, B, C and D form an orthocentric tetrad (see Fig. 7.3). If we evaluate the expression

$$(A - D + B - C)^2 - (A - D - B + C)^2,$$

our rules for the inner product show that this can be evaluated as if it were an ordinary algebraic expression, and we have

$$(A - D + B - C)^2 - (A - D - B + C)^2 = 4(A - D) \cdot (B - C) = 0,$$

since AD is perpendicular to BC.
Again

$$(A - D + B - C)^2 - (A + D - B - C)^2 = 4(B - D) \cdot (A - C) = 0,$$

and

$$(A + D - B - C)^2 - (A - D - B + C)^2 = 4(C - D) \cdot (B - A) = 0,$$

since BD is perpendicular to AC, and CD is perpendicular to AB. Hence

$$(A + B - C - D)^2 = (A - B + C - D)^2 = (A - B - C + D)^2.$$

We know that for any vector P, $(-P)^2 = (P)^2$. If we multiply each vector inside the parentheses by -1, we have the equalities:

$$(A + B - C - D)^2 = (-A - B + C + D)^2 = (A - B + C - D)^2$$
$$= (-A + B - C + D)^2 = (A - B - C + D)^2$$
$$= (-A + B + C - D)^2.$$

If we introduce the point $N = (A + B + C + D)/4$, these equalities can be written in the form:

$$[N - (C + D)/2]^2 = [N - (A + B)/2]^2 = [N - (B + D)/2]^2$$
$$= [N - (A + C)/2]^2 = [N - (B + C)/2]^2$$
$$= [N - (A + D)/2]^2.$$

If we now recall that the distance $|PQ|$ between two points P and Q is given by the formula

$$|PQ|^2 = (P - Q) \cdot (P - Q) = (P - Q)^2,$$

the six equalities we have just obtained show that the point N is equidistant from the six midpoints of the points A, B, C and D, taken in pairs. These midpoints therefore lie on a circle, center N. In Exercise 7.4 a method is suggested by which it may be shown that this circle also passes through the intersections of AD and BC, of BD and CA and of CD and AB. These 9 points, intimately associated with a triangle ABC, justify the name 'nine-point circle' for the circle center N. We shall come across it again, in §12.1

Exercises

7.1 From the formula

$$|PQ|^2 = (P - Q)^2,$$

deduce the cosine formula for the triangle OPQ.

7.2 If A, B, C and D are any set of four coplanar points, and A' is the midpoint of BC, B' is the midpoint of CA and C' is the midpoint of AB, prove that

$$(D - A') \cdot (C - B) + (D - B') \cdot (A - C) + (D - C') \cdot (B - A) = 0.$$

7.3 Prove that if the joins of the midpoints of AB and CD and of the midpoints of BC and AD are equal in length, then AC is perpendicular to BD, and conversely, where A, B, C and D are any set of four coplanar points.

7.4 Show that the join of the midpoints of AD and BC, where A, B, C and D form an orthocentric tetrad, passes through the center N of the circle through the six points described above in §7.3, and that therefore this circle also passes through the intersection of AD and BC.

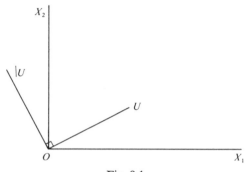

Fig. 8.1

8.1   The supplement of vectors in a plane

If U be a vector in the Euclidean plane (Fig. 8.1), we define the *supplement* of U to be the vector obtained from U by rotating it, with unchanged magnitude, through a right angle in the positive direction, from OX_1 towards OX_2. We denote this vector by $|U$.

If $U = (a_1, a_2)$, so that as a complex number $U = a_1 + ia_2$, the complex number which corresponds to the supplement of U is obtained from U by multiplication by i, so that

$$|U = -a_2 + ia_1.$$

A similar relation holds if we write $U = a_1E_1 + a_2E_2$, in terms of the unit vectors E_1 and E_2, and use the facts that $|E_1 = E_2$, and $|E_2 = -E_1$, and that the supplementation process is distributive over addition. We then have

$$|U = -a_2E_1 + a_1E_2,$$

and

$$||U = -a_1E_1 - a_2E_2 = -U,$$

as we should expect.

We now have the relations

$$|(k_1U_1 + k_2U_2) = k_1|U_1 + k_2|U_2.$$

The supplement of the zero vector is again the zero vector, and any linear relation, such as

$$k_1 U_1 + k_2 U_2 + k_3 U_3 = 0$$

becomes

$$k_1 |U_1 + k_2| U_2 + k_3 |U_3 = |0 = 0,$$

when we take the supplement. Since $||U = -U$, the linear relation

$$|(k_1 U_1 + k_2 U_2 + k_3 U_3) = 0,$$

becomes

$$||(k_1 U_1 + k_2 U_2 + k_3 U_3) = |0 = 0,$$

and this is

$$-(k_1 U_1 + k_2 U_2 + k_3 U_3) = 0,$$

so that we may replace a given linear relation which has been supplemented by the original linear relation.

There is a connexion between supplements and inner products. If $U = (a_1, a_2)$, and $V = (b_1, b_2)$, then $|V = (-b_2, b_1)$, and $|U = (-a_2, a_1)$, and whereas

$$U \cdot |V = -a_1 b_2 + a_2 b_1,$$

$$V \cdot |U = -a_2 b_1 + a_1 b_2 = -U \cdot |V.$$

This is, of course, evident geometrically.

8.2   Example

A square is described with a given side AB (Fig. 8.2). Prove that if P be the center of the square, then

$$P = (A + B)/2 \pm |(A - B)/2,$$

the sign depending on the orientation of the square.

We can reach the center of the square by moving half the way from B towards A, and at the midpoint of AB turning through a right angle and moving through a length $\overline{AB}/2$. This is expressed in the vector form of the point P, and both cases are shown in Fig. 8.2.

8.3   Orthologic triangles

If a triangle PQR (Fig. 8.3) has its sides QR, RP and PQ respectively perpendicular to DA, DB and DC, where A, B, C and D are a coplanar tetrad of points, then there is a point F such that FP, FQ and FR are respectively perpendicular to BC, CA and AB.

The triangles PQR and ABC are called *orthologic triangles* if this reciprocal relationship exists between them. We have already indicated one proof of this theorem, by rotation of one of the triangles through a right angle, when we are back at the consideration of a pair of reciprocal triangles (Exercise 6.1, and §6.1). We make use of the idea of supplements to prove the theorem on orthologic triangles, using practically the same algebra that was used in §6.1.

Let $D = xA + yB + zC$, where $x + y + z = 1$.
Then since QR is perpendicular to DA, we have

$$Q - R = |k_1(A - D),$$

and we also have

$$R - P = |k_2(B - D), \text{ and } P - Q = |k_3(C - D),$$

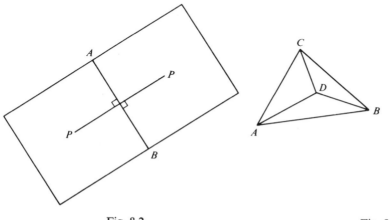

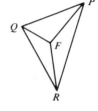

Fig. 8.2   Fig. 8.3

since RP is perpendicular to DB, and PQ is perpendicular to DC. On adding these equations, we see that

$$0 = |[k_1A + k_2B + k_3C - (k_1 + k_2 + k_3)(xA + yB + zC)].$$

We may remove the supplemental $|$ from this equation, and as in the proof in §6.1, since A, B and C are not collinear, and the sum of the coefficients of A, B and C in this equation is zero, we must have each coefficient equal to zero. Hence

$$k_1 = (k_1 + k_2 + k_3)x, \; k_2 = (k_1 + k_2 + k_3)y, \; k_3 = (k_1 + k_2 + k_3)z.$$

Now consider the point $F = xP + yQ + zR$, where $x + y + z = 1$. Then

$$P - F = y(P - Q) + z(P - R)$$
$$= k_2k_3|(C - D) + k_3k_2|(D - B)$$
$$= k_2k_3|(C - B),\,.$$

and therefore $P - F$ is perpendicular to $B - C$, that is, PF is perpendicular to BC. Similarly QF is perpendicular to CA, and RF is perpendicular to AB.

Exercises

8.1 If P, Q, R and S be the centers of squares described externally on the sides AB, BC, CD and DA respectively of any quadrilateral $ABCD$, prove that the segments PR and QS are perpendicular and equal in length.

8.2 If A, B are two distinct points, and $C = tA + (1 - t)B + k|(B - A)$, where t and k are fixed real numbers, prove that the triangle ABC is always similar to a fixed triangle, however A and $B \neq A$ are chosen.

8.3 Similar triangles BCX, CAY and ABZ are described on the sides of a triangle ABC. Show that the centroids of ABC and XYZ are the same point.

8.4 Prove that the centroids of equilateral triangles described externally on the sides of any triangle form an equilateral triangle. (Compare Exercise 16.1.)

9.1 The exterior product of two vectors

The vector space V we have been working in so far is a two-dimensional space over the field of the real numbers. We now construct a vector space, which we denote by Λ^2, over the field of the reals. We call this space **the space of 2-vectors on** V.

It consists of all sums

$$\Sigma x_i(u_i \wedge v_i),$$

where the x_i are real numbers, and the u_i, v_i are vectors in V which are subject to the following rules for the 'wedge' product $\wedge$:

$$(x_1 u_1 + x_2 u_2) \wedge v = x_1(u_1 \wedge v) + x_2(u_2 \wedge v),$$
$$u \wedge (y_1 v_1 + y_2 v_2) = y_1(u \wedge v_1) + y_2(u \wedge v_2),$$
$$u \wedge u = 0,$$

and

$$u \wedge v + v \wedge u = 0.$$

In other words the wedge product (also called the *exterior* product) of two vectors is distributive over addition, but non-commutative. The rules given above are not independent. From

$$(u + v) \wedge (u + v) = 0,$$

we deduce that

$$u \wedge u + u \wedge v + v \wedge u + v \wedge v = 0,$$

which leads to

$$u \wedge v + v \wedge u = 0.$$

Since the wedge product $u \wedge v$ is a member of a vector space over the field of real numbers, the equation $ku \wedge v = 0$, where the vector on the right is the zero vector of the space Λ^2, leads to the two possibilities: either $k = 0$, or $u \wedge v = 0$.

If u and v are dependent, so that $bv = cu$, where, say, $b \neq 0$, then

$$u \wedge v = u \wedge (b^{-1}cu) = bc^{-1}(u \wedge u) = 0.$$

If this is not the case, we shall soon prove that $u \wedge v \neq 0$.

9.2   Geometrical interpretation of exterior product

In Fig. 9.1 let $u = x_1 E_1 + x_2 E_2$, $v = y_1 E_1 + y_2 E_2$, where E_1, E_2 are unit vectors parallel to OX_1, OX_2 respectively. Then

$$u \wedge v = (x_1 E_1 + x_2 E_2) \wedge (y_1 E_1 + y_2 E_2)$$

$$= (x_1 y_2 - x_2 y_1) E_1 \wedge E_2.$$

The expression $x_1 y_2 - x_2 y_1$ is equal to twice the area of the triangle formed by the origin and the points (x_1, x_2), (y_1, y_2), and it is a positive real number if the orientation of the triangle is as shown, from OX_1 towards OX_2.

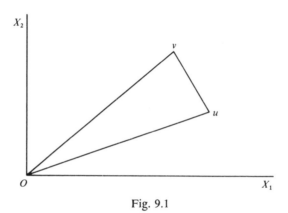

Fig. 9.1

The equation for $u \wedge v$ shows also that the exterior product of the basis vectors $E_1 \wedge E_2$ is a basis for the space of 2-vectors on V. This means that the dimension of Λ^2 is equal to one. Since we do not wish the vector space Λ^2 to consist of only the zero vector, we assume that $E_1 \wedge E_2 \neq 0$. It then follows that $u \wedge v = 0$ if and only if u and v are linearly dependent, which means that $b(x_1, x_2) = c(y_1, y_2)$, where b, c are not both zero.

If we evaluate $u \wedge v \wedge w$, where u, v and w are all in V, we obtain a sum $\Sigma x_i y_j z_k (E_i \wedge E_j \wedge E_k)$, where i, j, k are either 1 or 2. Since $E_i \wedge E_i = 0$, and $E_j \wedge 0 = 0$, and $0 \wedge E_k = 0$, it follows that the triple exterior product $u \wedge v \wedge w$ is always zero.

9.3   A criterion for collinearity

Theorem   *The points A, B and C are collinear if and only if*

$$A \wedge B + B \wedge C + C \wedge A = 0.$$

Proof. If $A = x_1E_1 + x_2E_2$, $B = y_1E_1 + y_2E_2$, $C = z_1E_1 + z_2E_2$, evaluation of $A \wedge B + B \wedge C + C \wedge A$ gives us

$$[(x_1y_2 - x_2y_1) + (y_1z_2 - y_2z_1) + (z_1x_2 - z_2x_1)]E_1 \wedge E_2 = 0,$$

and therefore an equivalent criterion is the determinantal expression

$$\begin{vmatrix} x_1 & x_2 & 1 \\ y_1 & y_2 & 1 \\ z_1 & z_2 & 1 \end{vmatrix} = 0,$$

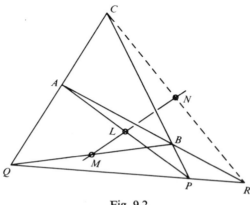

Fig. 9.2

which we know is a necessary and sufficient condition for the collinearity of the points (x_1, x_2), (y_1, y_2) and (z_1, z_2).

Exercises

9.1 Prove that $u \wedge |v = (u \,.\, v)E_1 \wedge E_2$.

9.2 Prove Thm 9.3 by the use of vector methods.

9.3 Prove Thm 9.3 from the consideration that the given criterion leads to a certain triangle having zero area.

9.4 Deduce from Thm 9.3 that if A, B and C are vectors, and the endpoints determine a triangle of area $\triangle$, then $A \wedge B + B \wedge C + C \wedge A = 2\triangle E_1 \wedge E_2$.
    Show that an equivalent expression for $2\triangle E_1 \wedge E_2$ is $(A - C) \wedge (B - C)$.

9.4 Application of the criterion

Let A, B, P, Q be four points such that no set of three of the lines AB, BP, PQ and QA are concurrent (Fig. 9.2). These four lines are said to form a *complete quadrilateral*.

If AQ, BP meet in C, AB, PQ in R, the lines AP, BQ, CR are *diagonals*. We prove that the midpoints of the diagonals of a complete quadrilateral $ABPQ$ are collinear.

This theorem goes back to Newton. Let L, M and N be the midpoints of AP, BQ, and CR respectively. We wish to prove that the points L, M and N are collinear. Since they are midpoints, we have

$$2L = A + P, 2M = B + Q, 2N = C + R.$$

Hence

$$4L \wedge M = (A + P) \wedge (B + Q) = A \wedge B + A \wedge Q + P \wedge B + P \wedge Q,$$
$$4M \wedge N = (B + Q) \wedge (C + R) = B \wedge C + B \wedge R + Q \wedge C + Q \wedge R,$$
$$4N \wedge L = (C + R) \wedge (A + P) = C \wedge A + C \wedge P + R \wedge A + R \wedge P.$$

We add these equations, and note that since A, B and R are collinear we have $A \wedge B + B \wedge R + R \wedge A = 0$, that since A, Q and C are collinear we have $A \wedge Q + Q \wedge C + C \wedge A = 0$, and finally that since P, B and C are collinear we have $P \wedge B + B \wedge C + C \wedge P = 0$, and since P, Q and R are collinear we have $P \wedge Q + Q \wedge R + R \wedge P = 0$. This exhausts the sum on the right-hand side, and leads to $L \wedge M + M \wedge N + N \wedge L = 0$, which proves that the points L, M and N are collinear.

Another proof will be given in §11.1.

9.5 The parallel case of Pappus' theorem

Theorem *If A, B, C and A', B', C' are collinear triads of points, and if BC' is parallel to $B'C$ and CA' is parallel to $C'A$, then the line AB' is parallel to the line $A'B$. (See Fig. 9.3.)*

Proof. We obtain the equation

$$(A - B') \wedge (A' - B) + (B - C') \wedge (B' - C) + (C - A') \wedge (C' - A)$$
$$= (B' \wedge C' + C' \wedge A' + A' \wedge B') - (B \wedge C + C \wedge A + A \wedge B)$$

by simple evaluation of the left-hand side, and using the fundamental identities $A \wedge A' + A' \wedge A = B \wedge B' + B' \wedge B = C \wedge C' + C' \wedge C = 0$. Since A, B and C are collinear, and also A', B' and C' are collinear, the theorem of § 9.3 shows us that both terms on the right-hand side are zero. Since BC' is parallel to $B'C$, $B - C'$ is parallel to $B' - C$, and so $(B - C') \wedge (B' - C)$ is zero. Similarly since CA' is parallel to $C'A$, $(C - A') \wedge (C' - A)$ is zero, and so we end up with $(A - B') \wedge (A' - B)$ zero, which means that $A - B'$ and $A' - B$ are dependent, or AB' is parallel to $A'B$.

Remark. The Pappus Theorem says that the opposite sides of the hexagon $AB'CA'BC'$, the vertices A, B, C lying on one line, and the vertices A', B', C' lying on another line, intersect in three collinear points. Theorem 9.5 deals with the case when the three collinear points are on the line at infinity. Pascal's Theorem (Cor. Thm III, §78.2) is a generalization of the theorem to the case when the six vertices of the hexagon lie on any conic. In applications of the Pappus Theorem care has to be taken that the hexagon considered does have alternate vertices on one of two lines, and that the intersections of opposite sides are considered. The criss-cross pattern of the formulation of the theorem is easy to remember (Fig. 9.4).

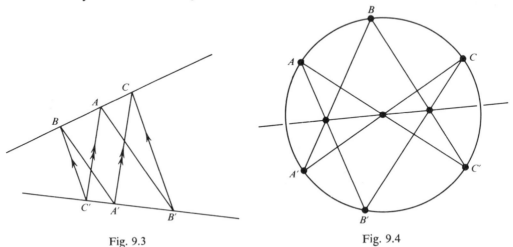

Fig. 9.3   Fig. 9.4

Exercise

9.5 A line drawn through the vertex A of a parallelogram $ABCD$ cuts CB in P and CD in Q. A line through C cuts AB in R and AD in S. Prove that PR and QS are parallel.

10.1 A necessary and sufficient condition for the concurrency of three lines

Thm 9.3 gives us a necessary and sufficient condition for the endpoints of the three position vectors A, B and C to be collinear, namely

$$B \wedge C + C \wedge A + A \wedge B = 0.$$

It is useful to have a criterion for three lines, each passing through one vertex of a given triangle, to be concurrent. Of course, we have the Ceva Theorem (§4.3), but this involves the intercepts cut on the sides of the given triangle, and we want a criterion which involves the vectors along the given lines only.

   Suppose then that we have three non-collinear points P, Q and R (Fig. 10.1), given by three position vectors P, Q and R, and that through the point P, and originating at P there passes a free vector P^*, through the point Q, and originating at Q there passes the free vector Q^*, and through the point R and originating at R there passes the free vector R^*. These free vectors P^*, Q^* and R^* determine lines through P, Q and R and we seek conditions for these three lines to be concurrent. We replace the free vectors P^*, Q^* and R^* by vectors P', Q' and R', along the same lines and also originating at the points P, Q and R respectively. With this explanation of the terms involved we have:

10.2   The theorem itself

Theorem   *A necessary and sufficient condition for the lines of action of the vectors* P^*, Q^* *and* R^* *to be concurrent is the existence of vectors* P', Q' *and* R' *such that*

$$P \wedge P' + Q \wedge Q' + R \wedge R' = 0, \text{ where } P' + Q' + R' = 0.$$

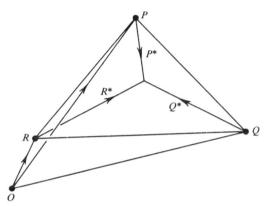

Fig. 10.1

Proof. Necessity. We suppose that the lines of action of the vectors P^*, Q^* and R^* intersect in a point W. Then we may write (Thm 4.2)

$$W = xP + yQ + zR, \text{ where } x + y + z = 1.$$

Since P^*, Q^* and R^* are along the lines PW, QW and RW respectively, we may choose $P' = x(W - P)$, $Q' = y(W - Q)$ and $R' = z(W - R)$. We then find that $P' + Q' + R' = (x + y + z)W - xP - yQ - zR = 0$, and

$$P \wedge P' + Q \wedge Q' + R \wedge R' = (Px + Qy + Rz) \wedge W = W \wedge W = 0.$$

Sufficiency. We are given the existence of P', Q' and R' satisfying the conditions. Let the line of action of P^* through P intersect the line of action of Q^* through Q in the point W. We may find x and y so that

$$P' = x(W - P), \quad Q' = y(W - Q),$$

since by hypothesis P' is along P^* and Q' is along Q^*. Since

$$P \wedge P' + Q \wedge Q' + R \wedge R' = 0,$$

$$xP \wedge (W - P) + yQ \wedge (W - Q) + R \wedge R' = 0.$$

Also

$$P' + Q' = (x + y)W - (Px + Qy) = -R',$$

so that

$$Px + Qy = (x + y)W + R',$$

and on substituting in

$$xP \wedge (W - P) + yQ \wedge (W - Q) + R \wedge R' = (Px + yQ) \wedge W + R \wedge R' = 0,$$

we obtain the equation

$$[(x + y)W + R'] \wedge W + R \wedge R' = 0,$$

which simplifies to

$$R' \wedge W + R \wedge R' = 0,$$

which is

$$R' \wedge (W - R) = 0.$$

Hence $R' = z(W - R)$, for some definite z, and since R' is along R^*, the line of action of R^* passes through W.

Remark. The theorem continues to hold if W is not a finite point. If P' is parallel to Q', the condition $P' + Q' + R' = 0$ shows that R' is also parallel to P' and to Q', and so P^*, Q^* and R^* are all parallel. If we are given that P^*, Q^* and R^* are all parallel, we can prove the necessity of the given conditions thus:

We take $P' = xU$, $Q' = yU$, $R' = zU$, where U is a vector in the given parallel direction, and since we wish $P' + Q' + R'$ to be zero, we have $x + y + z = 0$. If we look at

$$P \wedge P' + Q \wedge Q' + R \wedge R'$$
$$= xP \wedge U + yQ \wedge U + zR \wedge U = [x(P - R) + y(Q - R)] \wedge U,$$

on substituting for z, we note that since P, Q and R are *not* collinear, the vectors $P - R$ and $Q - R$ are *not* along the same line, that is, they are independent. Since any vector in the plane can be expressed in terms of two independent vectors, we can choose x and y, which are at our disposal, so that $x(P - R) + y(Q - R) = U$, and then $[x(P - R) + y(Q - R)] \wedge U = 0$.

10.3   Applications of the concurrency conditions

If ABC is any triangle, the medians are parallel to the vectors

$$P' = A - (B + C)/2, \quad Q' = B - (C + A)/2, \quad R' = C - (A + B)/2.$$

It is clear that $P' + Q' + R' = 0$, and

$$A \wedge P' + B \wedge Q' + C \wedge R' = 0,$$

so that the medians are concurrent.

If we now consider the perpendiculars from A, B and C respectively onto BC, CA and AB, these are parallel to

$$P' = |(B - C), \quad Q' = |(C - A), \quad R' = |(A - B),$$

and $P' + Q' + R' = 0$, and if we consider

$$A \wedge |(B - C) + B \wedge |(C - A) + C \wedge |(A - B),$$

and use the result of Exercise 9.1, that $u \wedge |v = (u \cdot v)E_1 \wedge E_2$, we are led to consider the expression

$$A \cdot (B - C) + B \cdot (C - A) + C \cdot (A - B),$$

and this is zero, so that the three perpendiculars are concurrent.

Finally, the theorem on orthologic triangles (§8.3) is easily proved by these methods. The triangle PQR has its sides QR, RP and PQ respectively perpendicular to DA, DB and DC. Hence

$$(Q - R) \cdot (D - A) = (R - P) \cdot (D - B) = (P - Q) \cdot (D - C) = 0.$$

If we add these three equations, we obtain the equation

$$(Q - R) \cdot A + (R - P) \cdot B + (P - Q) \cdot C = 0,$$

which can be written

$$P \cdot (C - B) + Q \cdot (A - C) + R \cdot (B - A) = 0.$$

But by the result of Exercise 9.1 this equation can also be written

$$P \wedge |(C - B) + Q \wedge |(A - C) + R \wedge |(B - A) = 0.$$

Hence if $P' = |(C - B)$, $Q' = |(A - C)$, and $R' = |(B - A)$, we not only have $P' + Q' + R' = 0$, but $P \wedge P' + Q \wedge Q' + R \wedge R' = 0$. Hence vectors in the direction of P', Q' and R' respectively through the vertices P, Q and R of triangle PQR are concurrent. That is, perpendiculars to BC, CA and AB respectively through the points P, Q and R are concurrent, which is what we wish to prove.

Exercises

10.1 Prove that if triangles $A_1A_2A_3$ and $B_1B_2B_3$ are orthologic, and C_1, C_2 and C_3 divide the segments A_1B_1, A_2B_2 and A_3B_3 in the same ratio, then the triangle $C_1C_2C_3$ is orthologic both to $A_1A_2A_3$ and to $B_1B_2B_3$.

10.2 If the Exercise above has been proved by the use of Thm 10.2, consider the following: Let D_1 be the point of intersection of perpendiculars from B_1 onto A_2A_3, from B_2 onto A_3A_1 and from B_3 onto A_1A_2, and let H be the orthocenter of triangle $A_1A_2A_3$. Prove that the perpendicular from C_1 onto A_2A_3 meets D_1H in a point D′ which divides D_1H in the same ratio that C_1 divides B_1A_1. Deduce that perpendiculars from C_1 onto A_2A_3, from C_2 onto A_3A_1 and from C_3 onto A_1A_2 all pass through D′.

11.1   The exterior product of three vectors

We have seen that the exterior product of three vectors of our two-dimensional vector space is always zero. It is useful to consider the exterior product of three vectors lying in a *three*-dimensional vector space. The rules we shall use will be an evident extension of those used already. Besides the two distributive laws, we also have

$$u \wedge v = -v \wedge u,$$

and

$$u \wedge u = 0,$$

where u, v are any two vectors of the three-dimensional vector space. We are still interested in the Euclidean plane, and we obtain a useful tool for dealing with the area

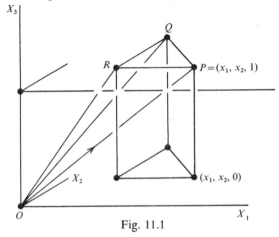

Fig. 11.1

of a triangle in this plane if we suppose that our vectors in a three-dimensional space have their endpoints on a fixed plane, the plane $X_3 = 1$. (See Fig. 11.1.)

Let $P = (x_1, x_2, 1) = x_1 E_1 + x_2 E_2 + E_3$, where E_1, E_2 and E_3 are unit vectors parallel to the coordinate axes OX_1, OX_2, OX_3. Then we note that the foot of the perpendicular from the endpoint P onto the plane $OX_1 X_2$ is just the point (x_1, x_2) in this plane. If we have two other vectors

$$Q = (y_1, y_2, 1) = y_1 E_1 + y_2 E_2 + E_3,$$

and

$$R = (z_1, z_2, 1) = z_1 E_1 + z_2 E_2 + E_3,$$

and we calculate

$$P \wedge Q \wedge R = (x_1 E_1 + x_2 E_2 + E_3) \wedge (y_1 E_1 + y_2 E_2 + E_3) \wedge (z_1 E_1 + z_2 E_2 + E_3),$$

we see that this exterior product is, in the first place

$$[(x_1 y_2 - x_2 y_1)E_1 \wedge E_2 + (x_1 - y_1)E_1 \wedge E_3 + (x_2 - y_2)E_2 \wedge E_3]$$
$$\wedge (z_1 E_1 + z_2 E_2 + E_3),$$

and if we admit the exterior multiplication of E_i, E_j and E_k, where i, j and k are 1, 2 or 3, we naturally have

$$E_i \wedge E_j \wedge E_i = -(E_j \wedge E_i) \wedge E_i = -E_j \wedge (E_i \wedge E_i) = -E_j \wedge 0 = 0,$$

whereas if i, j and k are all distinct,

$$E_i \wedge E_j \wedge E_k = -E_i \wedge E_k \wedge E_j \neq 0,$$

and the basis for exterior multiplication in the three-dimensional vector space is the product $E_1 \wedge E_2 \wedge E_3$. We now see that

$$P \wedge Q \wedge R = [(x_1y_2 - x_2y_1) + (x_2z_1 - x_1z_2) + (y_1z_2 - y_2z_1)]E_1 \wedge E_2 \wedge E_3$$

$$= \begin{vmatrix} x_1 & x_2 & 1 \\ y_1 & y_2 & 1 \\ z_1 & z_2 & 1 \end{vmatrix} E_1 \wedge E_2 \wedge E_3,$$

and we recognize that the determinantal expression is precisely twice the area of the triangle formed by the points P, Q and R in the plane $X_3 = 1$ (or by the feet of the perpendiculars from P, Q and R onto the plane OX_1X_2).

We now have another method for considering the areas of triangles in a plane. We can suppose the triangles to lie in the plane $X_3 = 1$, and use the exterior product of vectors issuing from $(0, 0, 0)$ with their endpoints in this plane. Since a necessary and sufficient condition for three points to be collinear is that the area of the triangle formed by the points be zero, we now have another criterion for the collinearity of three points, namely $L \wedge M \wedge N = 0$, if the points are given by L, M and N, instead of $L \wedge M + M \wedge N + N \wedge L = 0$.

11.2   Applications

We once again consider the Theorem of Menelaus (§4.1). We have a triangle ABC (Fig. 11.2), with a line cutting BC, CA and AB in the points L, M and N respectively. If we take $L = xB + x'C$, $M = yC + y'A$, and $N = zA + z'B$, where

$$x + x' = y + y' = z + z' = 1,$$

then since L, M and N are collinear, we have

$$L \wedge M \wedge N = (xB + x'C) \wedge (yC + y'A) \wedge (zA + z'B) = 0,$$

and on expansion, this reduces to

$$(xyz + x'y'z')A \wedge B \wedge C = 0.$$

Since A, B and C are not collinear, we deduce that $xyz + x'y'z' = 0$, and this is the Theorem of Menelaus.

Again, let us consider the theorem of §9.4, that the midpoints of the diagonals of a complete quadrilateral $ABPQ$ are collinear (Fig. 11.3).

We note that

$$A \wedge B \wedge R = P \wedge Q \wedge R = A \wedge Q \wedge C = P \wedge B \wedge C = 0,$$

since the corresponding triads of points are collinear. Hence

$$8L \wedge M \wedge N = (A + P) \wedge (B + Q) \wedge (C + R)$$
$$= (A \wedge B \wedge C - Q \wedge P \wedge C) - (A \wedge R \wedge Q - P \wedge B \wedge R)$$

after reduction. Now each expression in parentheses is proportional to the area of the quadrilateral $ABPQ$. Hence $L \wedge M \wedge N = 0$, and the midpoints are collinear.

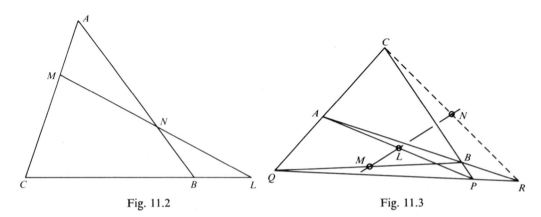

Fig. 11.2   Fig. 11.3

11.3   Example

We now give an application of the concept of the exterior product of three vectors to the area of triangles. Let ABC be a given triangle, and let D, E and F be points on BC, CA and AB respectively dividing the segments BC, CA and AB in the same ratio, $k:1$. We wish to find the ratio *area of triangle DEF : area of triangle ABC*.

The points D, E and F are given by the equations

$$(1 + k)D = B + kC, (1 + k)E = C + kA, (1 + k)F = A + kB.$$

Hence, forming the exterior product:

$$(1 + k)^3 D \wedge E \wedge F = (B + kC) \wedge (C + kA) \wedge (A + kB)$$
$$= (1 + k^3)A \wedge B \wedge C$$

after reduction, all other exterior products being zero. We know that

$$2 \text{ area triangle } DEF (E_1 \wedge E_2 \wedge E_3) = D \wedge E \wedge F,$$

and

$$2 \text{ area triangle } ABC (E_1 \wedge E_2 \wedge E_3) = A \wedge B \wedge C,$$

and therefore

$$\text{area triangle } DEF : \text{area triangle } ABC = 1 + k^3 : (1 + k)^3.$$

Exercises

11.1 Prove that if A_1, B_1 and C_1 are variable points on the sides BC, CA and AB of triangle ABC such that triangle $A_1B_1C_1$ is of constant area, then if AA_1, BB_1 and CC_1 be divided at A_2, B_2 and C_2 in the same fixed ratio $k:1$, triangle $A_2B_2C_2$ will have a constant area.

11.2 If A', B' and C' be the midpoints of BC, CA and AB in a triangle ABC, show that $8A' \wedge B' \wedge C' = 2A \wedge B \wedge C$, and deduce that:

$$\text{area triangle } A'B'C' = \text{area triangle } ABC/4.$$

11.3 If the lengths of the sides of a triangle ABC be denoted by a_0, b_0 and c_0, and the angles by $\hat{A}$, $\hat{B}$ and $\hat{C}$, show that the foot of the perpendicular D' from the vertex A onto the side BC is given by the equation $a_0 D' = (b_0 \cos \hat{C})B + (c_0 \cos \hat{B})C$. Deduce that the area of the triangle formed by the feet of the perpendiculars from the vertices of triangle ABC onto the opposite sides is

$$2 \cos \hat{A} \cos \hat{B} \cos \hat{C} \ (\text{area triangle } ABC).$$

11.4 If L be the point of contact with BC of the inscribed circle of triangle ABC, show that

$$a_0 L = (s_0 - c_0)B + (s_0 - b_0)C,$$

where $2s_0 = a_0 + b_0 + c_0$. Deduce a formula for the area of the triangle formed by the points of contact of the inscribed circle with the sides.

11.5 Obtain the result of the Example above by using the result of Exercise 9.4 for the area of a triangle, involving the exterior product of pairs of vectors.

11.6 Let ABC be a given triangle, and let D, E, F be points on BC, CA and AB respectively dividing the segments BC, CA and AB in the ratio $k_1:1$, $k_2:1$ and $k_3:1$ respectively. Show that

$$\text{area of triangle } DEF : \text{area of triangle } ABC = 1 + k_1k_2k_3 : (1 + k_1)(1 + k_2)(1 + k_3).$$

11.7 Deduce from the preceding Exercise that two triangles whose vertices lie on the sides of a given triangle at equal distances from their midpoints are equal in area. (If DEF is one triangle, and $D'E'F'$ the other, and the midpoints are A', B', C', we are given that

$$\overline{DA'} = \overline{A'D'}, \quad \overline{EB'} = \overline{B'E'} \quad \text{and} \quad \overline{FC'} = \overline{C'F'}.)$$

11.8 D, E, F are points on the sides BC, CA and AB of a triangle ABC such that $D = q_1B + p_1C$, $E = q_2C + p_2A$ and $F = q_3A + p_3B$, where

$$p_1 + q_1 = p_2 + q_2 = p_3 + q_3 = 1.$$

The lines AD, BE and CF form a triangle PQR. Prove that

$$(p_2 + q_2q_3)P = q_3p_2A + p_3p_2B + q_3q_2C.$$

Find expressions for Q and R in terms of A, B and C, and deduce that the ratio of the areas of the triangles PQR and ABC is

$$(p_1p_2p_3 - q_1q_2q_3)^2 : (p_1 + q_1q_2)(p_2 + q_2q_3)(p_3 + q_3q_1).$$

II

CIRCLES

The circle, in various guises, occurs in many branches of mathematics. In this chapter we shall consider those properties which have the habit of appearing in different contexts. Our methods will be mixed. We shall use Euclidean geometry or analytical geometry, according to our needs at the time. We begin with the consideration of certain one-to-one mappings of the plane onto itself, and the product of these mappings. (See 0.11.) Later on we study inversion, which is only one-to-one onto if we cut out a point of the plane, or better still, add a point to the plane.

12.1 The nine-point circle

We have already encountered this circle (§7.3). It is a circle which passes through nine points intimately connected with any given triangle *ABC*. (We shall not be using vector methods, and we shall use the customary notation for the points connected with a

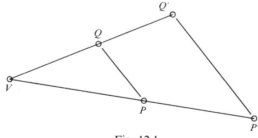

Fig. 12.1

triangle). The nine-point circle is the first really exciting one to appear in any course on Euclidean geometry which goes far enough.

To prove the existence of the nine-point circle of a triangle, we make use of *a one-to-one mapping* of the Euclidean plane onto itself. This mapping is called *a central dilatation*. We choose a point *V*, and call it *the center of the dilatation* (Fig. 12.1).

If *P* is any point in the plane its image, or map *P'* is found by choosing the unique point *P'* which lies on *VP*, is on the same side of *V* as is *P*, and is such that $\overline{VP'} = k\overline{VP}$, where *k* is a fixed real positive number. If *P* = *V*, we define *V'* to be *V*.

This mapping is evidently one-to-one. It is also called *a central similarity*. It will occur again later (§41.5) as one of the fundamental mappings of the Euclidean plane onto itself. Here we are only interested in the case when k is positive, and when the point P moves on a circle, center O (Fig. 12.2). If O' is the map of O, then the triangles VOP and $VO'P'$ are similar, so that $\overline{O'P'} = k\overline{OP}$. Since O is a fixed point, and $|OP| = r$, the radius of the circle on which P moves, we deduce that P' moves on a circle, center O' of radius kr.

Let us now recall some properties of a triangle ABC which we wish to use (Fig. 12.3). We know that the perpendiculars AD', BE' and CF' from the vertices A, B and C onto

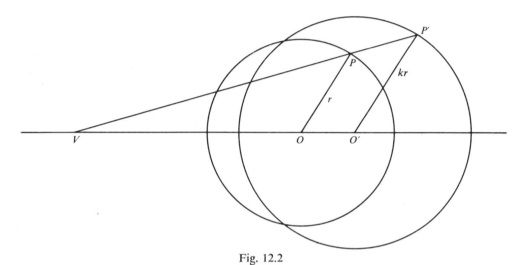

Fig. 12.2

the opposite sides all pass through a point H, the orthocenter (§7.2). If AD' is produced to cut the circumcircle of triangle ABC again at D, it is easy to see that $\overline{HD'} = \overline{D'D}$, since $\sphericalangle BCD = \sphericalangle BAD$ (angles on the same arc), and $\sphericalangle BCH$ is complementary to $\sphericalangle B$, and is therefore equal to $\sphericalangle BAD$. Hence the triangles HCD', DCD' are congruent, and $\overline{HD'} = \overline{D'D}$.

Now consider the central dilatation which has vertex H, and for which $k = 1/2$. This maps every point P on P', where $\overline{HP'} = \overline{HP}/2$. Let P move around the circumcircle of triangle ABC. We know that P' will move on a circle of half the radius, with center bisecting HO, where O is the center of the circumcircle of triangle ABC. This circle will pass through the midpoints of HA, HB and HC, since A, B and C are on the circumcircle, and also through the points D', E' and F', since we have shown that D' is the midpoint of HD. We have therefore found *six* points connected with the triangle ABC which lie on a circle.

Now, a circle is uniquely determined if we are given *three* non-collinear points through which it must pass. Consider the triangle BHC. Its orthocenter is A. By what we have just proved, there is a circle which passes through D', E' and F', the feet of the perpendiculars from H, B and C onto the opposite sides BC, HC and HB, and through the midpoints of AH, AB and AC, A being the orthocenter of triangle HBC. But there is only one circle through the points D', E' and F'. Hence the circle connected with triangle ABC which passes through D', E' and F' and the midpoints of HA, HB

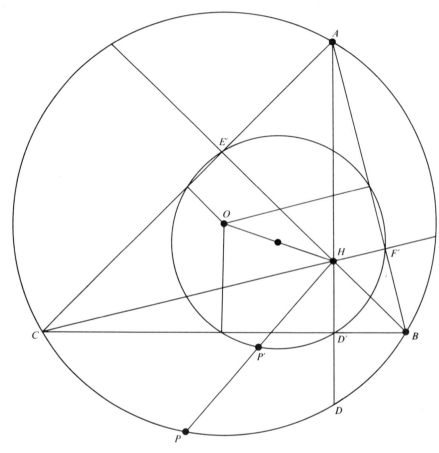

Fig. 12.3

and HC also passes through the midpoints of AB and AC. By considering the triangle HCA we see that this same circle also passes through the midpoint of BC. We have therefore established the existence of a circle which passes through the midpoints of the sides of triangle ABC, the midpoints of HA, HB and HC, and through the feet of the perpendiculars from the vertices of ABC onto the opposite sides. We also know that its center bisects OH, and that its radius is half that of the circumcircle.

This is a first example of the use of a mapping to establish a theorem in geometry. We shall come across other examples fairly soon.

Exercises

12.1   Let AO (where O is the circumcenter of triangle ABC) intersect the circumcircle again in Y. Prove that $BHCY$ is a parallelogram, and deduce that the circle through six points found above also passes through the midpoints of BC, CA and AB.

12.2   Let A', B', C' be the midpoints of BC, CA and AB, and let L', M' and N' be the midpoints of HA, HB and HC. Using the theorem that the join of the midpoints of two sides of a triangle is parallel to the third side prove that both $B'C'M'N'$ and $C'A'N'L'$ are rectangles, and deduce that $A'L'$, $B'M'$ and $C'N'$ are three diameters of one circle. Prove that this circle passes through the feet of the perpendiculars, D', E' and F' from A, B and C onto the sides BC, CA and AB, and so complete another proof of the existence of the nine-point circle.

12.3   We know that N, the center of the nine-point circle, bisects OH (in the notation of §12.1). Deduce from Exercise 12.1 that if A' is the midpoint of BC, then since A' bisects HY and O bisects AY, $\overline{AH} = 2\overline{OA'}$. Now join O to H and show that the centroid of ABC, which is at the point G on AA', where $\overline{AG} = 2\overline{GA'}$, must lie on OH, and $\overline{HG} = 2\overline{GO}$. (The line which contains the circumcenter, centroid, nine-point center and orthocenter of triangle ABC is called the *Euler line* of triangle ABC.)

12.4   If L' is the midpoint of AH, show that $AL'A'O$ is a parallelogram, so that $|A'L'| = |OA| =$ radius of circumcircle of ABC, and therefore, since A' and L' both lie on the nine-point circle, $A'L'$ must be a diameter of the nine-point circle. (Compare Exercise 12.2.)

12.5   On the arc $A'D'$ of the nine-point circle, take the point X one-third the way from A' to D'. Take similar points Y and Z on the arcs $B'E'$ and $C'F'$. Show that the triangle XYZ is equilateral. (Hint: the triangle $X'Y'Z'$ which corresponds to XYZ under the mapping, center H, which maps the circumcircle onto the nine-point circle, must be equilateral.)

12.6   If A', B', C' be the respective midpoints of the sides BC, CA and AB of a triangle ABC, show that O, the circumcenter of ABC, is the orthocenter of $A'B'C'$. From the parallel case of the Desargues Theorem (§5.1) show that AA', BB', CC' pass through a point G. (This, is of course, the centroid of ABC). There is a dilatation, center G, which maps ABC onto $A'B'C'$. Show that H maps onto O, and deduce that H, G and O are collinear, and $\overline{HG} = 2\overline{GO}$. (See Exercise 12.3.) Deduce also that $\overline{AH} = 2\overline{OA'}$.

12.7   Show that O is the incenter of the triangle PQR produced by drawing tangents to the circumcircle of ABC at the vertices. Show also that H is the incenter of the triangle $D'E'F'$, where D' is the foot of the perpendicular from A onto BC, etc. Prove that the sides of PQR and $D'E'F'$ are respectively parallel. Deduce that the joins $D'P$, $E'Q$, $F'R$ pass through a point V, the center of the dilatation which maps $D'E'F'$ on PQR, and that V lies on OH, the Euler line of ABC. Deduce further that since the circumcenter of $D'E'F'$ is the nine-point center of ABC, the circumcenter of PQR lies on the join of V to the nine-point center of ABC, that is, on the Euler line of ABC.

13.1   The centers of similitude of two circles

If V is a center for a dilatation which lies outside a circle $\mathscr{C}$ of center O (see Fig. 13.1), we know already that the image $\mathscr{C}'$ of $\mathscr{C}$ is a circle center O', where $\overline{VO'} = k\overline{VO}$, and that the radius of $\mathscr{C}' = kr$, where r is the radius of $\mathscr{C}$. The points P and P' lie on the same side of V, the radii OP and $O'P'$ are parallel, and $\overline{VO}:\overline{VO'} = 1:k = $ radius of $\mathscr{C}$: radius of $\mathscr{C}'$. If P, Q are distinct points on $\mathscr{C}$, and P', Q' the image points, the chord PQ is parallel to the chord $P'Q'$. If Q moves towards P on $\mathscr{C}$, the point Q' moves towards P' on $\mathscr{C}'$. Hence each of the two tangents from V to $\mathscr{C}$ will be a tangent to $\mathscr{C}'$, and V is the intersection of two common tangents to $\mathscr{C}$ and to $\mathscr{C}'$, each tangent having $\mathscr{C}$ and $\mathscr{C}'$ on the same side of it. We call such tangents *external* common tangents.

If V does not lie outside $\mathscr{C}$, all the statements made above continue to be true, but there are no tangents from V to $\mathscr{C}$, and therefore none from V to $\mathscr{C}'$. (see Fig. 13.2).

These situations can be reproduced with any two given circles. Let us suppose that

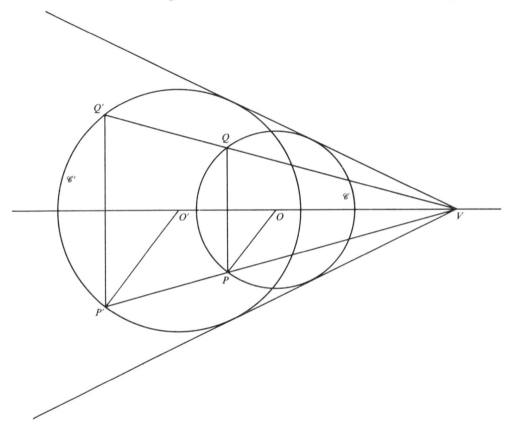

Fig. 13.1

we are given two circles $\mathscr{C}$ and $\mathscr{C}'$, with distinct centers O and O'. Draw parallel radii OP and $O'P'$ *in the same sense* through O and O' respectively. Then the join of the points P and P' will intersect the line of centers OO' of the two circles in a fixed point V (unless the circles are of equal radii), and $\overline{VO} : \overline{VO'}$ = radius of $\mathscr{C}$: radius of $\mathscr{C}'$. This point V is called *the external center of similitude* of $\mathscr{C}$ and $\mathscr{C}'$. It divides OO' externally in the ratio of the radii of $\mathscr{C}$ and $\mathscr{C}'$. We see that $\mathscr{C}$ can always be mapped onto $\mathscr{C}'$ by a central dilatation, vertex V.

If, in our description of central dilatations, we choose the image point P' so that P and P' are on *opposite* sides of V (Fig. 13.3), the map of a circle is still a circle, the

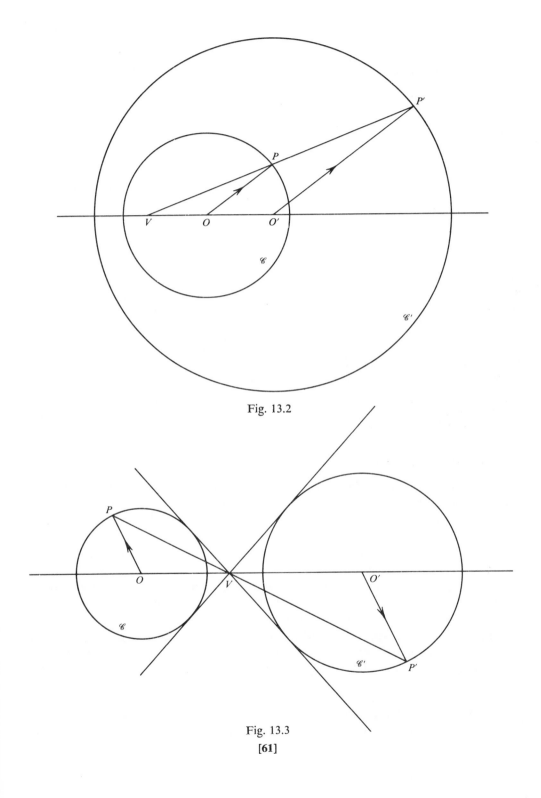

Fig. 13.2

Fig. 13.3

[61]

center of the circle $\mathscr{C}'$ is the map O' of the center O of $\mathscr{C}$, and this time V divides OO' internally in the ratio of the radii of $\mathscr{C}$ and $\mathscr{C}'$, and OP and $O'P'$ are parallel but in opposite directions. If V lies outside $\mathscr{C}$, the tangents from V to $\mathscr{C}$ are also tangents to $\mathscr{C}'$, but $\mathscr{C}$ and $\mathscr{C}'$ lie on opposite sides of each common tangent. The tangents are called *internal common tangents*, or *transverse* common tangents.

Again, if we are given any two circles $\mathscr{C}$ and $\mathscr{C}'$ with centers O and O' the above situation can be reproduced. Draw parallel radii OP and $O'P'$ in *opposite senses* through O and O' respectively. Then the join of the points P and P' will intersect the line of centers OO' in a fixed point V, and $|VO|:|VO'| = $ radius of $\mathscr{C}$:radius of $\mathscr{C}'$. This point V is called *the internal center of similitude of $\mathscr{C}$ and $\mathscr{C}'$*. It divides OO' *internally* in the ratio of the radii of $\mathscr{C}$ and $\mathscr{C}'$.

Exercises

13.1 Show that if the circles $\mathscr{C}$ and $\mathscr{C}'$ are concentric, the common center must be considered to be both the internal and external center of similitude. Where are the internal and external centers when:
(a) $\mathscr{C}$ and $\mathscr{C}'$ touch *externally*, lying on opposite sides of the tangent at the point of contact and:
(b) $\mathscr{C}$ and $\mathscr{C}'$ touch *internally*, lying on the same side of the tangent at the point of contact?

13.2 If we have three circles in the plane, no two with equal radius, and the centers forming a triangle, prove that the six centers of similitude (two from each pair of circles) lie in sets of three on four lines. (Two proofs are possible, one using the Desargues Theorem (§5.2), and the other the Menelaus Theorem (§4.1.)) The four lines are called *axes of similitude of the three circles*.

13.3 In the preceding Exercise show that the join of two internal centers of similitude always passes through an external center of similitude. What deduction can you make about the join of the points of contact of a circle which touches two other circles externally?

14.1   Reflexions in a line

Before introducing the mapping of the punctured plane (the plane from which one point has been removed) called *inversion*, which can be thought of as the process of reflexion in a given circle, let us consider the mapping of the Euclidean plane defined by (orthogonal) *reflexion* in a given line l.

In Fig. 14.1 if P is any point in the plane, N the foot of the perpendicular from P onto l, and PN is produced to P', where $|NP'| = |PN|$, so that P and P' are on opposite sides of l and at an equal distance from the line, then P' is called the *reflexion* or *geometrical image* of P in l. It is clear that P is the reflexion of P', and that the mapping given by $P \to P'$ is a one-to-one mapping of the plane onto itself.

The points of l are mapped onto themselves, and are therefore the *fixed* (or *invariant*) points of the mapping. There are no other fixed points. If we denote the operation of reflexion in l by the symbol M_l, and write $PM_l = P'$, then since $P'M_l = P$, we have

$$PM_lM_l = P(M_l)^2 = P,$$

so that $(M_l)^2$ is *the identity mapping*, which maps every point P on itself.

Now consider the successive operations of reflexion in two lines l and m which intersect at a point O and make an acute angle A with each other (Fig. 14.2). If we reflect P in l, obtaining P^*, and then reflect P^* in m, obtaining the point P', we have

$$PM_l = P^*, \qquad P^*M_m = P',$$

so that

$$PM_lM_m = P'.$$

We can replace *the product of two reflexions* by a *rotation* about O.

14.2   Product of reflexions in intersecting lines

Theorem   *The product of reflexions M_lM_m is equivalent to a rotation about O, the intersection of l and m, through an angle $2A$ from l towards m.*

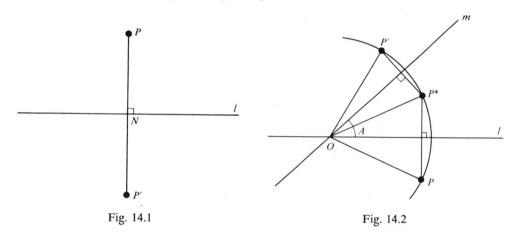

Fig. 14.1   Fig. 14.2

Proof. It is clear that $|OP| = |OP^*| = |OP'|$, so that the mapping from P to P' is a rotation about O. The fact that $\angle POP' = 2\hat{A}$, the direction of rotation being from l towards m is also clear when P is taken as shown in Fig. 14.2. The reader should also examine the cases when P is within the acute angle formed by l and m, and within the other obtuse angle formed by l and m, and the acute angle formed by the extension of l and m. Since this is a theorem of real Euclidean geometry, in which orientation plays an important part, these different cases have to be considered. Another proof will emerge in a later treatment, when complex numbers will be used (§50.2).

A *rotation* of the plane about a point O through an angle B in the sense from OX_1 towards OX_2 will be symbolized by $R(O, \hat{B})$ (See Fig. 14.3) A rotation is evidently a one-to-one mapping of the plane onto itself, with only one fixed point, the point O itself. If P is mapped on P', then $|OP| = |OP'|$, and $\angle POP' = B$, the sense of rotation being from OX_1 towards OX_2.

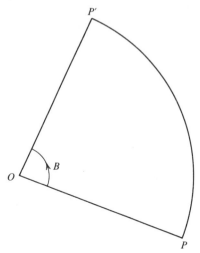

Fig. 14.3

14.3   Product of reflexions in parallel lines

Theorem   *If the lines l and m are parallel, and at a distance d apart, the product of reflexions M_lM_m can be replaced by a translation from l towards m and through a distance 2d. (See Fig. 14.4.)*

Proof. Here again the proof is clear when P is not between the lines l and m, but it should be verified for the case when P is between the two lines.

A *translation* is also a one-to-one mapping of the plane onto itself. If D is now a given vector (see Fig. 14.5), the mapping given by the vector equation $P' = P + D$ is a

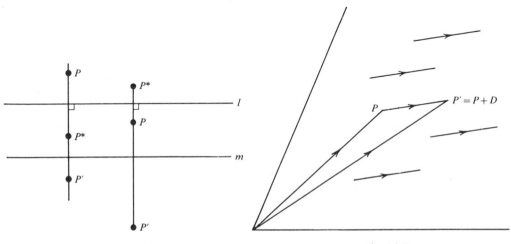

Fig. 14.4   Fig. 14.5

translation of every point of the plane in a direction and through a distance given by the vector D. If D is not zero, there are no fixed points under a translation. If there is one fixed point, $D = 0$, and every point is fixed, the mapping being then $P' = P$, the identity mapping.

The following Exercises are converses to the theorems just proved, and the results to be proved will be used later.

Exercises

14.1 Prove that a rotation $R(O, 2\hat{A})$ of angle $2\hat{A}$ about a point O can always be replaced by an equivalent product of reflexions in two lines intersecting at O at an angle A, and that either of the two lines can be chosen arbitrarily through O, the other then being fixed.

14.2 Prove that a translation through a distance $|D|$ given by a vector D can always be replaced by an equivalent product of reflexions in two parallel lines at a distance $|D|/2$ apart and perpendicular to D, and either one of the lines can be chosen arbitrarily, the other then being fixed.

14.4 Example

If we follow a rotation of the plane through an angle of 90° about a point C by a rotation, in the same sense, through 90° about a point B, then the result is equivalent to a rotation of 180° about the center of a square having BC for a side.

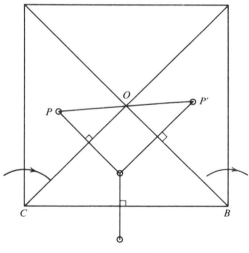

Fig. 14.6

Let O be the center of the square on BC as one side which is such that the given rotations are from CO towards CB (Fig. 14.6). Then the rotation of 90° about C is equivalent to the product of reflexions $M_{CO}M_{CB}$, since the angle between CO and CB

is 45°. Hence $R(C, 90°) = M_{CO}M_{CB}$. The rotation of 90° about B can be expressed as the product of reflexions $M_{BC}M_{BO}$, so that

$$R(C, 90°)R(B, 90°) = M_{CO}M_{CB}M_{BC}M_{BO},$$

and since $M_{CB}M_{BC}$ is equivalent to a repeated reflexion in BC, which is the identity, we have

$$R(C, 90°)R(B, 90°) = M_{CO}M_{BO},$$

and a product of reflexions in OC and OB is equivalent to a rotation through $2(\angle COB) = 180°$ about O.

Exercise

14.3 *ACPQ* and *BARS* are squares on the sides *AC* and *BA* of a triangle *ABC*. If *B* and *C* remain fixed, whilst *A* varies, prove that the line *PS* always passes through a fixed point.

15.1 Applications of reflexions to minimum problems

Let P, Q be two given points on the same side of a line l, and in the same plane with l (Fig. 15.1). If X be a point on l, it is required to find the position of X on l when the sum $|PX| + |XQ|$ is a minimum.

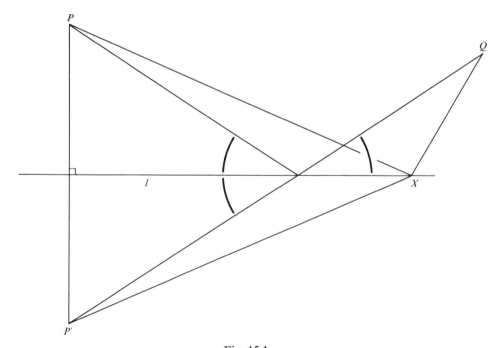

Fig. 15.1

If P' be the geometric image of P in l, then for all X on l, $|PX| = |P'X|$, and $|PX| + |XQ| = |P'X| + |XQ|$. We know that two sides of a triangle added together always exceed the length of the third side, unless the vertices are collinear, so that for the triangle QXP' we can say $|P'X| + |XQ| \geqslant |P'Q|$, with equality if and only if the points Q, X and P', in this order, lie on a line. Since $|PX| + |XQ| = |P'X| + |XQ| \geqslant |P'Q|$, and P' and Q are fixed points, we see that the minimum is attained when the points P', X and Q are on a line, in this order. When this is the case, we see that the minimum path from P to Q is that of a ray of light *reflected* at l. In other words, a ray of light emanating from P and reflected by l to Q takes the shortest path from P to Q via l.

In §24.2, after the necessary preliminaries on inversion, we shall obtain another important law of optics, Snell's law for refraction of light between two media, as another important illustration of the minimum principle for the path of a ray of light.

Exercise

15.1 $\mathscr{C}$ and $\mathscr{C}'$ are two coplanar circles, on the same side of a line AB in the same plane. Show how to find a point V on AB which is such that a tangent from V to $\mathscr{C}$ makes an angle with AB which is equal to the angle made with BA by a tangent from V to $\mathscr{C}'$.

16.1 Product of rotations

Theorem   *The product of two rotations* $R(O_1, \hat{A})$, $R(O_2, \hat{B})$ *where* $O_1 \neq O_2$ *is equivalent to a rotation* $R(O_3, \hat{A} + \hat{B})$. *If* $\hat{A} + \hat{B} = n(360°)$, *where* $n = 0, \pm 1, \pm 2, \ldots$, *the product is equivalent to a translation.* (See Fig. 16.1.)

Proof. A rotation about O_1 can be effected by a reflexion in a line through O_1 followed by a reflexion in the line O_1O_2 (Exercise 14.1). A rotation about O_2 can be effected by a reflexion in the line O_2O_1 followed by a reflection in a line through O_2. In the product of reflexions, the reflexions in O_1O_2 give the identity mapping, and we are left with a reflexion in a line through O_1 followed by a reflexion in a line through O_2. If these lines intersect in a point O_3, the two reflections are equivalent to a rotation about O_3 through an angle which is $2(\hat{A} + \hat{B})/2 = \hat{A} + \hat{B}$. If the lines are parallel, which is the case when $\hat{A} + \hat{B} = n \cdot (360°), n = 0, \pm 1, \pm 2, \ldots$, the two reflexions are equivalent to a *translation* (Thm 14.3). Note that in Fig. 16.1 both $\hat{A}$ and $\hat{B}$ are greater than $90°$, and the equivalent rotation about O_3 through $(\hat{A} + \hat{B})$ is greater than $180°$.

It is, of course, evident that when $O_1 = O_2$ the above theorem still holds, but the translation is the zero translation when $\hat{A} + \hat{B} = n(360°)$. On the other hand, when $O_1 \neq O_2$, and if $\hat{A} + \hat{B} = n(360°)$, but $\hat{A} \neq 0°$, the translation is not the zero translation, since the two parallel lines in which the sequence of reflexions take place are at a non-zero distance apart.

Exercises

16.1 Equilateral triangles are constructed on the sides of a triangle ABC and exterior to the triangle. Prove that the centroids O_1, O_2 and O_3 of these triangles also form an equilateral triangle. (Hint: consider the product of rotations $R(O_3, 120°)$, $R(O_1, 120°)$, $R(O_2, 120°)$ applied to A. Show that this product is equivalent to a translation. This has no fixed points unless it is the identity. Proceed . . . ..)

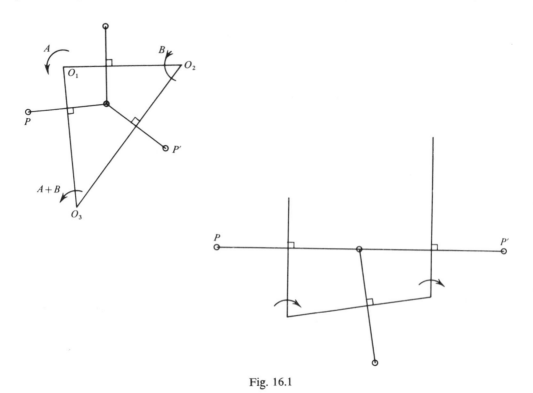

Fig. 16.1

16.2 On the sides of a convex quadrilateral $ABCD$ squares are constructed, all exterior to the quadrilateral. If the centers of the squares are M_1, M_2, M_3 and M_4, show that $|M_1M_3| = |M_2M_4|$, and that M_1M_3 is perpendicular to M_2M_4 (Compare Exercise 8.1) (Hint: consider the product of rotations about M_1, M_2, M_3 and M_4, each through 90°).

17.1 Another application of reflexions

Given an acute-angled triangle ABC (Fig. 17.1), we seek to find the inscribed triangle UVW (with U on BC, V on CA and W on AB) with the minimum perimeter.

If we keep U and V fixed we should expect, by the result of §15.1, that if there is a minimum it will occur when VW and UW make equal angles with BA and AB respectively, and that we can go all around the triangle UVW with this kind of argument, and end up with WV and UV making equal angles with CA and AC, and VU and WU

making equal angles with *BC* and *CB* respectively. The first question we ask is: is there such a triangle? As we shall show, there is, but it is a far cry from the existence of such a triangle to the proof that it is the minimum triangle we are seeking. Let *P*, *Q* and *R* be the feet of the perpendiculars from *A*, *B* and *C* onto the opposite sides *BC*, *CA* and *AB* of triangle *ABC* (Fig. 17.2). We prove that $\sphericalangle QRH = \sphericalangle HRP$. The points *Q*, *H*, *R* and *A* are concyclic, since the angles *HQA* and *HRA* are both right

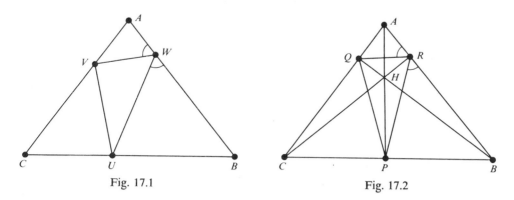

Fig. 17.1   Fig. 17.2

angles. Hence the angles *QRH* and *QAH*, on the same arc, are equal. But $\sphericalangle QAH$ is the complement of $\hat{C}$. The angle *QBP* is also the complement of $\hat{C}$, and since *P*, *H*, *R* and *B* are on the circle with diameter *BH*, angles *HRP* and *HBP* are equal, being on the same arc. Hence angles *QRH*, *HRP* are equal, and therefore *QR* and *PR* are equally inclined to *BA* and *AB* respectively. Similarly for the angles made by the other sides of triangle *PQR* with the other sides of triangle *ABC*.

We now prove that if *UVW* is any triangle inscribed in triangle *ABC*, then the perimeter of $UVW \geqslant$ perimeter of *PQR*. The proof uses reflexions, and was discovered by H. A. Schwarz (1843–1921).

We consider the triangles *PQR* and *UVW* simultaneously (Fig. 17.3), and first reflect

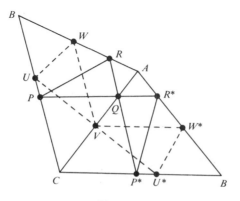

Fig. 17.3

the triangle ABC, with the two inscribed triangles, in the side AC. The point Q remains fixed, and if $P \to P^*$ and $R \to R^*$, we see that the points P, Q and R^* are collinear, as are the points R, Q, P^*. This is because of the equality of the angles made by QP with AC and QR with CA. We also note that the triangle P^*QR^* is congruent to triangle PQR, and that in consequence the length $|PR^*| = |PQ| + |QR|$. The triangle ABC, after reflexion in AC, has been merely rotated about AC, out of the plane and back again into the plane. The side BC is rotated clockwise through an angle $2\hat{C}$ into its new position.

We now continue to reflect, reflecting the triangle arising from ABC after its reflexion in AC in the side AB of this new triangle, and then reflecting the resulting triangle in

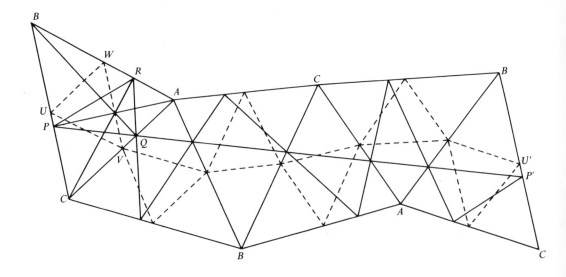

Fig. 17.4

its side BC, then again in AC and finally in AB (Fig. 17.4). The figure obtained shows the procedure rather better than a complete verbal description.

If we add up the total angle through which BC has been turned by the five reflexions, we see that it is turned through $2\hat{C}$ by the first reflexion, clockwise, through $2\hat{B}$ clockwise by the second reflexion, it remains unturned by the third (which is in itself), it is turned counterclockwise through $2\hat{C}$ by the fourth reflexion, and in the final reflexion it is turned counterclockwise through $2\hat{B}$. Hence the total angle of rotation of BC is zero, and so in the final position BC is parallel to its original position.

If we examine the various positions of the vertices of PQR after the reflexions, we have seen already that P, Q and R^* are collinear, and if P' be the final position of P, on the final position of BC, we see that the line segment PP' is made up of six pieces which are, in turn, equal to the three sides of triangle PQR, so that $|PP'| = 2$(perimeter of triangle PQR). If U' be the final position of U, the broken line from U to U' is similarly made up of the three sides of triangle UVW, each taken twice. It is also clear that the straight line UU' is parallel to the line PP', and since the initial and final positions of BC are parallel to each other, we have

$$|\text{Broken line } UU'| = 2(\text{perimeter of triangle } UVW)$$

$$\geqslant \text{straight line segment } |UU'|$$

$$= \text{line segment } |PP'|$$

$$= 2(\text{perimeter of triangle } PQR).$$

We have therefore proved that the triangle PQR is the inscribed triangle of minimum perimeter inscribed in triangle ABC.

18.1   The power of a point with respect to a circle

Before introducing inversion, there are some preliminary results in connection with circles which we cannot assume to be universally known. The equation of a circle can be written in the form:

$$C(X, Y) \equiv X^2 + Y^2 + 2gX + 2fY + c = 0.$$

This will be called the *normalized form*, since the coefficient of $X^2 + Y^2$ is unity. On completing the squares in x and y we have

$$(x + g)^2 + (y + f)^2 = g^2 + f^2 - c,$$

if (x, y) is a point on the circle, so that we deduce, if our coordinates are rectangular cartesian coordinates, that the center of the circle is the point $(-g, -f)$, and the radius r is given by $r^2 = g^2 + f^2 - c$.

Suppose now that we have a point $P = (x, y)$, and we substitute the coordinates of P in $C(X, Y)$. We then obtain a real number $C(x, y)$ which depends on the relation of P to the circle. Since

$$C(x, y) = (x + g)^2 + (y + f)^2 - (g^2 + f^2 - c),$$

and $(x + g)^2 + (y + f)^2 = $ square of the distance of P from the center, we see that $C(x, y) > 0$ if P is outside the circle, $C(x, y) = 0$ if P is on the circle, and $C(x, y) < 0$ if P lies inside the circle. We call $C(x, y)$ *the power of P with respect to the circle* $C(X, Y) = 0$. There is an important geometrical interpretation of the power of P with respect to a given circle.

18.2   Geometrical interpretation of the power of a point

Theorem   *Let $P = (x, y)$ be a given point and $\mathscr{C}$ a circle with normalized equation $C(X, Y) = 0$. If a line through P intersects the circle in the points Q and R, $C(x, y) = $*

$\overline{PQ} \cdot \overline{PR}$. *In particular, if PT is a tangent to the circle at T, then $C(x, y) = \overline{PT}^2$. (See Fig. 18.1.)*

Proof. We use polar coordinates with origin at P, and if (x', y') is the point Q, we write

$$x' = x + r \cos \theta, \ y' = y + r \sin \theta,$$

where $r = \overline{PQ}$, and $\overrightarrow{PQ}$ makes an angle θ with the OX-axis.

Since we are assuming that $C(x', y') = 0$, we have

$$(x + r \cos \theta)^2 + (y + r \sin \theta)^2 + 2g(x + r \cos \theta)$$
$$+ 2f(y + r \sin \theta) + c = 0,$$

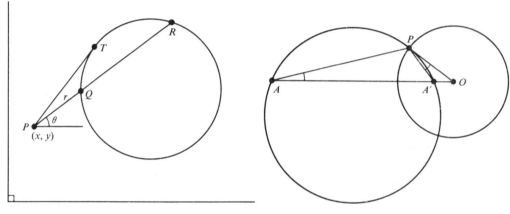

Fig. 18.1   Fig. 18.2

which, on expansion, becomes the equation in r:

$$r^2 + 2r(x \cos \theta + y \sin \theta + g \cos \theta + f \sin \theta) + C(x, y) = 0.$$

The product of the roots of this equation is $C(x, y)$, since the coefficient of r^2 is unity, and therefore

$$\overline{PQ} \cdot \overline{PR} = C(x, y).$$

18.3   The Apollonius circle of two points

A and A' are two given distinct points, and a point P moves so that the ratio of the lengths $|PA| : |PA'| = \lambda$, a constant. We prove that the locus of P is a circle, center O on AA', and of radius k, where $k^2 = |OA| \cdot |OA'|$.

We also prove the converse, that given any circle of radius k and center O, if we choose any two distinct points A and A' on a diameter on the same side of O, where $|OA| \cdot |OA'| = k^2$, then if P is any point on the circle, the ratio $|PA| : |PA'| =$ constant.

The circle is called *an Apollonius circle* with regard to the two points A and A'.

For any given position of P, draw the circle through P, A and A', and let O be the point where the tangent at P to this circle meets AA' (Fig. 18.2). Then the angle $A'PO =$ the angle in the alternate segment $=$ angle PAA'. Hence the triangles AOP and POA' are similar, having a common angle at O, and angle $A'PO =$ angle PAO. We may therefore write

$$|AP|:|PA'| = |OP|:|OA'| = |AO|:|PO| = \lambda,$$

and therefore

$$(|AO|:|PO|)(|OP|:|OA'|) = \lambda^2 = |AO|:|OA'|.$$

This shows that O, which lies outside the segment AA', is a fixed point on AA'. Since also $|OP|^2 = |OA| \cdot |OA'|$, the length $|OP|$ is constant, and therefore P moves on a circle, center on AA', and radius k, where $k^2 = |OA| \cdot |OA'|$.

We shall see in §20.1 that this means that A, A' are *inverse points* in the Apollonius circle.

If $\lambda = 1$, the Apollonius circle is the perpendicular bisector of the segment AA'.

To prove the converse, since $|OP|^2 = |OA| \cdot |OA'|$, we have $|OP|:|OA'| = |OA|:|OP|$, and therefore the triangles AOP, POA' are similar, having a common angle at O, and the ratio of the sides about the angle equal. If

$$|AP|:|PA'| = |OP|:|OA'| = |AO|:|PO| = \lambda,$$

then $(|OP|:|OA'|)(|AO|:|PO|) = \lambda^2 = |AO|:|OA'|$, and since A, A' and O are given points, this shows that λ^2, and therefore λ is a fixed positive number. Hence $|AP|:|PA'|$ = constant, as P moves round the circle.

If a line be given, we naturally choose A and A' as mirror images in the line, and then the line is the Apollonius circle for the points A and A', and $\lambda = 1$.

Exercises

18.1 If through a center of similitude of two circles (§13.1) we draw a line which cuts one of them in the points R, R' and the other in the corresponding points S, S', then $\overline{OR} \cdot \overline{OS'} = \overline{OR'} \cdot \overline{OS} =$ constant, where O is the center of similitude.

18.2 $\mathscr{C}$ and $\mathscr{D}$ are given circles, and $\mathscr{E}$ is any circle which intersects both of them. Let the points of intersection of $\mathscr{C}$ and $\mathscr{E}$ be P, Q, and the points of intersection of $\mathscr{D}$ and $\mathscr{E}$ be R and S, and suppose that $V = PQ \cap RS$. Prove that the power of V with respect to $\mathscr{C} =$ the power of V with respect to $\mathscr{D}$, and that a circle center V, of suitable radius, will intersect $\mathscr{C}$ and $\mathscr{D}$ orthogonally, provided that V lies outside both circles.

18.3 $\mathscr{C}$ and $\mathscr{D}$ are two circles external to each other, and $\mathscr{E}$ and $\mathscr{F}$ are circles which touch both $\mathscr{C}$ and $\mathscr{D}$ externally in the points R, R' and S, S' respectively. Show that RR' intersects SS' at an external center of similitude of $\mathscr{C}$ and $\mathscr{D}$, and that this point has the same power with respect to $\mathscr{E}$ and $\mathscr{F}$. (As we shall say later (§27.3), *it lies on the radical axis of $\mathscr{E}$ and $\mathscr{F}$.*)

18.4 Show that the Apollonius circles for two given points A and A' form a non-intersecting system of circles for varying λ, one circle of the system, and only one, passing through any given point of the plane. Show that any circle of the system is orthogonal to any circle through the points A and A'.

18.5 If the bisectors of the angle APA' meet the line AA' at X and Y, show that the Apollonius circle of P with respect to A and A' is the circle on XY as diameter.

18.6 If Q, R, S, T are four distinct points and $P = QR \cap ST$, and if $\overline{PQ} \cdot \overline{PR} = \overline{PS} \cdot \overline{PT}$, prove that the four points Q, R, S, T lie on a circle. (Compare Thm 18.2.)

19.1 Harmonic division

Let A, A' X, Y be collinear points, with the relation

$$|OA| \cdot |OA'| = |OX|^2 = |OY|^2,$$

where O is the midpoint of the segment XY. We prove that X and Y divide the segment AA' internally and externally in the same ratio, and that A, A' divide the segment XY internally and externally in the same ratio.

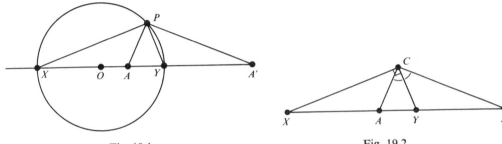

Fig. 19.1   Fig. 19.2

Let P be a point on the circle on XY as diameter (Fig. 19.1). Then since $|OP|^2 = |OX|^2 = |OY|^2$, we have $|OP|^2 = |OA| \cdot |OA'|$. By the preceding Section, this circle is therefore an Apollonius circle for the points A and A', and therefore the locus of a point which moves so that $|PA|:|PA'| = $ constant. Since X and Y are points on this circle, we have

$$|XA|:|XA'| = |YA|:|YA'|.$$

This means that Y and X divide the segment AA' internally and externally in the same ratio.

If we use *signed segments*, we write this relation in the form

$$\overline{XA}:\overline{XA'} = -\overline{YA}:\overline{YA'}.$$

Since this can also be written

$$\overline{XA}:\overline{YA} = -\overline{XA'}:\overline{YA'},$$

or

$$\overline{AX}:\overline{AY} = -\overline{A'X}:\overline{A'Y},$$

we have deduced that *A, A' divide the segment XY internally and externally in the same ratio.*

The pair of points (A, A') has a special relationship to the pair of points (X, Y). They are said to be *harmonic conjugates*, and the two pairs are said to form *a harmonic range*.

Harmonic ranges occur frequently in Euclidean geometry, a simple example arising when we bisect the angle C of a triangle ACA' both internally and externally (Fig. 19.2). The points of intersection X, Y of the bisectors with AA' form a harmonic range with (A, A'), since $|AX|:|XA'| = |AC|:|CA'| = |AY|:|YA'|$, using unsigned lengths.

Exercises

19.1 If V be the midpoint of AA', prove that $|VX| \cdot |VY| = |VA|^2 = |VA'|^2$.

19.2 If X and Y are kept fixed, whilst A' moves off to infinity, prove that A moves towards the midpoint of XY.

20.1 Inversion

We are now ready to define *inversion*. We assume that we are given a circle Σ (Fig. 20.1), center O, and radius k, and that given any point P in the plane of Σ we find a point P' on OP such that $\overline{OP} \cdot \overline{OP'} = k^2$. Then P' is said to be *the inverse of P in the*

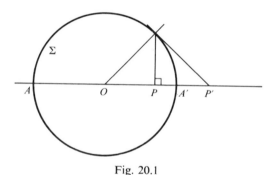

Fig. 20.1

circle Σ, or *with respect to* the circle Σ, and Σ is called *the circle of inversion*, and O is called *the center of inversion*. We allow inversion in a virtual circle (§0.9) of radius ik also, in which case we have $\overline{OP} \cdot \overline{OP'} = -k^2$, so that the inversion is the composition of an inversion in a real circle of radius k and a point-reflection center O.

Our definition of inversion does not hold for the center of inversion O. We consider our Euclidean plane as the plane of the complex numbers, and introduce a further point, the point ∞. We define the inverse of O to be the point ∞, and the inverse of the point ∞ to be the center of inversion O. We now have a one-to-one map of our extended plane, which we call *the inversive plane*. We note that if P' is the inverse of P, then P is the inverse of P', so that inversion, like reflexion in a line, is a transformation of period two. That is, if the inversion in Σ be denoted by the symbol T_Σ, $T_\Sigma^2 = $ Identity.

Inversion is sometimes called reflexion in a circle, and we shall see that there is a strong resemblance between the two processes, inversion in a circle and reflexion in a line.

From the results of the preceding Section we observe that if the diameter OP meets Σ in the points A, A', the points P, P' divide the segment AA' internally and externally in the same ratio.

Points P on the circle of inversion Σ map into themselves under inversion in Σ, since the definition gives us $\overline{OP} = \overline{OP'}$. We are referring to ordinary inversion, of course, and the following treatment applies to ordinary inversion. Inversion in a virtual circle will arise naturally in Chapter III and subsequently, when inversion is represented as a mapping of Euclidean space of three dimensions.

20.2   A property of inverse points

Theorem   *If $\mathscr{C}$ is any circle which passes through a pair of points P, P' inverse in the circle Σ, then $\mathscr{C}$ is orthogonal to Σ.* (See Fig. 20.2.)

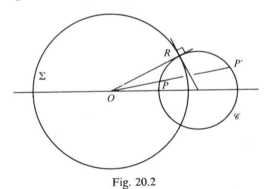

Fig. 20.2

Proof. We know from §18.1 that $\overline{OP} \cdot \overline{OP'}$ is the power of O with respect to the circle $\mathscr{C}$, and that if OT is tangent at T to $\mathscr{C}$ then $\overline{OP} \cdot \overline{OP'} = \overline{OT}^2$. But, by the definition of inversion, $\overline{OP} \cdot \overline{OP'} = \overline{OR}^2$, where R is a point of intersection of $\mathscr{C}$ and Σ. Hence $|OR| = |OT|$. Now a circle center O and radius $|OT|$ intersects $\mathscr{C}$ in two points, and these are the points of contact of tangents from O to $\mathscr{C}$. Hence OR is a tangent at R to $\mathscr{C}$. Therefore the circles $\mathscr{C}$ and Σ cut at right angles at their points of intersection.

Exercises

20.1   If $\mathscr{C}$ and $\mathscr{D}$ are circles with respective centers C and D which intersect at a point R, show that the tangents at R to the two circles are perpendicular if and only if each tangent passes through the center of the other circle.

20.2   Prove the converse to Thm 20.2: if $\mathscr{C}$ and $\mathscr{D}$ are two circles which cut orthogonally, then any diameter of $\mathscr{C}$ cuts $\mathscr{D}$ in a pair of points which are inverse in $\mathscr{C}$.

20.3   Inversion in a line

Let us suppose that we mark the point A on the circle Σ (Fig. 20.3), fix the diameter through A, and allow the center O of Σ to move off to infinity along the diameter. Since any pair of inverse points (P, P') on the diameter are harmonic conjugates with respect to A and A', as A' moves off to infinity with O the points P, P' become *image points* in the line through A to which Σ more and more nearly approximates as its radius tends towards infinity. We therefore *define* the process of inversion in a line to be *reflexion* in that line (see §14.1), and we shall see that this notion is in agreement with all subsequent developments.

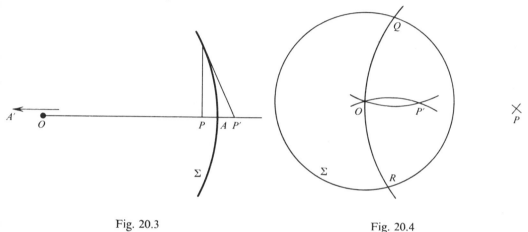

Fig. 20.3   Fig. 20.4

20.4   A construction for inverse points

We give a construction, using only a pair of compasses, which enables us to find the inverse of a given point P in a given circle Σ, center O (Fig. 20.4). The construction is easy to remember, since there is very little we can do with a pair of compasses, given a circle Σ, its center, and a point P. We put the point of the compasses on P, extend until the radius is $|PO|$, and draw a circle, center P, which cuts Σ in the points Q and R. We now put the point of the compasses on Q, extend until the radius is $|QO|$, and draw a circle center Q and radius $|QO|$. Finally, put the point of the compasses on R, and with the same radius draw a circle. This will intersect the circle center Q and radius $|QO|$ in O and another point P'. This is the inverse of P in the circle Σ.

By symmetry the point P' is on the line OP. The angle QOP is always an acute angle, and therefore the point P' is on the same side of O as P. The triangles PQO and QOP' are both isosceles, and share a common base-angle, the angle QOP. Hence, since they are similar triangles,

$$|OP|:|OQ| = |OQ|:|OP'|,$$

so that $\overline{OP} . \overline{OP'} = \overline{OQ^2}$, and P, P' are inverse in Σ.

This construction may break down (see Exercise 20.3), and the difficulty may be overcome with the help of the next construction.

20.5   Extending a segment with a pair of compasses

Given two distinct points A and B, we can find a point C on the line AB which is such that B is the midpoint of AC, using a pair of compasses only (Fig. 20.5). We merely

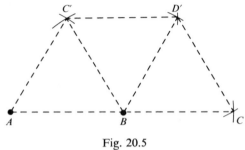

Fig. 20.5

use the compasses to draw equilateral triangles. We find C' so that ABC' is equilateral, then D' so that $BC'D'$ is equilateral, and then C so that $BD'C$ is equilateral. Since three angles of 60° add to 180°, it is clear that C is the desired point. We may evidently find a point Q on AB such that $\overline{AQ} = n\overline{AB}$, where n is any positive integer.

Exercises

20.3   When does the construction for the inverse of a given point P in a given circle Σ, center O, break down? How may P' be found, all the same, using a pair of compasses only?

20.4 A, B are given points. Using a pair of compasses only, show how the midpoint of the segment AB may be determined.

20.5   A circle Σ, center O is given, and with center P and radius $|PO|$ a circle is drawn to cut Σ in Q and R. Why must the angle QOP be acute? (This was stated in the proof of the construction for inverse points given in §20.4.)

21.1   Inverses of lines and circles

If a point P moves on a curve $\mathscr{C}$, and P', the inverse of P in a circle Σ moves on a curve $\mathscr{C}'$, the curve $\mathscr{C}'$ is called *the inverse of* $\mathscr{C}$. It is evident that $\mathscr{C}$ is the inverse of $\mathscr{C}'$ in Σ.

If we keep the center O of Σ fixed, and change the radius of inversion from k to k', we obtain an inverse curve $\mathscr{C}''$ which is related to $\mathscr{C}'$ by a central dilatation, center O. For if $\overline{OP}.\overline{OP'} = k^2$, and $\overline{OP}.\overline{OP''} = k'^2$, $\overline{OP''}:\overline{OP'} = k'^2:k^2$, and so P'' and P' are related in a central dilatation, center O (§12.1). Hence $\mathscr{C}''$ is similar to $\mathscr{C}'$, and *the radius of inversion merely alters the size of curves, and not their shape*, the center of inversion remaining fixed.

We shall show that the inverse of a line l which does *not* pass through the center of inversion is a circle which *does* pass through the center, and that the inverse of a circle $\mathscr{C}$ which does not pass through the center of inversion is a circle $\mathscr{C}'$ which likewise does not pass through the center of inversion.

We shall use algebraic methods, although geometrical methods are also available. If the center O of inversion be chosen as the origin of rectangular cartesian coordinates, and $P = (r, \theta)$ in polar coordinates, then $P' = (r', \theta)$, where $r = k^2/r'$, so that $r \cos \theta = k^2 r' \cos \theta/r'^2$, $r \sin \theta = k^2 r' \sin \theta/r'^2$, which leads to the simple formulae connecting $P = (X, Y)$ with $P' = (X', Y')$:

$$X = k^2 X'/(X'^2 + Y'^2), \quad Y = k^2 Y'/(X'^2 + Y'^2), \tag{1},$$

with exactly similar formulae if we interchange the primed and unprimed symbols.

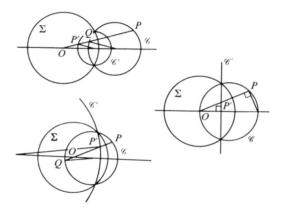

Fig. 21.1

The transform of the equation

$$C(X, Y) \equiv X^2 + Y^2 + 2gX + 2fY + c = 0$$

is, after multiplication all through by $(X'^2 + Y'^2)$:

$$c(X'^2 + Y'^2) + 2k^2(gX' + fY') + k^4 = 0.$$

If $c \neq 0$, which is the condition that the circle $\mathscr{C}$ should *not* pass through the origin, which is the center of inversion, then the inverse $\mathscr{C}'$ is also a circle which does not pass through the center of inversion.

The center of $\mathscr{C}$ is $(-g, -f)$, and the center of $\mathscr{C}'$ is $(-gk^2/c, -fk^2/c)$. (See Fig. 21.1.)

Note that this is *not* the inverse of the center of $\mathscr{C}$. We determine the inverse of the center of $\mathscr{C}$ later (§23.3). The radius R' of $\mathscr{C}'$ is given by the equation:

$$R'^2 = \frac{k^4(g^2 + f^2) - k^4}{c^2} = \frac{k^4(g^2 + f^2 - c)}{c^2} = \frac{k^4 R^2}{c^2} = \frac{k^4 R^2}{(g^2 + f^2 - R^2)^2},$$

where R = radius of $\mathscr{C}$. If we put $d^2 = g^2 + f^2$, this is the square of the distance of the center of $\mathscr{C}$ from O, and we obtain the useful formula:

$$R'^2(d^2 - R^2)^2 = k^4R^2, \tag{2}$$

This gives us the radius of the inverse circle $\mathscr{C}'$ in terms of the radius of the original circle, and the distance of the center of inversion from the center of the original circle $\mathscr{C}$.

If $c = 0$, the circle $\mathscr{C}$ passes through $(0, 0)$, the center of inversion, and the inverse is not a circle, but *a line*:

$$2(gX' + fY') + k^2 = 0.$$

Conversely, the inverse of a line which does *not* pass through the center of inversion is a circle which *does* pass through the center of inversion.

We now see that our lines consist of the sets of points they contain when considered to be immersed in a Euclidean plane, plus the special point ∞ with which we completed the Euclidean plane to form the inversive plane. It is the inverse of this point which is the center of any given inversion. It is clear from the definition of inversion that the inverses of points P lying on a line through the center of inversion continue to lie on the same line. Hence lines and circles invert into lines, or circles. If we regard a line as a special kind of circle, we can say that inversion is a *circular transformation*, since the set of all lines and circles in the plane is mapped onto itself.

Exercises

21.1 Show that the inverse of a circle through the center of inversion is a line parallel to the tangent to the circle at the center of inversion.

21.2 Two lines through a point P intersect at an angle A. Show that the inverse of the lines with respect to a point $O \neq A$ is a pair of circles through O and P' which intersect at an angle A.

21.3 From the indications given in Fig. 21.1, give a geometrical proof of the theorem that a circle through the center of inversion inverts into a line, and the theorem that a circle not through the center of inversion inverts into a circle which does not contain the center of inversion.

21.4 In the preceding section we saw that if the center of $\mathscr{C}$ is $(-g, -f)$, the center of $\mathscr{C}'$ is $(-gk^2/c, -fk^2/c)$. Deduce that if O, the center of inversion, lies outside $\mathscr{C}$, then the center of $\mathscr{C}$ and the center of $\mathscr{C}'$ lie on the same side of O, but if O lies inside $\mathscr{C}$, the centers lie on opposite sides of O.

21.5 Writing the formula (2) in the form $R'/R = k^2/|c|$, show that O is a center of similitude of $\mathscr{C}$ and $\mathscr{C}'$. Show further that O is an external center of similitude if O lies outside $\mathscr{C}$ (and therefore outside $\mathscr{C}'$), and an internal center of similitude if O lies inside $\mathscr{C}$ (and therefore inside $\mathscr{C}'$).

21.6 Show that there is a linear relation between the equations

$$\Sigma \equiv X^2 + Y^2 - k^2 = 0, \quad C \equiv X^2 + Y^2 + 2gX + 2fY + c = 0,$$

and

$$C' \equiv X^2 + Y^2 + 2k^2(gX + fY)/c + k^4/c = 0,$$

namely

$$k^2C \equiv cC' + (k^2 - c)\Sigma.$$

(In §27.1, we shall see that this identity means that *the circle $\mathscr{C}$, its inverse $\mathscr{C}'$ and the circle of inversion Σ are coaxal*.)

21.7 *ABC* is a given triangle, and tangents to the circumcircle of *ABC* at *A*, *B* and *C* form a triangle *A'B'C'*. If *O* be the circumcenter of *ABC*, show that if α, β, γ are the respective midpoints of *BC*, *CA* and *AB*, then the pairs of points (α, A'), (β, B') and (γ, C') are inverse in the circumcircle of *ABC*, center *O*. Deduce that the circumcircle of *A'B'C'* has its center on the line which joins the point *O* to the center of the nine-point circle of *ABC*, the Euler line of triangle *ABC*.

21.8 If, in Fig. 21.2, *ABCD* is a rhombus ($|AB| = |BC| = |CD| = |DA|$) and *O* is equidistant from its opposite vertices *A* and *C*, prove that *O*, *B*, *D* are collinear, and that $\overline{OB} . \overline{OD} = |OA|^2 - |AB|^2$. Deduce that if all the lines shown in the figure are rigid bars, and the points are freely moving joints, then if *O* be kept fixed, the points *B* and *D* describe figures inverse in a fixed circle center *O*. If *D* be constrained to move on a fixed circle which passes through *O*, prove that *B* will describe a portion of

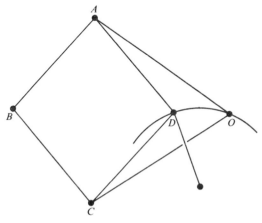

Fig. 21.2

a line. (This simple apparatus, named after the inventor, Peaucellier, was the first mechanical method for converting circular motion into straight line motion. Watts, the inventor of the steam engine, needed such a mechanism, but had to approximate to linear motion.)

21.9 Deduce from the converse of the theorem of the Apollonius circle (§18.2) that if *A*, *B*, *C*, *D* are four points which are such that there is a circle through *A* and *B* in which *C*, *D* are inverse points, then there is also a circle through *C* and *D* in which *A*, *B* are inverse points. (Such a set of points is called an *orthocyclic* set.)

21.10 Show that the composition of inversions in two concentric circles, center *O*, is equivalent to a dilatation center *O*, and that conversely, any given dilatation center *O* is equivalent to such a composition of inversions.

22.1 The angle of intersection of two circles

As a preliminary towards proving that inversion is a conformal mapping, that is a mapping which does not change the angle of intersection of two curves, we investigate the angle of intersection of two circles given by their equations

$$C_1(X, Y) \equiv X^2 + Y^2 + 2g_1X + 2f_1Y + c_1 = 0,$$
$$C_2(X, Y) \equiv X^2 + Y^2 + 2g_2X + 2f_2Y + c_2 = 0.$$

If O_1, O_2 be their respective centers (see Fig. 22.1), and P a point of intersection, we know that the radius vectors O_1P, O_2P are respectively perpendicular to the tangents at P to the two circles, so that we may take the angle O_1PO_2 as the measure of the angle of intersection of $\mathscr{C}_1$ and $\mathscr{C}_2$. Using the cosine formula,

$$|O_1O_2|^2 = |O_1P|^2 + |O_2P|^2 - 2|O_1P||O_2P| \cos (O_1PO_2).$$

Expressing this in terms of the equations, we have

$$2R_1R_2 \cos \theta = (g_1{}^2 + f_1{}^2 - c_1) + (g_2{}^2 + f_2{}^2 - c_2) - |O_1O_2|^2,$$

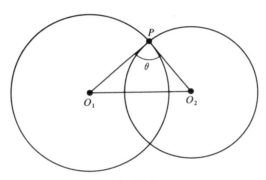

Fig. 22.1

where

$$|O_1O_2|^2 = (g_1 - g_2)^2 + (f_1 - f_2)^2,$$

so that

$$2R_1R_2 \cos \theta = 2g_1g_2 + 2f_1f_2 - c_1 - c_2, \tag{1}$$

gives us the angle θ of intersection of the two circles. As a consequence we have *the condition for orthogonality*

$$2g_1g_2 + 2f_1f_2 - c_1 - c_2 = 0. \tag{2}$$

Now, suppose that the circles given above are inverted with respect to a circle Σ, center at $(0, 0)$. Then we obtain the circles

$$C_1'(X, Y) \equiv c_1(X^2 + Y^2) + 2k^2(g_1X + f_1Y) + k^4 = 0,$$

and

$$C_2'(X, Y) \equiv c_2(X^2 + Y^2) + 2k^2(g_2X + f_2Y) + k^4 = 0.$$

In applying formula (1) to these circles, we must remember that they are not in normalized form, with coefficient of $(X^2 + Y^2)$ unity. If we evaluate the expression

$$2g_1'g_2' + 2f_1'f_2' - c_1' - c_2',$$

we obtain

$$2k^4(g_1g_2 + f_1f_2)/c_1c_2 - k^4(1/c_1 + 1/c_2),$$

and since formula (2) of §21.1 gives us

$$R_1'/R_1 = k^2/|c_1|, \text{ and } R_2'/R_2 = k^2/|c_2|,$$

we see that

$$\frac{2g_1g_2 + 2f_1f_2 - c_1 - c_2}{2R_1R_2} = \frac{(2g_1'g_2' + 2f_1'f_2' - c_1' - c_2')}{2R_1'R_2'} \frac{|c_1c_2|}{c_1c_2}.$$

If θ' be the angle of intersection of $\mathscr{C}_1'$ and $\mathscr{C}_2'$, this shows that

$$\cos\theta' = \pm\cos\theta.$$

Since we did not actually specify whether we were choosing the acute or obtuse angle for the angle of intersection of $\mathscr{C}_1$ and $\mathscr{C}_2$, we have proved:

22.2   Inversion preserves angles

Theorem   *Inversion is a mapping which does not alter the angle of intersection of two circles.*

In particular, orthogonal circles invert into orthogonal circles (or lines), and circles which touch invert into circles which touch. In the first case $\cos\theta = 0$, and in the second case $\cos\theta = \pm1$.

Remark. In the proof given above, we have tacitly assumed that neither of the circles $\mathscr{C}_1$, $\mathscr{C}_2$ passes through the center of inversion. But the theorem we have proved is true even if the circles invert into lines. For example, suppose that $\mathscr{C}_1$ and $\mathscr{C}_2$ both pass through O, and are orthogonal. Then $c_1 = c_2 = 0$, and both circles invert into lines. These lines are perpendicular, since the orthogonality condition

$$2g_1g_2 + 2f_1f_2 = 0$$

for the circles is precisely the condition that the lines they invert into be orthogonal.

22.3   Example

We have a given circle $\mathscr{C}$, and two distinct points P and Q. How many circles can we draw to pass through P and Q and to touch $\mathscr{C}$?

Let us suppose that we have such a circle $\mathscr{D}$ (Fig. 22.2). If we invert the whole figure with respect to P as center of inversion, we obtain a circle $\mathscr{C}'$ from $\mathscr{C}$ which is still touched by the inverse of $\mathscr{D}$. But the inverse $\mathscr{D}'$ of $\mathscr{D}$ is a line through the inverse Q' of Q. We now ask: how many lines through a given point Q' touch a given circle $\mathscr{C}'$? We know that there are two. Inverting these lines with respect to P, we obtain *two* circles through P and Q which touch the circle $\mathscr{C}$.

We recall that when a line cuts a circle at a point P, the angle between the line and the circle at P is measured by the angle between the tangents to the two curves at P

(Fig. 22.3). But the tangent to a line at a point P on the line is the line itself. The angle is therefore that between the line and the tangent to the circle at P. It follows that if a line cuts a circle orthogonally, *the line must be a diameter of the circle*. Since we can draw only one diameter of a given circle $\mathscr{C}'$ through a given point Q' which is not the center of $\mathscr{C}'$, we see that through two given points P, Q we can draw a unique circle which is orthogonal to a given circle $\mathscr{C}$. All we do is to invert with respect to P, as above.

If there is more than one orthogonal circle, there is an infinity.

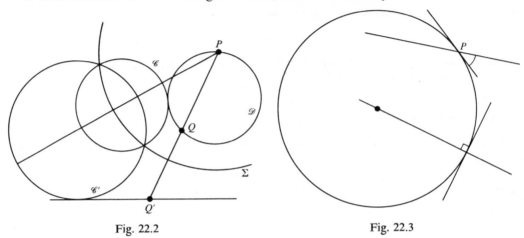

Fig. 22.2   Fig. 22.3

Exercises

22.1 If each of two circles which pass through two given points P, Q is orthogonal to a given circle $\mathscr{C}$, prove that all circles through P, Q are orthogonal to $\mathscr{C}$, that P, Q must lie on a diameter of $\mathscr{C}$, and that P, Q are inverse points with respect to the circle $\mathscr{C}$.

22.2 Prove that circles which touch each other at a point O invert into parallel lines under inversion with respect to O. How does this fit in with the conformal property of inversion?

22.3 A circle $\mathscr{C}$ and a point O lying on a line l are given. Show that through O and touching l there is either no circle orthogonal to $\mathscr{C}$, one circle orthogonal to $\mathscr{C}$, or an infinity of circles orthogonal to $\mathscr{C}$, depending on the relative positions of $\mathscr{C}$, O and l.

22.4 If the circle $\mathscr{C}$ inverts into the circle $\mathscr{C}'$, and $\mathscr{C}$ intersects the circle of inversion Σ at P and Q, show that $\mathscr{C}'$ also passes through P and Q, and that the circle of inversion Σ bisects the angle at P and Q between $\mathscr{C}$ and $\mathscr{C}'$.

22.5 The circle $\mathscr{C}$ and the circle Σ are orthogonal, and $\mathscr{C}$ is inverted in Σ. Prove that not only is $\mathscr{C}$ inverted into itself, but all points *inside* $\mathscr{C}$ invert into points inside $\mathscr{C}$. What can we say about the inverse of Σ in Σ?

22.4   Another look at the angle of intersection

From the viewpoint of differential geometry, the conformal property of inversion is very clear, and it is worth while indicating how Thm 22.2 can be proved if we assume that the tangent at a point P of a curve $\mathscr{C}$ is the limit of the chord PQ, Q being another point on the curve $\mathscr{C}$ which tends towards P in the limit (Fig. 22.4). If the inverse of P with respect to a center of inversion O be P', and the inverse of Q be Q', then since $\overline{OP}\,.\,\overline{OP'} = \overline{OQ}\,.\,\overline{OQ'} = k^2$, the points P, Q, P' and Q' are concyclic, so that $\measuredangle QPO = \measuredangle OQ'P'$, and in the limit we see that the angle made by the tangent at P to the curve $\mathscr{C}$ with $\overline{PO}$ is equal to the angle made by the tangent at P' to the curve $\mathscr{C}'$ with $\overline{OP'}$.

Hence curves which touch at P invert into curves which touch at P'. We now deduce

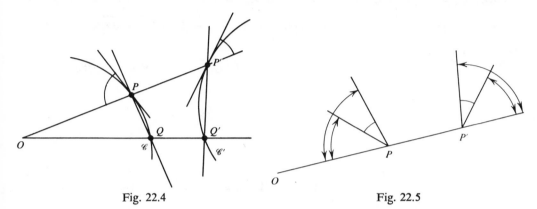

Fig. 22.4   Fig. 22.5

very readily that inversion is an indirect conformal transformation (Fig. 22.5), that is, that *the orientation of the angle of intersection is changed by the mapping, the measure of the angle being preserved.*

From this point of view we do not have to consider special cases, when circles invert into lines. We shall need part of the algebra of §22.1 later, which is one reason for giving the algebraic treatment. Both points of view are rewarding.

We can now proceed with the manifold applications of inversion, since all we need are the fundamental results of §§ 21.1 and 22.2. But there are also three special results which we shall need later, and we give them here.

23.1 The effect of inversion on length

A segment inverts into a circular arc (Fig. 23.1), but we may compare the *length* of the segment PQ with that of the segment $P'Q'$. If O is the center of inversion, the triangles OPQ, $OQ'P'$ are similar, so that

$$|P'Q'|:|PQ| = |OP'|:|OQ| = |OP|\,.\,|OP'|:|OP|\,.\,|OQ| = k^2:|OP|\,.\,|OQ|$$

so that

$$|P'Q'| = k^2|PQ|:|OP|\,.\,|OQ|$$

We also have

$$|PQ| = k^2|P'Q'|:|OP'|\,.\,|OQ'|.$$

23.2   The cross-ratio of points on a line

If A, B, C and D be four collinear points, an important function of the points, which we shall investigate in more detail later (§53.3) is the *cross-ratio*. This is *a ratio of ratios*, and is dependent on the order in which the four points are taken. If we write $\{AB, CD\} = \overline{AC}/\overline{AD} : \overline{BC}/\overline{BD} = \overline{AC} \cdot \overline{BD}/\overline{AD} \cdot \overline{BC}$, all that we need here is the proof that if we invert with respect to a center O on the line of the points, then $\{A'B', C'D'\} = \{AB, CD\}$.

Since $\overline{AC} = \overline{OC} - \overline{OA}$, $\overline{BD} = \overline{OD} - \overline{OB}$, $\overline{AD} = \overline{OD} - \overline{OA}$, and $\overline{BC} = \overline{OC} - \overline{OB}$,

$$\{AB, CD\} = (\overline{OC} - \overline{OA})(\overline{OD} - \overline{OB})/(\overline{OD} - \overline{OA})(\overline{OC} - \overline{OB}),$$

and

$$\overline{OC} = k^2/\overline{OC'}, \ \overline{OA} = k^2/\overline{OA'}, \ \overline{OD} = k^2/\overline{OD'}; \text{ and } \overline{OB} = k^2/\overline{OB'};$$

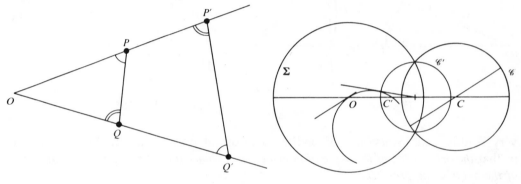

Fig. 23.1   Fig. 23.2

substitution gives us

$$\{AB, CD\} = (\overline{OC'} - \overline{OA'})(\overline{OD'} - \overline{OB'})/(\overline{OD'} - \overline{OA'})(\overline{OC'} - \overline{OB'}),$$

so that

$$\{AB, CD\} = \{A'B', C'D'\}.$$

23.3   The inverse of the center of a circle

We have already warned the reader that the center C of a circle $\mathscr{C}$ does not invert into the center of the inverse circle $\mathscr{C}'$. To find the point it does invert into, we note that lines through C cut the circle $\mathscr{C}$ orthogonally (Fig. 23.2). These lines invert into circles through the center of inversion O, and through the inverse C' of the center of $\mathscr{C}$. But we know that if two or more circles through two points cut a given circle orthogonally, these points are inverse points in the given circle (Exercise 22.1). Therefore O and C' are inverse points in $\mathscr{C}'$, the inverse of $\mathscr{C}$.

Hence, *to obtain the inverse of the center of $\mathscr{C}$, we find the inverse of O, the center of inversion, in the inverse circle $\mathscr{C}'$.*

The case when $\mathscr{C}'$ is a line is of importance (Fig. 23.3). Lines through the center of $\mathscr{C}$ cut the circle $\mathscr{C}$ orthogonally, and therefore invert into circles through O and C' which cut the line $\mathscr{C}'$ orthogonally. Since the line $\mathscr{C}'$ must be a diameter of all these circles, O and C' must be mirror images of each other in the line $\mathscr{C}'$. This confirms our previous remark (§20.3) that *to obtain the inverse of a point in a line, we find the reflexion of the point in the line.*

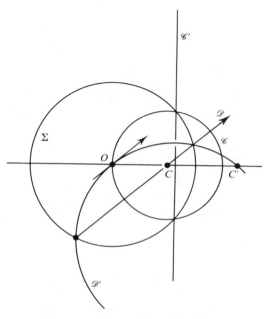

Fig. 23.3

23.4 Two celebrated theorems

Feuerbach proved, using trigonometry, that the nine-point circle of a triangle ABC touches the inscribed circle and the three escribed circles of the triangle. We give a proof by inversion. It will be recalled that the nine-point circle of triangle ABC passes through the midpoints A', B', C' of the sides BC, CA and AB and through the feet D', E' and F' of the perpendiculars from the vertices A, B and C onto the opposite sides.

Let I be the center of the incircle, and I' the center of the excircle which touches BC externally (Fig. 23.4). The two circles we are considering have three common tangents, the sides of triangle ABC, and the fourth common tangent meets the line II' on BC, at the point K, say, and meets AC at E and AB at F. The points A and K are external and internal centers of similitude for the two circles, and therefore divide the segment II' externally and internally in the same ratio. That is, the points I, I' and A, K are pairs in a harmonic range. If we drop perpendiculars from these points onto the line BC, the feet of the perpendiculars also form a harmonic range. Let X, X' be the respective feet of the perpendiculars from I and I'. These are the respective points of contact of the incircle with BC, and the excircle with BC. The points X, X' are therefore harmonic conjugates with respect to D' and K. It is a simple matter to show that $|CX'| = |BX| = (|AB| + |BC| - |CA|)/2$, so that A' is the midpoint not only of BC but also of XX'. From the properties of harmonic ranges (§19.1), we may therefore

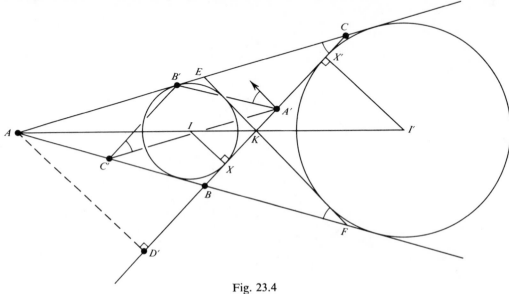

Fig. 23.4

say that $|A'K| \cdot |A'D'| = |A'X|^2 = |A'X'|^2$. We are going to invert with respect to a circle center A', and radius $|A'X|$, and we see that the point D' inverts into the point K.

The circle of inversion, by choice, is orthogonal to both the incircle and the excircle we are considering, having its center on a tangent, and passing through the point of contact, for each circle. Therefore both the incircle and the excircle invert into themselves. The nine-point circle passes through the center of inversion A', and therefore inverts into a line, and since the nine-point circle also passes through D', which inverts into K, the line passes through K. We wish to show that the inverse is the tangent EKF, which will prove Feuerbach's Theorem.

We know that the inverse of a circle with respect to a point A' on it is a line which is parallel to the tangent to the circle at the point A'. We must therefore show that the tangent to the nine-point circle at A' is parallel to the line EKF.

The nine-point circle passes through A', B' and C', and the tangent to it at A', by the alternate segment theorem for circles, makes an angle equal to the angle $B'C'A'$

with $A'B'$, but on the side of $A'B'$ which is opposite to the side on which C' lies. This angle, since $C'B'CA'$ is a parallelogram, is equal to the angle ECK, and this, by symmetry, is equal to the angle KFB. Since $A'B'$ is parallel to BA, the tangent to the nine-point circle at A' is parallel to EKF, and therefore the inverse of the nine-point circle touches both the incircle and the excircle. Since these inverted into themselves, the nine-point circle itself touches both circles. Similarly, it touches the other two escribed circles.

Our second theorem is mentioned by Pappus as being already ancient in his time (towards the end of the third century, A.D.). Let X, Y, Z be three collinear points

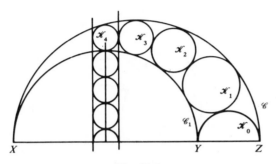

Fig. 23.5

(Fig. 23.5), with Y between X and Z, and let $\mathscr{C}, \mathscr{C}_1, \mathscr{K}_0$ denote semicircles, all lying on the same side of XZ, on XZ, XY and YZ as diameters. Let $\mathscr{K}_1, \mathscr{K}_2, \mathscr{K}_3, \ldots$ denote circles touching $\mathscr{C}$ and $\mathscr{C}_1$, with $\mathscr{K}_1$ also touching $\mathscr{K}_0$, with $\mathscr{K}_2$ also touching $\mathscr{K}_1$, with $\mathscr{K}_3$ also touching $\mathscr{K}_2$, and so on. If the radius of $\mathscr{K}_n$ be r_n, and the distance of the center of $\mathscr{K}_n$ from XZ be h_n, then

$$h_n = 2nr_n.$$

Let t_n be the length of the tangent from X to the circle $\mathscr{K}_n$. The figure shows the case $n = 4$. We invert in the circle center X, radius t_n. Then $\mathscr{K}_n$ inverts into itself, since the circle of inversion is orthogonal to it, and $\mathscr{C}$ and $\mathscr{C}_1$ invert into a pair of lines which are both parallel to the tangent shared by the two circles at X, that is into a pair of lines each perpendicular to XZ. The semicircle $\mathscr{K}_0$ inverts into a semicircle touching both lines, the circle $\mathscr{K}_1$ inverts into a circle touching both lines, so does the circle $\mathscr{K}_2, \mathscr{K}_3$ and so on. Also the inverse of $\mathscr{K}_1$ touches the inverse of $\mathscr{K}_0$, the inverse of $\mathscr{K}_2$ touches the inverse of $\mathscr{K}_1$, the inverse of $\mathscr{K}_3$ touches the inverse of $\mathscr{K}_2$, and so on. The radii of all the inverses, of $\mathscr{K}_0, \mathscr{K}_1, \mathscr{K}_2, \mathscr{K}_3, \ldots$ are therefore the same, and equal to r_n. Therefore $h_n = 2nr_n$.

24.1   An extension of Ptolemy's theorem

The triangle inequality, that two sides of a triangle added together give a sum not less than the length of the third side, is a fundamental inequality in Euclidean geometry.

Let us see what happens if we invert it. We shall obtain the following

Theorem   *If A, B, C and D are any four coplanar points, then*

$$|AB| \cdot |CD| + |AD| \cdot |BC| > |AC| \cdot |BD|,$$

unless A, B, C and D lie, in this order, on a circle or a line. In the latter case the inequality becomes an equality.

The equality is usually designated as *Ptolemy's Theorem*.

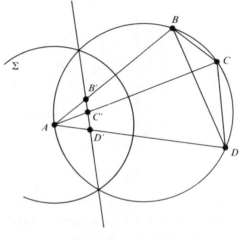

Fig. 24.1

Proof. Given four points A, B, C and D, we invert with respect to A (see Fig. 24.1). Let B', C' and D' be the respective inverses of B, C and D. Then

$$|B'C'| + |C'D'| > |B'D'|$$

unless C' lies on the segment $B'D'$ between B' and D', when the inequality becomes an equality

$$|B'C'| + |C'D'| = |B'D'|.$$

Using the results of §23.1, which give the effect of inversion on length, the inequality becomes

$$\frac{|BC|}{|AB| \cdot |AC|} + \frac{|CD|}{|AC| \cdot |AD|} > \frac{|BD|}{|AB| \cdot |AD|},$$

or

$$|AB| \cdot |CD| + |AD| \cdot |BC| > |AC| \cdot |BD|,$$

with equality if and only if C' lies on the segment $B'D'$ between B' and D', in which case A, B, C and D lie on a circle in this order, or on a line, in this order. Then we have

$$|AB| \cdot |CD| + |AD| \cdot |BC| = |AC| \cdot |BD|,$$

which is Ptolemy's Theorem.

This is naturally as powerful a theorem as the triangle inequality. We apply it to give a proof of an inequality which a number of texts declare cannot be proved by elementary methods.

24.2   Fermat's principle and Snell's law

We have seen (§15.1) that there is a principle of least time involved in the law of reflexion of a ray of light which is reflected from a point P to a point Q by means of a line

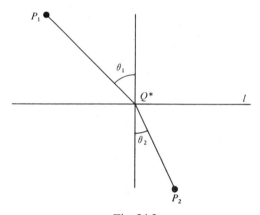

Fig. 24.2

(mirror) l. The minimum distance from P to Q via l is actually the path which obeys the physical law of reflexion. Suppose that we now have a line l bounding two media with different physical properties, so that the speed of light is v_1 in the one medium and v_2 in the other (Fig. 24.2). A ray of light starts out from a point P_1 in the first medium, is refracted where it meets l, and passes through a given point P_2 in the second medium, on the other side of l. Fermat's Principle, which is a Principle of Least Time, asserts that the time taken by the ray of light is a minimum. In other words, if Q^* is the point where the ray of light hits l, then

$$|P_1Q^*|/v_1 + |P_2Q^*|/v_2$$

is a minimum, so that if Q is any other point on the line l,

$$|P_1Q|/v_1 + |P_2Q|/v_2 > |P_1Q^*|/v_1 + |P_2Q^*|/v_2.$$

Snell's Law asserts that for the point Q^*,

$$\sin \theta_1 / \sin \theta_2 = v_1/v_2,$$

where θ_1, θ_2 are the angles of incidence and refraction at the point Q^*, that is the angles made by the incident ray and the refracted ray with the normal to l at Q^*. We prove that Fermat's Principle and Snell's Law are equivalent, so that Q^*, the point given by Snell's Law, is the minimum point which satisfies the Fermat Principle.

Draw the circle through the points P_1, Q^* and P_2 (Fig. 24.3), and let the perpendicular

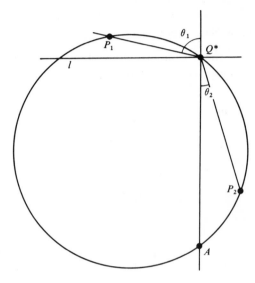

Fig. 24.3

to l at Q^* intersect the circle again at the point A. If R is the radius of the circle,

$$|AP_1| = 2R \sin(\pi - \theta_1) = 2R \sin \theta_1, \text{ and } |AP_2| = 2R \sin \theta_2,$$

so that we may write

$$|AP_1| = k/v_2, \ |AP_2| = k/v_1,$$

for a constant k.

Applying the Ptolemy Theorem to the four points P_1, Q^*, P_2 and A, which lie on a circle in the given order, we have

$$|P_1P_2| \cdot |AQ^*| = |P_1Q^*| \cdot |AP_2| + |P_2Q^*| \cdot |AP_1|,$$

whereas if $Q \neq Q^*$ is any point on l, the extension of the Ptolemy Theorem proved in §24.1 gives us the inequality

$$|P_1P_2| \cdot |AQ| < |P_1Q| \cdot |AP_2| + |P_2Q| \cdot |AP_1|.$$

If we substitute for $|AP_1|$ and $|AP_2|$, we obtain the equality

$$k(|P_1Q^*|/v_1 + |P_2Q^*|/v_2) = |P_1P_2| \cdot |AQ^*|,$$

and the inequality

$$k(|P_1Q|/v_1 + |P_2Q/v_2) > |P_1P_2| \cdot |AQ|.$$

Since $|AQ| > |AQ^*|$, AQ^* being perpendicular to l, we have

$$|P_1Q|/v_1 + |P_2Q|/v_2 > |P_1Q^*|/v_1 + |P_2Q^*|/v_2,$$

which establishes the Fermat Principle for a ray satisfying Snell's Law.

Exercises

24.1 Ptolemy used the theorem named after him to evaluate $\sin(A + B)$ and $\cos(A + B)$ in terms of $\sin A$, $\cos A$, $\sin B$ and $\cos B$. Can you imitate his procedures?

24.2 In Fig. 24.3, if we take Q at the other intersection ($\neq Q^*$) of l with the circle, the points P_1, P_2, A and Q lie on a circle, and the Ptolemy Theorem applies. Does this invalidate the proof given in §24.2? If not, why not?

24.3 An equilateral triangle BCD is described on the side BC of a given triangle ABC so that D lies in the opposite half-plane to A. Prove that unless the point P is at the intersection $\neq D$ of AD with the circumcircle of BCD, we have the inequality $|PA| + |PB| + |PC| > |AD|$.

24.3 The Fermat problem

As a final application of the extension of the Ptolemy theorem we give a solution of the problem proposed by Fermat in the seventeenth century: A, B and C are three given points (see Fig. 24.4) Find a point P in the plane of the three given points such that $|PA| + |PB| + |PC|$ is a minimum.

At least two of the angles of triangle ABC are acute. Let them be $\hat{B}$ and $\hat{C}$. On BC, and in the opposite half-plane to A, construct an equilateral triangle BCD, and its circumcircle $\mathscr{C}$.

By the extension of Ptolemy's theorem, unless P lies on the circle $\mathscr{C}$ so that the order of the points B, P, C and D is as described,

$$|BP| \cdot |CD| + |PC| \cdot |DB| > |PD| \cdot |BC|.$$

Since $|CD| = |DB| = |BC|$, this becomes $|PB| + |PC| > |PD|$. Therefore

$$|PA| + |PB| + |PC| > |PA| + |PD|.$$

Unless P lies on AD, we have $|PA| + |PD| > |AD|$. Hence, unless P is at the intersection $\neq D$ of AD with the circle $\mathscr{C}$,

$$|PA| + |PB| + |PC| > |AD|.$$

Let the other intersection of AD with $\mathscr{C}$ be P'. If P' is between A and D, both the in-equalities written above become equalities, and we have

$$|P'A| + |P'B| + |P'C| = |AD|.$$

Therefore if $P \neq P'$,

$$|PA| + |PB| + |PC| > |P'A| + |P'B| + |P'C|,$$

so that P' is the point which solves the Fermat problem.

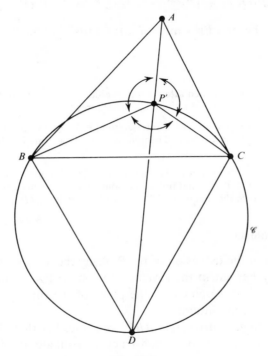

Fig. 24.4

Exercises

24.4 Show that the above solution is valid as long as $\hat{A} \leqslant 120°$. If this condition is satisfied, show that each side of triangle ABC subtends $120°$ at the point P'. Show also that if $\hat{A} > 120°$, A is the required point.

24.5 ABC is an acute-angled triangle, and equilateral triangles $A'BC$, $B'CA$ and $C'AB$ are drawn away from the opposite vertices on the sides BC, CA and AB respectively. Prove that the line segments AA', BB' and CC' are equal, concurrent, and inclined at $60°$ to each other.

24.6 ABC is an acute-angled triangle, P a point inside the triangle. The triangle APB is rotated through $60°$ about B, and $P \to P'$, $A \to C'$. Show that ABC' and PBP' are equilateral triangles, and that

$$|AP| + |BP| + |CP| = |C'P'| + |P'P| + |PC|.$$

Since C and C' are fixed points, where is P when the left-hand sum is a minimum? Show that when this is the case the angles subtended at P by BC and AB and AC are each $120°$.

25.1   A fixed point theorem

Theorem *If $\mathscr{C}$ and $\mathscr{D}$ are two non-intersecting circles, any circle $\mathscr{E}$ which cuts them both orthogonally passes through two fixed points on their line of centers.* (See Fig. 25.1.)

Proof. We anticipate the result of §29.4 which indicates that we may choose our co-ordinate axes so that the equations of the given circles $\mathscr{C}$ and $\mathscr{D}$ are respectively:

$$X^2 + Y^2 + 2gX + p^2 = 0, \quad X^2 + Y^2 + 2g'X + p^2 = 0.$$

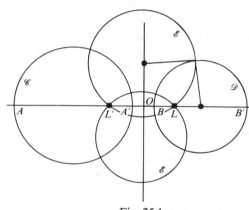

Fig. 25.1

The X-axis is the line of centers, and the Y-axis is a special line related to both circles called *the radical axis*. This line does not intersect either circle, as we see on putting $X = 0$ in both equations.

If $\mathscr{E}$ is the circle with equation

$$X^2 + Y^2 + 2kX + 2lY + c = 0,$$

the orthogonality condition (Eqn (2), §22.1) applied to $\mathscr{E}$ and $\mathscr{C}$ and $\mathscr{E}$ and $\mathscr{D}$ is

$$2gk = p^2 + c, \quad 2g'k = p^2 + c.$$

Since $g \neq g'$, the circles $\mathscr{C}$ and $\mathscr{D}$ being distinct circles, we have $k = 0$, and $c = -p^2$. Hence the equation of $\mathscr{E}$ is

$$X^2 + Y^2 + 2lY - p^2 = 0,$$

an equation in which l is the only variable parameter. This shows that a circle which cuts both $\mathscr{C}$ and $\mathscr{D}$ orthogonally must have its center on the radical axis (the Y-axis in this case) of the two circles, and such a circle cuts the line of centers, $Y = 0$, in the points given by $X^2 - p^2 = 0$, that is in the two fixed points $(p, 0)$ and $(-p, 0)$.

If we call these fixed points L and L', and the points where the line of centers cuts $\mathscr{C}$ and $\mathscr{D}$ respectively A, A' and B, B', we have

$$|OL|^2 = |OL'|^2 = p^2 = \overline{OA} \cdot \overline{OA'} = \overline{OB} \cdot \overline{OB'},$$

since if we put $Y = 0$ in the equation of each circle we obtain the equations $X^2 + 2gX + p^2 = 0$, $X^2 + 2g^1X + p^2 = 0$, which show that $\overline{OA} \cdot \overline{OA'} = p^2 = \overline{OB} \cdot \overline{OB'}$. From §19.1 it follows that L, L' are harmonic conjugates with respect to A, A' and to B, B'. Hence we have:

Corollary. Two non-overlapping pairs of points on a line have a pair of common harmonic conjugates.

This pair of common harmonic conjugates is unique. For if M, M' be another pair of harmonic conjugates to A, A' and B, B', and O' be the midpoint of MM', we should have $|O'M|^2 = |O'M'|^2 = \overline{O'A} \cdot \overline{O'A'} = \overline{O'B} \cdot \overline{O'B'}$, and so the point O' would have the same *power* (§18.1) with respect to the circles $\mathscr{C}$ and $\mathscr{D}$. Such a point must lie on the radical axis of $\mathscr{C}$ and $\mathscr{D}$ (§27.3) and is therefore uniquely defined. Hence $O' = O$, and thence M, M' coincide with L, L'.

25.2   The inverse of a circle and a pair of inverse points

Theorem *If P and Q are points inverse in a circle $\mathscr{C}$, inversion in any circle center O maps P, Q on points P', Q' inverse in the inverse $\mathscr{C}'$ of $\mathscr{C}$.*

Proof. All circles through P and Q are orthogonal to $\mathscr{C}$ (Thm 20.2). Hence after inversion all circles through P' and Q' are orthogonal to $\mathscr{C}'$. One such circle is the line $P'Q'$, the inverse of circle POQ. Hence $P'Q'$ is a diameter of $\mathscr{C}'$, and since circles through P' and Q' are orthogonal to $\mathscr{C}'$, the points P' and Q' are inverse points in $\mathscr{C}'$.

Remark. The proof holds with only minor modifications if $\mathscr{C}'$ is a line. Then P' and Q' are mirror images in the line.

Exercises

25.1 $\mathscr{C}$ is a circle which cuts a line l. From a point O on l a tangent OT is drawn to $\mathscr{C}$, and with center O and $|OT|$ as radius a circle is drawn which is orthogonal to $\mathscr{C}$. Show that this circle does *not* intersect the diameter of $\mathscr{C}$ which is perpendicular to l.

25.2 Let $\mathscr{C}$ and $\mathscr{D}$ be two circles which intersect in the points A and B. By inversion with respect to A as center show that the operations of inversion in distinct intersecting circles $\mathscr{C}$ and $\mathscr{D}$ are commutative if and only if $\mathscr{C}$ and $\mathscr{D}$ are orthogonal. (If inversion in $\mathscr{C}$ is represented by $T_{\mathscr{C}}$, and in $\mathscr{D}$ by $T_{\mathscr{D}}$, we wish to show that if P is any point, $PT_{\mathscr{C}}T_{\mathscr{D}} = PT_{\mathscr{D}}T_{\mathscr{C}}$ if and only if $\mathscr{C}$ and $\mathscr{D}$ are orthogonal.)

25.3 Show that four mutually orthogonal circles can be inverted into a circle $\mathscr{C}$, two perpendicular diameters l and m of $\mathscr{C}$, and a concentric circle of radius r', where $r'^2 + r^2 = 0$, r being the radius of

$\mathscr{C}$, so that this last circle is a virtual circle (§0.9). If the process of inversion in four mutually orthogonal circles be denoted by T_1, T_2, T_3, T_4, show that

$$P(T_1 T_2 T_3 T_4) = P,$$

for any point P. (This is the first encounter with inversion in a virtual circle, defined in §20.1.)

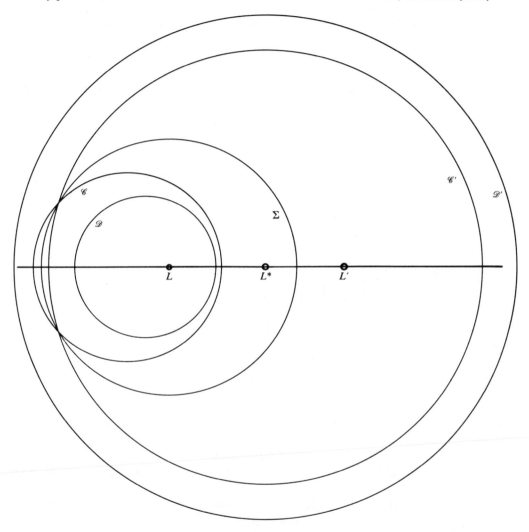

Fig. 25.2

25.3   Circles into concentric circles

Theorem   *Any two non-intersecting circles $\mathscr{C}$ and $\mathscr{D}$ may be inverted simultaneously into a pair of concentric circles.* (See Fig. 25.2.)

Proof. By Thm 25.1 there exists a pair of points L, L' on the line of centers of $\mathscr{C}$ and $\mathscr{D}$ such that all circles orthogonal to $\mathscr{C}$ and to $\mathscr{D}$ pass through L and L', and, of course, such that all circles through L and L' are orthogonal to $\mathscr{C}$ and to $\mathscr{D}$. We invert with respect to one of these points, say L. Let L^* be the inverse of L'. Then since all circles through L and L' are orthogonal to $\mathscr{C}$ and to $\mathscr{D}$, all lines through L^* are orthogonal to $\mathscr{C}'$ and $\mathscr{D}'$, the inverses of $\mathscr{C}$ and of $\mathscr{D}$. Hence L^* is the common center of $\mathscr{C}'$ and $\mathscr{D}'$.

25.4   The Steiner porism

We are given two circles $\mathscr{C}$ and $\mathscr{D}$, one lying inside the other (Fig. 25.3), and we begin to draw circles which touch both $\mathscr{C}$ and $\mathscr{D}$. Beginning with one such circle $\mathscr{C}_1$, the next

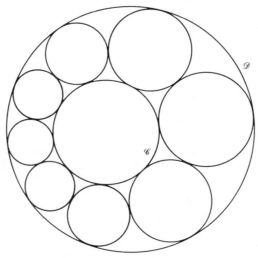

Fig. 25.3

circle we draw, $\mathscr{C}_2$, touches $\mathscr{C}$ and $\mathscr{D}$ and also $\mathscr{C}_1$. Then we draw $\mathscr{C}_3$ to touch $\mathscr{C}$ and $\mathscr{D}$ and $\mathscr{C}_2$, and so on, $\mathscr{C}_i$ touching $\mathscr{C}$ and $\mathscr{D}$ and $\mathscr{C}_{i-1}$. We ask whether there exists a finite chain of circles $\mathscr{C}_1$, $\mathscr{C}_2$, . . ., $\mathscr{C}_{n-1}$, $\mathscr{C}_n$, where each circle of the chain touches $\mathscr{C}$ and $\mathscr{D}$ and the two numerically adjacent circles, $\mathscr{C}_i$ touching $\mathscr{C}_{i-1}$ and $\mathscr{C}_{i+1}$, and $\mathscr{C}_n$ itself touching $\mathscr{C}_{n-1}$ and $\mathscr{C}_1$?

If there is such a chain does it depend on where we start with $\mathscr{C}_1$, or may we draw $\mathscr{C}_1$ in any position, as long as it touches $\mathscr{C}$ and $\mathscr{D}$?

The answer to this problem is simple if we invert $\mathscr{C}$ and $\mathscr{D}$ into two concentric circles, which we may do by Thm 25.3, since they do not intersect. Our problem is reduced to the case when $\mathscr{C}'$ and $\mathscr{D}'$ are concentric. It is clear now that a chain only exists when there is a special relationship between n and the radii of $\mathscr{C}'$ and $\mathscr{D}'$, but that when this relationship *does* exist, since the whole configuration can be rotated about the common center of $\mathscr{C}'$ and $\mathscr{D}'$, it is independent of the position of the initial circle of the chain.

Hence the problem of inscribing a chain of circles as described above has *no* solution or it has *an infinity* of solutions. This is why it is described as a porism. It was first formulated by Steiner.

25.5   The Steiner formula

It is of interest to obtain the formula given by Steiner which relates the radii of $\mathscr{C}$ and $\mathscr{D}$, the distance between their centers, and n, in the case when a chain of n circles can be inscribed to touch each other and $\mathscr{C}$ and $\mathscr{D}$.

Let us suppose that a common diameter of $\mathscr{C}$ and $\mathscr{D}$ intersects them in the points

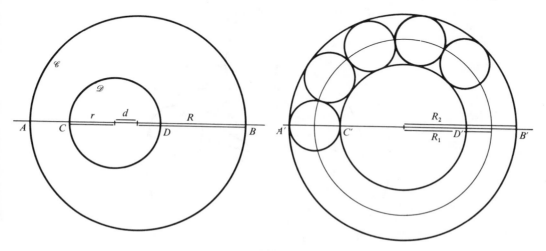

Fig. 25.4

A, B and C, D respectively (Fig. 25.4), and that after inversion with respect to L, which is a point on the common diameter, the points invert into A', B', and C', D', which will be the points of intersection of the same diameter with the concentric circles $\mathscr{C}'$ and $\mathscr{D}'$. Suppose that the radii of $\mathscr{C}$ and $\mathscr{D}$ are R and r, and the distance between their centers is d. Let the radii of the concentric circles $\mathscr{C}'$ and $\mathscr{D}'$ be R_1 and R_2. We use the result of §23.2, that $\{AB, CD\} = \{A'B', C'D'\}$, that is, that

$$\overline{AC}.\overline{BD}/\overline{AD}.\overline{BC} = \overline{A'C'}.\overline{B'D'}/\overline{A'D'}.\overline{B'C'}.$$

The left-hand side is readily seen to equal

$$(R - d - r)(R + d - r)/(R - d + r)(R + d + r) = [(R - r)^2 - d^2]/[(R + r)^2 - d^2],$$

and the right-hand side is

$$(R_1 - R_2)^2/(R_1 + R_2)^2.$$

Equating these two expressions, a little reduction leads to

$$(R - r)^2 - d^2 = Rr(R_1/R_2 + R_2/R_1 - 2).$$

Now if a chain of n circles can be inscribed to touch $\mathscr{C}'$ and $\mathscr{D}'$ (Fig. 25.5), we have the simple relation

$$(R_2 - R_1)/(R_2 + R_1) = \sin(\pi/n),$$

so that

$$R_2/R_1 = (1 + \sin(\pi/n))/(1 - \sin(\pi/n)),$$

and substitution gives us

$$(R - r)^2 - d^2 = 4Rr\tan^2(\pi/n),$$

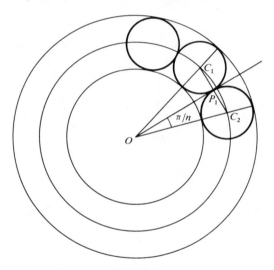

Fig. 25.5

for the relation between the radii of circles $\mathscr{C}$ and $\mathscr{D}$ and the distance between their centres which has to be satisfied for a chain of n circles to be inscribed, touching both circles.

Exercises

25.4 Show that two circles $\mathscr{C}$ and $\mathscr{D}$ intersecting at B and C are orthogonal if and only if there exist two circles touching $\mathscr{D}$ at B and C respectively which also touch each other at A, where A is any point on $\mathscr{C}$ distinct from B and C.

25.5 $\mathscr{C}$ and $\mathscr{D}$ are two distinct circles. Find the locus of a point P which is such that two circles, tangent both to $\mathscr{C}$ and $\mathscr{D}$ are also tangent to each other at P. (There are three cases to examine: $\mathscr{C}$ and $\mathscr{D}$ intersect: $\mathscr{C}$ and $\mathscr{D}$ touch, and $\mathscr{C}$ and $\mathscr{D}$ do not intersect.)

25.6 If the conditions for a chain of circles touching $\mathscr{C}$ and $\mathscr{D}$ are satisfied (The Steiner Porism,

§25.4), show that the mutual points of contact of circles of the chain lie on a circle, and that this circle serves for inverting $\mathscr{C}$ into $\mathscr{D}$.

25.7 Continuing with Exercise 25.2, show that the operations of inversion in $\mathscr{C}$ and $\mathscr{D}$ cannot be commutative if $\mathscr{C}$ and $\mathscr{D}$ touch, or if $\mathscr{C}$ and $\mathscr{D}$ do not intersect.

26.1 The problem of Apollonius

Given three circles $\mathscr{C}_1$, $\mathscr{C}_2$ and $\mathscr{C}_3$, how many circles can be drawn to touch all of them?

This is an example of a problem in which the variation or specialization of the given conditions (the circles in this case) may produce a change in the *number* of distinct

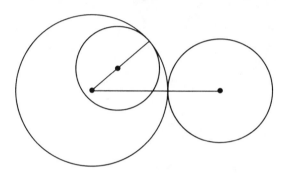

Fig. 26.1

solutions. A case encountered in the beginnings of Euclidean geometry is when the three circles are lines, and we consider how many circles can be drawn to touch the sides of a given triangle. There are *four* such circles, the incircle and the three excircles. Later we shall consider the matter algebraically (§38.1) and we shall see that the maximum *finite* number of solutions to the Apollonius problem is *eight*. Specialization of the circles to lines has halved the number of solutions.

In this section we shall examine the different possible cases of the Apollonius problem. If the circles $\mathscr{C}_1$, $\mathscr{C}_2$ and $\mathscr{C}_3$ are concentric, there are no real circles touching all three of them. We should stress, perhaps, that it is always *real* solutions we are seeking. (See §38.1.)

The condition for the contact of two circles can always be expressed thus:

(Distance between centers) = Sum or difference of radii,

according to the kind of contact, external or internal.

If the two circles be $\mathscr{C}$, $\mathscr{C}'$ (Fig. 26.1), given in the usual form (§22.1), this condition becomes

$$(2gg' + 2ff' - c - c')^2 = 4(g^2 + f^2 - c)(g'^2 + f'^2 - c'),$$

expressing the condition that if θ be the angle of intersection, $\cos \theta = \pm 1$. Suppose now that the circle $\mathscr{C}'$ is a *point-circle*, that is a circle of zero radius, so that $g'^2 + f'^2 - c' = 0$. Then the condition for contact between $\mathscr{C}$ and a point-circle is

$$2gg' + 2ff' - c - c' = 0,$$

which is the condition that the point-circle itself, which is merely the point $(-g', -f')$, should lie on the circle $\mathscr{C}$, which is

$$g'^2 + f'^2 - 2gg' - 2ff' + c = 0.$$

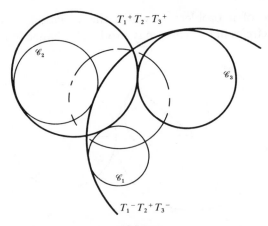

Fig. 26.2

This is a condition we could have reached intuitively, but it is satisfactory to have an algebraic verification. (In §37.1, we shall see that a similar result is true when a point-circle cuts another circle *orthogonally*.)

If the three given circles are each specialized to a point, the condition for a circle $\mathscr{C}$ to touch them is the condition that $\mathscr{C}$ should pass through the three given points. There is only one circle through three points, so we have a case when there is only *one* solution to the problem of Apollonius.

We shall set some special cases, which yield two, three, four, five and six solutions, as Exercises. In the meantime let us see why the number *eight* appears to be the correct number of solutions when the given circles are not specialized. We assume that there is at least one circle which touches the three given circles. Now suppose that we denote a circle (Fig. 26.2) which touches $\mathscr{C}_1$ externally by the symbol $T_1{}^+$, and one which touches $\mathscr{C}_1$ internally by the symbol $T_1{}^-$. Then a combination of the three symbols such as

$$T_1{}^+T_2{}^-T_3{}^+$$

describes a circle which touches $\mathscr{C}_1$ externally, $\mathscr{C}_2$ internally, and $\mathscr{C}_3$ externally. There are *eight* combinations of a triad of plus or minus symbols, namely $+++$, $++-$, $+--$, $+-+$, $-++$, $-+-$, $---$, $--+$, and therefore if there is a unique circle which corresponds to a triad of the symbols (for instance, a unique circle which touches all of the given circles externally), we expect a total of eight contact circles in all, since there seems no reason, if the circles are not specialized, for one kind of contact to be favoured more than another.

This sounds a rather unmathematical argument, but it can be made to depend on mathematical reasoning. We observe that the distinction between external and internal contact of a circle $\mathscr{C}$ with $\mathscr{C}_1$ vanishes if $\mathscr{C}_1$ becomes a point circle, and we expect, if our reasoning has anything in it, that there are only four circles which pass through a point and touch two given circles $\mathscr{C}_2$ and $\mathscr{C}_3$. In fact, inversion with respect to the

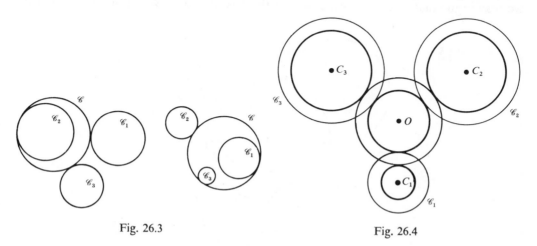

Fig. 26.3   Fig. 26.4

point $\mathscr{C}_1$ shows this at once, since the circles $\mathscr{C}_2'$ and $\mathscr{C}_3'$ have four common tangents, and these arise from the circles through $\mathscr{C}_1$ which touch $\mathscr{C}_2$ and $\mathscr{C}_3$.

We may think that the case in which $\mathscr{C}_1$, $\mathscr{C}_2$ and $\mathscr{C}_3$ all lie outside each other is more general than the case when, say, $\mathscr{C}_2$ and $\mathscr{C}_3$ lie inside $\mathscr{C}_1$. But if we invert with respect to $\mathscr{C}_1$, this goes into itself, and if $\mathscr{C}_2$ and $\mathscr{C}_3$ lie outside $\mathscr{C}_1$ they invert into circles inside $\mathscr{C}_1$. Again, if we have a circle $\mathscr{C}$ (Fig. 26.3) which touches $\mathscr{C}_1$ externally, $\mathscr{C}_2$ internally, and $\mathscr{C}_3$ externally, and we invert with respect to $\mathscr{C}$, the case $T_1{}^+T_2{}^-T_3{}^+$ does turn into $T_1{}^-T_2{}^+T_3{}^-$, since $\mathscr{C}$ inverts into itself, and $\mathscr{C}_1$, which lies outside $\mathscr{C}$, inverts into a circle inside $\mathscr{C}$, the circle $\mathscr{C}_2$ which lies inside $\mathscr{C}$ inverts into a circle outside $\mathscr{C}$, and $\mathscr{C}_3$, which lies outside $\mathscr{C}$, inverts into a circle inside $\mathscr{C}$. Hence if we can show that there is one circle which satisfies the condition $T_1{}^+T_2{}^+T_3{}^+$, and one which satisfies, say, the condition $T_1{}^-T_2{}^+T_3{}^+$, inversions will prove that circles satisfying the other six conditions exist, since we may evidently permute the numbers 1, 2 and 3 in these considerations. (As soon as we have established the existence of a circle Σ orthogonal to $\mathscr{C}_1$, $\mathscr{C}_2$ and $\mathscr{C}_3$ (§28.2), inversion in Σ maps each $\mathscr{C}_i$ onto itself, and can interchange the signs for any given circle condition, but see the remark below.)

We choose our circles $\mathscr{C}_1$, $\mathscr{C}_2$ and $\mathscr{C}_3$ (Fig. 26.4) so that the radii of $\mathscr{C}_2$ and $\mathscr{C}_3$ are equal, and so that the center of $\mathscr{C}_1$ is equidistant from the centers of $\mathscr{C}_2$ and $\mathscr{C}_3$. We also choose the radius of $\mathscr{C}_1$ to be less than that of $\mathscr{C}_2$ or $\mathscr{C}_3$. If there is a circle $\mathscr{C}$ of center O which touches the three given circles externally, decreasing the radii of the three circles by the same amount enables us to construct a circle center O which still touches the diminished circles externally. All we have to do is to increase the radius of $\mathscr{C}$ by the amount the other three radii are decreased. Eventually we arrive at a point circle center C_1, and two equal circles centers C_2 and C_3. If we can draw a circle through C_1 to touch the equal circles centers C_2 and C_3 externally, then we can move back to the circle touching the three original circles externally. If there is only one such circle, we may argue that in a more general case there is only one. We naturally invert with

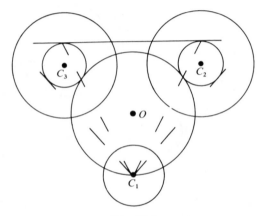

Fig. 26.5

respect to C_1 (Fig. 26.5), but we choose as circle of inversion the circle center C_1 which is orthogonal to both circles $\mathscr{C}_2$ and $\mathscr{C}_3$. Since their centers are at equal distances from C_1, and their radii are equal, this is possible. After inversion therefore, the circles centers C_2 and C_3 are unchanged, but the circle through C_1 which touches them has inverted into a common tangent. It is clear that it is the external common tangent shown, the other common tangents inverting into circles through C_1 which do not have external contact with both of the circles $\mathscr{C}_2$ and $\mathscr{C}_3$.

To show that there is one circle satisfying the condition $T_1^-T_2^+T_3^+$ is left as an Exercise.

Remark. Nothing in the above discussion indicates that the eight circles which touch three given circles *always* conform to the pattern we have described. For example, if the centers of $\mathscr{C}_1$, $\mathscr{C}_2$ and $\mathscr{C}_3$ are collinear, the circle orthogonal to the three circles is the line of centers, and inversion in this circle is equivalent to reflexion in the line. It is clear that if there is a $T_1^+T_2^+T_3^+$ contact circle, reflexion in the line of centers produces another contact circle *of the same type*, and similarly all existing types of contact circles occur in pairs. What we have shown above is that there is a case (or there are cases) where our heuristic argument is valid.

Exercises

26.1 Show that a variation of the method given above enables us to assert that there is a unique circle satisfying the condition $T_1^- T_2^+ T_3^+$.

26.2 Discuss the types of contact circle given by the other common tangents to the reduced circles $\mathscr{C}_2$ and $\mathscr{C}_3$ in the method for finding a circle with three external contacts described above. In combination with the method of the first Exercise above, are all the eight contact circles produced by the two methods?

26.3 $\mathscr{C}_1$ and $\mathscr{C}_2$ are point-circles, and $\mathscr{C}_3$ is a proper circle. Show that there are two contact circles, touching all three circles.

26.4 $\mathscr{C}_1$ is a point-circle, and $\mathscr{C}_2$ and $\mathscr{C}_3$ are proper circles which touch each other. Show there are three solutions to the problem of Apollonius.

26.5 $\mathscr{C}_1$, $\mathscr{C}_2$ and $\mathscr{C}_3$ are proper circles, and $\mathscr{C}_1$ touches both $\mathscr{C}_2$ and $\mathscr{C}_3$, but these latter circles do not touch. Show there are four solutions to the Apollonian problem, or five if we count $\mathscr{C}_1$ in the solutions.

26.6 $\mathscr{C}_1$, $\mathscr{C}_2$ and $\mathscr{C}_3$ all touch each other. Show that there are two circles distinct from the given ones which touch the three circles, so that, since each of the given circles is a solution of the problem, there are five solutions in all.

26.7 $\mathscr{C}_1$ and $\mathscr{C}_2$ touch, and $\mathscr{C}_3$ is a proper circle which does not touch $\mathscr{C}_1$ or $\mathscr{C}_2$. Show that there are six distinct circles which touch the three given circles.

†**26.8** Can you find a case when there are seven distinct solutions? (None of the many papers on the problem of Apollonius mentions this case specifically. Can you prove that such a case is impossible?)

26.9 If $\mathscr{C}_1$ contains both $\mathscr{C}_2$ and $\mathscr{C}_3$, discuss the possible types for the eight contact circles.

26.10 l and m are two parallel tangents to a circle $\mathscr{C}$, and a further tangent n intersects l in P and m in Q. Prove that the circle on PQ as diameter contains the center of $\mathscr{C}$.

Deduce by inversion that if three equal circles $\mathscr{C}_1$, $\mathscr{C}_2$ and $\mathscr{C}_3$ pass through the same point O, $\mathscr{C}_1$ and $\mathscr{C}_2$ touching at O, then if $\mathscr{C}_1$ intersects $\mathscr{C}_3$ again at P, and $\mathscr{C}_2$ intersects $\mathscr{C}_3$ again at Q, PQ is a diameter of $\mathscr{C}_3$. (There is also a more direct and simple proof.)

---

† See Pedoe, D., '*The Missing Seventh Circle*', *Elemente der Mathematik*, **25,** Jan. 1970, p. 14–15.

III

COAXAL SYSTEMS OF CIRCLES

These are the simplest algebraic systems of circles, and we shall discuss them from the algebraic point of view, although they can also be investigated by means of pure geometry.

27.1 Pencils of circles

We have already come across pencils of lines. The equation of any line passing through the origin of coordinates in the (X, Y) plane may be written in the form $\lambda X + \mu Y = 0$, and this equation depends linearly on two distinct (and therefore independent) lines of the system, given by $X = 0$ and $Y = 0$. Such a system of lines is called a *pencil* of lines.

Similarly if $u \equiv aX + bY + c = 0$, $v \equiv a'X + b'Y + c' = 0$ are any two distinct lines, the line

$$\lambda u + \mu v = 0,$$

represents, for suitable λ, μ, any line through the intersection of $u = 0$ and $v = 0$, and the set of lines is called a *pencil* of lines.

The idea can be extended to circles. Let $\mathscr{C}$ and $\mathscr{C}'$ be two distinct circles, given in normalized form by the equations

$$C(X, Y) \equiv X^2 + Y^2 + 2gX + 2fY + c = 0,$$
$$C'(X, Y) \equiv X^2 + Y^2 + 2g'X + 2f'Y + c' = 0.$$

The system of circles $\lambda C + \mu C' = 0$, derived linearly from the two given circles, is also called a *pencil* of circles. The equation of a circle of the system in normalized form is

$$X^2 + Y^2 + 2X \frac{(\lambda g + \mu g')}{\lambda + \mu} + 2Y \frac{(\lambda f + \mu f')}{\lambda + \mu} + \frac{\lambda c + \mu c'}{\lambda + \mu} = 0.$$

If (x', y') is a point of intersection of the circles $\mathscr{C}$ and $\mathscr{C}'$, then

$$C(x', y') = C'(x', y') = 0,$$

and therefore, for all values of λ and μ,

$$\lambda C(x', y') + \mu C'(x', y') = 0.$$

Hence the real points of intersection of $\mathscr{C}$ and $\mathscr{C}'$, if any, lie on all circles of the pencil.

[106]

If $P = (x, y)$ is a point which does *not* lie on both circles $\mathscr{C}$ and $\mathscr{C}'$, we cannot have both $C(x, y) = 0$ *and* $C'(x, y) = 0$. Hence the equation $\lambda C(x, y) + \mu C'(x, y) = 0$ admits a unique solution for the ratio $\lambda : \mu$, so that we can say: *one circle of the pencil passes through any given point which does not lie on both circles $\mathscr{C}$ and $\mathscr{C}'$.* This is evident if $\mathscr{C}$ and $\mathscr{C}'$ intersect, of course. The algebraic proof holds for all cases.

27.2   Pencil of circles as a locus

If $P = (x, y)$ is any point in the plane, and a line through P cuts a given circle $\mathscr{C}$ in the points Q and R, we know that the product $\overline{PQ} \cdot \overline{PR}$ is independent of the particular line chosen through P, and that in fact

$$\overline{PQ} \cdot \overline{PR} = x^2 + y^2 + 2gx + 2fy + c,$$

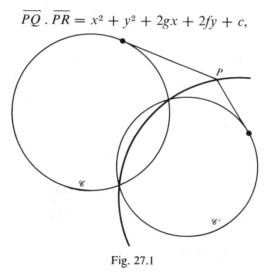

Fig. 27.1

where the equation of $\mathscr{C}$, in normalized form, is

$$C(X, Y) \equiv X^2 + Y^2 + 2gX + 2fY + c = 0.$$

We defined this product as *the power of P* with respect to the circle $\mathscr{C}$ (§18.1), and we know that if P is outside the circle, it is equal to the square of the tangent from P to the circle. If P lies on the circle it is zero, and if P lies inside the circle it is negative. Since a circle of the pencil $\lambda C + \mu C' = 0$ may be written in the form

$$\lambda/\mu = -C'(x, y)/C(x, y)$$

for points (x, y) which do not lie on the circle $\mathscr{C}$, and since λ and μ are fixed for the particular circle of the pencil under consideration, we may interpret the given circle of the pencil as *the locus of a point which moves so that the ratio of its powers with respect to two given circles $\mathscr{C}$ and $\mathscr{C}'$ is a constant.* (See Fig. 27.1.)

As we vary the value of the constant, we may obtain all circles of the pencil.

Exercises

27.1 Show that the power of a point P with respect to a point-circle, center A, is equal to $|PA|^2$. Hence determine the locus of a point P which moves so that the ratio of its distances $|PA|:|PA'|$ from two given points A, A' is constant.

27.2 Determine the locus of the previous Exercise by bisecting the angle APA' internally and externally, and proving that the points where the bisectors meet the line AA' are independent of P. Show also that the points A, A' are inverse points for the locus (called *an Apollonius circle* with regard to A and A'. See §18.3).

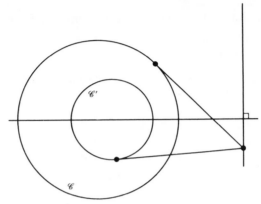

Fig. 27.2

27.3 QR is a chord of a circle $\mathscr{C}$ which subtends a right angle at a fixed point L. If P be the midpoint of QR, show that the power of P with respect to the circle $\mathscr{C}$ is equal to $-|PL|^2$. Deduce that the locus of P is a circle in the pencil determined by $\mathscr{C}$ and the point-circle L.

27.3 The radical axis of two circles

In the pencil of circles $\lambda C + \mu C' = 0$, a particularly important member is obtained by taking $\lambda/\mu = -1$. It is, in fact, a *line*:

$$C(X, Y) - C'(X, Y) \equiv 2(g - g')X + 2(f - f')Y + c - c' = 0,$$

and this line is called the *radical axis* of the circles $\mathscr{C}$ and $\mathscr{C}'$.

We know that it is the locus of a point P which moves so that P has equal powers with respect to $\mathscr{C}$ and $\mathscr{C}'$ (Fig. 27.2). If the two circles intersect in the points A and B, the radical axis, as a member of the pencil $\lambda C + \mu C' = 0$, will pass through A and B, and is then *the common chord* of the two circles. But it exists as a real line if $\mathscr{C}$ and $\mathscr{C}'$ do *not* intersect, and *intersects neither circle in this case*.

Exercises

27.4 Prove that the radical axis of $\mathscr{C}$ and $\mathscr{C}'$ is perpendicular to the line joining the centers of the two circles, (*a*) from general considerations, and (*b*) algebraically.

27.5 If $\mathscr{C}$ and $\mathscr{C}'$ do not intersect, prove that the midpoints of the segments determined by the points of contact of each of the four common tangents lie on a line.

27.6 If $\mathscr{C}$ and $\mathscr{C}'$ do not intersect, and have centers C and C', and radii r and r', show that the radical axis meets CC' in the point V which is such that $|VC|^2 - |VC'|^2 = r^2 - r'^2$. Deduce that V is the midpoint of CC' if and only if $r = r'$. [Since $\mathscr{C}$ and $\mathscr{C}'$ are supposed to be distinct, we cannot have one lying inside the other when $r = r'$. Again, if one lies inside the other, V cannot lie between C and C'.]

27.7 What happens to the radical axis of two concentric circles, for which $g = g', f = f'$, but $c \neq c'$?

27.4   Justification of the term: coaxal system

We now show that any two distinct circles of a pencil of circles $\lambda C + \mu C' = 0$ have the same radical axis as have $\mathscr{C}$ and $\mathscr{C}'$. This will explain the term 'coaxal system', which is used for a pencil of circles.
Let

$$C_1 \equiv \lambda_1 C + \mu_1 C' = 0,$$
$$C_2 \equiv \lambda_2 C + \mu_2 C' = 0,$$

be two distinct circles of the pencil. Then $\lambda_1/\mu_1 \neq \lambda_2/\mu_2$, so that $\lambda_1\mu_2 - \lambda_2\mu_1 \neq 0$. To find the radical axis of $\mathscr{C}_1$ and $\mathscr{C}_2$, we have to *normalize* both equations, then subtract one from the other. The normalized equations are

$$(\lambda_1 + \mu_1)^{-1}(\lambda_1 C + \mu_1 C') = 0,$$

and

$$(\lambda_2 + \mu_2)^{-1}(\lambda_2 C + \mu_2 C') = 0,$$

and on subtracting one equation from the other we easily obtain

$$(\lambda_1\mu_2 - \lambda_2\mu_1)(C - C') = 0,$$

and we know that $C - C' = 0$ is the radical axis of $\mathscr{C}$ and $\mathscr{C}'$, so we have proved our result.

Again, if $\mathscr{C}$ and $\mathscr{C}'$ intersect, the result is intuitive geometrically, since the radical axis of any two circles which pass through two points A and B is the line AB. But in the non-intersecting case this argument cannot be used.

Exercise

27.8 Deduce from the above result, and prove directly also, by algebraic means, that the centers of the circles of a coaxal system move on a line perpendicular to the radical axis. What is the exceptional case?

27.5   Determination of a coaxal system

We prove now that our system of circles $\lambda C + \mu C' = 0$, which we derived linearly from two distinct circles $\mathscr{C}$ and $\mathscr{C}'$, is *equally derivable from any two distinct circles of the system.* This is, of course, a property of a straight line, obtained from two distinct points in the first instance. The points of the line can be obtained from any two distinct points of the line. The algebra in both cases is exactly the same. Later (§36.2), when we shall represent circles by points in space of three dimensions, coaxal systems will be represented by lines in space of three dimensions.

With the notation of the previous section, let

$$C_1 \equiv \lambda_1 C + \mu_1 C' = 0, \qquad C_2 \equiv \lambda_2 C + \mu_2 C' = 0$$

be two distinct circles of the coaxal system, so that $\lambda_2\mu_1 - \lambda_1\mu_2 \neq 0$. The system formed by $k_1 C_1 + k_2 C_2 = 0$ contains the circles

$$k_1(\lambda_1 C + \mu_1 C') + k_2(\lambda_2 C + \mu_2 C') = 0,$$

which is the same as

$$(\lambda_1 k_1 + \lambda_2 k_2)C + (\mu_1 k_1 + \mu_2 k_2)C' = 0.$$

We wish to show that this system of circles contains all the circles of the system $\lambda C + \mu C' = 0$, and no additional ones!

If we are given λ and μ, we can solve the linear equations

$$k_1\lambda_1 + k_2\lambda_2 = \lambda,$$

$$k_1\mu_1 + k_2\mu_2 = \mu$$

for k_1 and k_2 uniquely, since $\lambda_2\mu_1 - \lambda_1\mu_2 \neq 0$. On the other hand, if k_1 and k_2 are given, λ and μ are defined uniquely by the above linear equations. We have therefore shown that every circle of $k_1 C_1 + k_2 C_2 = 0$ is a member of the system $\lambda C + \mu C' = 0$, and every circle of $\lambda C + \mu C' = 0$ is a member of the system $k_1 C_1 + k_2 C_2 = 0$. The two systems of circles are therefore identical.

Exercises

27.9 If two circles $\mathscr{C}_1$ and $\mathscr{C}_2$ of a given coaxal system $\lambda C + \mu C' = 0$ do not intersect, prove that no two circles of the system intersect.

27.10 Prove that the coaxal system $\lambda C + \mu C' = 0$, where $C \equiv 3(X^2 + Y^2) - 2X - 4Y + 3 = 0$, and $C' \equiv 3(X^2 + Y^2) - 4X - 2Y + 3 = 0$, contains two real point-circles, and that the coaxal system may also be written in the form $k_1[(X - 1)^2 + Y^2] + k_2[X^2 + (Y - 1)^2] = 0$.

Show that the condition for a circle of the system to be real is $k_1 k_2 \leqslant 0$. Where are the centers of real circles of the system situated?

28.1 A theorem on radical axes

Theorem *The three radical axes determined by taking pairs of circles from three given circles are either concurrent or parallel.*

This is a simple theorem with far-reaching consequences. One proof uses the notion that the radical axis of two circles $\mathscr{C}_1$ and $\mathscr{C}_2$ is the locus of points from which equal tangents may be drawn to $\mathscr{C}_1$ and to $\mathscr{C}_2$. If the radical axis of $\mathscr{C}_1$ and $\mathscr{C}_2$ intersects the radical axis of $\mathscr{C}_2$ and $\mathscr{C}_3$ in the point V, then equal tangents may be drawn from V to $\mathscr{C}_1$ and $\mathscr{C}_2$, and to $\mathscr{C}_2$ and $\mathscr{C}_3$ (or, more generally, the power of V with respect to $\mathscr{C}_1$ and $\mathscr{C}_2$ is the same, and the power of V with respect to $\mathscr{C}_2$ and $\mathscr{C}_3$ is the same). Hence V is a point which has the same power with respect to $\mathscr{C}_1$ and to $\mathscr{C}_3$. Hence V lies on the radical axis of $\mathscr{C}_1$ and $\mathscr{C}_3$ also (Fig. 28.1).

If the first two radical axes are parallel, the join of the centers of $\mathscr{C}_1$ and $\mathscr{C}_2$ must be parallel to the join of the centers of $\mathscr{C}_2$ and $\mathscr{C}_3$. Since there is only one parallel to a

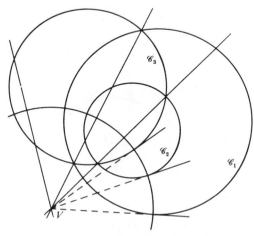

Fig. 28.1

given line through a given point, the centers of the three circles must be collinear, and therefore the radical axis of $\mathscr{C}_1$ and $\mathscr{C}_3$, which is perpendicular to the line of centers of $\mathscr{C}_1$, $\mathscr{C}_3$, must be parallel to the other two radical axes.

An algebraic proof is very simple, and this we now give:

Proof. If the equations of the three circles are normalized, the radical axes are

$$L_{12} \equiv C_1 - C_2 = 0,$$
$$L_{23} \equiv C_2 - C_3 = 0.$$

Since

$$L_{12} + L_{23} \equiv C_1 - C_3 = 0 \equiv L_{13},$$

the radical axis L_{13} of $\mathscr{C}_1$ and $\mathscr{C}_3$ passes through the intersection of L_{12} and L_{23}, if they have an intersection, or L_{13} is parallel to both lines, if they are parallel.

28.2   The circle orthogonal to three given circles

The theorem of the preceding Section shows us that there is at least one point, finite or at infinity, which has the same power with respect to three given circles $\mathscr{C}_1$, $\mathscr{C}_2$, $\mathscr{C}_3$

(Fig. 28.2). If the three circles belong to the same coaxal system, every point on the common radical axis has this property. If the circles are not in the same coaxal system there is a unique point with this property. It is called *the radical center* of the three circles. If it lies outside one, and therefore each of the three circles, and the point be designated as V, and the common power as $|VT|^2$, a circle radius $|VT|$ and center V will cut each of the three circles $\mathscr{C}_i$ orthogonally. If the radical center lies inside one of

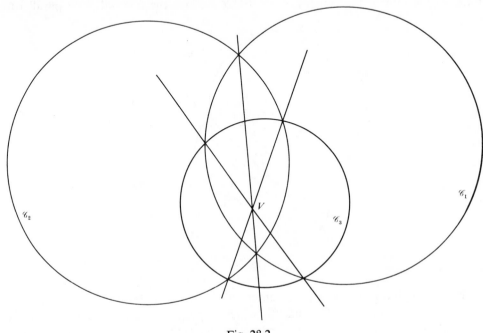

Fig. 28.2

the circles, and therefore inside each of them, and the common power is $-|VT|^2$, the common orthogonal circle is a *virtual* circle, with equation

$$(X - a)^2 + (Y - b)^2 + r^2 = 0,$$

where $r^2 = |VT|^2$, and the coordinates of V are (a, b). Algebraically, this virtual circle satisfies the condition for orthogonality with each of the circles $\mathscr{C}_i$.

The common orthogonal circle, if taken as circle of inversion, inverts each circle $\mathscr{C}_i$ into itself, and, in particular, inverts any circle $\mathscr{C}$ which touches the three circles into a distinct circle $\mathscr{C}'$ which touches the three circles. We assume that $\mathscr{C}$ is not also orthogonal to the common orthogonal circle. Hence we expect the circles which touch the three given circles to fall into pairs, mutually inverse in the common orthogonal circle. (See §26.1 and §38.1.) The consideration of inversion in a virtual circle arises quite naturally in this context.

When the centers of the three circles $\mathscr{C}_i$ are collinear, the common orthogonal circle is a line, the line of centers, and the radical center is at infinity. Inversion in the common orthogonal circle is reflexion in the line of centers in this case.

Exercises

28.1 If $\mathscr{C}_1$, $\mathscr{C}_2$ are two non-intersecting circles, and $\mathscr{C}_3$ intersects $\mathscr{C}_1$ in the points P, Q and $\mathscr{C}_2$ in the points R, S, prove that the point $PQ \cap RS$ lies on the radical axis of $\mathscr{C}_1$ and $\mathscr{C}_2$.

28.2 A circle drawn through two given points P, P' intersects a given circle $\mathscr{C}$ in the points Q and R. Prove that the lines QR form a pencil, that is, they either all pass through a point, or they are all parallel to each other. Hence prove that two circles through P and P' touch $\mathscr{C}$, and show how to construct them.

28.3 Prove the following construction, due to Gergonne, for a circle which touches three given circles $\mathscr{C}_1$, $\mathscr{C}_2$ and $\mathscr{C}_3$. Draw the circle which is orthogonal to the three given circles. (See §28.2.) Draw the three chords of intersection of this circle with $\mathscr{C}_1$, $\mathscr{C}_2$ and $\mathscr{C}_3$ respectively. From the points where these chords meet an axis of similitude of the three circles $\mathscr{C}_1$, $\mathscr{C}_2$, $\mathscr{C}_3$ (See Exercise 13.2) draw pairs of tangents to $\mathscr{C}_1$, $\mathscr{C}_2$ and $\mathscr{C}_3$. Then the two circles drawn through these six points of contact will touch $\mathscr{C}_1$, $\mathscr{C}_2$ and $\mathscr{C}_3$. (We shall see later, in §38.1, an algebraic verification of the theorem that the circles touching three given circles $\mathscr{C}_1$, $\mathscr{C}_2$ and $\mathscr{C}_3$ are inverse in pairs in the common orthogonal circle of the three given circles. But since circles are invariant under inversion in a circle orthogonal to them, the circles $\mathscr{C}_1$, $\mathscr{C}_2$ and $\mathscr{C}_3$ invert into themselves under inversion in their common orthogonal circle, and a circle which touches them at A, B and C say inverts into a circle which touches them at A', B' and C', where the primes denote points inverse to A, B and C in the common orthogonal circle. Hence circle ABC, the orthogonal circle, and circle $A'B'C'$ are in a coaxal system (Exercise 21.6). Therefore, by Thm 28.1, the radical axes of $\mathscr{C}_1$ and each of the three coaxal circles we are discussing meet in a point. Two of the radical axes are tangents at A and A' to $\mathscr{C}_1$, and the third is the common chord of $\mathscr{C}_1$ and the common orthogonal circle. Finally, if P is the point of intersection of these three lines, P must be on an axis of similitude (See Exercise 13.2) of the three given circles, because $|PA| = |PA'|$ shows that P lies on the radical axis of circles ABC, $A'B'C'$, and by Exercise 18.3 we know that if two circles touch two others the radical axis of either pair passes through a center of similitude of the other pair, and since circles ABC, $A'B'C'$ touch $\mathscr{C}_1$, $\mathscr{C}_2$ and $\mathscr{C}_3$, the radical axis of the two touching circles passes through three centers of similitude: it is an axis of similitude of $\mathscr{C}_1$, $\mathscr{C}_2$ and $\mathscr{C}_3$.)

28.4 Since three circles have four axes of similitude, the above construction yields eight Apollonian circles (compare §26.1). Draw them.

28.3 Another proof of the nine-point circle theorem

In §12.1 we saw that if A, B and C are the vertices of a triangle, D', E' and F' the feet of the perpendiculars from the vertices onto the opposite sides, and α, β, γ the midpoints of BC, CA and AB respectively, then there is a circle which passes through these six points and also the three midpoints of HA, HB and HC, where H is the orthocenter, the intersection of AD', BE' and CF' (Fig. 28.3). We remark that $|CD'| \cdot |CB| = |CE'| \cdot |CA|$, since a circle on AB as diameter passes through D' and E'. We deduce that $|C\alpha| \cdot |CD'| = |C\beta| \cdot |CE'|$, by halving each side of the equation. Hence the points α, D', β and E' are concyclic. Let them lie on a circle $\mathscr{C}_3$. Similarly the points α, D', γ, F' are concyclic. Let them lie on a circle $\mathscr{C}_2$. If the circles $\mathscr{C}_3$ and $\mathscr{C}_2$ do not coincide, they have the chord of intersection $\alpha D'$, which is the line BC. The points γ, F', β, E' are concyclic. Let them lie on a circle $\mathscr{C}_1$. If $\mathscr{C}_1 \neq \mathscr{C}_2$, their chord of intersection is AB, and if $\mathscr{C}_1 \neq \mathscr{C}_3$, their chord of intersection is AC. Hence, if the circles $\mathscr{C}_1$, $\mathscr{C}_2$ and $\mathscr{C}_3$ are distinct, their radical axes, taking the circles in pairs, form the sides of a triangle. This contradicts Thm 28.1. Hence the points α, β, γ, D', E' and F' all lie on one circle. As in §12.1, by applying this result to the triangle BHC, we show that this circle also contains the midpoints of HA, HB and HC.

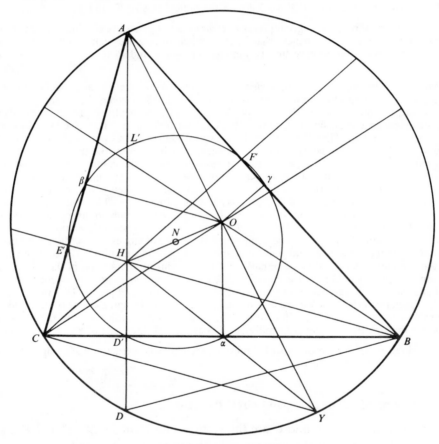

Fig. 28.3

28.4   Clerk Maxwell's derivation of reciprocal figures

We use the notation of §6.1. Let ABC be a triangle, and D a point not on the sides. The triangle PQR is such that QR is parallel to DA, RP is parallel to DB and PQ is parallel to DC. We wish to show the existence of a point F such that BC is parallel to FP, CA is parallel to FQ and AB is parallel to FR.

Maxwell, who was a geometer as well as a great physicist, drew circles with centers at the points A, B, C and D respectively (Fig. 28.4), and remarked that the chord of intersection of each pair of circles is perpendicular to the line joining the centers, and that the chords of intersection of the pairs which can be taken from three circles pass through one point (Thm 28.1). Hence the four circles constructed have common

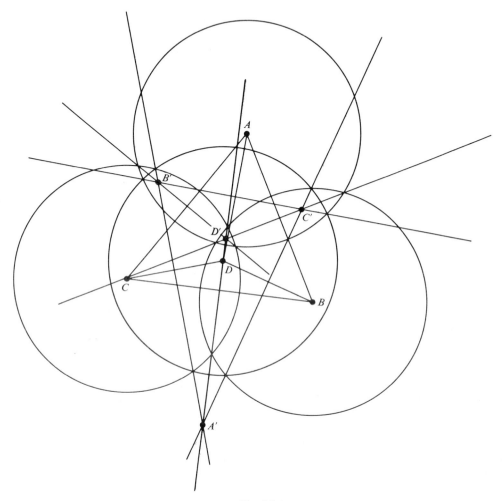

Fig. 28.4

chords which pass in threes through four points A', B', C' and D', where $B'C'$ is perpendicular to DA, $C'A'$ is perpendicular to DB, $A'B'$ is perpendicular to DC, and $D'A'$ is perpendicular to BC, $D'B'$ is perpendicular to CA and $D'C'$ is perpendicular to AB.

If we now rotate the figure $A'B'C'D'$ through a right angle, the figure reciprocal to $ABCD$ is in existence! The triangle $A'B'C'$ is similar to the given triangle PQR, and the point F is immediately obtained from the point D' which corresponds to it in triangle

$A'B'C'$. Through P we draw a line parallel to BC, and through Q a line parallel to CA. The intersection F of these two lines will be such that FR is parallel to AB.

29.1   Canonical form for the equation of a coaxal system

We know that if a coaxal system contains two circles with distinct centers, then the centers of all the circles of the system move on the line containing these two centers, and that this line is perpendicular to the radical axis, which is a proper (finite) line in

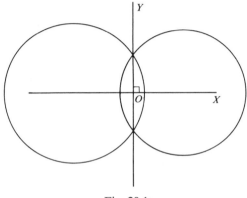

Fig. 29.1

the Euclidean plane. If our coaxal system is defined by two *concentric* circles, all circles of the system have the same center, and there is no finite radical axis. We shall see later (§30.1) that a pencil of lines has to be considered as a coaxal system also, but for the moment we consider only a coaxal system of proper circles (that is, not lines), which contains at least two circles with distinct centers. Of course, every such system does contain a line, the radical axis.

    The line of centers and the radical axis are perpendicular, and suggest a natural system of coordinates, that is, a system independent of the original system, and attached *geometrically* to the coaxal system. By this, we mean a system of coordinates which is *invariant* under the change from one orthogonal system of cartesian coordinates to another. We should prove systematically that under a change of axes the coaxal system $\lambda C + \mu C' = 0$ changes into the coaxal system $\lambda C_T + \mu C'_T = 0$, where $C_T = 0$ is the transform of $C = 0$, and $C'_T = 0$ is the transform of $C' = 0$, that the line of centers is transformed into the line of centers, and the radical axis into the radical axis; but this is something that can be left to the reader, although it must be pointed out.

    Let us therefore take the line of centers of our given coaxal system as our new X-axis (see Fig. 29.1), and the radical axis of the system as our new Y-axis. Then in the new coordinates, the original circles $\mathscr{C}$ and $\mathscr{C}'$ will have the equations

$$C_T \equiv X^2 + Y^2 + 2gX + c = 0,$$

$$C'_T \equiv X^2 + Y^2 + 2g'X + c' = 0,$$

since their centers are on the X-axis. The radical axis is given by

$$C_T - C'_T = 2(g - g')X + c - c' = 0,$$

and since this is now $X = 0$, we must have $c = c'$. We can define the coaxal system by means of the special circle $X = 0$, and the special (distinct) circle $g'C_T - gC'_T = (g' - g)(X^2 + Y^2 + c) = 0$, so that the equation of the coaxal system we are investigating is, in the new coordinates

$$(X^2 + Y^2 + c) + 2kX = 0;$$

that is

$$X^2 + Y^2 + 2kX + c = 0,$$

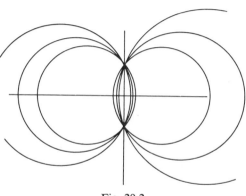

Fig. 29.2

where k *varies* through all real values, and c is *a constant* fixed for the particular coaxal system.

By saying that this is a *canonical* (or *normal*) form for coaxal systems, we imply that this equation always represents a coaxal system, and that *any* coaxal system of circles which does not consist of concentric circles may be transformed to have such an equation, by a suitable choice of axes.

We now distinguish between the different types of coaxal system, obtaining new canonical forms for the distinct kinds of system.

29.2   The intersecting type of coaxal system

The circles of the coaxal system all cut the radical axis in the same two points (Fig. 29.2), given by $X = 0$, $Y^2 + c = 0$. If $c = -p^2$, these points are real, and all circles of the system pass through the two fixed points $(0, p)$ and $(0, -p)$. The circle of the system which has minimum radius is the one with its center at $(0, 0)$. Any coaxal system which consists of intersecting circles may be written in the form

$$X^2 + Y^2 + 2kX - p^2 = 0,$$

by a suitable choice of axes of coordinates.

29.3   The tangent type of coaxal system

When $p = 0$, the circles of the system all touch the radical axis at the same point (Fig. 29.3). Hence any system of coaxal circles which consists of circles which touch each other at a point may be written in the form

$$X^2 + Y^2 + 2kX = 0,$$

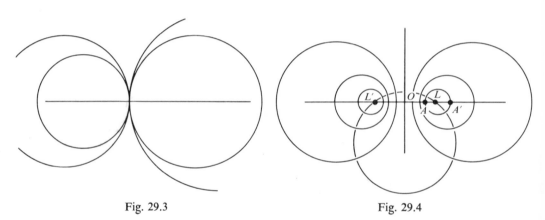

Fig. 29.3   Fig. 29.4

by a suitable choice of coordinate axes. Since there is no constant coefficient in this equation, all such systems are indistinguishable from each other.

29.4   The non-intersecting type of coaxal system

If c is positive, $c = p^2$, the circles of the coaxal system do not intersect the radical axis, and no two circles of the system intersect, since if they did, all circles of the system would pass through the points of intersection, which would lie on the radical axis (Fig. 29.4). Any system of non-intersecting circles may therefore be written in the form

$$X^2 + Y^2 + 2kX + p^2 = 0,$$

by a suitable choice of coordinate axes. Writing the equation as

$$(X + k)^2 + Y^2 = k^2 - p^2,$$

we see that the center is the point $(-k, 0)$, and that the radius is $(k^2 - p^2)^{1/2}$. We deduce that for real circles we must have $k^2 \geqslant p^2$, and therefore:

No circle of the system has its center between the points $(-p, 0)$ and $(p, 0)$. The circles with centers at these points are point-circles, of zero radius.

This is the distinguishing feature of the non-intersecting type of coaxal system. If we put $p = 0$, we are in the intermediate case of contact circles, when the system contains one point-circle.

The points $(-p, 0)$ and $(p, 0)$ are sometimes called *limiting points* of the coaxal system. If a circle of the system cuts the X-axis in the points A, A', we have, for the coordinates of A, A':

$$X^2 + 2kX + p^2 = 0,$$

so that $\overline{OA} \cdot \overline{OA'} = p^2$. If we denote the limiting points by L, L',

$$\overline{OA} \cdot \overline{OA'} = |OL|^2 = |OL'|^2,$$

and therefore L and L' are *harmonic conjugates* with respect to A and A'. Hence (§19.1) if O' is the midpoint of AA', the center of the circle we are considering, $\overline{O'L} \cdot \overline{O'L'} = |C'A|^2 = |O'A'|^2$, so that L and L' are *inverse points for any circle of the system*. If we look back at Thm 25.1, Corollary, we can now state the theorem:

Theorem *If A, A' and B, B' are two non-overlapping pairs of points on a line, cut by two non-intersecting circles which have their centers on the line, then the unique pair of harmonic conjugates determined by the two pairs of points are the limiting points of the coaxal system determined by the two circles.*

29.5   Polar systems of coaxal circles

The points L and L' are inverse points for every circle of the coaxal system we have been discussing (see Fig. 29.5), and therefore any circle which passes through L and L' cuts every circle of the non-intersecting system orthogonally (Thm 20.2). This is immediately verified algebraically, since circles through L and L' are given by the equation $X^2 + Y^2 + 2k'Y - p^2 = 0$, and the criterion for two circles to be orthogonal (§22.1, (2))

$$2g_1g_2 + 2f_1f_2 - c_1 - c_2 = 0$$

becomes, in this case,

$$2k(0) + 2(0)k' - p^2 + p^2 = 0.$$

Hence, every circle of the coaxal system

$$X^2 + Y^2 + 2kX + p^2 = 0$$

is orthogonal to every circle of the system (also a coaxal system)

$$X^2 + Y^2 + 2k'Y - p^2 = 0,$$

of circles through the point-circles of the first system.

Similarly, to every intersecting system of coaxal circles there is associated a system of non-intersecting coaxal circles, the limiting points of which are the points of intersection of all circles of the intersecting system. The equations

$$X^2 + Y^2 + 2kX - p^2 = 0,$$

and

$$X^2 + Y^2 + 2k'Y + p^2 = 0,$$

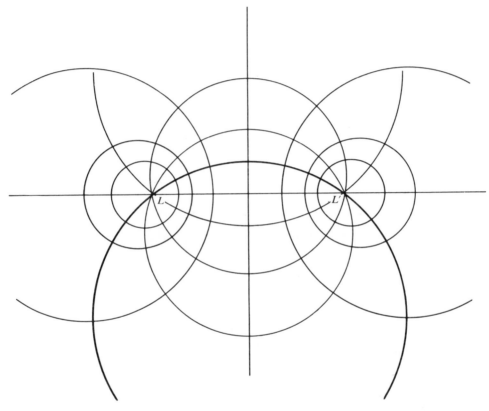

Fig. 29.5

demonstrate this theorem. If one system consists of circles which touch each other at a point, the other system is of the same kind. The equations

$$X^2 + Y^2 + 2kX = 0,$$

$$X^2 + Y^2 + 2k'Y = 0,$$

show two such systems.

Pairs of coaxal systems which are such that every circle of the one intersects every circle of the other orthogonally will be called *polar systems*. (They are also called *orthogonal* systems, and *conjugate* systems.) The reason for this terminology will be clearer when we represent coaxal systems by lines in space of three dimensions (§37.2). A special case of such systems is afforded by the usual grid system $X = k$, $Y = k'$. One line of each of the families of lines $X = k$ and $Y = k'$ passes through an arbitrary point of the plane. We have seen that one circle of a given coaxal system passes through any point of the plane which does not lie on all circles of the system (§27.1). It is to be expected that polar coaxal systems should appear in many branches of mathematics, and they do!

Exercises

29.1   Prove algebraically that the condition that the circle $\mathscr{C}$ be orthogonal to the point-circle with center at (x_1, y_1) is that the point (x_1, y_1) lie on $\mathscr{C}$.

29.2   Prove algebraically, using the canonical forms of §29.1, that if a circle $\mathscr{C}$ is orthogonal to two distinct members of a coaxal system then it is a member of the polar coaxal system, and is therefore also orthogonal to every member of the coaxal system.
   If a circle passes through both limiting points of a non-intersecting coaxal system it is orthogonal to every circle of the system. How does this fit in with the first part of this question?

29.3   Prove that a circle which is orthogonal to one circle of an intersecting coaxal system and has its center on the radical axis of this system must be a member of the polar coaxal system.

29.4   Justify the following method for the construction of a non-intersecting coaxal system: Let $\mathscr{C}$ be any fixed circle through two distinct points L, L'. Let O be any point on LL' from which a tangent OT may be drawn to $\mathscr{C}$. With center O and radius $|OT|$, draw a circle. Then this circle is a member of the non-intersecting coaxal system with limiting points L and L', and all circles of the system may be drawn in this way.

29.5   Construct a system of coaxal circles and the polar system, using the method of the preceding Exercise. (If done neatly, the result is pleasing: aesthetic pleasure plays an important part in the development of geometrical skills.)

29.6   Deduce from Exercise 29.4 that a circle $\mathscr{C}$ and a pair of inverse points P, P' in $\mathscr{C}$ define a coaxal system, in the sense that $\mathscr{C}$ is in the coaxal system defined by the point-circles P and P'.

30.1   The inverse of a coaxal system

If we invert an intersecting coaxal system with common points A and B with respect to a center of inversion which is neither at A or B, we obtain a system of circles through the inverse points A' and B'. This is a coaxal system. If we invert with respect to A, say, we obtain a pencil of lines through B'. If the circles of the original system touch at A, inversion with respect to A will produce a system of parallel lines. Hence, if we wish to have the theorem: *the inverse of a coaxal system is also a coaxal system*, we must admit a pencil of intersecting or parallel lines into the domain of our coaxal systems.

If the system we are inverting is a non-intersecting system of circles, only one circle of the system passes through the center of inversion, so that the inverse system contains only one line. The polar system inverts into a proper coaxal system or a pencil of lines. In the first case we can argue that since orthogonality is preserved by inversion, Exercise 29.2 indicates that the non-intersecting coaxal system inverts into a non-intersecting coaxal system. In the second case the vertex of the pencil is the common center of the system we have inverted, and we accept the fact that a system of concentric circles is a coaxal system. If the pencil of lines is one of parallel lines, the non-intersecting system inverts into the set of perpendicular parallels, and the original non-intersecting system must have consisted of circles touching at a point.

If it is remembered that the canonical forms of §29.1 do not apply to systems of concentric circles, but that these are coaxal systems, and that the polar system to a set of concentric circles consists of the pencil of lines through the common center, all systems which contain circles are allowed for.

Exercises

30.1   Investigate the inverse of a coaxal system algebraically, using the formula of §21.1.

30.2   The radical axis of a coaxal system cuts a given line in the point V, and a circle of the system cuts the line at the points P, Q. Show that the product $\overline{VP} \cdot \overline{VQ}$ is the same for all circles of the system which cut the line. Deduce that two circles of the system *touch* the line. What happens if the coaxal system is one of concentric circles?

30.3 P' is the inverse of a given point P in the circle $\mathscr{C}$ of a given coaxal system. Show that the locus of P', as $\mathscr{C}$ varies in the coaxal system, is the unique circle through P of the polar coaxal system.

30.4 $\mathscr{C}$ and $\mathscr{C}'$ are two given circles. Show that a circle can always be found in the coaxal system defined by $\mathscr{C}$ and $\mathscr{C}'$ such that $\mathscr{C}$ and $\mathscr{C}'$ are inverse to each other in this circle.

30.5   If we denote by M the transformation of the inversive plane which is the product of inversions in circles $\mathscr{C}$ and $\mathscr{C}'$, show that the same mapping is obtained if we take the product of inversions in two other suitable circles $\mathscr{D}$ and $\mathscr{D}'$ of the coaxal system determined by $\mathscr{C}$ and $\mathscr{C}'$, and that one of the two circles $\mathscr{D}$, $\mathscr{D}'$ may be chosen arbitrarily in the coaxal system. Show also that M leaves all circles of the coaxal system polar to that determined by $\mathscr{C}$ and $\mathscr{C}'$ invariant.

31.1   Geometry of the compasses

Anyone who has attempted some accurate geometrical drawings will have found that the so-called "straight" edge is rarely straight, whereas a good pair of compasses seem to offer a more accurate means of construction than a ruler. Is it possible to dispense with a straight edge in Euclidean constructions? We have to be able to:

(i) find the intersection of two given circles, each given by its center and radius,
(ii) find the intersection of a line given by two distinct points with a circle given by its center and radius,
(iii) find the intersection of two lines, each given by a pair of distinct points.

We can naturally carry out (i) with a pair of compasses only. We bring inversion into the scheme of things! We saw in §20.4 that we can find the inverse of a given point P in a given circle Σ, center O, using compasses only, Exercise 20.3 being the problem of what to do when the construction breaks down. To show how (ii) can be carried out with compasses only, let P, Q be the points which determine the line, and let $\mathscr{C}$ be the given circle, center V. Invert in $\mathscr{C}$. Then $\mathscr{C}$ inverts into itself, but the line

PQ inverts into the circle *P'Q'V*. We can find *P'* and *Q'*. If we can find the center of a circle, given three points on the circle, using compasses only, we can draw the circle *P'Q'V*, find the two points *X'* and *Y'* where this circle intersects $\mathscr{C}$, and invert these two points in $\mathscr{C}$ to find the points *X*, *Y* where the line *PQ* intersects the circle $\mathscr{C}$.

To carry out (iii), using compasses only, invert with respect to any suitable circle, center *O*. If the first line *PQ* inverts into the circle *P'Q'O*, and the second line *RS*

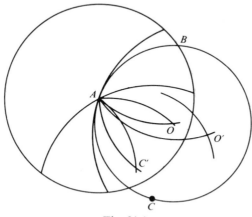

Fig. 31.1

inverts into the circle *R'S'O*, we know that we can draw these two circles, using compasses only. If they intersect in the points *O* and *Z'*, the inverse *Z* of *Z'* is the point of intersection of the lines *PQ* and *RS*.

Hence all Euclidean constructions can be carried out with a pair of compasses only if we can determine the center of a circle through three given points *A*, *B* and *C*.

31.2   Determination of the center of the circle through three points

Let the points be *A*, *B* and *C* (Fig. 31.1). We invert with respect to the circle center *A* and radius $|AB|$. If *C'* be the inverse of *C* in this circle, the circle *ABC* inverts into the line *BC'*, since *B* inverts into itself. We construct this point *C'*, using compasses only, of course. Now, by §23.3, the center of circle *ABC* inverts into the inverse of the center of inversion, which is *A*, in the inverse of circle *ABC*, which is the line *BC'*. We can determine this point, which is the *reflexion* of *A* in the line *BC'*. All we have to do is to find the other intersection of circles center *B* and radius $|BA|$, and center *C'* and radius $|C'A|$. These two circles intersect at *A* and another point. Call this other point *O'*. It is the reflexion of *A* in the line *BC'*, and the inverse of the center of circle *ABC*. Hence if we find the inverse of *O'* in the circle of inversion, we obtain the center *O* of the circle *ABC*.

All construction lines are shown in the drawing of Fig. 31.1. The reader is advised to carry out the construction for himself. The test of the accuracy of the drawing comes when we put the point of the compasses on O, extend the compasses to A, say, and see whether the circle we draw does indeed pass through B and C.

Many a bright student discovers this geometry of the compasses for himself, and is dashed to find that it was fully investigated, without the benefit of inversion, by Mascheroni in *La Geometria del Compasso*, published in Pavia in 1797, and even earlier by Mohr in *Euclides Danicus*, published in 1672.

Exercise

31.1 Given a circle $\mathscr{C}$, justify the following construction, using compasses only, for the determination of the center O of the circle:

With center at any point A on $\mathscr{C}$ draw a circle to cut $\mathscr{C}$ at B and C. Find the geometrical image O' of A in the line BC. Then the inverse of O' in the circle center A is the required center of $\mathscr{C}$.

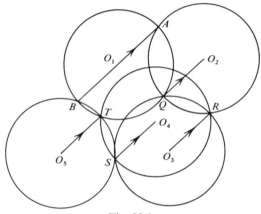

Fig. 32.1

32.1 Constructions with a disk

Besides the compass geometry of Mohr–Mascheroni, there are other possibilities, such as using a straight edge only, or using a straight edge in combination with a circle and its center in order to construct points which would seem to be elusive without the use of both straight edge and compasses. Constructions with a straight edge only belong to projective geometry (Chapter VII). We shall not pursue the various possibilities here, but an interesting construction suggested by Paul Pargas (*Am. Math. Monthly* (67), 1960, p. 292) is worth discussing.

We use a given circular disk $\mathscr{D}$ to draw circles, and assume that the disk can be placed on any two distinct points P and Q (not too far apart, of course), and that a circle can then be drawn around the circumference of $\mathscr{D}$ to contain both P and Q. There are two such circles. Now, mark a point A on one of the circles we have drawn with the help of $\mathscr{D}$. By drawing circles only with the help of $\mathscr{D}$, how can we determine the point on this circle which is diametrically opposite to A?

We use the evident theorem that if two equal circles (see Fig. 32.1), centers O_1 and

O_2 intersect at the points A and Q, then the lines O_1A and QO_2 are parallel and in the same direction. If the circle center O_1 be the given circle, use the disk $\mathscr{D}$ to draw a circle through A and any other point Q on the given circle. Let its center be O_2. Use $\mathscr{D}$ to draw a circle, center O_3, through Q, cutting the circle center O_2 in R. Use $\mathscr{D}$ to draw a circle, center O_4, through R, and let this circle cut the circle center O_3 in S, and the circle center O_1 in T. Use $\mathscr{D}$ to draw the other circle through S and T. Then if this circle, center O_5 cuts the circle center O_1 in B as well as T, the point B is diametrically opposite to A, because

$$\overrightarrow{O_1A} = \overrightarrow{QO_2} = \overrightarrow{O_3R} = \overrightarrow{SO_4} = \overrightarrow{O_5T} = \overrightarrow{BO_1}.$$

Exercises

32.1 l, m are two parallel tangents to a circle $\mathscr{C}$, and a further tangent n intersects l at P and m at Q. Prove that the circle on PQ as diameter contains the center of $\mathscr{C}$. Deduce, by inversion, that if three equal circles $\mathscr{C}_1$, $\mathscr{C}_2$ and $\mathscr{C}_3$ pass through the same point O, and $\mathscr{C}_1$ and $\mathscr{C}_2$ touch at O, then if $\mathscr{C}_3$ intersects $\mathscr{C}_1$ again at P and $\mathscr{C}_2$ again at Q, the line PQ is a diameter of $\mathscr{C}_3$. (This has already appeared as Exercise 26.10. The deduction can be proved directly and simply, without the use of inversion. It is needed for the next Exercise.)

32.2   A circle is drawn with the disk $\mathscr{D}$ described above, and a point A is marked on it. Show how the disk may be used to draw an equal circle which touches the given circle at A. (The result of Exercise 32.1 may be assumed.)

IV

THE REPRESENTATION OF CIRCLES BY POINTS IN SPACE OF THREE DIMENSIONS

The equation of a circle in normalized form depends on three parameters, and it is natural to enquire whether there is a useful mapping of circles onto the points of Euclidean space of three dimensions. We give such a mapping, and we shall see that this leads to valuable geometric insights into the geometry of circles. We begin, however, with a rapid survey of the few simple geometrical properties of E_3, Euclidean space of

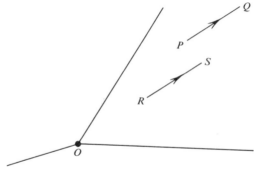

Fig. 33.1

three dimensions, which we wish to use. Because of the possibility of extension to Euclidean and affine space of n dimensions, we use coordinates (X_1, X_2, X_3) here instead of (X, Y, Z), but in §36.1 we shall return to (X, Y, Z) coordinates.

33.1 Vectors in Euclidean and affine space of three dimensions

Let $P = (x_1, x_2, x_3)$ and $Q = (y_1, y_2, y_3)$ be two points in Euclidean or affine space of three dimensions, where the coordinates are taken with reference to a given system of Cartesian coordinates, origin O (Fig. 33.1). Then the directed segment or *vector* $\overrightarrow{PQ}$ is defined to be the ordered triad of real numbers

$$\overrightarrow{PQ} = (y_1 - x_1, y_2 - x_2, y_3 - x_3).$$

Two vectors $\overrightarrow{PQ}$ and $\overrightarrow{RS}$ are thought of as being equivalent when PQ is parallel to RS, when the length of PQ = length of RS, and when the direction of motion from

[126]

If OP and OQ be regarded as two adjacent sides of a parallelogram, then $P + Q$ is represented by the diagonal of the parallelogram which passes through O.

Again, the vector rP is a dilatation of the vector P from the origin, and rP is in the same direction as $\overrightarrow{OP}$ if r is positive, and in the opposite direction if r is negative.

If two free vectors, represented by the bound vectors P and Q, are parallel, then $P = rQ$, for some real number r.

The parallelogram law for the sum of two bound vectors shows at once that the free vector $\overrightarrow{QP}$ is represented by the bound vector $P - Q$. This result is often used (Fig. 33.4).

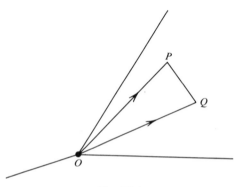

Fig. 33.4

33.3 Linear dependence

Our bound vectors, with the rules for addition and multiplication by a scalar given above, form a vector space over the reals. The laws governing such a space are exactly those given in §1.3.

Vectors $P_1, P_2, \ldots, P_n$ are said to be *linearly dependent* if real numbers $k_1, k_2, \ldots, k_n$, *not all zero*, exist so that

$$k_1 P_1 + k_2 P_2 + \ldots + k_n P_n = 0,$$

where 0 stands for the zero vector. If a set of vectors is not dependent, it is a set of *independent* vectors. In our space, the unit vectors

$$E_1 = (1, 0, 0), \ E_2 = (0, 1, 0), \ E_3 = (0, 0, 1)$$

are clearly independent, and any vector can be expressed in terms of them, since

$$P = (x_1, x_2, x_3) = x_1 E_1 + x_2 E_2 + x_3 E_3.$$

Such a set of vectors is said to form *a basis* for the vector space, and it is easily proved that any set of three independent vectors form a basis.

33.4   Points on a line

Theorem   *If C is any point on the line determined by two distinct points A and B, then we may always write*

$$C = (1 - t)A + tB,$$

where the ratio of the real numbers $t : 1 - t$ is equal to the ratio $\overline{AC} : \overline{CB}$ of the signed lengths of the segments AC and CB.

Proof. This is exactly the same as in §3.1. Here again we use the same symbol for the endpoints of a vector and the vector itself. The equation which connects C with A and B is equivalent to three equations between the coordinates of the points.

An immediate deduction from this theorem is that $C = kA + k'B$, where $k + k' = 1$, is a point on the line joining the points A and B. On the other hand, if $k + k' \ne 1$, the point C is still in the plane determined by the lines OA and OB, and all such bound vectors $\overrightarrow{OC}$ are given by $C = kA + k'B$, where k and k' take all real values.

Since we may write $kA + k'B - C = 0$, it follows that *any three coplanar bound vectors through O are linearly dependent.* It is equally simple to prove that *any three non-zero bound linearly dependent vectors through O are coplanar.*

As is always the case with the linear dependence of vectors, the zero vector has the property of making any set of vectors which contains it a linearly dependent set, since

$$k \,.\, 0 + 0 \,.\, P_1 + 0 \,.\, P_2 + \ldots + 0 \,.\, P_n = 0,$$

and our theorems, to be geometrically significant, refer to non-zero vectors.

33.5   Points in a plane

In §33.3 we remarked that any set of three independent vectors through O form a basis for the vectors through O. An immediate consequence is the theorem that *any set of four bound vectors through O is a linearly dependent set.* A set is dependent if a subset is dependent, so that if three of the vectors are coplanar, and therefore dependent, the four vectors are dependent. Suppose that the four vectors are A, B, C and D, and that the three vectors A, B and C are not coplanar. The plane (ABC) of the three endpoints cannot contain the origin O, for if it did, the three lines OA, OB and OC would be in one plane, the plane (ABC) (see Fig. 33.5). Hence the line of the vector OD intersects the plane (ABC) in a point $\ne O$, or is parallel to it,

Let P be the point of intersection of the line of the vector OD with the plane (ABC) (Fig. 33.6). (The case when the line of OD is parallel to the plane (ABC) is considered in Exercise 33.4.) We cannot have the joins AP, BP and CP parallel to BC, CA and AB respectively, so let us assume that CP intersects the side AB of triangle ABC in the point E. By Thm 33.4 we may write $P = kE + k'C$, where $k + k' = 1$. We may also write $E = qB + q'A$, where $q + q' = 1$. Hence

$$P = k(qB + q'A) + k'C = kq'A + kqB + k'C,$$

and $kq' + kq + k' = k(q + q') + k' = k + k' = 1$. We therefore have:

Theorem I   *Any point in the plane* (ABC), *where* A, B *and* C *are independent vectors, may be written as* $P = k_1A + k_2B + k_3C$, *where* $k_1 + k_2 + k_3 = 1$.
(Compare Thm 4.2.)

Example. If A, B and C are three independent vectors, the midpoint of BC is $(B + C)/2$, and the centroid of triangle ABC is again $(A + B + C)/3$. If D is a vector with end-

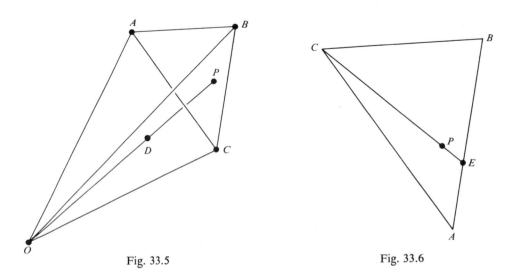

Fig. 33.5   Fig. 33.6

point D not in the plane (ABC), the point which divides the join of D to the centroid of ABC in the ratio $3:1$ is

$$[3(A + B + C)/3 + D]/4 = (A + B + C + D)/4.$$

The symmetry of this result shows that the four such joins are concurrent.

We are assuming that the lines AB and CD do not intersect, for if they did, the point D would lie in the plane (ABC). If we use the theorem that the four vectors A, B, C and D are linearly dependent, and write

$$k_1A + k_2B + k_3C + k_4D = 0,$$

we may assume that none of the coefficients k_i is zero, since $k_4 = 0$ would imply that the plane (ABC) contains O, and so on, and we may assume that none of the planes (ABC), (ABD), (ACD) and (BCD) contains the origin. Writing this equation in the form

$$(k_1A + k_2B)/(k_1 + k_2) = -(k_3C + k_4D)/(k_1 + k_2),$$

where we assume, for the moment, that $k_1 + k_2 \neq 0$, the left-hand side of the equation represents a point P on the line AB, and the right-hand side represents a point in the plane formed by OC and OD. The line OP intersects AB, since P lies on AB, and it also intersects CD, since OP and CD lie in the same plane (Fig. 33.7). Hence, with modifications when lines are parallel, we have the theorem:

Theorem II *Given two skew (non-intersecting and non-parallel) lines AB and CD in E_3, a unique transversal may be drawn through a given point O to intersect both lines.*

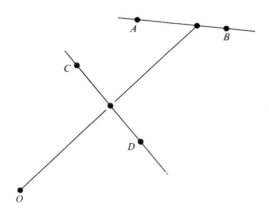

Fig. 33.7

Exercises

33.1 In §33.5 there is the statement: "we cannot have the joins AP, BP and CP parallel to BC, CA and AB respectively", where ABC is a triangle, and P a point in its plane. What theorem in real Euclidean geometry assures us of the truth of this statement?

33.2 If the relation of dependence between A, B, C and D in the proof of Thm II, §33.5, be such that $k_1 + k_2 = 0$, show that a line through O parallel to AB intersects CD, so that the theorem is still valid.

33.3 In Thm II, §33.5 why is the transversal to AB and CD through O unique? Show that if AB intersects CD, there is either one or an infinity of transversals. Show also that if AB and CD do not intersect, the transversal is the line of intersection of the planes (OAB) and (OCD). Does this statement suggest an alternative proof of the theorem? Is it true if AB and CD intersect?

33.4 In the proof of Thm I, §33.5 we assumed that P is a finite point in the plane (ABC). Prove that if OP is parallel to the plane (ABC), we may always write $P = k_1A + k_2B + k_3C$, where $k_1 + k_2 + k_3 = 0$.

33.5 If A, B, C and D are given vectors, show that a necessary and sufficient condition for the points A, B, C and D to be coplanar is the existence of real numbers k_1, k_2, k_3, k_4, not all zero, such that $k_1 + k_2 + k_3 + k_4 = 0$, and $k_1A + k_2B + k_3C + k_4D = 0$.

33.6 If A, B, C and D are given vectors, and the points A, B, C and D are not coplanar, show that the equations $k_1A + k_2B + k_3C + k_4D = 0$, $k_1 + k_2 + k_3 + k_4 = 0$ imply that $k_1 = k_2 = k_3 = k_4 = 0$.

33.7 If A, B, C and D form a proper tetrahedron, show that the joins of the midpoints of opposite edges, such as AB and CD, pass through the same point.

33.8 Referring to §11.1, let A, B and C be three bound vectors with endpoints in the plane $X_3 = 1$, and let P be any point in the plane (ABC). Writing $P = k_1A + k_2B + k_3C$, where $k_1 + k_2 + k_3 = 1$, show that

$$P \wedge B \wedge C = k_1(A \wedge B \wedge C), \quad P \wedge C \wedge A = k_2(B \wedge C \wedge A),$$

$$P \wedge A \wedge B = k_3(C \wedge A \wedge B),$$

so that the numbers k_1, k_2 and k_3 may be regarded as ratios of areas. If P moves in the plane (ABC) on a line parallel to BC, what is the relation between k_1, k_2 and k_3 during the motion? What is the relation if P moves on a line through the vertex C?

34.1   Inner products in Euclidean space of three dimensions

If $P = (x_1, x_2, x_3)$, and $Q = (y_1, y_2, y_3)$ are points in E_3, where we now assume that the coordinate axes are mutually perpendicular, the inner product of the vectors $\overrightarrow{OP}$, $\overrightarrow{OQ}$ is defined as:

$$\overrightarrow{OP} . \overrightarrow{OQ} = x_1y_1 + x_2y_2 + x_3y_3.$$

As in §7.1, the inner product obeys the usual distributive laws. If we use the symbols P, Q, R, . . . for position vectors,

$$P . (Q + R) = P . Q + P . R,$$

and

$$(Q + R) . P = Q . P + R . P,$$

where of course

$$P . Q = Q . P.$$

Again

$$P . (kQ) = k(P . Q),$$

for any real number k.

Once more we write $P . P = P^2$. If P and Q are now once again the end-points of vectors $\overrightarrow{OP}$, $\overrightarrow{OQ}$, the vector $\overrightarrow{OQ} - \overrightarrow{OP}$ represents the directed line-segment $\overrightarrow{PQ}$, and the formula for $\overrightarrow{OP} . \overrightarrow{OQ}$ shows that when $P = Q$, $\overrightarrow{OP} . \overrightarrow{OP} = \overrightarrow{OP^2}$ gives the *length* of the vector $\overrightarrow{OP}$ squared, which we write $|OP|^2$. Hence

$$|PQ|^2 = (\overrightarrow{OQ} - \overrightarrow{OP})^2 = |OQ|^2 + |OP|^2 - 2\overrightarrow{OP} . \overrightarrow{OQ}.$$

But since the cosine formula for triangle OPQ also gives us

$$|PQ|^2 = |OP|^2 + |OQ|^2 - 2|OP||OQ| \cos \angle POQ,$$

we can identify the inner product thus:

$$\overrightarrow{OP} . \overrightarrow{OQ} = |OP||OQ| \cos \angle POQ.$$

Once again, we have the important result that $\overrightarrow{OP}$ and $\overrightarrow{OQ}$ are perpendicular if and only if $\overrightarrow{OP} \cdot \overrightarrow{OQ} = 0$, neither vector being the zero vector.

We may derive the formula for the angle between two vectors in terms of the *direction-cosines* of the two vectors. Since $\overrightarrow{OP} = x_1E_1 + x_2E_2 + x_3E_3$, and $\overrightarrow{OQ} = y_1E_1 + y_2E_2 + y_3E_3$, where E_1, E_2, E_3 are unit vectors along the coordinate axes,

$$\overrightarrow{OP} \cdot E_1 = x_1, \text{ since } E_1 \cdot E_2 = E_1 \cdot E_3 = 0,$$

and also

$$\overrightarrow{OP} \cdot E_1 = |OP| \cos \hat{A}_1,$$

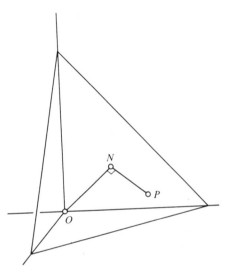

Fig. 34.1

with similar results for the other axes, where $\hat{A}_1$ is the angle made by $\overrightarrow{OP}$ with the positive direction of the X_1-axis. The *cosines*, $\cos \hat{A}_1$, $\cos \hat{A}_2$ and $\cos \hat{A}_3$ associated with the vector $\overrightarrow{OP}$ are called its *direction-cosines*. Since $\overrightarrow{OP^2} = |OP|^2$, we have the fundamental identity

$$\cos^2 \hat{A}_1 + \cos^2 \hat{A}_2 + \cos^2 \hat{A}_3 = 1,$$

and if $\overrightarrow{OQ}$ has the direction-cosines $\cos \hat{B}_1$, $\cos \hat{B}_2$ and $\cos \hat{B}_3$,

$$\cos \sphericalangle POQ = \cos \hat{A}_1 \cos \hat{B}_1 + \cos \hat{A}_2 \cos \hat{B}_2 + \cos \hat{A}_3 \cos \hat{B}_3.$$

We make use of the orthogonality condition to obtain *the equation of a plane*.

Let N be the foot of the perpendicular from the origin onto the plane (Fig. 34.1), and let $P = (x_1, x_2, x_3)$ be any point in the plane. The vector $\overrightarrow{PN}$ is perpendicular to the vector $\overrightarrow{ON}$. If $N = (a_1, a_2, a_3)$, the representative of the vector $\overrightarrow{PN}$ which passes through O has the end-point $(a_1 - x_1, a_2 - x_2, a_3 - x_3)$, and since this is perpendicular to $\overrightarrow{ON}$,

$$a_1(a_1 - x_1) + a_2(a_2 - x_2) + a_3(a_3 - x_3) = 0,$$

so that the points (x_1, x_2, x_3) satisfy *a linear equation*

$$a_1X_1 + a_2X_2 + a_3X_3 = \text{constant}$$

as they move in a plane.

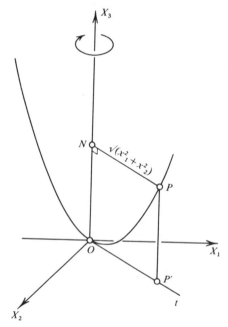

Fig. 34.2

There is one other elementary result we shall need later. If the foot of the perpendicular from $P = (x_1, x_2, x_3)$ onto the plane (OX_1X_2) be P' (Fig. 34.2), the coordinates of P' are $(x_1, x_2, 0)$. The distance of P' from $OX_3 = (x_1{}^2 + x_2{}^2)^{1/2}$, and this is also the distance of P from the axis OX_3.

Suppose that N is the foot of the perpendicular from P onto OX_3, and we have the relation $|PN|^2 = |ON|$. This gives us, in a fixed plane through the axis OX_3, a parabola. In three dimensions, we have the surface obtained by rotating this parabola about the axis OX_3, and its equation is derived from

$$|PN|^2 = x_1{}^2 + x_2{}^2 = |ON| = x_3,$$
$$\text{and is} \quad X_1{}^2 + X_2{}^2 - X_3 = 0.$$

This *paraboloid of revolution* plays a fundamental part in our subsequent development of the representation of circles by points in E_3.

35.1   The polar plane of a point

Let us call the paraboloid of revolution whose equation we have just obtained Ω. Through a given point $P = (a_1, a_2, a_3)$ we draw lines intersecting the surface Ω (Fig. 35.1). If Q, R be the points of intersection of a line through P with Ω, we wish to find

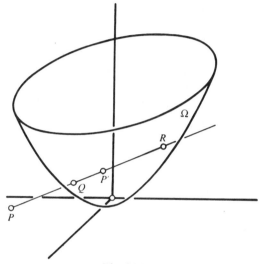

Fig. 35.1

the harmonic conjugate of P with respect to Q and R, and to determine the locus of this harmonic conjugate P' as the line through P varies. We shall show that the locus is a plane.

We saw in §19.1 that if P' be the harmonic conjugate of P with respect to Q and R, then Q and R are also harmonic conjugates with respect to P and P'. If the coordinates of P' are (x_1, x_2, x_3), and

$$Q = \left(\frac{a_1 + kx_1}{1 + k}, \quad \frac{a_2 + kx_2}{1 + k}, \quad \frac{a_3 + kx_3}{1 + k} \right)$$

the coordinates of R will merely have $-k$ instead of k, corresponding to the fact that Q and R divide the segment PP' internally and externally in the same ratio. Now, the point Q lies on Ω, and its coordinates therefore satisfy the equation

$$X_1{}^2 + X_2{}^2 - X_3 = 0.$$

On substitution in this equation, and simplification, we obtain an equation for k:

$$k^2(x_1{}^2 + x_2{}^2 - x_3) + k(2a_1x_1 + 2a_2x_2 - a_3 - x_3) + a_1{}^2 + a_2{}^2 - a_3 = 0. \quad (1)$$

This equation determines the values of k which give Q and R, and since the sum of these values is zero, the coefficient of k must be zero. Hence

$$2a_1x_1 + 2a_2x_2 - x_3 - a_3 = 0 \qquad (2)$$

if the points (a_1, a_2, a_3) and (x_1, x_2, x_3) are divided harmonically by the points Q and R in which their join cuts Ω. Such pairs of points are said to be *conjugate with respect to* Ω. We note that if $P = (a_1, a_2, a_3)$ be kept fixed, *the locus of points conjugate to P is a plane*, which is called *the polar plane of P with respect to* Ω. The point P is called *the pole of the plane*, and it is easy to prove that every plane has a unique pole with respect to Ω.

The symmetry of the conjugacy relation between the points (a_1, a_2, a_3) and (x_1, x_2, x_3) shows that if *the polar plane of P passes through the point P', the polar plane of P' passes through P*. This is also immediate from the definition of conjugacy. If P lies on Ω, one root of equation (1) is zero. If the other root is also zero, the line PP' has no further intersection with Ω other than P, and in this case P' satisfies equation (2). But P' lies in the tangent plane to Ω at P in this case. Hence the polar plane of P is the tangent plane at P when P lies on Ω.

Let us see what we deduce from equation (1) when P does not lie on Ω, but P' is a point on the intersection with Ω of the polar plane of P, given by equation (2). It is convenient to write the equation as a quadratic in $p = k^{-1}$, so that since $x_1^2 + x_2^2 - x_3 = 0$, the equation is

$$p(2a_1x_1 + 2a_2x_2 - x_3 - a_3) + p^2(a_1^2 + a_2^2 - a_3) = 0.$$

One root of this equation in p is zero, showing that P' is an intersection of the line PP' with Ω, as we expect. But if P' lies not only on Ω, but in the polar plane of P, the coefficient of p in the above equation, which is the constant term in the equation after division by p, is also zero, and therefore the line PP' has no other intersection with Ω other than P'. This means that PP' *touches* Ω at P'. Conversely, a similar argument shows that a line through P touching Ω at P' touches Ω precisely at the points P' where the polar plane of P cuts Ω. Hence:

The tangent cone of lines from a point P to Ω touches Ω along the points of a plane, the polar plane of P with respect to Ω.

We should stress here that all these theorems on polar planes and contacts are true for any quadric surface, that is for any surface given by a polynomial equation of the second degree in (X_1, X_2, X_3).

Finally, the condition that equation (1) should have *equal* roots in k, which means that the line PP' touches Ω, is, writing $(x_1, x_2, x_3) = (X_1, X_2, X_3)$,

$$(2a_1X_1 + 2a_2X_2 - X_3 - a_3)^2 - 4(a_1^2 + a_2^2 - a_3)(X_1^2 + X_2^2 - X_3) = 0.$$

This is, of course, the equation of the cone of lines through P which touch Ω along the polar plane of P with respect to Ω (Fig. 35.2). We note that the equation

$$(2a_1X_1 + 2a_2X_2 - X_3 - a_3)^2 - \lambda(X_1^2 + X_2^2 - X_3) = 0$$

for varying values of λ represents a system of quadric surfaces which have *ring-contact* with Ω along its plane curve of intersection with the polar plane of P with respect to Ω.

To prove this we remark that if $U = 0$ and $V = 0$ are any two planes, the quadric surface

$$UV - \lambda(X_1{}^2 + X_2{}^2 - X_3) = 0$$

passes through the plane curves of intersection of $U = 0$ and Ω, and $V = 0$ and Ω. For a point (x_1, x_2, x_3) which satisfies, say, $U = 0$ and lies on Ω satisfies the given equation. When $U = 0$ and $V = 0$ are the same plane, it is clear that the quadric

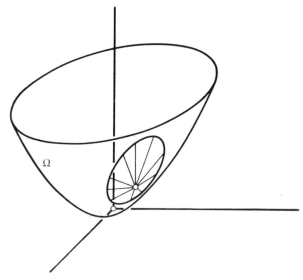

Fig. 35.2

touches Ω along its intersection with $U = 0$. (The discussion and results of this section are applicable to a quadric Ω in affine space of three dimensions, but we naturally do not then call it a paraboloid of revolution.)

Exercises

35.1 Two edges AB, CD of a tetrahedron $ABCD$ are perpendicular. Prove that the distance between the midpoints of AC and BD is equal to the distance between the midpoints of AD and BC.

35.2 $ABCD$ is a tetrahedron in which AB is perpendicular to CD, and AC is perpendicular to BD. Prove that AD is perpendicular to BC.

35.3 The equation of a plane in E_3 is $p_1X_1 + p_2X_2 + p_3X_3 + p_4 = 0$. Show that if $p_3 = 0$, the axis OX_3 is parallel to the plane. Comparing the equation of this plane with that of the polar plane of (a_1, a_2, a_3), which is $2a_1X_1 + 2a_2X_2 - X_3 - a_3 = 0$, show that any plane in E_3 which is not parallel to the axis OX_3 has a unique pole, given by the equations

$$2a_1/p_1 = 2a_2/p_2 = -1/p_3 = -a_3/p_4.$$

35.4   Show that the polar plane with respect to Ω of the point $(1 - t)A + tB$, where A and B are vectors in E_3, is the plane $(1 - t)P(A) + tP(B) = 0$, where $P(A) = 0$ is the polar plane of A, and $P(B) = 0$ is the polar plane of B with respect to Ω. Deduce that the polar planes of points on AB always pass through the line given by the equations $P(A) = P(B) = 0$, and that the polar planes of points on this line always contain the line AB.

35.5   If we call the lines which are the subject of the preceding Exercise *polar lines*, show that if l, m are two polar lines, and l intersects Ω at the points X and Y, then m is the intersection of the tangent planes to Ω at X and Y. When does the polar plane of a point A contain the point A? Show that polar lines l and m intersect if and only if they are tangent lines to Ω at the point of intersection.

35.6   Justify the following construction for a tetrahedron $ABCD$ which is such that every face is the polar plane with respect to Ω of the opposite vertex (such a tetrahedron is called a *self-polar* tetrahedron). Take any point A which does not lie on Ω, and let π be the polar plane of A. This does not

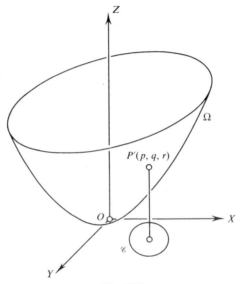

Fig. 36.1

contain A. Let B be any point in π which does not lie on Ω, and let ρ be its polar plane. This contains the point A. Let the line of intersection of π and ρ be l, and choose on l any point C which does not lie on Ω. The polar plane of C contains A and B and intersects l in a point D. Then $ABCD$ is a self-polar tetrahedron.

36.1   The representation of circles by points of E_3

We assume that we have a circle $\mathscr{C}$ in normalized form, with equation:

$$C(X, Y) \equiv X^2 + Y^2 - 2pX - 2qY + r = 0,$$

so that the center is the point (p, q), and the radius R is given by the formula $R^2 = p^2 + q^2 - r$. We set up a system of orthogonal cartesian axes OX, OY and OZ for E_3, and we suppose that all the circles we shall consider lie in the OXY plane. We map the circle $\mathscr{C}$ on the point (p, q, r) of E_3 (Fig. 36.1).

Then if $P' = (p, q, r)$, the foot of the perpendicular, P, from P' onto the OXY-plane is the center of the circle $\mathscr{C}$. We may say that the point representing the circle $\mathscr{C}$ is *vertically above* the center of the circle $\mathscr{C}$. As r increases, the radius of $\mathscr{C}$ decreases. When we arrive at the point (p, q, r) for which $r = p^2 + q^2$ (we suppose that p and q are being kept fixed for the moment), the circle $\mathscr{C}$ is *a point-circle*, that is a circle with real center but of zero radius. *Hence all point-circles are mapped onto the points of the surface*

$$\Omega \equiv X^2 + Y^2 - Z = 0,$$

in E_3. We saw in §34.1 that this surface is a paraboloid of revolution. If we project any point (x', y', z') of Ω downward onto the OXY-plane, we obtain a point-circle, center (x', y') which is represented by the point (x', y', z') of Ω. The correspondence between the points of Ω and the point-circles of the OXY-plane is one-to-one.

If we keep (p, q) fixed and suppose that $r = p^2 + q^2$, so that $\mathscr{C}$ is a point-circle, center (p, q), if we increase r, the representative point (p, q, r) lies *inside* the surface Ω. When this is the case, the circle $\mathscr{C}$ has a real center (p, q), but R^2 is negative, and so there are no real points on the circle. We shall call such circles *virtual circles*. Our circles $\mathscr{C}$ always have equations with real coefficients, and we do not wish to use the term *imaginary circles* for such circles, even though we may be unable to draw them. They are represented by real points in E_3, which lie inside Ω. (See the remarks in §0.9.)

We have set up a one-to-one correspondence between the proper circles of a real Euclidean plane and the points of real Euclidean space of three dimensions. Point-circles are in one-to-one correspondence with the points of a paraboloid of revolution $\Omega \equiv X^2 + Y^2 - Z = 0$, and virtual circles, those with real centers but with no real points on them, are mapped on points inside Ω.

36.2   The representation of a coaxal system

If $C' \equiv X^2 + Y^2 - 2p'X - 2q'Y + r' = 0$ be the normalized equation of another circle $\mathscr{C}'$, the coaxal system defined by the circles $\mathscr{C}$ and $\mathscr{C}'$ is the system of circles given by the equations

$$\lambda C + \mu C' = 0,$$

where λ and μ vary through all real values (§27.1). To represent the circles of this coaxal system, we must first normalize the equation. We obtain the equation

$$X^2 + Y^2 - \frac{2(\lambda p + \mu p')X}{\lambda + \mu} - \frac{2(\lambda q + \mu q')Y}{\lambda + \mu} + \frac{\lambda r + \mu r'}{\lambda + \mu} = 0.$$

The representative point is therefore

$$\left(\frac{\lambda p + \mu p'}{\lambda + \mu}, \; \frac{\lambda q + \mu q'}{\lambda + \mu}, \; \frac{\lambda r + \mu r'}{\lambda + \mu} \right).$$

Now $\mathscr{C}$ is represented by the point (p, q, r) of E_3, and $\mathscr{C}'$ is represented by the point (p', q', r'). The point which represents the circle $\lambda C + \mu C' = 0$ is a point on the line joining the points which represent $\mathscr{C}$ and $\mathscr{C}'$ (§33.4). As λ and μ vary, all points of the line are covered. Hence we have the fundamental result:

The circles of a coaxal system are represented by the points of a line in E_3.

Since a line is determined by any two points on it, provided they are distinct, we once more obtain the result of §27.5:

A coaxal system is derivable from any two distinct circles of the system (Fig. 36.2).

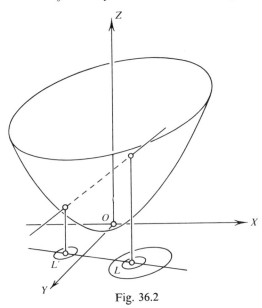

Fig. 36.2

The line representing the coaxal system may or may not intersect the paraboloid Ω in real points. If it does, the coaxal system contains circles of zero radius, and is of the non-intersecting kind (§29.4). If the system is of the intersecting kind (§29.2), the representative line does not intersect Ω. When the line touches Ω, there is only one circle in the system of zero radius, and the system of coaxal circles consists of circles which touch each other at a point (§29.3).

Exercises

36.1 What is the representation in E_3 of a system of concentric circles, with center at the fixed point (p, q) of the (X, Y)-plane? What corresponds geometrically in the representation in E_3 to the fact that a system of concentric circles always contains one circle of zero radius?

36.2 The representation of circles by points of E_3 does not allow for circles of infinite radius (lines). Where is the point which represents the special circle of the coaxal system $\lambda C + \mu C' = 0$ obtained by taking $\lambda = -\mu$? Is it correct to say that there is a one-to-one correspondence between the circles of the coaxal system $\lambda C + \mu C' = 0$ and the points of E_3?

36.3   If two lines l and m in E_3 intersect in a point P', what can you say about the coaxal systems represented by the lines? If each system is composed of intersecting circles, the one of circles through the points P, Q, and the other of circles through the points R, S, what can you say about the points P, Q, R and S when the lines l and m intersect?

36.4   Show that a line l which lies in the (X, Y)-plane represents a coaxal system of circles which pass through the point $(0, 0)$. Where is the other common point of intersection? What kind of system is represented by l if l passes through $(0, 0)$?

36.5   Show that the coaxal system determined by the circles

$$C \equiv X^2 + Y^2 - 2pX - 2qY - k^2 = 0, \quad C' \equiv X^2 + Y^2 - 2p'X - 2q'Y - k^2 = 0,$$

is always of the intersecting type, with distinct points of intersection for $k \neq 0$.

36.6   If L and L' be the limiting points (zero circles) of a coaxal system, there are no real circles of the system with centers between L and L'. How is this seen immediately from the representation in E_3? Show that for real circles in the coaxal system

$$k_1[(X - p)^2 + (Y - q)^2] + k_2[(X - p')^2 + (Y - q')^2] = 0$$

we must have $k_1 k_2 \leqslant 0$.

36.3   Deductions from the representation

V is a point in E_3 which does not lie on l or m, two non-intersecting, non-parallel lines. If a line n through V intersects l, it must lie in the plane $[Vl]$. If the line n intersects m, it must lie in the plane $[Vm]$. Two planes in E_3 which are not identical intersect in a line, when they are not parallel. Since l and m do not intersect, they cannot lie in a plane with V. Since the planes $[Vl]$ and $[Vm]$ certainly intersect in V, they are not parallel. Hence the planes $[Vl]$ and $[Vm]$ intersect in a line, and this line n intersects l, since it lies in $[Vl]$, and it intersects m, since it lies in $[Vm]$. There is therefore a transversal through V to both lines l and m. If there were more than one, the two transversals would determine a plane, which would contain both l and m, and these would intersect, contrary to hypothesis. Hence there is a unique transversal through V to l and m. We proved this in §33.5, using vectors. (See Exercise 36.10.)

We see what this theorem implies in the geometry of coaxal systems in the plane (OXY). The line l is the representative of a coaxal system. Let us suppose that l does not intersect Ω, so that the coaxal system is of the intersecting type. Let the points common to all circles of the system be P and P'. Similarly, let the coaxal system of which the line m is the representative be an intersecting type, all the circles of the system passing through the points Q and Q'. Since l and m do not intersect, *there is no circle common to both coaxal systems*. Hence the points P, P' and Q, Q' do not lie on a circle. *They are not concyclic.*

The point V represents a circle $\mathscr{C}$, say. If we draw lines through V which do not intersect Ω, we are mapping intersecting coaxal systems of which $\mathscr{C}$ is a member. Each line through V determines a pair of points R, R' on $\mathscr{C}$, and all circles of the coaxal system of which the line is a map pass through R and R'.

Suppose that the transversal through V to l and m does not intersect Ω. Then it determines a pair of points R, R' on $\mathscr{C}$ which are concyclic with P, P' and with Q, Q'. The circle through R, R' and P, P' is mapped on the point in which the transversal through V intersects l, and the circle through R, R' and Q, Q' is mapped on the point in which the transversal through V meets m.

We have proved the following plane theorem:

P, P', Q, Q' are four points which are not concyclic and do not lie on a given circle $\mathscr{C}$.

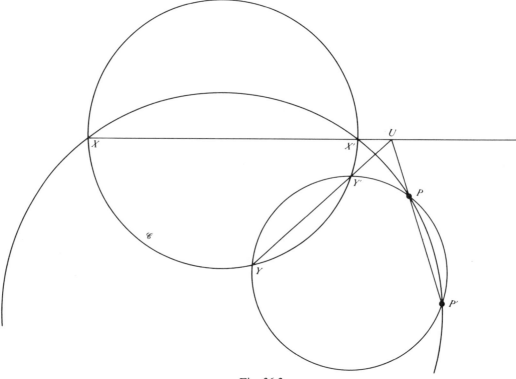

Fig. 36.3

There is a unique pair of points R, R' on $\mathscr{C}$ such that P, P', R, R' are concyclic, and Q, Q', R, R' are concyclic.

We have to face the possibility that the pair of points R, R' on $\mathscr{C}$ may not be real. If the unique transversal from V, which represents $\mathscr{C}$, to the lines l and m which represent the coaxal systems of circles through P, P' and Q, Q' respectively, intersects Ω, the coaxal system of circles of which $\mathscr{C}$ is a member will be of the non-intersecting kind, and we shall end up with a circle through P, P' and a circle through Q, Q' which are in a coaxal system which contains $\mathscr{C}$.

A proof of the above theorem which does not use the representation in E_3 is also instructive. If we draw circles through P, P' to intersect the given circle $\mathscr{C}$ in the points X, X', the chord XX' passes through a fixed point U, or is always parallel to a fixed *direction.* In other words, *the chords of intersection with $\mathscr{C}$ form a pencil.* This is Exercise 28.2. The proof depends on the theorem that the radical axes of three circles, taken in pairs, are concurrent or parallel (Thm 28.1). If we draw two circles through P, P'

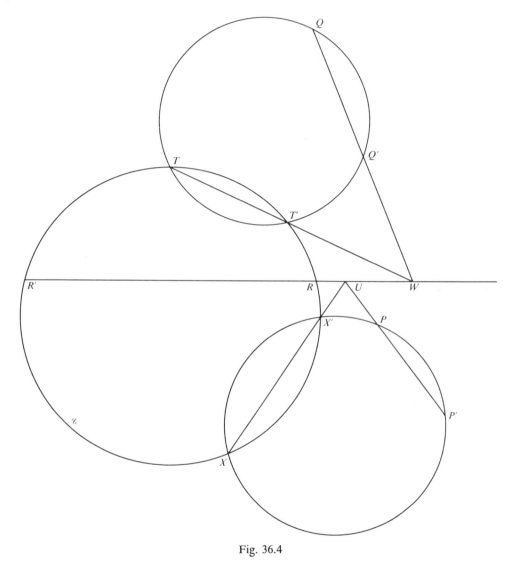

Fig. 36.4

(Fig. 36.3), the one intersecting $\mathscr{C}$ in X, X' and the other intersecting $\mathscr{C}$ in Y, Y', the radical axes are PP', XX' and YY', and these are either concurrent at a point U, or parallel. The point U is determined by the intersection of PP' and one circle through PP', and is therefore a fixed point for all other chords of intersection.

We note that any chord through U cuts $\mathscr{C}$ in a pair of points Z, Z' which are con-cyclic with P, P', since a circle through P, P' and Z must also intersect $\mathscr{C}$ in Z'.

Now draw the circles through Q, Q' (Fig. 36.4), intersecting $\mathscr{C}$ in points T, T'. All chords TT' form a pencil, and therefore pass through a fixed point W or are parallel to a fixed direction. Again, any chord of $\mathscr{C}$ through W gives a pair of points on $\mathscr{C}$ concyclic with Q and Q'. The pair of points R, R' on $\mathscr{C}$ we are looking for lie on a line through U and also on a line through W, and therefore they are *the intersection of the line UW with the circle $\mathscr{C}$.*

This line may not intersect the circle $\mathscr{C}$.

Exercises

36.7 Let
$$C_1 \equiv X^2 + Y^2 - 2p_1X - 2q_1Y + k^2 = 0,$$
$$C_2 \equiv X^2 + Y^2 - 2p_2X - 2q_2Y + k^2 = 0, \quad C_3 \equiv X^2 + Y^2 - 2p_3X - 2q_3Y + k^2 = 0$$

be three distinct circles. If $\mathscr{C}'$ be any circle of the coaxal system determined by $\mathscr{C}_1$ and $\mathscr{C}_2$, we consider the zero circles of the coaxal system determined by $\mathscr{C}_3$ and $\mathscr{C}'$. Show that as $\mathscr{C}'$ varies, all these circles of zero radius have their centers on a circle, *a*) from the remark that the point $(0, 0)$ has the same power with respect to all the circles we are considering, and (*b*) by considering the representation of the circles in E_3.

36.8 Assuming the theorem that four mutually skew lines in E_3 possess in general just two trans-versals, which intersect all four lines (Thm III of §83.1), deduce a theorem about coaxal systems in E_2.

36.9 Show that two coaxal systems of circles in the plane OXY whose maps in E_3 are parallel lines are coaxal systems with the same radical axis.

36.10 Looking more closely at §36.3, show that if V lies in the plane through l which is parallel to m, then the line $[Vl] \cap [Vm]$ intersects l but is parallel to m.

36.11 If the transversal through V to l and m in §36.3 intersects l but is parallel to m, show that the interpretation given still holds, but the points Q, Q', R, R, lie on a line, not a proper circle.

37.1 Orthogonal circles and conjugate points

The angle of intersection between the two circles
$$C_1 \equiv X^2 + Y^2 - 2p_1X - 2q_1Y + r_1 = 0,$$
$$C_2 \equiv X^2 + Y^2 - 2p_2X - 2q_2Y + r_2 = 0,$$

is given by the formula (§22.1, (1))
$$4(R_1^2R_2^2)\cos^2\theta = (2p_1p_2 + 2q_1q_2 - r_1 - r_2)^2,$$

or
$$4(p_1^2 + q_1^2 - r_1)(p_2^2 + q_2^2 - r_2)\cos^2\theta = (2p_1p_2 + 2q_1q_2 - r_1 - r_2)^2. \tag{1}$$

We first examine the case of *contact*, when $\cos^2\theta = 1$. The equation
$$4(p_1^2 + q_1^2 - r_1)(p_2^2 + q_2^2 - r_2) = (2p_1p_2 + 2q_1q_2 - r_1 - r_2)^2$$

is the condition that the quadratic in k:

$$k^2(p_1{}^2 + q_1{}^2 - r_1) + k(2p_1p_2 + 2q_1q_2 - r_1 - r_2) + (p_2{}^2 + q_2{}^2 - r_2) = 0$$

should have equal roots. When this is the case, the line of points

$$\left(\frac{kp_1 + p_2}{k + 1}, \frac{kq_1 + q_2}{k + 1}, \frac{kr_1 + r_2}{k + 1} \right)$$

joining the point (p_1, q_1, r_1) to (p_2, q_2, r_2) *touches* the paraboloid of revolution Ω.

Since the intersections of this line with Ω give the limiting points of the coaxal system determined by $\mathscr{C}_1$ and $\mathscr{C}_2$, we have the fairly intuitive result:

A necessary and sufficient condition that two circles touch is that the limiting points of the coaxal system they determine should coincide.

We now consider the case of the circles $\mathscr{C}_1$ and $\mathscr{C}_2$ intersecting orthogonally. If $\cos \theta = 0$, equation (1) is simply

$$2p_1p_2 + 2q_1q_2 - r_1 - r_2 = 0.$$

This is the condition that the quadratic in k given above should have the sum of its roots equal to zero, so that if one intersection of the line P_1P_2 with Ω be $kP_1 + P_2$, where of course $P_1 = (p_1, q_1, r_1)$, $P_2 = (p_2, q_2, r_2)$, then the other intersection is $-kP_1 + P_2$. The points of intersection are *harmonic conjugates* with respect to P_1 and P_2, and P_1, P_2 are *conjugate points* with respect to the quadric Ω. (See §35.1.)

Hence *orthogonality implies conjugacy, and conversely.*

If the points P_1, P_2 are conjugate with respect to Ω, the right-hand side of equation (1) is zero, and $\cos^2 \theta = 0$ if neither of the factors on the left-hand side is zero. Hence the converse of our stated theorem is certainly true for two circles, neither of which is a point-circle. Suppose that $\mathscr{C}_1$ is a point-circle, and that the conjugacy condition holds. Then both sides of equation (1) are zero, whatever the value of $\cos^2 \theta$. But if $\mathscr{C}_1$ is a point-circle, $r_1 = p_1{}^2 + q_1{}^2$, and the conjugacy condition can be written:

$$2p_1p_2 + 2q_1q_2 - p_1{}^2 - q_1{}^2 - r_2 = 0,$$

or

$$p_1{}^2 + q_1{}^2 - 2p_2p_1 - 2q_2q_1 + r_2 = 0,$$

which is the condition that the center of the point-circle $\mathscr{C}_1$ should lie on the circle $\mathscr{C}_2$. We deduce that *if we are considering a system of circles which cut a given circle $\mathscr{C}_2$ at a given angle θ, the point-circles which have their centers on $\mathscr{C}_2$ always form part of the system.*

Now, the point-circles which have their centers on $\mathscr{C}_2$ are represented in E_3 by the points in which lines drawn parallel to OZ from points on $\mathscr{C}_2$ intersect Ω (Fig. 37.1). These points (x, y, z) on Ω satisfy the condition of being conjugate with respect to $P_2 = (p_2, q_2, r_2)$, and therefore lie in the plane

$$2p_2X + 2q_2Y - Z - r_2 = 0.$$

This is the polar plane of P_2, the point which represents the circle $\mathscr{C}_2$. Hence we have proved that *the projection of the conic onto the OXY-plane in which the polar plane of a point in E_3 intersects Ω is precisely the circle which is represented by the point.*

Conversely, if we erect parallels at the points (x, y) of the circle $\mathscr{C}_2$ to the OZ-axis, each parallel meets Ω in a unique point (x, y, z) for which $z = x^2 + y^2$, and points on the curve of intersection of the parallels with Ω satisfy the equation

$$Z - 2p_2 X - 2q_2 Y + r_2 = 0,$$

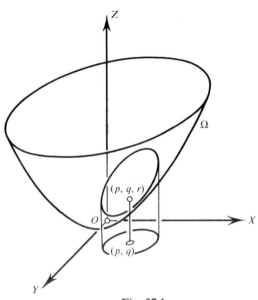

Fig. 37.1

which is the equation of a plane, the polar plane of (p_2, q_2, r_2), the point which represents $\mathscr{C}_2$, with respect to Ω.

We deduce that *four points (x_i, y_i) $(i = 1, 2, 3, 4)$ lie on a circle if and only if their representations $(x_i, y_i, x_i^2 + y_i^2)$ on Ω lie in a plane.*

37.2   Polar lines and polar coaxal systems

Let A, B be two distinct points in E_3 which determine a line l, and suppose that the line of intersection of the polar planes of A and B with respect to Ω is a line m. Let C, D be two distinct points on m. Then since the polar plane of A contains C, the polar plane of C contains A. Similarly the polar plane of B contains C and therefore the polar plane of C contains B. Similar reasoning shows that the polar plane of D contains A and also B, so that the line l is the intersection of the polar planes of C and D, where C and D are any two points on m. Hence the polar plane of every point of m contains

P to Q is the same as that from R to S. All this is expressed thus: if $R = (z_1, z_2, z_3)$, and $S = (t_1, t_2, t_3)$, then $\overrightarrow{PQ} \equiv \overrightarrow{RS}$ if and only if the ordered triad of real numbers defining $\overrightarrow{PQ}$ is the same as the ordered triad $(t_1 - z_1, t_2 - z_2, t_3 - z_3)$ which defines $\overrightarrow{RS}$.

The fact that this definition does define an equivalence relation is immediately verified. As in §1.1, there is always one representative of an equivalence class of vectors with its first number triad at $(0, 0, 0)$. Vectors issuing from the origin will be called *bound vectors*, and these are representative of *free vectors*, which do not necessarily issue from $(0, 0, 0)$. (Fig. 33.2.)

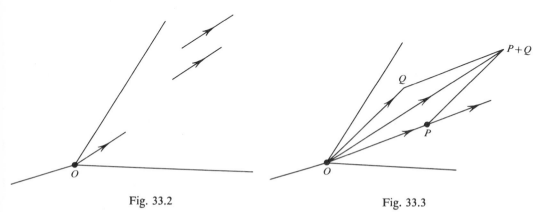

Fig. 33.2   Fig. 33.3

33.2   The sum of two vectors

Our operations on *vectors*, which is the more usual name for directed segments, will always be on bound vectors, with the origin as the first number triad, $(0, 0, 0)$. Since this is to be kept fixed, we may speak of a bound vector as one triad of ordered real numbers

$$P = (x_1, x_2, x_3),$$

and the bound vector $\overrightarrow{OP}$ is given by this ordered triad (see Fig. 33.3). We may also call the vector $\overrightarrow{OP}$ the *position vector* of the point P. If $Q = (y_1, y_2, y_3)$, we define the *sum* of the two position vectors by the formula

$$P + Q = (x_1 + y_1, x_2 + y_2, x_3 + y_3),$$

and we form a *scalar multiple* of a vector P by the real number r thus:

$$rP = (rx_1, rx_2, rx_3).$$

The geometric interpretation of these rules is exactly as in two dimensions (or in space of n dimensions). We see that bound vectors are added according to *the parallelogram law:*

l, and therefore the polar plane of every point of l contains every point of m. The two lines l and m, which are in this special mutual relationship, are called *polar lines* with respect to Ω.

Algebraically, the polar planes of points on a line form a *pencil* of planes, and so intersect in a line (Exercise 35.4).

Suppose that the line l intersects Ω in the points A and B. The polar plane of a point on Ω is the tangent plane to Ω at the point. Hence the tangent planes to Ω at A and at B intersect in m. Let us consider what system of circles is represented in E_3 by the points of a plane tangent to Ω at a point P'. Let P' represent the circle $\mathscr{C}$, which is a point-circle, since P' is on Ω. Any point in the tangent plane to Ω at P' is conjugate with respect to P', and therefore represents a circle which is orthogonal to $\mathscr{C}$. But circles orthogonal to a point-circle contain the center of the point-circle. We therefore deduce that *the points of a plane tangent to Ω at a point P' represent the circles of the OXY-plane which pass through the point which is the projection of P' onto the OXY-plane*.

Hence the line l, if it intersects Ω at the points A and B, is associated with a coaxal system of circles, represented by the polar line m, which pass through the projections of the two fixed points A and B onto the OXY-plane. All these circles are orthogonal to the circles of the coaxal system represented by the line l, and the two fixed points are the point-circles, or limiting points of the coaxal system represented by the line l.

We have thus found a simple geometrical interpretation of the coaxal systems which we have called *polar* coaxal systems (§29.5). Since a non-intersecting type of coaxal system is polar to an intersecting type, we deduce that *if l intersects Ω in two distinct points, its polar line m will not intersect Ω.*

37.3   Bundles of circles

We have considered the set of circles linearly dependent on two given distinct circles. Let us now consider the set of circles linearly dependent on the three circles (themselves linearly independent):

$$C_i \equiv X^2 + Y^2 - 2p_iX - 2q_iY + r_i = 0 \qquad (i = 1, 2, 3).$$

In our representation this set of circles, called a *bundle* of circles, is mapped onto the points of the plane determined by the three points (p_i, q_i, r_i), $(i = 1, 2, 3)$. This plane has a unique pole P, and the point P represents a circle which is orthogonal to each of the circles $\mathscr{C}_i$ and to every circle of the bundle. If the point P lies outside the paraboloid Ω, the circle represented by P is a real circle, and the bundle consists of all circles orthogonal to a given circle. It is then called *a hyperbolic bundle*. Such a system will appear later, in our discussion of a model of a hyperbolic non-Euclidean geometry (§56.1). If the point P lies on Ω, the common orthogonal circle is a point-circle, and all the circles of the bundle pass through the center. The bundle is then called *a parabolic bundle*. We now consider the case when the point P lies *inside* Ω, so that the common orthogonal circle is a virtual circle. Let $P = (x', y', z')$, where we know that $x'^2 + y'^2 < z'$.

The circles of the bundle can be obtained thus. Take any plane π through P, and project the intersection of π with Ω onto the OXY-plane. These planes have the equation:

$$a(X - x') + b(Y - y') + c(Z - z') = 0,$$

and they project into the circles

$$a(X - x') + b(Y - y') + c(X^2 + Y^2 - z') = 0.$$

We can choose one such plane section which projects into a circle which has the point (x', y') as center in the OXY-plane. It is simply:

$$-2x'(X - x') - 2y'(Y - y') + Z - z' = 0,$$

and its projection is the circle

$$-2x'(X - x') - 2y'(Y - y') + X^2 + Y^2 - z' = 0,$$

which is the circle

$$(X - x')^2 + (Y - y')^2 = z' - x'^2 - y'^2.$$

All circles of the bundle (now called an *elliptic bundle*) *cut this circle at the ends of diameters.* The planes whose intersections give the circles of the bundle, by projection, all pass through (x', y', z'), and therefore intersect any plane through this point in a line which passes through (x', y', z'). On projection the circles corresponding to the plane sections have a chord of intersection which passes through the point (x', y') in the OXY-plane. We have found a circle of the bundle which has this point as center. Therefore all circles of the bundle intersect this fixed circle at the ends of a diameter.

The preceding arguments are based on the assumption that the point P, the pole of the plane determined by the points (p_i, q_i, r_i), is a finite point. If, and only if, the plane is parallel to the OZ-axis, the pole is not a finite point. But in this case the centers of the three circles $\mathscr{C}_i$ lie on a line, and this line, regarded as a circle, is orthogonal to the three given circles and all circles of the bundle defined by the given circles, and so the bundle is a hyperbolic bundle.

Exercises

37.1 Show that two polar lines project onto the OXY-plane into two perpendicular lines.

37.2 Three planes in E_3 intersect in a point. If the three planes chosen are the tangent planes to Ω at the three points P', Q' and R' on Ω, what circle is represented by the point of intersection of the three tangent planes? What is its relation to the point-circles represented by P', Q' and R' respectively?

37.3 Let A, B, and C be three points in E_3 which determine a plane which is not parallel to the OZ-axis, and therefore has the equation

$$p_1 X + p_2 Y + p_3 Z + p_4 = 0 \qquad (p_3 \neq 0).$$

By Exercise 35.3, this plane has a unique pole, given by the equations

$$2a_1/p_1 = 2a_2/p_2 = -1/p_3 = -a_3/p_4.$$

Show that this point (a_1, a_2, a_3) in E_3 represents the unique circle which is orthogonal to the three circles represented by the points A, B and C, and that the equation of the circle is

$$p_1X + p_2Y + p_3(X^2 + Y^2) + p_4 = 0.$$

37.4   If $ABCD$ is a self-polar tetrahedron with respect to Ω, show that the four circles represented by these points are mutually orthogonal, and that their centers form an orthocentric tetrad (a triangle and its orthocenter: see §7.2).

37.5   If $ABCD$ is a self-polar tetrahedron, the lines AB and CD are polar lines with respect to Ω. If AB intersects Ω in two real points which are distinct, CD does not intersect Ω. Prove that a coaxal system of the intersecting type, represented by CD, contains no virtual circles, so that we may assume that both C and D lie outside Ω. Then prove that if A lies outside Ω, B must lie inside. From the deduction that at least one vertex of a self-polar tetrahedron must lie inside Ω, prove that if four circles are mutually orthogonal, at least one must be virtual.

37.6   Verify algebraically that every circle of the bundle

$$a(X - x') + b(Y - y') + c(X^2 + Y^2 - z') = 0$$

intersects the fixed circle of the bundle

$$-2x'(X - x') - 2y'(Y - y') + X^2 + Y^2 - z' = 0$$

at the ends of a diameter of this fixed circle.

38.1   Circles which cut three given circles at equal angles

In §37.1 we found a formula connecting two circles $\mathscr{C}_1$ and $\mathscr{C}_2$ which intersect at an angle θ or $\pi - \theta$. It was

$$4(p_1^2 + q_1^2 - r_1)(p_2^2 + q_2^2 - r_2)\cos^2\theta = (2p_1p_2 + 2q_1q_2 - r_1 - r_2)^2.$$

This formula implies that the points in E_3 which map circles which intersect the given circle $\mathscr{C}_1$ at an angle θ or $\pi - \theta$ fill the quadric surface

$$4(p_1^2 + q_1^2 - r_1)(X^2 + Y^2 - Z)\cos^2\theta = (2p_1X + 2q_1Y - Z - r_1)^2.$$

Towards the end of §35.1 we hinted that a quadric surface with this type of equation touches Ω along its intersection with the plane

$$2p_1X + 2q_1Y - Z - r_1 = 0.$$

But it is not essential to grasp this geometrical fact for complete understanding of what follows. (Such a quadric is said to have *ring-contact* with Ω.)

We have three given circles, $\mathscr{C}_1$, $\mathscr{C}_2$ and $\mathscr{C}_3$, and we wish to consider those circles which cut each of the three circles at an angle θ or $\pi - \theta$. The points representing these circles lie on the three quadrics

$$4(p_i^2 + q_i^2 - r_i)(X^2 + Y^2 - Z)\cos^2\theta = (2p_iX + 2q_iY - Z - r_i)^2$$
$$(i = 1, 2, 3).$$

For the sake of brevity we write these three equations as

$$4(X^2 + Y^2 - Z)k_i \cos^2 \theta = X_i^2 (i = 1, 2, 3), \tag{1}$$

where $X_i = 0$ is the polar plane of (p_i, q_i, r_i) and $k_i = p_i^2 + q_i^2 - r_i$.

If we equate the common value of $X^2 + Y^2 - Z$ in the three equations, we obtain the system of equations

$$\frac{X_1^2}{k_1} = \frac{X_2^2}{k_2} = \frac{X_3^2}{k_3}. \tag{2}$$

We note that these equations do not depend on $\cos \theta$. If we take them in pairs, the set (2) is equivalent to the set of linear equations

$$\begin{cases} \sqrt{k_2}X_1 \pm \sqrt{k_1}X_2 = 0, \\ \sqrt{k_3}X_1 \pm \sqrt{k_1}X_3 = 0. \end{cases}$$

(If our three circles $\mathscr{C}_1$, $\mathscr{C}_2$, $\mathscr{C}_3$ all contain real points, k_1, k_2, k_3 are all $\geqslant 0$.) Taking all the possible combinations of sign, we have four sets of two linear equations in X, Y and Z. Since two planes in E_3 intersect in a line, the solutions of the set of equations (2) lie on four lines. The point given by $X_1 = X_2 = X_3 = 0$ automatically satisfies each equation of the four sets. Hence all four lines of solutions pass through the point which is the intersection of the polar planes of the three points (p_i, q_i, r_i) $(i = 1, 2, 3)$. This point represents the circle which is orthogonal to the three circles $\mathscr{C}_1$, $\mathscr{C}_2$ and $\mathscr{C}_3$.

Since the common value of $X^2 + Y^2 - Z$, derived from the equations (1), gave us the set (2), we expect that each of the three quadrics given by the equations (1) will intersect each of the four lines given by the set (2) in the same pair of points, and this set of eight points is the set of common solutions of the three equations given by (1). In fact, if (X', Y', Z') be a point which satisfies the equation

$$4(X^2 + Y^2 - Z)k_1 \cos^2 \theta = X_1^2,$$

and also the equations

$$X_1^2/k_1 = X_2^2/k_2 = X_3^2/k_3,$$

we obtain, besides the equation

$$4(X'^2 + Y'^2 - Z')k_1 \cos^2 \theta = X_1^2,$$

the equations

$$4(X'^2 + Y'^2 - Z')k_2 \cos^2 \theta = X_2^2,$$

and

$$4(X'^2 + Y'^2 - Z')k_3 \cos^2 \theta = X_3^2.$$

For a given θ, the three quadrics intersect in eight points, which lie by twos on four lines which are independent of θ. We have therefore proved the theorem:

There are eight circles which intersect three given circles at a given angle θ or $\pi - \theta$.

As the angle θ varies, the circles vary in four coaxal systems with a common circle, the circle orthogonal to the three given circles.

When $\theta = 0$ or π, we have the solution of the Apollonius problem (§26.1).

39.1 The representation of inversion

Let A and A' be inverse points in a circle $\mathscr{C}$. To represent this transformation in E_3 we do not make use of the metrical definition of inversion (§20.1). We use the theorem that if we consider the points A and A' as point-circles, and consider the coaxal system

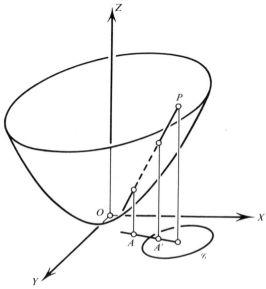

Fig. 39.1

they define, the circle $\mathscr{C}$ for which they are inverse points is a member of this coaxal system (Exercise 29.6). This is clear from the fact that A and A' lie on a line through the center of $\mathscr{C}$, and every circle through A and A' is orthogonal to $\mathscr{C}$. In fact, the second fact is enough to establish our theorem. When we represent the circle $\mathscr{C}$ by a point in E_3 (Fig. 39.1), and the points A and A' are represented by points on Ω, the three points we obtain will be *collinear*, representing coaxal circles. Hence we have an extremely simple construction for the inverse of a curve $\mathscr{D}$ in the circle $\mathscr{C}$:

Erect perpendiculars to the plane OXY at the points of $\mathscr{D}$, and let $\mathscr{E}$ be the curve in which these perpendiculars intersect Ω. The cone of lines joining P (the point which represents the circle $\mathscr{C}$) to the points of $\mathscr{E}$ intersects Ω again in a curve $\mathscr{E}'$. Project the points of $\mathscr{E}'$ down onto the plane OXY: the resulting curve $\mathscr{D}'$ is the inverse of $\mathscr{D}$ in the circle $\mathscr{C}$.

If P, the point which represents $\mathscr{C}$, lies inside Ω, the construction is carried out in exactly the same way, and gives the inverse of $\mathscr{D}$ in a virtual circle $\mathscr{C}$.

We remark that if a line through P intersects Ω in the points Q and R, these points are harmonic conjugates with respect to P and the point P' in which the line intersects the polar plane of P with respect to Ω. Since P and the polar plane of P can be regarded as fixed, we can extend the mapping, given by taking inverse points, to the whole of E_3, by joining any point X of E_3 to P, finding the point in which PX intersects the polar plane of P, and if this be P', finding the harmonic conjugate of X with respect to the pair of points P, P'. Such a mapping of E_3 onto itself is called *a harmonic homology* (see §76.2), and P and the points in the polar plane are all fixed points under this mapping. The points of Ω are interchanged by this mapping, and so Ω *maps into itself*, but not pointwise.

Returning to the method in E_3 for finding inverses, we know that the inverse of a circle is a circle, if the original circle does not pass through the center of inversion. We also know that erecting perpendiculars to the plane OXY at the points on a circle gives us a plane curve of intersection with Ω. Hence we see that if we form the quadric cone of lines which join P to the points in which a plane intersects Ω, this quadric cone of lines will intersect Ω again in a plane section. We also know that a circle, its inverse and the circle of inversion are in the same coaxal system (Exercise 21.6). Hence the poles of the two plane sections are collinear with P, since these poles represent the circles obtained by projecting the plane sections down onto the OXY-plane. The complete theorem we have proved is therefore:

A quadric cone which intersects Ω in one plane section intersects it again in a plane section, and the poles of the two plane sections are collinear with the vertex of the cone.

This theorem is true for any quadric. It is interesting to see that in this special case it arises naturally from the process of inversion.

Exercises

39.1   The circle $\mathscr{D}$ passes through the center V of the circle $\mathscr{C}$. Show that if we carry out the process of inversion of $\mathscr{D}$ in $\mathscr{C}$ in E_3, the further plane of intersection of the quadric cone with Ω will be parallel to OZ, so that the inverse of $\mathscr{D}$ is a line, not a circle.

39.2   The circle $\mathscr{D}$ is orthogonal to the circle $\mathscr{C}$. Show that if we erect perpendiculars to the plane OXY along the points of $\mathscr{D}$, the plane section in which these lines intersect Ω contains the point which represents $\mathscr{C}$. Hence show that a circle orthogonal to $\mathscr{C}$ inverts into itself in $\mathscr{C}$.

39.3   A plane which contains the point (p, q, r) has the equation

$$a(X - p) + b(Y - q) + c(Z - r) = 0.$$

Deduce that the circle

$$a(X - p) + b(Y - q) + c(X^2 + Y^2 - r) = 0$$

is always orthogonal to the circle

$$X^2 + Y^2 - 2pX - 2qY + r = 0.$$

39.4   In our investigation of circles which intersect three given circles at angles θ or $\pi - \theta$, why does inversion prepare us for the discovery that the solutions, in the representation in E_3, lie in pairs on lines through the point which represents the circle orthogonal to the three given circles?

39.5 Why is it clear geometrically that the solutions of the Apollonian problem, the circles which *touch* three given circles, are represented by the common intersections of three quadric *cones*, each of which has ring-contact with Ω? (What can you say about the circles of zero radius in the coaxal system defined by two circles which touch?)

39.6 It was suggested in Exercises 25.2 and 25.7 that the operations of inversion in two given distinct circles are commutative if and only if the circles are orthogonal. Let U and V be the representative points for the two given circles, and S a section of Ω by a plane through the line UV. Let P be any point on S, P_1 the other intersection of S with the line UP, and P_2 the other intersection of S with VP. Let P_{12} be the other intersection of VP_1 with S, and let P_{21} be the other intersection of UP_2 with S (which is a conic). Show that the representation of inversion given in §39.1 indicates that the operations of inversion in the circles represented by U and V are commutative if and only if $P_{12} = P_{21}$. (We shall see later that this implies that U and V are conjugate with respect to S, and since this is any section of Ω through the line UV, that U and V are conjugate with respect to Ω, and therefore represent orthogonal circles. See Exercise 78.14.)

40.1 An algebra of circles

We shall conclude this chapter on circles by developing an algebra of circles, and we shall use this algebra to prove an attractive theorem, first discovered by Descartes, which gives the algebraic relation between the radii of four circles which have mutual contact.

We wish to combine circles linearly. We have been doing this for some time in the present chapter, using the representation of circles by points in E_3 to obtain a geometric picture of the results. For our present purpose it is more convenient to work with the polar planes of the representative points. That is, if the circle $X^2 + Y^2 - 2pX - 2qY + r = 0$ is represented by the point (p, q, r) in E_3, it is also represented by the plane $Z - 2pX - 2qY + r = 0$, and this plane is obtained by erecting parallels to the OZ-axis along the points of the circle, and taking the intersection of these lines with the paraboloid of revolution

$$\Omega \equiv X^2 + Y^2 - Z = 0.$$

These points of intersection lie in a plane, which has the equation just given, and this plane is the polar plane of the point (p, q, r) with respect to the quadric.

The advantage of the representation by plane sections is that there is no difficulty in representing improper circles, that is circles of infinite radius, which appear as lines, radical axes in fact, as soon as we consider coaxal systems of circles. Thus the circle

$$0(X^2 + Y^2) + aX + bY + c = 0$$

has no *point* representation, but it is represented perfectly well in the plane section representation. In fact the representative plane has precisely the same equation. If we erect parallels to the OZ-axis along the points of the line, which has the equations

$$aX + bY + c = 0 = Z$$

in E_3, we obtain the plane

$$aX + bY + c = 0,$$

which is a plane parallel to the OZ-axis.

We cannot, however, represent all radical axes in this simple way. If we have two concentric circles, with centers at the point (p, q), their equations are

$$C_1 \equiv X^2 + Y^2 - 2pX - 2qY + r_1 = 0,$$
$$C_2 \equiv X^2 + Y^2 - 2pX - 2qY + r_2 = 0,$$

and

$$C_1 - C_2 \equiv r_1 - r_2 = 0.$$

The radical axis of two concentric circles is not a finite line in the OXY-plane, and the plane representation in E_3 can only be *the plane at infinity*. This is represented by the apparently paradoxical equation

$$\theta = 0,$$

where θ is a non-zero constant. We could avoid such apparent paradoxes by using homogeneous coordinates (see §80.1), but we shall not have any special difficulty in using this special plane section of Ω, and we shall find it very useful to do so. We shall normalize the constant in the next section.

We can talk about the linear dependence or independence of planes in E_3, and it is clear that we can find *four* independent planes, and that any five given planes must be linearly dependent. For instance, the four planes

$$X = 0,\ Y = 0,\ Z + 1 = 0,\ Z - 1 = 0$$

are independent. If we try to find constants k_i ($i = 1, \ldots, 4$) such that

$$k_1 X + k_2 Y + k_3(Z + 1) + k_4(Z - 1) \equiv 0,$$

we find that $k_1 = k_2 = k_3 + k_4 = k_3 - k_4 = 0$, so that all the k_i are zero. We note that an equivalent statement is that the four circles in the OXY-plane given by

$$X = 0,\ Y = 0,\ X^2 + Y^2 + 1 = 0,\ X^2 + Y^2 - 1 = 0$$

are linearly independent. Two of these circles are improper circles, and one is a virtual circle, containing no real points.

If we are given four linearly independent planes, with equations

$$\pi_i = 0 \quad (i = 1, \ldots, 4)$$

in E_3, the addition of the special plane $\theta = 0$ to this set makes the five planes dependent, and we can therefore always write, for suitable x_i,

$$x_1\pi_1 + x_2\pi_2 + x_3\pi_3 + x_4\pi_4 \equiv \theta. \tag{1}$$

If the plane sections $\pi_i = 0$ represent the circles $C_i = 0$, this equation is equivalent to the equation

$$x_1 C_1 + x_2 C_2 + x_3 C_3 + x_4 C_4 \equiv 0 \qquad (2)$$

and we move from one identity to the other by simply interchanging Z and $X^2 + Y^2$.

40.2   An inner product for two circles

Let $\mathscr{C}_1$, $\mathscr{C}_2$, $\mathscr{C}_3$ be three circles with equations

$$C_i = a_i(X^2 + Y^2) - 2p_i X - 2q_i Y + r_i = 0 \quad (i = 1, 2, 3).$$

We note that these are only normalized equations if $a_i = 1$, but on the other hand if $a_i = 0$ the circle $\mathscr{C}_i$ is a line, and if $a_i = p_i = q_i = 0$, $r_i \neq 0$, then $\mathscr{C}_i$ is our special circle $\theta = 0$.

We define an *inner product* $[C_1 . C_2]$ for the circles $\mathscr{C}_1$, $\mathscr{C}_2$ thus:

$$[C_1 . C_2] = \tfrac{1}{2}[2p_1 p_2 + 2q_1 q_2 - r_1 a_2 - r_2 a_1].$$

It is clear that $[C_1 . C_2] = [C_2 . C_1]$, and we note that the inner product is linear in the coefficients of the equation of each circle. Hence

$$[C_1 . (kC_2)] = [(kC_1) . C_2] = k[C_1 . C_2],$$

and since $C_2 + C_3$ is obtained by adding the expressions for C_2 and C_3, thus:

$$C_2 + C_3 = (a_2 + a_3)(X^2 + Y^2) - 2(p_2 + p_3)X - 2(q_2 + q_3)y + r_2 + r_3,$$

it is immediately verified that

$$[C_1 . (C_2 + C_3)] = [C_1 . C_2] + [C_1 . C_3].$$

This *distributive law* can evidently be extended, so that

$$[C_1 . (C_2 + C_3 + C_4)] = [C_1 . C_2] + [C_1 . C_3] + [C_1 . C_4],$$

and so on, and it is valid for any of the circles, proper or improper, which may arise. In particular, we can assert that

$$[\theta . (C_1 + C_2 + C_3 + C_4)] = \sum_{i=1}^{4} [\theta . C_i]$$

for any given circles C_1, C_2, C_3, C_4, and our special circle $\theta = 0$.

The geometrical interpretation of the inner product for $\mathscr{C}_1$, $\mathscr{C}_2$ is immediate if we take $a_1 = a_2 = 1$, so that the circles are normalized, when we find that

$$[C_1 . C_2] = \tfrac{1}{2}(2p_1 p_2 + 2q_1 q_2 - r_1 - r_2),$$

which we recognize as being equal to the expression

$$\tfrac{1}{2}(R_1{}^2 + R_2{}^2 - D^2),$$

where R_1, R_2 are the respective radii of the two circles, and D is the distance between their centers. (§22.1.)

It follows that $[C_1 . C_2] = 0$ if the circles $\mathscr{C}_1$ and $\mathscr{C}_2$ are orthogonal, and in fact $[C_1 . C_2] = 0$ for non-normalized circles which are orthogonal, in the special cases when one or both circles are lines, and either $a_1 = 0$, or $a_2 = 0$, or both $a_1 = a_2 = 0$.

We are especially interested in the cases when the normalized circles $\mathscr{C}_1$, $\mathscr{C}_2$ touch each other. If they touch externally, so that $D = R_1 + R_2$,

$$[C_1 . C_2] = \tfrac{1}{2}[R_1^2 + R_2^2 - (R_1 + R_2)^2] = -R_1 R_2,$$

and if they touch internally, so that $R_1 - R_2 = \pm D$, we see that

$$[C_1 . C_2] = +R_1 . R_2.$$

If the circles are *concentric*, so that $D = 0$,

$$[C_1 . C_2] = (R_1^2 + R_2^2)/2,$$

and if we find the inner product of a circle with itself,

$$[C_1 . C_1] = R_1^2.$$

Finally, if we *weight* given circles $\mathscr{C}_i$, by multiplying C_i by the number k_i, it is clear that for the weighted circles:

$$[C_1 . C_1] = k_1^2 R_1^2, \qquad [C_1 . C_2] = -k_1 k_2 R_1 R_2$$

when C_1, C_2 touch externally, and

$$[C_1 . C_2] = k_1 k_2 R_1 R_2$$

if they touch internally. We shall apply this weighting procedure later, choosing $k_1 = 1/R_1$, $k_2 = 1/R_2$, and then, for the weighted circles,

$$[C_1 . C_1] = 1, [C_1 . C_2] = -1, \text{ and } [C_1 . C_2] = +1$$

in the three respective cases we have discussed.

Before we apply this inner product to our algebra of circles, let us consider any given circle $\mathscr{C}_1$ and the special circle $\theta = 0$, which has the equation

$$\theta \equiv 0(X^2 + Y^2) + 0X + 0Y + r = 0,$$

where $r \neq 0$. The constant r is at our disposal. If we form

$$[C_1 . \theta] = \tfrac{1}{2}(2p_1(0) + 2q_1(0) - r_1(0) - ra_1) = -(ra_1)/2,$$

we see that if we choose $r = -2$, then for *any normalized circle* $\mathscr{C}$ we always have $[C . \theta] = 1$.

Again, the inner product

$$[\theta . \theta] = 0.$$

We now have the formal apparatus for proving theorems.

40.3   A theorem of Descartes

Theorem   *If four circles $\mathscr{C}_i$ ($i = 1, \ldots, 4$) in a plane touch each other externally, each circle touching the three other circles, and $\varepsilon_i = 1/R_i$, where R_i is the radius of the circle $\mathscr{C}_i$, then*

$$2(\varepsilon_1{}^2 + \varepsilon_2{}^2 + \varepsilon_3{}^2 + \varepsilon_4{}^2) = (\varepsilon_1 + \varepsilon_2 + \varepsilon_3 + \varepsilon_4)^2.$$

Proof. We assume that the four circles are first normalized and then weighted by multiplying each of them by its curvature $\varepsilon_i = 1/R_i$. Since each pair of circles C_i, C_j touch externally, we have $[C_i \, . \, C_j] = -1$, and since $[C_i \, . \, \theta] = 1$ when C_i is normalized, we now have $[C_i \, . \, \theta] = \varepsilon_i$.

There is still a linear relation of the form (2) of §40.1 between the weighted circles C_i and the special circle θ. We write once more

$$x_1 C_1 + x_2 C_2 + x_3 C_3 + x_4 C_4 \equiv \theta.$$

Forming the inner product of both sides of this identity with C_1, and using the distributive formula, we have

$$x_1[C_1 \, . \, C_1] + x_2[C_1 \, . \, C_2] + x_3[C_1 \, . \, C_3] + x_4[C_1 \, . \, C_4] = [C_1 \, . \, \theta],$$

which is

$$x_1 - x_2 - x_3 - x_4 = \varepsilon_1$$

by the relations given above for circles with external contact.

Similarly, forming the inner product with C_2, then C_3 and then C_4, we obtain the equations

$$-x_1 + x_2 - x_3 - x_4 = \varepsilon_2,$$

$$-x_1 - x_2 + x_3 - x_4 = \varepsilon_3,$$

and

$$-x_1 - x_2 - x_3 + x_4 = \varepsilon_4.$$

We now form the inner product of both sides of the identity with θ, and obtain the equation

$$x_1[\theta \, . \, C_1] + x_2[\theta \, . \, C_2] + x_3[\theta \, . \, C_3] + x_4[\theta \, . \, C_4] = [\theta \, . \, \theta],$$

which is

$$x_1 \varepsilon_1 + x_2 \varepsilon_2 + x_3 \varepsilon_3 + x_4 \varepsilon_4 = 0.$$

If we square each of the first four equations and add, we find that

$$\Sigma \varepsilon_i{}^2 = 4\Sigma x_i{}^2,$$

and if we multiply the equations in pairs, and then add, we find that

$$\Sigma \varepsilon_i \varepsilon_j = 4\Sigma x_i x_j.$$

If we now substitute the values of the ε_i given by the first four equations in the last equation, we find that

$$\Sigma x_i^2 - 2\Sigma x_i x_j = 0.$$

Hence, substituting for Σx_i^2 and for $\Sigma x_i x_j$, we end up with

$$\Sigma \varepsilon_i^2 - 2\Sigma \varepsilon_i \varepsilon_j = 0,$$

which can be rewritten as

$$2(\varepsilon_1^2 + \varepsilon_2^2 + \varepsilon_3^2 + \varepsilon_4^2) = (\varepsilon_1 + \varepsilon_2 + \varepsilon_3 + \varepsilon_4)^2.$$

This theorem has an interesting history. It can be found in the correspondence of Descartes with the Princess Elizabeth (Oeuvres de Descartes, par Adam et Tannery, Vol. IV (Paris, 1901, p. 63)). The mother of the Princess was a daughter of James 1st of England, and married to the King of Bohemia. The Princess lived from 1618–1680. She corresponded with Descartes on philosophical matters, as well as mathematics. It was rediscovered in 1936 by F. Soddy, and given great publicity. Soddy sent a statement of the theorem to the English science journal 'Nature', together with a poem entitled 'The kiss precise', the idea being that circles which touch are osculating. Soddy was a very distinguished scientist, had worked with Rutherford, and was a man of great vitality and varied interests. There is an extension of the Descartes theorem to the case of $n + 2$ spheres in external contact in Euclidean space of n dimensions, when the formula is

$$n(\Sigma \varepsilon_i^2) = (\Sigma \varepsilon_i)^2.$$

This extension can be proved with precisely the same algebra that we have used above. (See Pedoe, (**15**)).

This extension was stated, without proof, by T. Gosset in 1937, a year after Soddy had stated the theorem for two and for three dimensions, without proof. Gosset added a verse to the Soddy poem to cover the n-dimensional case. The references are: *Nature*, 137 (1936), p. 1021, and *Nature*, 139 (1937), p. 62. Coxeter (**3**) gives a verse of the Soddy poem.

The exercises which follow are exercises on the algebra of circles. The reader may have noticed that our algebra could have been developed without any geometrical introduction, since the equations of circles can be added, and the equation (2) of §40.1 obtained without any geometrical justification. The constant θ is then merely a constant, and does not represent anything special. The inner product of two circles proceeds as before, and can be applied to both sides of any identity involving linear sums of multiples of circles. But we think that the geometrical motivation is worth giving, and it is reassuring that we can even explain the curious $[\theta \cdot \theta] = 0$ phenomenon. For a proper circle $\mathscr{C}_1$ we saw that $[C_1 \cdot C_1] = R_1^2$, when the circle is normalized. We may say that the circle θ is given by the section of the plane at infinity with our paraboloid of revolution Ω. But this paraboloid can be shown to touch the plane at infinity, and so our special circle θ is a point-circle, that is, one with zero radius!

Exercises

40.1 If $\mathscr{C}_i(i = 1, 2, 3, 4)$ are four mutually orthogonal circles, of respective radii $R_i = 1/\varepsilon_i$, prove that

$$\varepsilon_1{}^2 + \varepsilon_2{}^2 + \varepsilon_3{}^2 + \varepsilon_4{}^2 = 0.$$

40.2 If the three circles $\mathscr{C}_1$, $\mathscr{C}_2$ and $\mathscr{C}_3$ are in the same coaxal system show that

$$\begin{vmatrix} R_1{}^2 & [C_1 \cdot C_2] & [C_1 \cdot C_3] \\ [C_2 \cdot C_1] & R_2{}^2 & [C_2 \cdot C_3] \\ [C_3 \cdot C_1] & [C_3 \cdot C_2] & R_3{}^2 \end{vmatrix} = 0.$$

40.3 If $\mathscr{C}_4$, with radius R_4, is the unique circle orthogonal to the three circles $\mathscr{C}_1$, $\mathscr{C}_2$ and $\mathscr{C}_3$, show that R_4 is given by the determinantal equation

$$\begin{vmatrix} R_1{}^2 & [C_1 \cdot C_2] & [C_1 \cdot C_3] & 1 \\ [C_2 \cdot C_1] & R_2{}^2 & [C_2 \cdot C_3] & 1 \\ [C_3 \cdot C_1] & [C_3 \cdot C_2] & R_3{}^2 & 1 \\ 1 & 1 & 1 & -1/R_4{}^2 \end{vmatrix} = 0.$$

40.4 By taking the circles $\mathscr{C}_1$, $\mathscr{C}_2$ and $\mathscr{C}_3$ to be point-circles, show that the result of the previous Exercise leads to the standard formula for the radius of the circumcircle of a triangle ABC:

$$R = a_0 b_0 c_0 / 4 \triangle,$$

where a_0, b_0, c_0 are the sides of the triangle, and $\triangle$ is the area.

40.5 In the proof of the Descartes Theorem (§40.3), it is assumed that the four circles touch each other externally. Show that if $\mathscr{C}_2$ lies inside $\mathscr{C}_1$, so do $\mathscr{C}_3$ and $\mathscr{C}_4$, and the correct relation between the curvatures is obtained by writing $-\varepsilon_1$ instead of ε_1 in the formula.

40.6 Obtain the relation which holds in E_3 for five spheres which touch each other externally.

V

MAPPINGS OF THE EUCLIDEAN PLANE

In this chapter we propose to investigate the fundamental bijective (one-to-one onto) mappings of the Euclidean plane, using algebraic methods, and, in particular, complex numbers. Some of these mappings have already appeared in purely geometric guise (§12.1, §14.1). Some of the proofs given previously may seem to depend on the acceptance of the correctness of diagrams, and it is important to show that algebraic proofs are also possible. The algebra of complex numbers is very useful in this connection, but we shall try not to lose sight of the geometrical meaning of theorems and, in fact, it will be evident in many cases that the algebra is motivated by the geometry.

In §0.4 we gave the properties of complex numbers which we shall be using. We saw that we may regard the Euclidean plane as a representative of the field of complex numbers, the point (a, b) being associated with the complex number $a + ib$. When the Euclidean plane is regarded in this way we shall refer to it as *the Gauss plane*. Complex numbers will be denoted by lower-case letters, such as z, a, b, etc., and although we have just mapped the point (a, b), where a, b are real numbers, onto the complex number $z = a + ib$, there will be no misunderstanding if we talk of the mapping $z' = az$, where $|a| = 1$, it being understood that a is a complex number in this case. We shall naturally talk of 'the point z' instead of 'the point of the Euclidean plane associated with the complex number z'.

41.1 Translations

Let b be a fixed complex number, and consider the mapping

$$z' = z + b.$$

It is clear that every point z is moved through a fixed distance and in a fixed direction, given by the complex number b regarded as a vector. Such a mapping is called a *translation* (see Fig. 41.1). If $b = 0$, the mapping is the identity mapping.

41.2 Rotations

The mapping $z' = (\cos \varphi + i \sin \varphi)z$, leaves the modulus of z unchanged, but increases its amplitude by φ, and therefore moves the point representing z around the circle, center at the origin, which passes through the point z, and through a fixed angle φ. This is *a rotation about the origin* (see Fig. 41.2).

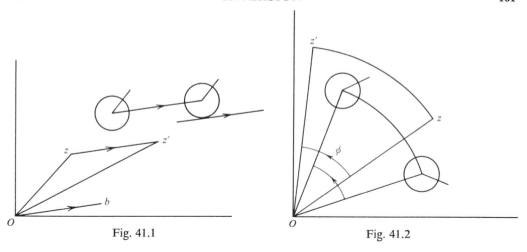

Fig. 41.1   Fig. 41.2

This mapping may also be written $z' = az$, where $|a| = 1$, so that $a = \cos \varphi + i \sin \varphi$. If $a = 1$, we obtain the identity mapping.

41.3 Reflexions

The mapping $z' = \bar{z}$ is a reflexion in the X_1-axis (Fig. 41.3). If we require a reflexion in the line which passes through the origin, and makes an angle α with the X_1-axis, it is given by $z' = (\cos 2\alpha + i \sin 2\alpha)\bar{z}$, since we have at once $|z| = |z'|$, and $am\ z' + am\ z =$

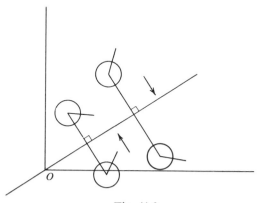

Fig. 41.3

2α, so that the given line always bisects the angle between the joins of the origin to the points z and z'. The difference between the mappings when the map is a rotation and when it is a reflexion should be noted.

41.4 Inversion

This is not a mapping of the Gauss plane, but of the plane with the center of inversion removed (Fig. 41.4). We shall overcome this difficulty in the next chapter, but in the meantime we remark that $z' = 1/z$ is equivalent to inversion in the unit circle, center at the origin, followed by reflexion in the X_1-axis. For if $z = r(\cos\theta + i\sin\theta)$,

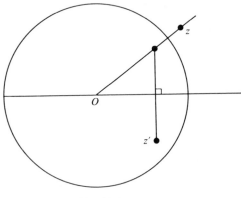

Fig. 41.4

$z^{-1} = r^{-1}(\cos(-\theta) + i\sin(-\theta))$, and the inverse of z is the point $r^{-1}(\cos(\theta) + i\sin(\theta))$. If we wish to obtain the inverse of z directly, we write $z' = 1/\bar{z}$.

41.5 Central dilatations

We can combine mappings (Fig. 41.5). Let us consider the mapping $z' = az$, where

$$a = |a|(\cos\alpha + i\sin\alpha).$$

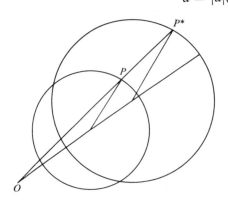 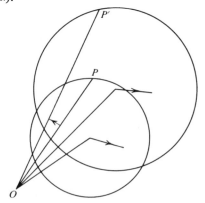

Fig. 41.5

Then $z = r(\cos \theta + i \sin \theta)$ is mapped on

$$z' = |a|r(\cos (\theta + \alpha) + i \sin (\theta + \alpha)).$$

The point P representing z is stretched to the point P^*, where $\overline{OP^*} = |a|\overline{OP}$, (that is, stretched in a fixed ratio), and OP^* is then rotated about O through the fixed angle α to OP', where P' represents z'.

We note that the order in which the two operations, stretching and rotation, are carried out does not affect the final result. That is, they are *commutative* operations. If a is real, so that $\alpha = 0$, the mapping is called either a *central similarity*, or a *central dilatation*, or merely a *magnification* or *enlargement*, but if rotation is also involved, it is called a *dilative rotation* about O, or a *spiral similarity*. Each of these terms is fairly easy to remember. We encountered central dilatations in §12.1, reflexions in a line in §14.1, rotations in the same section, and inversions in §20.1. Translations, of course, are implicit in the equivalence relation defined for vectors in §1.1. A free vector is obtained from a bound vector by means of a translation.

All of the mappings mentioned above can be used to solve certain types of geometrical problem. We have already seen the power of inversion in the §§ 20.1–onwards, and we have used reflexions to solve certain problems and to prove special theorems. The Exercises below can be solved by the use of the mappings we have discussed. We omit Exercises which need inversion.

Exercises

41.1 Two circles $\mathscr{C}_1$ and $\mathscr{C}_2$ are given, and a line l. Find a point A on $\mathscr{C}_1$ and a point B on $\mathscr{C}_2$ such that AB is parallel to l, and the distance $AB = d$, where d is given. (What is the map of a given circle under a translation?)

41.2 Two circles $\mathscr{C}_1$ and $\mathscr{C}_2$ are given, and a line l. Construct squares $ABCD$ with A and C on l, with B on $\mathscr{C}_1$, and D on $\mathscr{C}_2$. (What is the reflexion of a given circle in a given line?)

41.3 Show how to construct an equilateral triangle which has one vertex at a given point A, and one vertex on each of two given lines l and m. (What locus do we obtain if we rotate a given line through a given angle A about a given point?)

41.4 A point A and lines l and m are given. Show how to construct a square $ABCD$, with vertex A at A, vertex B on l, and vertex C on m. (What is the locus of l under the spiral similarity, center A, given by the equation $z' = az$, where $|a| = \sqrt{2}$, and am $a = +\pi/4$, or $-\pi/4$?)

41.5 Show that a central dilatation, $z' = pz$, where p is real, maps a line onto a parallel line. Show also that if $z'' = qz'$ be another central dilatation with the same center, this maps the point z' onto the point $z'' = qpz$. Show that if the two mappings be combined, with the second first, we obtain $z'' = pqz$ as a result of the composition, so that the final result is independent of the order in which the mappings are combined. Use this result to obtain the parallel case of the Pappus Theorem of §9.5. (The points A, B, C lie on l, and the points A', B', C' lie on m. Given that AC' is parallel to CA', and CB' is parallel to BC', we wish to prove that AB' is parallel to BA'. Use central dilatations with center at $l \cap m$. Let $C = AT_1$, so that $A' = C'T_1$, and let $B = CT_2$, so that $C' = B'T_2$. Show that $B = AT_1T_2$, and $A' = B'T_2T_1$. From $T_1T_2 = T_2T_1$, deduce the theorem.)

42.1 Isometries

We note that the mappings: *translation, rotation* and *reflexion,* are all such that the distance between two points is unchanged by one of these mappings. If the two points are z and w, a translation gives

$$|z' - w'| = |z + b - (w + b)| = |z - w|,$$

a rotation gives

$$|z' - w'| = |(\cos \varphi + i \sin \varphi)z - (\cos \varphi + i \sin \varphi)w|$$
$$= |(\cos \varphi + i \sin \varphi)||(z - w)| = |z - w|,$$

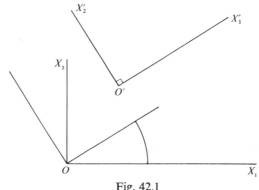

Fig. 42.1

and reflexion in a line through the origin, which we can write $z' = a\bar{z}$, where $|a| = 1$, gives

$$|z' - w'| = |a(\overline{z - w})| = |z - w|,$$

so that our assertion is proved in all three cases.

We know that inversion and central dilatations do change the distance between a pair of points.

Definition. An *isometry* of the Gauss plane is a one-to-one onto mapping of the plane such that if P is mapped on P' and Q is mapped on Q', the distance $PQ = $ distance $P'Q'$, whatever the points P and Q.

Such mappings are also called *congruent transformations,* or *rigid motions.* The mappings: *translation, rotation* and *reflexion* are all isometries, and compositions of them are naturally also isometries. Two important ones must be familiar to the reader.

42.2 Coordinate transformations

We assume that our coordinate axes are orthogonal cartesian, and we *rotate* them through an angle φ about the origin, and in a positive direction, from the X_1-axis towards the X_2-axis (Fig. 42.1).

We then *translate* the new position of the axes so that the original origin O falls on the point $O' = b$, where b is a complex number referred to the original axes. Calling the original axes OX_1, OX_2, call the new position of the axes $O'X_1'$, $O'X_2'$.

After the rotation the point which was originally z with respect to the original axes has become

$$z(\cos(-\varphi) + i\sin(-\varphi))$$

with respect to the axes rotated through φ about O, and the point b has become

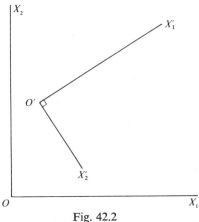

Fig. 42.2

$b(\cos(-\varphi) + i\sin(-\varphi))$ with respect to these axes. After the translation, the point z becomes, with respect to the axes $O'X_1'$, $O'X_2'$:

$$Z' = z(\cos(-\varphi) + i\sin(-\varphi)) - b(\cos(-\varphi) + i\sin(-\varphi)),$$

which we may write

$$Z' = az + c, \text{ where } |a| = 1.$$

The same point now has two different names, or aliases: z with respect to the original axes, and $Z' = az + c$ with respect to the new axes.

If we suppose that the coordinate axes remain unchanged, and consider the mapping $Z' = az + c$ ($|a| = 1$), we are not surprised to find that this is also an isometry. In fact, if z and w are mapped on Z' and W',

$$|Z' - W'| = |az + c - aw - c| = |a||z - w| = |z - w|.$$

In our transformation of coordinate axes above, we *preserved* the *orientation* of the axes: that is, the sense of the rotation from $O'X_1'$ towards $O'X_2'$ is the same as that of OX_1 towards OX_2. If, however, we *reverse* the direction of *one* of the new axes (Fig. 42.2), say $O'X_2$, the point z with respect to OX_1, OX_2 now has the form

$$Z' = a\bar{z} + c \qquad (|a| = 1)$$

with respect to the new axes $O'X_1'$, $O'X_2'$.

Once again, if we suppose the coordinate axes to remain unchanged, and consider the mapping $Z' = a\bar{z} + c$ ($|a| = 1$), we find that it is also an isometry, since

$$|Z' - W'| = |a\bar{z} + c - a\bar{w} - c| = |a||\bar{z} - \bar{w}| = |\overline{z - w}| = |z - w|.$$

We shall find that these two isometries of the Gauss plane, which arise in this natural way by coordinate transformations, are the fundamental isometries of the Gauss plane, and there are no others.

Exercises

42.1 The X_1 and X_2-axes are interchanged, so that if $z = a + ib$, $Z = b + ia$ after the transformation. Show that the following composition of coordinate transformations produces the above result: $Z_1 = iz = -b + ia$, $Z_2 = \bar{Z}_1 = -b - ia$, $Z = -Z_2 = b + ia$. Show also that the composition of these transformations is equivalent to $Z = i\bar{z}$.

42.2 Where are the new coordinate axes which induce the coordinate transformations

$$\text{(i) } Z = \bar{z} \qquad \text{(ii) } Z = i\bar{z} \qquad \text{(iii) } Z = -iz + 1?$$

42.3 Find the coordinate transformation produced if the new system has its origin at the point $1 + i$, the new negative X_1-axis passes through O, and the new positive X_2-axis intersects the original positive X_2-axis. In the same circumstances, find the transformation if the new positive X_2-axis intersects the old positive X_1-axis.

43.1   The main theorem on isometries of the Gauss plane

Theorem *The set $\mathcal{I}$ of all isometries of the Gauss plane is composed of two classes $\mathcal{I}_+$ and $\mathcal{I}_-$. The class $\mathcal{I}_+$ consists of all isometries of the form $z' = az + b$ ($|a| = 1$), and the class $\mathcal{I}_-$ of all isometries of the form $z' = c\bar{z} + d$ ($|c| = 1$).*

We recall that an isometry is a one-to-one onto mapping of the Gauss plane which preserves the distance between a pair of points, and that such a mapping is also called a *congruent* or *rigid* transformation, or *motion*, if the plane is thought of as being moved over itself.

We have already seen that the mappings $\mathcal{I}_+$ and $\mathcal{I}_-$ are such that distance is preserved, but we have not checked that both are one-to-one onto mappings. In each case this is immediate, since if $z' = az + b$, z' is uniquely defined by z, and $z = a^{-1}(z' - b)$, so that, since $a \neq 0$, z is uniquely defined by z'. If $z' = c\bar{z} + d$, z' is uniquely defined by z, and $\bar{z} = c^{-1}(z' - d)$, so that $z = \bar{\bar{z}}$ is uniquely defined by z'.

We have some auxiliary theorems to prove before we establish the main theorem, Thm 43.1.

43.2   An auxiliary theorem

Theorem *Given two pairs of points z_0, z_1 and w_0, w_1, where $|z_0 - z_1| = |w_0 - w_1| \neq 0$, there is just one mapping of type $\mathcal{I}_+$ and just one of type $\mathcal{I}_-$ which maps z_0 onto w_0 and z_1 onto w_1.*

Proof. All that we have to do is to determine the constants a, b for a mapping of the first kind, and the constants c, d for a mapping of the second kind. Since we must have $az_0 + b = w_0$, and $az_1 + b = w_1$, where $|a| = 1$, subtraction gives us $a(z_0 - z_1) = w_0 - w_1$, which defines a, since $z_0 - z_1 \neq 0$, and also gives us a value of a for which $|a| = 1$. Knowing the value of a, we substitute, and find that $b = w_0 - az_0$. Hence a and b are both uniquely determined.

The proof that c and d are uniquely determined is very similar.

One of the things which interests us for any given mapping of the Gauss plane is whether points which are collinear are mapped into points which are collinear.

Definition. A *collineation* of the Gauss plane is a mapping which preserves collinearity, so that straight lines are mapped into straight lines.

A collineation need not be an isometry, as we see from the example of a central dilatation. But if we consider isometries, we have:

43.3   Isometries are collineations

Theorem   *Every isometry of the Gauss plane is a collineation.*

Proof. In the real Euclidean plane we have a very powerful notion, derived from that of distance, namely the notion of a point lying on a line *between* two distinct points. If u, v and w are the complex numbers attached to three distinct collinear points, we know that v lies on the line joining the two other points, and between them, if and only if

$$|v - u| + |w - v| = |w - u|.$$

If under an isometry the points become u', v' and w' respectively, we have

$$|v' - u'| + |w' - v'| = |w' - u'|,$$

and since these points are also in the real Euclidean plane, the points represented by the complex numbers u', v' and w' must necessarily lie, in this order, on a line.

We deduce from this theorem that although Thm 43.2 tells us that there are at least two isometries which map z_0 on w_0 and z_1 on w_1, where $|z_0 - w_0| = |z_1 - w_1|$, each mapping maps a point of the line joining z_0 and z_1 onto a point of the line joining w_0 to w_1, and the effects of the mappings on points of the object line are identical.

43.4   Isometries and parallel lines

Theorem   *An isometry of the Gauss plane maps parallel lines onto parallel lines.* (See Fig. 43.1.)

Proof. If P is a point in the Gauss plane, and l a line which does not contain P, the minimum distance of P from the line is well defined, and given by PR, where R is the

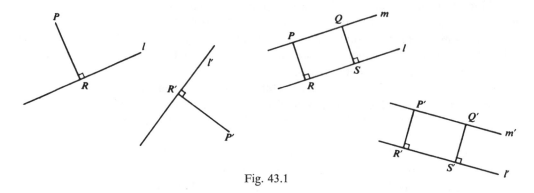

Fig. 43.1

foot of the perpendicular from P onto the line. Under an isometry, if P is mapped onto P', the line l onto a line l', and R onto R', the point R' lies on l', and $P'R'$ is the minimum distance of P' from l'. Hence $P'R'$ is perpendicular to l'.

Given two parallel lines l and m, take two distinct points P and Q on m, and let R and S be the respective feet of perpendiculars from P and Q onto the line l. Then

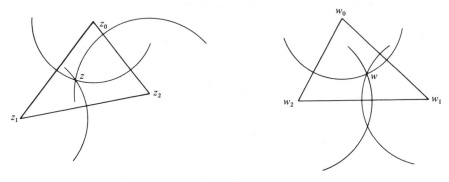

Fig. 43.2

$PR = QS$, and the isometry maps l and m into lines l' and m' which are likewise equidistant from each other, and therefore parallel.

Alternatively, and more rapidly, we may argue that if the maps l' and m' of l and m intersected at a point P', the object of P' would have to be a point lying on l and also on m, and by assumption these are parallel lines.

43.5   Determination of an isometry

Theorem   *An isometry of the Gauss plane is uniquely determined by the assignment of the congruent map of a given triangle.* (See Fig. 43.2.)

Proof. We are given three distinct non-collinear points z_0, z_1 and z_2, and their respective images w_0, w_1 and w_2, where

$$|z_i - z_j| = |w_i - w_j| \qquad (i, j = 0, 1, 2).$$

We wish to show that the image of a point z distinct from the three given points z_i is uniquely defined under an isometry which satisfies the given conditions.

We may argue that a point is uniquely determined if its distances from three given non-collinear points are given. For the assignment of its distances from two given points determines the point as the intersection of two distinct circles, which intersect in two points, and the point also lies on a third circle, center at the third given point.

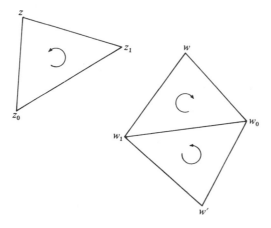

Fig. 43.3

This circle cannot pass through both points of intersection of the first two circles constructed, since its center would have to be collinear with the first two centers. Hence it passes through only one of the two points, and so the point is uniquely determined by its distances from three given non-collinear points.

Hence the image of z under an isometry is uniquely determined, since we are given the distances of z from the points z_i, and we are therefore also given the distances of the image w from the points w_i.

Another proof is indicated in Exercise 43.3.

43.6   An auxiliary theorem

Theorem   *There are precisely two isometries of the Gauss plane which map two given points z_0 and z_1 onto two given points w_0 and w_1, where $|z_0 - z_1| = |w_0 - w_1| \neq 0$.* (See Fig. 43.3.)

Proof. We have already found two isometries which do this, in Thm 43.2, one of type

$z' = az + b$ ($|a| = 1$), and the other of type $z' = c\bar{z} + d$ ($|c| = 1$). We show that there are not more than two.

By an argument similar to the one used in Thm 43.5, we know that if a point z is not on the line joining the points z_0 and z_1, its image point w is determined as the intersection of two circles, one center w_0 and of radius $|z - z_0|$, and the other center w_1 and radius $|z - z_1|$. Hence there are two possible images of z. Call these points w and w'. By the preceding theorem, the assignment of w_0, w_1 and w as image points of z_0, z_1 and z determines the isometry uniquely, and similarly the assignment of w_0, w_1 and w' as the image points of z_0, z_1 and z determines the isometry uniquely. Hence there are at most two isometries which map z_0 on w_0 and z_1 on w_1, and since we know two isometries which do this, there are exactly two, and these are of the form $z' = az + b$ ($|a| = 1$) and $z' = c\bar{z} + d$ ($|c| = 1$).

We have now proved our Main Thm 43.1, which we repeat here:

Main theorem on isometries of the Gauss plane. The set of all isometries $\mathscr{I}$ of the Gauss plane is composed of two classes $\mathscr{I}_+$ and $\mathscr{I}_-$. The class $\mathscr{I}_+$ consists of all isometries of the form $z' = az + b$ ($|a| = 1$), and the class $\mathscr{I}_-$ of all isometries of the form $z' = c\bar{z} + d$ ($|c| = 1$).

Remark. We point out once again that the isometry which maps the points of the line joining z_0 to z_1 onto the points of the line joining w_0 to w_1 is unique. The fact that two distinct isometries can effect the mapping is only visible when we consider the mappings of points not on the line joining z_0 to z_1. We notice that the sense of rotation (w_0, w_1, w) is opposite to the sense of rotation (w_0, w_1, w') in the proof of the theorem above. One isometry, $z' = az + b$ ($|a| = 1$) *preserves* orientation, and the other, $z' = c\bar{z} + d$ ($|c| = 1$) *reverses* orientation. Of course, we have not proved these statements as yet. The isometry $z' = az + b$ is called a *direct* isometry, and the isometry $z' = c\bar{z} + d$ is called an *indirect* (or *opposite*) isometry.

Exercises

43.1 Find the isometries of the Gauss plane which map the point i onto the point O, and the point O onto the point 1. Which isometry also maps the point 1 onto the point $1 + i$? What is the image of the point 1 under the other isometry? Compare the orientations of the two image triangles.

43.2 l and m are two parallel lines, the points A, B lie on l, the points C, D on m, $AB = CD$, and AD intersects BC in E. Show that $AE = ED$, and $BE = EC$. Under an isometry, the points map into A', B', C', D' and E', and $A'B' = C'D'$, and $A'E' = E'D'$, while $B'E' = E'C'$. Prove that $A'B'$ must be parallel to $C'D'$, and deduce Thm 43.4.

43.3 Give another proof of Thm 43.5, using the theorem that under an isometry parallel lines map onto parallel lines. (If z is not on a side of triangle $z_0z_1z_2$, draw parallels to z_0z_1 and z_0z_2 respectively through z, intersecting z_0z_2 and z_0z_1 respectively in z'' and z'. Prove that w is uniquely determined by a reversal of this construction.)

43.4   Show that in cartesian coordinates a direct isometry is given by equations of the form

$$x' = x \cos \alpha - y \sin \alpha + p,$$
$$y' = x \sin \alpha + y \cos \alpha + q,$$

whereas an indirect isometry is given by equations of the form

$$x' = x \cos \alpha + y \sin \alpha + p,$$
$$y' = x \sin \alpha - y \cos \alpha + q.$$

Show that the inverse mappings which give (x, y) in terms of (x', y') have similar forms.

44.1   Subgroups

The axioms which determine a group have already appeared in §0.10. Here we give some more of the theory, for the reader who is not too familiar with the subject.

A *subgroup* of a given group G is a group whose elements lie in G, and whose operations are the same as those of G. For example, the rational numbers form a group under the operation of addition, and so do the integers, and the latter is a subgroup of the former. The set of even integers is a subgroup of the set of integers, and therefore also a subgroup of the set of rationals.

If we consider the set of all non-zero complex numbers, the operation now being *multiplication*, they form a group, with identity element the complex number 1, since any non-zero complex number has an inverse which is a complex number, the product of two non-zero complex numbers is also a non-zero complex number, and multiplication is associative. Within this group if we consider those complex numbers whose modulus is equal to 1, this set forms a subgroup, since $|z_1 z_2| = |z_1||z_2|$, and $|1/z| = |z|^{-1}$, shows that the product of two unimodular numbers is unimodular, and the inverse of a unimodular number is unimodular.

44.2   Conditions for a subgroup

Theorem   *A non-empty subset H of a group G is a subgroup of G if and only if for every $a, b \in H$ (i) b^{-1} and $ab \in H$, or (ii) $ab^{-1} \in H$.*

Proof. If H is a subgroup, and $a, b \in H$, then $b^{-1} \in H$, and so both ab and $ab^{-1} \in H$. Conversely, if $ab^{-1} \in H$, then $aa^{-1} = \iota \in H$, and $\iota b^{-1} = b^{-1} \in H$, and $a(b^{-1})^{-1} = ab \in H$, so that (ii) proves that H is a group. If we assume (i), $ab^{-1} \in H$, and we are back at (ii).

44.3   Cosets

We assume that G is a group with H as a subgroup. Let $a \in G$ be kept fixed, and consider the set of all elements ha, where $h \in H$ and h varies. Then the set of elements ha, denoted by $\{ha\}$, and also by Ha, is called *a right coset with respect to the subgroup H*.
    In similar fashion we define *a left coset aH*.

As examples, using the group of rationals under addition, and the subgroup of integers, a right coset consists of the set $\{r + N\}$, where r is a fixed rational, and N is an integer. The group being commutative (Abelian), right and left-cosets coincide.

In this example it is clear that two cosets $\{r + N\}$, $\{s + N\}$ have no elements in common unless $r - s$ is an integer, in which case both cosets consist of the same elements.

To take an example in which the group operation is multiplication of complex numbers, we take H as the subgroup of unimodular complex numbers, G being the group of non-zero complex numbers. Then if z_0 is a fixed non-zero complex number, the right (= left)-coset Hz_0 consists of the set of complex numbers $\{zz_0\}$, where $|z| = 1$. If w_0 is another fixed complex number, the right coset $\{zw_0\}$ is distinct from $\{zz_0\}$, having no elements in common with it, unless $z_0w_0{}^{-1}$ is unimodular, when both sets are identical. For if $zz_0 = z'w_0$, then $z_0w_0{}^{-1} = z'z^{-1}$, and since both z' and z are unimodular, so is $z'z^{-1}$. On the other hand, if $z_0w_0{}^{-1}$ is unimodular, any element zz_0 of Hz_0 can be written

$$zz_0 = (zz_0w_0{}^{-1})w_0,$$

which is an element of Hw_0, since $zz_0w_0{}^{-1}$ is unimodular with z, and similarly, every element of Hw_0 is an element of Hz_0, so that the two sets are identical.

44.4   The identity of cosets

Theorem   *If Ha and Hb have one element in common, they coincide.*

Proof. Let $c \in Ha$ and $c \in Hb$. Then $c = ha = kb$, where $h, k \in H$. From $ha = kb$ we deduce that $h^{-1}k = ab^{-1} \in H$, since both $a, b \in H$. Also $a = h^{-1}k \, b \in Hb$, and therefore $a \in Hb$. It follows that $Ha \subseteq Hb$. Also $b = k^{-1}ha$, so that $b \in Ha$, and therefore $Hb \subseteq Ha$. Hence

$$Hb = Ha.$$

Corollary. The elements $a, b \in G$ lie in the same right coset of H if and only if $ab^{-1} \in H$.

Proof. Every element of G lies in a right coset. Thus $a \in Ha$ and $b \in Hb$, since H contains the unity element ι. If these cosets coincide, so that a and b lie in the same right coset, we have seen that $ab^{-1} \in H$. On the other hand, if $ab^{-1} \in H$, so that $ab^{-1} = h \in H$, $a = hb$, and the element $\iota a = a \in Ha$ is the same as the element hb of Hb, so that the cosets Ha and Hb coincide.

We consider another example, where the elements are vectors in a two-dimensional affine space, the operation is vector addition, and the subgroup consists of all vectors parallel to the X-axis (Fig. 44.1).

In this case we denote the cosets by the symbol $H + a$, and right cosets are the same as left cosets. The endpoints of the vectors of the set $H + a$ lie on the line through the endpoint of a parallel to the X-axis. It is clear in this case that the cosets Ha and Hb are distinct, and if they have one element in common they coincide. If they do, the vector $a - b$ is parallel to the X-axis, that is, $a - b$ belongs to the subgroup H.

44.5   Right and left cosets

Theorem   *If the number of right cosets with respect to a subgroup H is finite, then there is an equal number of left cosets, and conversely.*

Proof. If $h \in H$, then $(ah)^{-1} = h^{-1}a^{-1}$, and since H is a subgroup, h^{-1} runs through all the elements of H as h does so. Hence the set $\{h^{-1}a^{-1}\}$ may be written Ha^{-1}, which

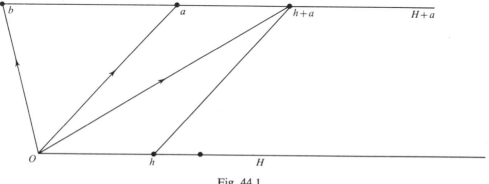

Fig. 44.1

means that the inverses of the left coset aH form a right coset Ha^{-1}. The converse is evidently true also. Since the elements of a group have a unique inverse, and cosets only have an element in common if they are identical, we see that there is a one-to-one correspondence between right and left cosets, and so, if the number of the one is finite, there is an equal number of the other.

We note that since $\iota H = H\iota = H$, the subgroup H itself is always a right or left coset, and since the cosets exhaust every element of the group, the number of distinct right or left cosets, if this is finite, called the *index* of H in G, is a useful indication of the number of elements in H, if this is finite. In fact, if the number of elements in G is finite and equal to N (this is called the *order* of G), then if the order of H is equal to n, we can show that every right coset of H contains n elements (Exercise 44.2), and since right cosets are disjoint and contain every element of G, the connexion between the index i of H and n, N is simply

$$ni = N.$$

Hence the order of a subgroup of a finite group (group of finite order) is a divisor of the order of the group.

Exercises

44.1 If in a group G with a subgroup H the relation $x \sim y$ means $xy^{-1} \in H$, where x, $y \in G$, show that $\sim$ is an equivalence relation, and that the equivalence classes of G defined by this relation are precisely the right cosets of G with respect to H. How would you obtain the left cosets?

44.2 If H is a finite subgroup of a group G, show that the number of distinct elements in any right coset Ha is always equal to the number of distinct elements in H.

45.1 Conjugate and normal subgroups

Theorem *If H is a subgroup of G, and $g \in G$, then the set of elements $K = g^{-1}Hg$ is a subgroup of G.*

Proof. If k, $h \in G$, then $g^{-1}kg$ and $g^{-1}hg$ are in K, and to show that K is a subgroup of G we must show that $(g^{-1}kg)(g^{-1}hg)^{-1} \in K$ (Thm 44.2). But $(g^{-1}hg)^{-1} = g^{-1}h^{-1}g$, and therefore

$$(g^{-1}kg)(g^{-1}hg)^{-1} = (g^{-1}kg)(g^{-1}h^{-1}g) = g^{-1}kh^{-1}g,$$

and $kh^{-1} \in H$, so that $g^{-1}kh^{-1}g \in K$.

Definition. The element $g^{-1}ag$ is called a *conjugate* of a, and $g^{-1}Hg$ is called *a subgroup conjugate* to H.

In general a subgroup conjugate to H is not contained in H. However, it may happen that H *coincides* with every one of its conjugates.

Definition. A subgroup which coincides with every one of its conjugates is called a *self-conjugate*, *invariant* or *normal* subgroup.

We shall see later (§46.4) that such subgroups are important geometrically. The definition says that $H = g^{-1}Hg$ for all elements $g \in G$. This implies

$$gH = Hg \quad \text{(all } g \in G\text{)},$$

which means that *as sets of elements* right and left cosets are identical.

Every group has two trivial normal subgroups, the group itself, and the group consisting of the identity element ι taken by itself, since $g^{-1}\iota g = g^{-1}g = \iota$.

In an Abelian group, multiplication is commutative, so that right and left cosets are identical, and therefore every subgroup is normal.

45.2 An equivalent definition of normal subgroup

Theorem *H is a normal subgroup of G if and only if $h \in H$ implies $g^{-1}hg \in H$, for all $g \in G$.*

Proof. Assume $g^{-1}hg \in H$. This means that $g^{-1}Hg \subseteq H$ for all $g \in G$. Since $g^{-1} \in G$, we also have

$$(g^{-1})^{-1}Hg^{-1} = gHg^{-1} \subseteq H, \quad \text{or} \quad H \subseteq g^{-1}Hg,$$

and therefore

$$H = g^{-1}Hg \quad \text{(all } g \in G\text{)},$$

which is the definition of a normal subgroup.

Conversely, if $H = g^{-1}Hg$, then $g^{-1}hg \in H$, for all $g \in G$. This is the criterion we shall use later when establishing that a given subgroup is normal.

45.3 Subgroups of index two

Theorem *A subgroup H of index two in G is always normal.*

Proof. Since $H\iota = \iota H = H$, H is both a right and left coset. Let $g \in G$ be an element of G which does not lie in H. Then g lies in the right coset Hg, which is distinct from H, and g also lies in the left coset gH, which is distinct from H. But since the index of H is two, each set Hg and gH covers all the elements of G which are not in H. Hence $gH = Hg$, for all $g \in G$, and so H is a normal subgroup.

45.4 Isomorphisms

We shall come across groups which appear to be different, but have the same abstract structure. We feel that the structure is the same when operations in one group can be imitated, step by step, in the other. The formal definition is as follows: Let G, G^* be two groups, the group operation in G being denoted by $\circ$, and that in G by $\circ^*$. Then the groups are said to be *isomorphic* if there exists a bijective mapping between the elements of G and G^* which is such that if a and b in G correspond to a^* and b^* in G^*, then $a \circ b$ in G corresponds to $a^* \circ^* b^*$ in G^*, for all elements a, b in G.

It is simple to prove that in this case the unity element ι of G corresponds to the unity element ι^* of G^* (Exercise 45.3), and that the inverse a^{-1} of a corresponds to the inverse a^{*-1} of a^* (Exercise 45.3).

We write $G \cong G^*$, for two isomorphic groups. This is *an equivalence relation* between groups (Exercise 45.1).

45.5 Automorphisms

An *automorphism* is an isomorphism of a group G with itself. There is the obvious identity automorphism, where every element is mapped onto itself. But there are also others.

Theorem *The mapping $\alpha : x' = g^{-1}xg$, where $g \in G$ is fixed, and $x \in G$, is an automorphism of G.*

Proof. The element x' is a uniquely defined element of G. Since $x = gx'g^{-1}$, and is uniquely defined, the mapping is one-to-one and onto G. If $y' = g^{-1}yg$, then $x'y' = (g^{-1}xg)(g^{-1}yg) = g^{-1}xyg = g^{-1}(xy)g$, and so $x'y'$ arises from the map of the element xy. The mapping is therefore an automorphism.

Definition. The mapping $\alpha : x' = g^{-1}xg$, is called an *inner* automorphism of G. All other automorphisms are called *outer* automorphisms.

Exercises

45.1 Prove that the relation of isomorphism between groups is an equivalence relation.

45.2 Prove that the group of the integers under addition is isomorphic to the group of the even integers under addition.

45.3 If G and G^* are two isomorphic groups, prove that the unity element ι of G corresponds to the unity element ι^* of G^*, and also that if a and a^* correspond, then so do a^{-1} and $(a^*)^{-1}$.

45.4 Show that the set of elements z in a group G which satisfy the relation $zg = gz$ for all elements $g \in G$ form a normal subgroup of G. (This subgroup is called *the center* of G.)

45.5 Show that the inner automorphisms $\alpha : x' = g^{-1}xg$, where $x \in G$, and $g \in H$, where H is a subgroup of G, form a group.

We now return to the one-to-one mappings of the Gauss plane, and consider their groups.

46.1 Translations

Theorem   *The translations of the Gauss plane form an Abelian group $\mathscr{T}$ which is isomorphic to the additive group of the complex numbers.*

Proof. Let $T_b : z' = z + b$ and $T_c : z' = z + c$ be two translations. The inverse of T_c is clearly given by $z' = z - c$, and if we call this translation T_c^{-1}, all that we need to show to prove the first part of the theorem is that $T_b T_c^{-1}$ is also a translation (§44.2), since the associative law holds for all mappings, and if we take $c = b$ we obtain the unity element, which is the identity mapping $z' = z$. The composition mapping $T_b T_c^{-1}$ is $z' = z + b - c$, and this is again a translation. Hence the set of translations form a group under composition, and it is clearly Abelian.

To prove the isomorphism, we merely map T_b onto the complex number b, and note that the translation $T_b T_c$ is given by $z' = z + b + c$, so that $T_b T_c = T_{b+c}$. Hence $T_b T_c$ is mapped onto $b + c$, and this establishes the isomorphism between T and the additive group of the complex numbers. Naturally, both groups are Abelian, this being a property conserved by an isomorphism.

In §41.5 we considered central dilatations, given by $z' = az$, where a is real, and dilative rotations, given by $z' = az$, where a is complex. When a is real the mapping is a stretch, in a fixed ratio, from the origin, and when a is complex, there is also a rotation about the origin. When $|a| = 1$, the mapping is a rotation only.

46.2   Dilative rotations

Theorem   *The dilative rotations of the Gauss plane about the origin form an Abelian group $\mathscr{D}$ which is isomorphic to the multiplicative group of the non-zero complex numbers.*

Proof. Composition of $D_a : z' = az$, and $D_b : z' = bz$ produces $z' = abz$. The inverse of D_a is $z' = a^{-1}z$, and the unity element is $z' = z$, when $a = 1$. If we set up the mapping $D_a \leftrightarrow a$, then $D_a D_b \leftrightarrow ab$, and so an isomorphism is established. Both groups are Abelian.

46.3   Central dilatations

Theorem   *The central dilatations form a subgroup $\mathscr{D}^*$ of $\mathscr{D}$. The rotations about the origin form a subgroup $\mathscr{R}_0$ of $\mathscr{D}$. Both $\mathscr{D}^*$ and $\mathscr{R}_0$ are Abelian.*

Proof. For central dilatations, the constant a is real, in $z' = az$, and $a \neq 0$. Then a^{-1} is real and not zero, and the product of two real numbers is real, and not zero if neither is zero. From these remarks it follows that the central dilatations form a subgroup of the group of dilative rotations.

For rotations, $z' = az$, where $|a| = 1$. If we compound this rotation with $z' = bz$, where $|b| = 1$, we obtain $z' = abz$, and since $|ab| = |a||b| = 1$, this is also a rotation. Again $|b^{-1}| = 1$, and so the rotations form a subgroup of the dilative rotations. As subgroups of an Abelian group, both $\mathscr{D}^*$, the subgroup of central dilatations, and $\mathscr{R}_0$, the subgroup of rotations, are Abelian.

We now consider the isometries of the Gauss plane, of which we know there are only two kinds (Thm 43.1). The one type, $\mathscr{I}_+$, is given by the formula $z' = az + b$ ($|a| = 1$), and the second type, $\mathscr{I}_-$ is given by the formula $z' = c\bar{z} + d$ ($|c| = 1$). We call the two types *direct* and *indirect* isometries.

46.4   Group property of isometries

Theorem   *The set $\mathscr{I}$ of isometries of the Gauss plane forms a group, with the subset $\mathscr{I}_+$ of direct isometries forming a normal (invariant) subgroup. The set $\mathscr{I}_-$ of opposite isometries forms a coset with respect to $\mathscr{I}_+$. Both $\mathscr{I}_+$ and $\mathscr{I}$ are non-Abelian.*

Before we prove this theorem, let us explain why normal subgroups are important geometrically.

Suppose that G is a group, and H a normal subgroup. Let F_1 be a geometrical figure which is mapped onto a geometrical figure F_2 by an element T of H (Fig. 46.1). (We are assuming that our groups are groups of geometrical mappings.) We write $F_2 =$

$F_1 . T$. Now let T^* be an element of G which maps F_1 onto F_1', and F_2 onto F_2'. Then
$F_2' = F_2 . T^* = F_1 . T . T^* = F_1' . (T^*)^{-1} . T . T^*$.

If H is a normal subgroup, the mapping $(T^*)^{-1} T . T^*$ is a mapping of H, and so
the property of being a map of a geometrical figure by an element of a normal subgroup
is left unchanged (invariant) under mappings by elements of the group G.

It will follow from the theorem we wish to prove, that if the triangles ABC, $A'B'C'$ are
related by a direct isometry, and we map ABC onto $A^*B^*C^*$, and $A'B'C'$ onto $A^{*'}B^{*'}C^{*'}$

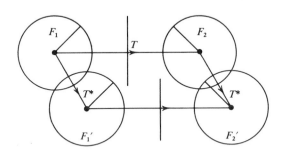

Fig. 46.1

by any isometry, then these last two triangles, $A^*B^*C^*$ and $A^{*'}B^{*'}C^{*'}$ are also related
by a direct isometry. This will be a corollary to Thm 46.4, which we now prove.

Proof. To show that the composition of two isometries is an isometry, we may use a
geometrical argument. Each isometry preserves the distance between two points, and
therefore so does their composition. The inverse of an isometry also preserves distances.
Hence $\mathscr{I}$ is a group.

If $I_1 : z' = az + b$, and $I_2 : z' = cz + d$ ($|a| = |c| = 1$) are two elements of I_+, the
composition is

$$I_1 I_2 : z' = c(az + b) + d = acz + bc + d.$$

Since $|ac| = 1$, $I_1 I_2 \in I_+$.

Again, I_1^{-1} is given by $z' = a^{-1}z - a^{-1}b$, and since $|a^{-1}| = 1$, we also have $I_1^{-1} \in \mathscr{I}_+$,
and therefore the set of direct isometries $\mathscr{I}_+$ is a subgroup of $\mathscr{I}$.

We may show that $\mathscr{I}_+$ is a normal subgroup either directly (Exercise 46.1) or by
using Thm 45.3. We therefore now prove that the set $\mathscr{I}_-$ forms a coset with respect to
$\mathscr{I}_+$.

We know that $I^* : z' = \bar{z}$ is an indirect isometry, and contained in $\mathscr{I}$. We have the
composition $I^* I_1 : z' = a\bar{z} + b$, and we know that all elements of $\mathscr{I}_-$ are of this form.
Hence $\mathscr{I}_- = I^* \mathscr{I}_+$, and so $\mathscr{I}_-$ is a left coset with respect to $\mathscr{I}_+$.

But Thm 43.1 tells us that $\mathscr{I}$ is completely made up of $\mathscr{I}_+$ and $\mathscr{I}_-$. Hence $\mathscr{I}$ is of
index two, and by Thm 45.3 we know that $\mathscr{I}_+$ is a normal subgroup.

Finally we show that $\mathscr{I}_+$, and therefore $\mathscr{I}$ are non-Abelian. If we take $I_1 : z' = az$

$(|a| = 1)$, and $T:z' = z + b$, we see that $I_1 . T:z' = az + b$, whereas $T . I_1:z' = a(z + b) = az + ab$, and for $a \neq 1$, $b \neq 0$, these compositions are different. Hence $\mathscr{I}_+$ is not Abelian, and therefore the group $\mathscr{I}$ of which it is a subgroup cannot be Abelian.

Corollary. ABC and $A'B'C'$ are triangles related by a direct isometry. If I_1 is any isometry, then the two triangles with vertices $(A, B, C) . I_1$ and vertices $(A', B', C') . I_1$ are also related by a direct isometry.

    We shall return to this concept of related triangles in the next section.

Exercises

46.1   If M be an isometry of the Gauss plane, and I_1 a direct isometry, show that $M^{-1}I_1M$ is a direct isometry, and deduce from Thm 45.2 that the direct isometries form a normal subgroup of the group of isometries (Thm 46.4).

46.2   Show that the mappings $x' = ax + b$ of the real line, where a and b are real and $a \neq 0$ form a group under composition of mappings. Prove that the elements of this group for which $a = 1$ form a normal subgroup. Is this subgroup Abelian? Is the group itself Abelian?

47.1   Similarity transformations

We have already extended the notion of central dilatation to that of dilative rotation (§41.5). We now extend the general notion of isometry.

Definition. A one-to-one onto mapping of the Gauss plane which is such that the distance between any given pair of points is changed in a fixed ratio is called *a similarity transformation*, or *similitude*.

    If the points u, v are mapped onto the points u', v', this means that $|u' - v'| = k|u - v|$, where $k > 0$ is fixed for the particular mapping. When $k = 1$, the definition overlaps that of an isometry.

    We saw that there are only two types of isometry in the Gauss plane, the one given by $z' = az + b$ $(|a| = 1)$, and the other given by $z' = c\bar{z} + d$ $(|c| = 1)$ (Thm 43.1). Let us examine the mappings $z' = az + b$, and $z' = c\bar{z} + d$, where now the complex numbers a and c are not restricted to have unit modulus. We shall see that each of these mappings is a similitude. Those of type $z' = az + b$ are called *direct* similitudes, and those of type $z' = c\bar{z} + d$ are called *opposite* similitudes.

    The geometrical significance of these types of similitude is not difficult to determine. Suppose that under a direct similitude $z' = az + b$ the points u and v are mapped onto the points u' and v'. Then $u' = au + b$, and $v' = av + b$, so that $u' - v' = a(u - v)$, and therefore $|u' - v'| = k|u - v|$, where $k = |a|$, and the vector joining the point v' to the point u' is obtained from the vector joining the point v to the point u by rotation through a fixed angle $= \arg a$ in the positive direction, from the X-axis towards the Y-axis (Fig. 47.1).

    If a triangle ABC is mapped onto a triangle $A'B'C'$ by a direct similitude, the sides

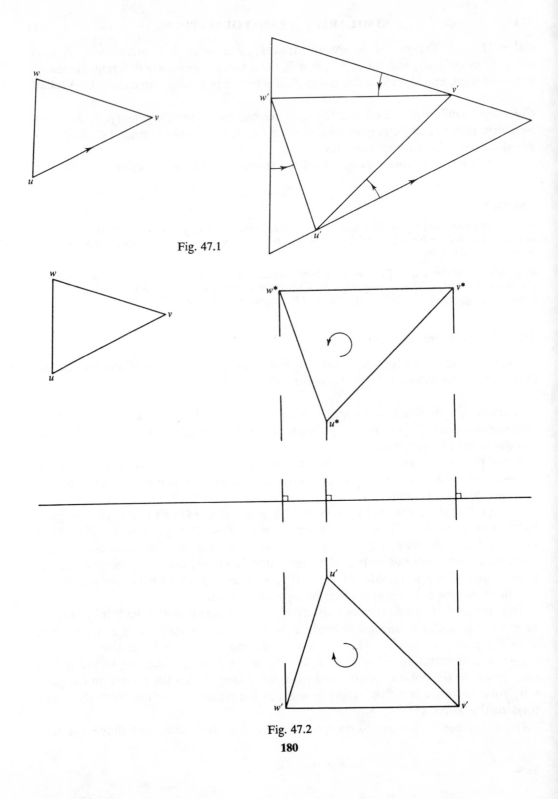

Fig. 47.1

Fig. 47.2

of $A'B'C'$ are therefore $= k$ (corresponding sides of ABC), and the sense of rotation around the vertices of the triangle ABC is the same for triangle $A'B'C'$. In other words, the two triangles are *equally oriented*. Such a pair of triangles are said to be *directly similar*, and it is no accident that this is brought about by a *direct* similitude.

Under an opposite similitude, $z' = c\bar{z} + d$, we have

$$\overline{u' - v'} = \bar{c}(u - v),$$

and although $|u' - v'| = k|u - v|$, where now $k = |c|$, the comparison between the vector joining v to u must be made with the geometrical reflexion of the vectors u' and v' in the X-axis (Fig. 47.2). This *reverses* orientation, and so the triangles ABC and $A'B'C'$ are *indirectly* similar, the sense of rotation around the vertices ABC being *opposite* to that around $A'B'C'$.

47.2 Main theorem for similitudes

Theorem *The set $\mathscr{S}$ of all similitudes of the Gauss plane is composed of two classes, $\mathscr{S}_+$ and $\mathscr{S}_-$. The class $\mathscr{S}_+$ consists of all similitudes of the form $z' = az + b$, and the class $\mathscr{S}_-$ of all similitudes of the form $z' = c\bar{z} + d$.*

The sequence of theorems which ends in the proof of the Main Theorem above runs step by step alongside the sequence given for the Main Theorem on Isometries (§43.1).

47.3 An auxiliary theorem

Theorem *Given two pairs of points z_0, z_1 and w_0, w_1, where $|z_0 - z_1| = k|w_0 - w_1| \neq 0$, there is just one mapping of type $\mathscr{S}_+$ and one of type $\mathscr{S}_-$ which maps z_0 onto w_0 and z_1 onto w_1.*

Proof. Exactly as in §43.2, except that $|a| = |c| = k$ in this case.

47.4 Similitudes are collineations

Theorem *Every similitude of the Gauss plane is a collineation.*

Proof. If u, v, w, which are collinear points, with v lying between u and w, are mapped onto u', v' and w' by a similitude, then $|u' - v'| = k|u - v|$, $|v' - w'| = k|v - w|$, and $|u' - w'| = k|u - w|$. Hence the equation

$$|u - v| + |v - w| = |u - w|$$

becomes the equation

$$|u' - v'| + |v' - w'| = |u' - w'|,$$

on multiplication by k^{-1}. But this is the condition that the points u', v' and w' should be collinear, with v' lying between u' and w'.

47.5 Similitudes and parallel lines

Theorem *A similitude of the Gauss plane maps parallel lines onto parallel lines.*

Proof. Exactly as in §43.4.

A similitude maps a triangle ABC onto a triangle $A'B'C'$ where

$$|B'C'| = k|BC|, \quad |C'A'| = k|CA| \quad \text{and} \quad |A'B'| = k|AB|.$$

Two such triangles are similar, in the ordinary sense of Euclidean geometry, and the enunciation of Thm 43.5 is therefore changed slightly.

47.6 Determination of a similitude

Theorem *A similitude of the Gauss plane is uniquely determined by the assignment of a map of a triangle which is similar to the given triangle.*

Proof. Exactly as in §43.5, with a variation along the lines of Exercise 43.1.

There has to be a slight change in the wording of Thm 43.6.

47.7 An auxiliary theorem

Theorem *There are precisely two similitudes of the Gauss plane which map two given points z_0 and z_1 onto two given points w_0 and w_1, where $|w_0 - w_1| = k|z_0 - z_1| \neq 0$.*

Proof. Along exactly the same lines as in §46.6.

This sequence of theorems leads to the Main Theorem on similitudes of §47.2.

47.8 Group property of similitudes

Theorem *The similitudes form a group $\mathscr{S}$, the direct similitudes forming a normal subgroup $\mathscr{S}_+$. The opposite similitudes form a coset $\mathscr{S}_-$ with respect to $\mathscr{S}_+$. Neither $\mathscr{S}$ nor $\mathscr{S}_+$ is Abelian. The group of isometries $\mathscr{I}$ is a normal subgroup of $\mathscr{S}$, and $\mathscr{I}_+$ is a normal subgroup of $\mathscr{S}_+$.*

Proof. Nothing new is involved in the proof, which runs along parallel lines to that in §46.4.

47.9 Condition for direct similarity

Theorem *The vertices z_1, z_2 and z_3 of a triangle are mapped by a direct similitude onto the corresponding vertices w_1, w_2 and w_3 of another triangle if and only if*

$$(z_2 - z_1)/(z_3 - z_1) = (w_2 - w_1)/(w_3 - w_1).$$

Proof. If there is a direct similitude

$$S : z' = az + b,$$

with $z_k S = w_k (k = 1, 2, 3)$ and $a \neq 0$, then

$$(w_2 - w_1)/(w_3 - w_1) = (az_2 - az_1)/(az_3 - az_1) = (z_2 - z_1)/(z_3 - z_1),$$

and so the necessity of the theorem is established.

To prove sufficiency, assume that we are given the equality of the given ratios, and consider the mapping

$$z' = \frac{w_3 - w_1}{z_3 - z_1} \; (z - z_1) + w_1.$$

This is a well-defined direct similitude, since the given conditions imply that $z_3 \neq z_1$, and $w_3 \neq w_1$. If we put $z = z_3$, then $z' = w_3 - w_1 + w_1 = w_3$. If we put $z = z_1$, the map is w_1, and if we put $z = z_2$, the map is

$$\frac{w_3 - w_1}{z_3 - z_1} \; (z_2 - z_1) + w_1 = \frac{w_3 - w_1}{w_3 - w_1} \; (w_2 - w_1) + w_1 = w_2,$$

using the given conditions. Hence we have found a direct similitude which effects the desired mapping, and the theorem is established.

47.10   Orientation

Part of the preceding theorem can be proved directly by considering the complex number represented by each of the given equal fractions, but the method given above is simpler. We have seen already that two triangles related by a direct similitude are directly similar, having the same orientation, and if the similitude is indirect, the triangles are indirectly similar (§47.1).

We have also seen that the concept of two triangles being related by a *direct isometry* is one which invariant under isometries (Corollary, Thm 46.4). This naturally extends to similitudes, and we have:

Theorem   *The points z_1, z_2 and z_3 are related to the points w_1, w_2 and w_3 by a direct similitude. If S is any direct or indirect similitude, then the triangles with vertices $z_i S$ and $w_i S$ ($i = 1, 2, 3$) are also related by a direct similitude.*

Proof. Follows directly from Thm 47.8, which proves that the direct similitudes form a normal subgroup in the group of all similitudes. The geometric content of this theorem is explained in §46.4.

Exercises

47.1   Prove that the relation, for two triangles, of one being the map of the other by a direct similitude is an equivalence relation.

47.2 Prove that a necessary and sufficient condition for the two triangles (z_1, z_2, z_3) and (w_1, w_2, w_3) to be directly similar is

$$\begin{vmatrix} 1 & 1 & 1 \\ z_1 & z_2 & z_3 \\ w_1 & w_2 & w_3 \end{vmatrix} = 0.$$

47.3 Prove that a necessary and sufficient condition for the triangle (z_1, z_2, z_3) to be equilateral is

$$z_1{}^2 + z_2{}^2 + z_3{}^2 - z_2 z_3 - z_3 z_1 - z_1 z_2 = 0.$$

47.4 Directly similar triangles BCX, CAY and ABZ are described on the sides of a triangle ABC. Show that the centers of mass of ABC and XYZ coincide.

47.5 If X, Y, Z are points on the sides BC, CA and AB of triangle ABC, such that

$$\overline{BX}/\overline{XC} = \overline{CY}/\overline{YA} = \overline{AZ}/\overline{ZB} = k,$$

and if ABC, XYZ are directly similar, prove that either $k = 1$, or both triangles are equilateral.

47.6 Prove that the square of an opposite isometry is a translation, and the square of an opposite similitude is equivalent to a translation following a dilatation from 0.

47.7 Give the proof of Thm 47.5, that a similitude of the Gauss plane maps parallel lines onto parallel lines.

47.8 Find the two similitudes which map the point $A = 0$ into the point $A' = 1 + i$, and the point $B = 1$ into the point $B' = 1 - i$. If the map of the point $C = i$ under the two similitudes be C' and C'', compare the orientations of the triangles $A'B'C'$ and $A'B'C''$ with the original triangle ABC.

We now return to consider translations and rotations from the point of view of group theory, and we shall also consider the connexion between direct isometries and rotations. We first introduce another concept of group theory.

Definition. A transformation group of the Gauss plane which contains an element which maps the origin 0 onto *any* given point z_0 is said to be a *transitive* group.

An equivalent statement is: there exists a transformation in the group which maps a given arbitrary point w onto another given arbitrary point w' (Exercise 48.1).

48.1 The group of translations

Theorem *The group $\mathscr{T}$ of translations of the Gauss plane is transitive, and a normal subgroup of the group $\mathscr{I}_+$ of direct isometries.*

Proof. $z' = z + z_0$ maps 0 onto z_0, and therefore the group $\mathscr{T}$ is transitive. Let $I_1 : z' = az + b$ $(|a| = 1)$ be a direct isometry, and let $T_d : z' = z + d$ be an element of $\mathscr{T}$. Then $I_1{}^{-1} : z' = a^{-1}(z - b)$ is the inverse of I_1, and if we calculate $I_1{}^{-1} T_d I_1$ we obtain

$$I_1{}^{-1} T_d I_1 : z' = a[a^{-1}(z - b) + d] + b = z + ad,$$

and therefore, since $z' = z + ad$ is a translation T_{ad}, we have

$$I_1{}^{-1} T_d I_1 \in \mathscr{T}$$

for any direct isometry I_1. Hence $\mathscr{T}$ is a normal subgroup of $\mathscr{I}_+$.

48.2   The fixed points of a mapping

We accept the fact that a rotation is an isometry which leaves one point of the plane unchanged, or fixed, namely, the center of the rotation. Invariant, or fixed points play an important part in the theory of mappings.

Definition. If α is a mapping of the Gauss plane, the point z is said to be an *invariant* or *fixed point* under α if $z\alpha = z$.

   A line l is said to be an *invariant line* if the points of l are mapped onto points of l

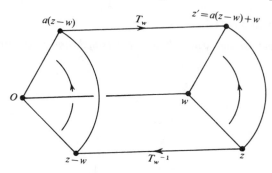

Fig. 48.1

under α. This does not mean that all points of l are fixed points. When this is the case, we shall say that l is *pointwise fixed.*

   For example, the mapping $z' = kz$, where k is real, maps every line through 0 onto itself, but 0 is the only fixed point on each line.

48.3   The fixed point of a direct isometry

Theorem   *Let I_1 be the direct isometry $z' = az + b$ ($|a| = 1$), where $a \neq 1$. Then I_1 has only one fixed point, given by $w = b(1 - a)^{-1}$, and I_1 can be written in the form $z' = a(z - w) + w$, which shows that I_1 is a rotation about w.*

Proof. The restriction $a \neq 1$ is to avoid the case of I_1 being a translation, which has no fixed points (see Fig. 48.1). If w is a fixed point under I_1, then

$$w = aw + b,$$

so that $w = b(1 - a)^{-1}$. Since $a \neq 1$, this is a well-defined point. If we subtract the equation for w from the equation of the mapping, we have

$$z' - w = a(z - w).$$

The geometrical meaning of this equation is clear. If we move the origin O to the point w, and write

$$Z' = z' - w, \quad Z = z - w,$$

which is a parallel transformation of axes, we have

$$Z' = aZ$$

for the direct isometry I_1. But this is a *rotation* about w. Conversely, a rotation about a point w is given by $z' - w = a(z - w)$, and is a direct isometry. This leads to

48.4 Direct isometries and rotations

Theorem *Every direct isometry which is not a translation is a rotation about the fixed point of the isometry.*

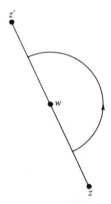

Fig. 48.2

We note that the complex number a is uniquely defined by the isometry, and conversely $Z' = aZ$ defines a direct isometry. We have therefore obtained a *standard*, *normal* or *canonical* form for direct isometries (all these terms are used). Any direct isometry not a translation can be transformed into the form $Z' = aZ$ by a parallel transformation of axes, and conversely a mapping of this form is a direct isometry.

If the mapping $z' = az$ is called R_a, and the translation $z' = z + w$ is called T_w, the inverse T_w^{-1} is $z' = z - w$, and

$$T_w^{-1} R_a T_w \text{ is } z' = a(z - w) + w$$

so that $I_1 = T_w^{-1} R_a T_w$, which is the group-theoretic expression of the statement made about the canonical form for a direct isometry.

Definition. A *half-turn* about a point w is a rotation about w through an angle π (see Fig. 48.2). Referred to w as origin, the equation of this mapping is $z' = -z$. Referred to O, the equation is $z' = -z + 2w$.

Intuitively, we feel that rotations about any point of the plane are similar to each other, and this is expressed in group-theoretical terms by the next theorem.

48.5 Rotation groups

Theorem *The set R_w of rotations about the point w form a group $\mathscr{R}_w$, and the groups $\mathscr{R}_w$, for all values of w, are isomorphic.*

Proof. We have just shown, in the preceding section, that every isometry which is a rotation about the point w can be written as $I_1 = T_w^{-1}RT_w$, where R is a rotation about the point O. Let $I_2 = T_w^{-1}R'T_w$ be another rotation about the point w. To show that the set of rotations about w form a subgroup of the group $\mathscr{I}$ of isometries, we have to show that $I_2 I_1^{-1}$ is also a rotation about w. But

$$I_2 I_1^{-1} = T_w^{-1}R'T_w(T_w^{-1}RT_w)^{-1} = T_w^{-1}R'T_wT_w^{-1}R^{-1}T_w = T_w^{-1}(R'R^{-1})T_w,$$

and since the rotations about O form a group (Thm 46.3), $R'R^{-1}$ is a rotation about O, and therefore $I_2 I_1^{-1}$ is a rotation about w.

We can set up a bijective mapping between rotations about O and rotations about the point w by mapping R onto $T_w^{-1}RT_w$. Calling this mapping β,

$$R\beta = T_w^{-1}RT_w = I_1,$$

and it is clearly bijective. Since we have just shown that

$$I_1 I_2 = (R\beta)(R'\beta) = (RR')\beta,$$

(by a slight modification of the algebra given above), we have shown that the mapping β is an isomorphism. Hence every group $\mathscr{R}_w \cong \mathscr{R}_O$, and so the $\mathscr{R}_w$ are isomorphic to each other.

Corollary. The rotation groups $\mathscr{R}_w$ form a complete set of conjugate subgroups of $\mathscr{R}_O$ within the group of all isometries of the Gauss plane.

Proof. In §45.1 the group $g^{-1}Hg$, where H is a subgroup of G, was defined as a subgroup conjugate to H within G. We have just seen that $\mathscr{R}_w = T_w^{-1}\mathscr{R}_O T_w$, so that $\mathscr{R}_w$ is a conjugate subgroup. We now wish to show that both $I_1^{-1}\mathscr{R}_O I_1$ and $I_2^{-1}\mathscr{R}_O I_2$, where I_1 is a direct and I_2 an indirect isometry, are rotation groups $\mathscr{R}_w$ for a particular point w.

Let $I_1: z' = az + b$ ($|a| = 1$) be a direct isometry, and let I_2 be $z' = c\bar{z} + d$ ($|c| = 1$). If the element R of $\mathscr{R}_O$ is $z' = kz$ ($|k| = 1$), then

$$I_1^{-1}RI_1 \text{ is } z' = a[k(a^{-1}(z - b)] + b = kz - bk + b,$$

whereas

$$T_b^{-1}RT_b \text{ is } z' = k(z - b) + b = kz - bk + b,$$

so that

$$I_1^{-1}RI_1 = T_b^{-1}RT_b,$$

and the elements of the conjugate subgroup $I_1^{-1}RI_1$ are identical with the elements of
the rotation group $\mathscr{R}_b$.
Again,

$$I_2^{-1}RI_2 \text{ is } z' = c[\bar{k}c^{-1}(z - d)] + d = \bar{k}z - \bar{k}d + d,$$

whereas

$$T_d^{-1}R'T_d \text{ is } z' = \bar{k}(z - d) + d = \bar{k}z - \bar{k}d + d,$$

where R' is the rotation $z' = \bar{k}z$. Here also the elements of the conjugate subgroup
$I_2^{-1}RI_2$ are identical with the elements of the rotation group $\mathscr{R}_d$. We have therefore
shown that rotation groups $\mathscr{R}_w$ are conjugate subgroups of $\mathscr{R}_o$ within the group of
isometries of the Gauss plane, and moreover that all conjugate subgroups of $\mathscr{R}_o$ within
the group of isometries are rotation groups.

Exercises

48.1 Prove that a transformation group G of the Gauss plane contains an element which maps the
origin 0 onto any assigned point z_0 if and only if it contains an element which maps a given arbitrary
point w onto another given arbitrary point w_0.

48.2 Which of the following transformation groups is transitive?
(a) $\mathscr{I}$, the group of isometries of the Gauss plane
(b) $\mathscr{T}$, the group of translations
(c) $\mathscr{R}_w$, the group of rotations about a point w
(d) $\mathscr{S}$, the group of similitudes
(e) $\mathscr{D}$, the group of dilative rotations about the origin.

48.3 Show that the product of a half-turn about the point w and a half-turn about the point w' ($\neq w$)
is a translation T_d, where $d = 2(w' - w)$.

48.4 Show that a translation T_d can always be regarded as a product of two half-turns about points
w and w', where $d = 2(w - w')$ and either of the points w, w' can be chosen arbitrarily.

48.5 Show that the product of the three half-turns $H_1H_2H_3$, where

$$H_i: z_i = -z_{i-1} + 2w_i \quad (i = 1, 2, 3),$$

gives the half-turn

$$z_3 = -z_0 + 2(w_1 - w_2 + w_3),$$

about the point w_4, where

$$w_4 - w_3 = w_1 - w_2.$$

48.6 Deduce, from the theorem that the product of three half-turns is a half-turn, that $H_1H_2H_3 = H_3H_2H_1$. (If H is a half-turn, $H^2 = $ Identity.)

49.1 The fixed points of indirect isometries

We have already seen, in §41.3, that the indirect isometry $z' = a\bar{z}$ ($|a| = 1$) is a reflexion
of the point z in the straight line through O which makes an angle (am a)/2 with the
X-axis. Hence the given indirect isometry has a whole line of fixed points, which serve
as a mirror for the reflexion. It is understood that the mirror is a special one, in which
reflexion is possible from either side. (Fig. 49.1).

We now consider the question of possible fixed points for the more general indirect isometry $z' = a\bar{z} + b\,(|a| = 1)$. If w is an invariant point, we have $w = a\bar{w} + b$, and on taking the conjugate of both sides of this equation, $\bar{w} = \bar{a}w + \bar{b}$. Hence

$$w = a(\bar{a}w + \bar{b}) + b = a\bar{a}w + a\bar{b} + b = w + a\bar{b} + b,$$

since $a\bar{a} = |a|^2 = 1$.

Hence $a\bar{b} + b = 0$ *is a necessary condition for the existence of a fixed point of the indirect isometry* $z' = a\bar{z} + b$.

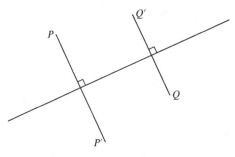

Fig. 49.1

The condition is also sufficient, since if it holds we can prove that $w = b/2$ is an invariant point. In fact

$$a\bar{w} + b = a\bar{b}/2 + b = (a\bar{b} + b)/2 + b/2 = b/2 = w.$$

We pursue this case further.

49.2   Indirect isometries as reflexions

Theorem   *An indirect isometry* $z' = a\bar{z} + b$ *has invariant points if and only if* $a\bar{b} + b = 0$. *If it has an invariant point, it is a line reflexion, and has a whole line of invariant points. Every line reflexion is an indirect isometry.*

Proof. We have proved the first part of this theorem in the preceding section. We assume that the given isometry has a fixed point w, so that $w = a\bar{w} + b$. Then if $z' = a\bar{z} + b$, we subtract the first equation from the second and have:

$$z' - w = a(\bar{z} - \bar{w}) = a(\overline{z - w}),$$

a form we have already encountered. If T_w be the translation $z' = z + w$, and M_a is the mapping $z' = a\bar{z}$, then the given indirect isometry can be written in the form

$$T_w^{-1}M_aT_w.$$

This is the usual group-theoretic form obtained by a parallel transformation of axes. Writing $Z = z - w$, $Z' = z' - w$, our indirect isometry is in the form:

$$Z' = a\bar{Z} \quad (|a| = 1)$$

and this is a mirror reflexion in the line through the point w which makes an angle $(\text{am } a)/2$ with the X-axis. Hence the indirect isometry is a line reflexion if it has any invariant points at all.

Conversely, suppose that a reflexion M in a line m be given (Fig. 49.2), let w be a point on m, and let M_a be the reflexion in the line l through O which is parallel to m. Then

$$M = T_w^{-1} M_a T_w$$

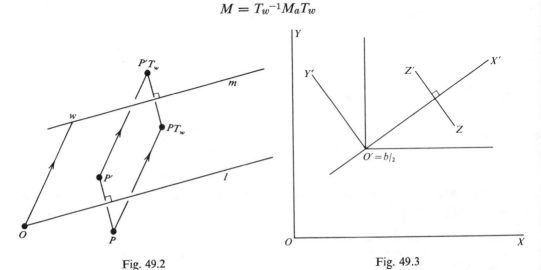

Fig. 49.2   Fig. 49.3

leaves invariant all the points of m. We know, from Thm 43.2, that there is just one direct isometry and one indirect isometry which map the points of the line m onto themselves. In this case the direct isometry is the identity mapping $z' = z$, and M is the uniquely determined indirect isometry which leaves the points of m, and only these points, invariant.

49.3 Canonical form for reflexions

Our geometrical intuition tells us that if we are given an indirect isometry with a fixed point, so that we have a line reflexion, we may shift the coordinate axes parallel to themselves to pass through one of the fixed points of the isometry, and then rotate the axes until the line of fixed points falls along the new X-axis. The equation of the isometry must then be $Z' = \bar{Z}$, and since any indirect isometry with fixed points can be put in this form, and since conversely this equation always represents an indirect isometry with fixed points, we have a normal, or canonical form for indirect isometries with fixed points (see Fig. 49.3).

We know that one fixed point of the mapping is the point $b/2$, the mapping being $z' = a\bar{z} + b$, and the line in which reflexion takes place makes an angle $(\operatorname{am} a)/2$ with the X-axis. A complex number with this amplitude is $\sqrt{a}$. Hence, if we follow the steps necessary for the change of coordinate axes (see §42.2), we find that $z = Z/\sqrt{\bar{a}} + b/2$ is the required transformation. Since $a\,\bar{a} = 1$, and therefore $\sqrt{a}\sqrt{\bar{a}} = 1$, we can simplify this transformation to $z = Z\sqrt{a} + b/2$. The mapping in the new co-ordinates is therefore

$$z' = Z'\sqrt{a} + b/2 = \overline{a(Z\sqrt{a} + b/2)} + b = \bar{Z}\sqrt{a} + b/2,$$

since $a\bar{b} + b = 0$, and so we are led to $Z' = \bar{Z}$, as we expected.

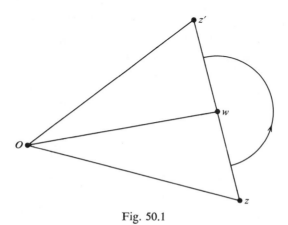

Fig. 50.1

Exercises

49.1   An indirect isometry $z' = a\bar{z} + b$ has invariant points if and only if $a\bar{b} + b = 0$, and it is then a line reflexion (Thm 49.2). By considering the image of the point 0, deduce that the point $b/2$ is a fixed point in this case. Prove also that the line in which reflexion takes place is at right angles to the join of 0 to the point b.

49.2   Show that the indirect isometry given by $z' = i\bar{z} + (1 - i)$ is a line reflexion, and carry out the coordinate transformation which leads to the equivalent transformation $Z' = \bar{Z}$.

50.1   Involutory isometries

A line reflexion M is evidently such that if $zM = z'$, then $z'M = z$, which leads to $zM^2 = z$, so that M^2 is the identity mapping. A mapping with this property is called an *involutory* mapping, and such a mapping divides the points of the Gauss plane into *pairs* (z, z'), each member of the pair being object to the other's image under the mapping: thus, $zM = z'$ and $z'M = z$.

Another involutory isometry of the Gauss plane is *a half turn about a point*. If the point be w (Fig. 50.1), the equation of this mapping is $z' - w = -(z - w)$, since the point z is rotated about the point w through an angle π. The equation shows that $z + z' = 2w$, as we expect, the midpoint of the join of z to z' being the fixed point w. It is clear that lines through the point w are left invariant by this mapping (Exercise 50.4). The following theorem shows that the examples just given exhaust the involutory isometries of the Gauss plane.

Theorem *Every involutory isometry of the Gauss plane is either a line reflexion, a half-turn or the identity.*

Proof. We consider the only two possible types of isometry of the Gauss plane, take the square, and equate to the identity. For a direct isometry $z' = az + b$ ($|a| = 1$), the square is $z' = a(az + b) + b = a^2z + ab + b$. If this is the identity, $z' = z$, then $a^2 = 1$, and $ab + b = 0$. Hence either $a = 1$, and $b = 0$, which gives the identity, or $a = -1$, and b is unrestricted, which gives the mapping $z' = -z + b$, or $z' - b/2 = -(z - b/2)$, which is a half-turn about the point $b/2$.

For the indirect isometry $z' = a\bar{z} + b$ ($|a| = 1$), the square is

$$z' = a(\overline{a\bar{z} + b}) + b = a\bar{a}z + a\bar{b} + b = z + a\bar{b} + b,$$

and if this is to be $z' = z$, we must have $a\bar{b} + b = 0$, which is the condition that the indirect isometry should be a *line reflexion* (Thm 49.2).

The above proof shows that the algebra of complex numbers can be very effective in proving certain types of theorem about direct and indirect isometries. In the following theorem, we obtain once again a theorem we have already encountered when we considered reflexions in lines from the geometrical point of view, that is, without the use of algebra (§14.2).

50.2 The composition of two reflexions

Theorem *The composition of reflexions in lines l and m results in* (i) *a translation if and only if the lines are parallel,* (ii) *a rotation about the point of intersection if and only if the lines intersect.*

Proof. Let the reflexion in l be given by $z' = a\bar{z} + b$, with $|a| = 1$ and $a\bar{b} + b = 0$, and the reflexion in m by $z' = c\bar{z} + d$, with $|c| = 1$, and $c\bar{d} + d = 0$. The composition of the two reflexions is given by

$$z' = c(\overline{a\bar{z} + b}) + d = \bar{a}cz + \bar{b}c + d.$$

The first thing we remark is that the sequence of the two reflexions is a *direct* isometry. There are two possibilities:
(i) $\bar{a}c = 1$. Then the composition $z' = z + \bar{b}c + d$ is a *translation*. From the condition $\bar{a}c = 1$, we have $a\bar{a}c = a$, or $c = a$, since $a\bar{a} = |a|^2 = 1$. Hence $(\text{am } a)/2 = (\text{am } c)/2$,

so that the lines l and m have the same slope with regard to the X-axis (§49.1). In this case l is parallel to m.

Conversely, if l is parallel to m, $a = c$ (since $|a| = |c| = 1$), and $\bar{a}c = 1$, and the composition of reflexions is a translation.

(ii) $\bar{a}c \neq 1$. Then the composition $z' = \bar{a}cz + \bar{b}c + d$ is a direct isometry which is *not* a translation, and therefore it is a rotation about the fixed point of the mapping (Thm 48.3). The fixed point is obtained by putting $z' = z$ in the equation of the mapping, and is the point $(\bar{b}c + d)/(1 - \bar{a}c)$. To prove that this is the point of intersection of l and m, we show that *both* reflexions leave this point invariant (Exercise 50.1). For the converse, we note that if l is not parallel to m, then $a \neq c$, and $\bar{a}c \neq 1$.

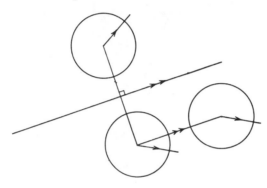

Fig. 50.2

50.3 Opposite isometries without fixed points

We have not yet investigated opposite isometries without fixed points. In this case the mapping is $z' = a\bar{z} + b$, and we do *not* have $a\bar{b} + b = 0$, although we still have $|a| = 1$. Since $z' = a\bar{z}$ is a line reflexion M_a, and $z' = z + b$ is a translation T_b, we may write the given opposite isometry as the composition M_aT_b, that is, as a line reflexion followed by a translation.

Such a mapping could be called a *reflexion glide*. It is also a *glide reflexion* if the order of composition does not affect the final mapping, that is, if $M_aT_b = T_bM_a$. For this to be the case

$$a(\overline{z + b}) = a\bar{z} + b,$$

which leads to $a\bar{b} = b$, or $a = b/\bar{b}$. From this we deduce that am $a = 2$ (am b), and since the direction of the line in which reflexion takes place is given by (am a)/2, we see that the translation T_b is then *parallel* to the line in which reflexion takes place.

Definition. The mapping which is a composition of reflexion in a line l followed by a translation *parallel* to the line l is called a *glide reflexion* (see Fig. 50.2). The operations are commutative.

It is an interesting fact that any reflexion glide can be shown to be equivalent to a glide reflexion, the new line in which reflexion takes place being parallel to the original line. We give an algebraic proof of this theorem, and indicate a geometric proof in Exercise 50.3.

Theorem   *A reflexion in a line l followed by a translation T_b results in an opposite isometry without invariant points, provided that l is not perpendicular to the position*

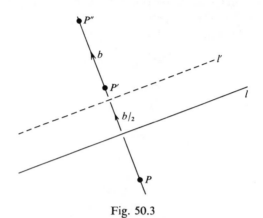

Fig. 50.3

vector of b. An opposite isometry without fixed points is equivalent to a glide reflexion, that is to a reflexion in a line followed by a translation parallel to the line.

Proof. If the mapping $M_a T_b$, which is expressed as $z' = a\bar{z} + b$, has fixed points, then $a\bar{b} + b = 0$. Then $\operatorname{am} a - \operatorname{am} b = \operatorname{am}(-b) = \operatorname{am} b \pm \pi$, and so $(\operatorname{am} a)/2 = \operatorname{am} b \pm \pi/2$. Since $\operatorname{am} a/2$ is the angle between the *X*-axis and the line *l*, it follows that *l is perpendicular to the vector b.*

If *l* is perpendicular to *b*, we can reverse the steps of the argument (remembering that $|a| = 1$), and find that $a\bar{b} + b = 0$, so that the opposite isometry we are considering is *a line reflexion* (Thm 49.2). This is also clear directly. We are reflecting in a line *l* (Fig. 50.3), and then moving the image point through a distance $|b|$ perpendicular to the line.

Moving the line *l* through a distance $|b|/2$ produces a line *l'* in which reflexion produces the same final image point. More precisely, let *u* be any point on *l*. Then the reflexion in *l* carries $u + b/2$ into the point $u - b/2$, and T_b carries this point into $u + b/2$, the original point. Hence the mapping has a line of invariant points.

We now prove the second part of the theorem, and consider $z' = a\bar{z} + b$ where $|a| = 1$ but $a\bar{b} + b \neq 0$. We may write

$$a\bar{z} + b = a\overline{(z - b/2)} + b/2 + (a\bar{b} + b)/2.$$

Writing $d = (a\bar{b} + b)/2 \neq 0$, the mapping can now be expressed as a composition of the translations $T_{b/2}$, T_d and the reflexion M_a, thus:

$$T_{b/2}{}^{-1}M_aT_{b/2}T_d,$$

since $zT_{b/2}{}^{-1}$ produces $z - b/2$, and $(z - b/2)M_a$ produces $a(\overline{z - b/2})$, and the subsequent translations $T_{b/2}, T_d$ add $b/2$ and d to this result. Now, the mapping $T_{b/2}{}^{-1}M_aT_{b/2}$ is merely a reflexion in a line which passes through the point $b/2$ and makes an angle (am $a)/2$ with the X-axis (§49.2). Hence our indirect isometry without fixed points is equivalent to a reflexion in such a line followed by a translation with vector d. We show that the vector d is parallel to the line in which reflexion takes place, that is we show that 2 am d = am a.

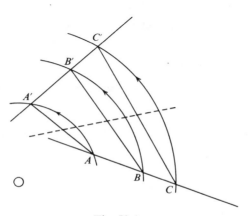

Fig. 50.4

We have $(d/|d|)^2 = d^2/dd = d/\bar{d} = (a\bar{b} + b)/(\bar{a}b + \bar{b}) = a$, and so am $(d/|d|)^2 =$ 2 am $d/|d|$ = 2 am d = am a, which completes the proof.

An indirect isometry is therefore either a line reflexion (Thm 49.2) or a glide reflexion. In the first case it has fixed points, in the second case it has no fixed points.

50.4   Hjelmslev's theorem

Theorem I *The points P on a line are mapped by a plane isometry onto the points P' of another line. Then the midpoints of the segments PP' either coincide, or they are distinct and collinear.* (Fig. 50.4)

Proof. A direct isometry is either a translation, or it is a rotation about a fixed point. In the first case the points $A, B, C, \ldots$ are mapped onto the points $A', B', C', \ldots$ of a parallel line, and the midpoints of segments AA', BB', CC', $\ldots$ lie on a line. In the case of a rotation through an angle $\neq \pi$, we have to show that the midpoints of AA', BB', CC', . . . lie on a line. If the angle of rotation is π, so that the mapping is a half-turn, the midpoints all coincide with the center of rotation.

If the isometry is indirect, it is either a reflexion in a line l, in which case it is clear that the midpoints of AA', . . . lie on l, or it is a glide-reflexion, a reflexion in l followed by a translation parallel to l, and again it is clear that the midpoints of AA', . . . lie on l. Hence the only part of the proof which is not immediate is the case when the isometry is a rotation about a point.

We can prove the theorem algebraically, and not only cover all cases, but generalize the theorem for isometries, leaving an extension of the theorem to similitudes for the reader in Exercise 50.2. Let the points A, B, C, . . . be represented by the complex numbers a, b, c, . . ., and the points A', B', C', by the complex numbers a', b', c', ... If c is the complex number which represents any point C on the line AB, we may write $c = ta + (1 - t)b$, where t is real, and the ratio of the real numbers $1 - t$: t is equal to the ratio of the signed segments $\overline{AC} : \overline{CB}$ (Thm 3.1). Let c' be the map of c under the direct isometry $z' = pz + q$ ($|p| = 1$). Then

$$c' = pc + q = p(ta + (1 - t)b) + q$$
$$= t[pa + q] + (1 - t)[pb + q]$$
$$= ta' + (1 - t)b'.$$

Now suppose that k, k' are real numbers, with $k + k' = 1$. Then

$$k'c + kc' = k'[ta + (1 - t)b] + k[ta' + (1 - t)b']$$
$$= t[k'a + ka'] + (1 - t)[kb' + k'b].$$

We suppose that k, k' are fixed. If $k'a + ka' \neq k'b + kb'$, the points $k'c + kc'$ move on a line, and are distinct for distinct values of t. If

$$k'a + ka' = k'b + kb',$$

then $$k'c + kc' = k'a + ka' = k'b + kb',$$

and all the points which divide the join CC' in the ration $k:k'$ coincide. The condition for this is $k'(a - b) = - k(a' - b')$, which can be expressed thus:

$$k'\overrightarrow{AB} = - k\overrightarrow{A'B'},$$

where A, B are any two points on the one line, A', B' the corresponding points on the other. The lines are then parallel, and since we are dealing with an isometry, we must also have $k = k'$.

The algebra for an opposite isometry is very similar, the crucial point being that t and $1 - t$ are real numbers, and therefore unchanged when we take the conjugate. We have therefore proved

Theorem II The points P on a line are mapped by an isometry onto the points P'

of another line. Then the points which divide the segments PP' in the fixed ratio $k:k'$ either coincide, or they are all distinct, collinear, and trace out a range similar to the ranges $\{P\}$ and $\{P'\}$. Coincidence is only possible if the given lines are parallel, and $k = k'$.

Remark The importance of the original Hjelmslev Theorem (Theorem I above) lies in the fact that it holds in *absolute* geometry, geometry in which the parallel axiom of Euclid is not invoked. See Eves **(5)**, Vol. I, p. 362. The extension of the theorem given as

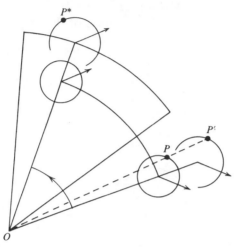

Fig. 50.5

Theorem II above, and suggested for similitudes in Exercise 50.2, has not appeared in print before, but Prof. Howard Eves has informed me that these extensions have also been known to him for some time.

50.5   Direct and opposite similitudes

We now consider the mappings $z' = az + b$, and $z' = c\bar{z} + d$, where for the first mapping $|a| \neq 1$, and for the second mapping $|c| \neq 1$. These are the direct and opposite (or indirect) similitudes proper, and we have already looked at some of their properties in §§ 47.1–9.

A *direct similitude always has a fixed point*. If w be a possible fixed point, then $w = aw + b$, which gives $w(1 - a) = b$. Since $a \neq 1$, this gives a unique value for w, and substitution shows that it is indeed a fixed point. On subtraction of $w = aw + b$ from $z' = az + b$, we have

$$z' - w = a(z - w),$$

and moving the coordinate axes parallel to themselves to pass through the fixed point reduces the equation of a direct similitude to the form $Z' = aZ$. Since $|a| \neq 1$ this is not just a *rotation* about the fixed point. If the fixed point be O (Fig. 50.5), the point P

is *stretched* to P', where $|OP'| = k|OP|$, $k = |a|$, and then P' is *rotated* about O through the fixed angle am a. It is clear that the order in which the two operations are carried out does not matter. We can call the mapping, as in §41.5, a *dilative rotation*, or a *spiral similarity*, and we have proved that *all proper direct similarities* $(|a| \neq 1)$ *are spiral similarities*.

Let us consider opposite similitudes $z' = c\bar{z} + d$ $(|c| \neq 1)$. *This mapping always has a fixed point*. If it be w, then $w = c\bar{w} + d$, from which $\bar{w} = \bar{c}w + \bar{d}$, $w = c(\bar{c}w + \bar{d}) + d$, which leads to $w(1 - c\bar{c}) = cd + d$. Since $c\bar{c} = |c|^2 \neq 1$, w is uniquely defined. On subtracting the equation for w from the equation of the mapping we have,

$$z' - w = c(\overline{z - w}),$$

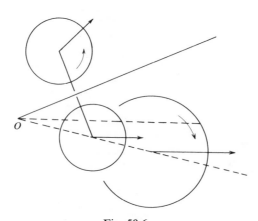

Fig. 50.6

and moving the origin to the fixed point gives the equation

$$Z' = c\bar{Z}$$

for the mapping.

This is a new kind of mapping, a *dilative-reflexion*. If O be the fixed point (Fig. 50.6), the point P is reflected in the line through O which makes an angle (am c)/2 with the X-axis, and the resulting point P' is stretched from O so that $|OP''| = k|OP'|$, where $k = |c|$. Here also, the operations *reflexion* and *stretching* are commutative. Hence *all opposite similitudes which are proper* $(|c| \neq 1)$ *are dilative reflexions*.

Exercises

50.1 Show that the point $(\bar{b}c + d)/(1 - \bar{a}c)$ is a fixed point both for the mapping $z' = a\bar{z} + b$, where $|a| = 1$ and $a\bar{b} + b = 0$, and for the mapping $z' = c\bar{z} + d$, where $|c| = 1$ and $c\bar{d} + d = 0$.

50.2 Let $r = tp + (1 - t)q$, where p and q are complex numbers, and t is real. Show that the image of the point r under any direct or opposite similitude is $r' = tp' + (1 - t)q'$, where p' is the image of p and q' the image of q. Generalize Hjelmslev's Theorem (§50.4).

50.3 Give a geometrical proof that a reflexion in a line l_1 followed by a translation parallel to a line l' can be replaced by reflexion in a line l^* parallel to l_1 followed by a translation parallel to l^*. (This is proved algebraically in §50.3.) (Hint: the translation can be replaced by successive reflexions in lines l_2 and l_3 perpendicular to l'. The total mapping is then effected by reflexions in l_1, l_2 and l_3, in succession. The reflexions in the first two lines can be replaced by reflexions in lines l_1' and l_2' through the intersection of l_1 and l_2, where l_2' is perpendicular to l_3. The reflexions in l_2' and l_3 can be replaced by reflexions in l_2^* and l_3^*, where l_2^* is parallel to l_1' and l_3^* is perpendicular to l_1', and therefore parallel to l_1. Reflexion in l_1' and then in l_2^* = translation.)

50.4 Show that the half-turn about the point w given by $z' = -z + 2w$ leaves lines through the point w invariant.

50.5 The equation of a line in the real Euclidean plane is $AX + BY + C = 0$, where A, B and C are real. Show that the equation in z is of the form

$$Ez + \bar{E}\bar{z} + F = 0,$$

where E is complex, and F is real.

50.6 The point z is mapped onto the point z' by the indirect isometry

$$z' = a\bar{z} + b \qquad (|a| = 1).$$

Show that the midpoint Z of the segment defined by z and z' moves on the line:

$$2Z(1 - \bar{a}) + 2\bar{Z}(1 - a) = b + \bar{b} - (a\bar{b} + \bar{a}b).$$

50.7 Verify that the line found in the preceding Exercise passes through the point $b/2$, and makes an angle φ with the X-axis, where $\tan \varphi = \tan(\theta/2)$, and am $a = \theta$.

50.8 Why is the result of Exercise 50.6 to be expected from Thm 49.2 and 50.3?

51.1   Line reflexions

Theorem *Every product of three line reflexions is either a line reflexion or a glide reflexion.*

Proof. Let the three reflexions be given by the mappings $z' = a_i\bar{z} + b_i$ $(i = 1, 2, 3)$. Since each mapping must have at least one fixed point, we also have the three conditions $a_i\bar{b}_i + b_i = 0$, but we shall not have to use these equations. The composition (or product) of the three mappings is given by

$$z' = a_3[\overline{a_2(a_1\bar{z} + b_1) + b_2}] + b_3 = a_1\bar{a}_2a_3\bar{z} + \bar{a}_2a_3b_1 + a_3\bar{b}_2 + b_3,$$

and this is an opposite isometry. Since we know (Thm 49.2, Thm 50.3) that an opposite isometry is either a line reflexion (when it has at least one fixed point) or a glide reflexion (when it has no fixed points), the theorem is proved.

We have already remarked (Thm 14.2) that a product of two reflexions is equivalent to a rotation about the point of intersection of the two lines in which we are reflecting, through an angle $2\hat{A}$, where $\hat{A}$ is the angle of intersection. If the lines are parallel, with no intersection, the product is equivalent to a translation (Thm 14.3). These theorems appeared again in Thm 50.2, in a form which stressed that the product is a translation if and only if the lines are parallel, and a rotation about the point of intersection if and only if the lines intersect.

Reflexions form the basis for a fairly new approach to geometry. *Aufbau der Geometrie aus dem Spiegelungsbegriff*, F. Bachmann (Springer, Berlin), is a recent text on the subject.

One of the important theorems in this approach is the condition that two given lines be perpendicular, in terms of reflexions in the lines. Thm 14.2 tells us that if the lines are perpendicular, the composition of reflexions is equivalent to a rotation through π about the point of intersection of the lines, and in this case the order of reflexion in the lines is immaterial to the final result: the operations are *commutative*. Conversely, if the operations are commutative, either the lines coincide, or if they are distinct they cannot be parallel, since reflexion in a line l followed by reflexion in a parallel line m is equivalent to a translation in the opposite sense to the translation given by reflexion in the line m followed by reflexion in the line l. The lines must therefore intersect, and if they do the condition for commutativity is

$$2\hat{A} = 2(\pi - \hat{A}),$$

where $\hat{A}$ is the angle of intersection, which gives us $\hat{A} = \pi/2$, as desired. Let us see how an algebraic proof of this theorem looks. (We made use of this result in Exercise 25.2.)

51.2   Condition for two lines to be perpendicular

Theorem   *Two distinct lines l and m are perpendicular to each other if and only if the product of the reflexions in l and m is involutory, and not the identity.*

Remark. An involutory mapping M was defined in §50.1 as being such that $M^2 =$ Identity. This is equivalent to the statement that $M = M^{-1}$. If M is the product of reflexions in l and m, so that $M = M_l M_m$, then the condition $M = M^{-1}$ is equivalent to $M_l M_m = M_m^{-1} M_l^{-1} = M_m M_l$, since a line reflexion is involutory, and equal to its own inverse. Conversely, if the product of reflexions is commutative, so that $M_l M_m = M_m M_l$, then $M = M_l M_m = M^{-1}$, and the product of the reflexions is involutory.

Proof. Let l be perpendicular to m, and let w be the point of intersection (see Fig. 51.1). From §49.2 we know that we can write the reflexion in l as $T_w^{-1} M_a T_w$, and the reflexion in m as $T_w^{-1} M_c T_w$, where T_w is the translation $z' = z + w$, M_a is a reflexion $z' = a\bar{z}$, and M_c is a reflexion $z' = c\bar{z}$, the lines l and m making angles (am a)/2 and (am c)/2 respectively with the X-axis. The product is

$$M = T_w^{-1} M_a T_w T_w^{-1} M_c T_w = T_w^{-1} M_a M_c T_w,$$

and this is to be proved involutory. Hence all we have to show is that the product $M_a M_c$ is involutory, where M_a is $z' = a\bar{z}$ and M_c is $z' = c\bar{z}$, so that the product is $z' = a\bar{c}z$.

Since l is perpendicular to m,

$$(\text{am } a)/2 - (\text{am } c)/2 = \pm \pi/2,$$

and therefore

$$\text{am } (a/c) = \pm \pi.$$

Since $|a| = |c| = 1$, $a/c = \cos \pi \pm i \sin \pi = -1$, and so $a = -c$. Hence M is given by

$$z' = c[a\overline{(z - w)}] + w = \bar{a}cz - \bar{a}cw + w = -z + 2w.$$

This is *a reflexion in the point* w, and therefore involutory, and not the identity.

To prove the converse of the theorem, we suppose that M is involutory, and not the

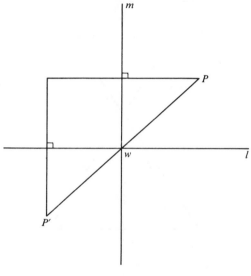

Fig. 51.1

identity. By Thm 14.3, l and m are not parallel, since if they were, the product of reflexions in l and m would be a non-zero translation, and this is not involutory. Let l and m intersect in the point w. As we have seen, we may write M in the form

$$z' = \bar{a}cz - \bar{a}cw + w.$$

This is a direct isometry, and since we are assuming that it is involutory, Thm 50.1 tells us that it must be a point-reflexion (half-turn) or the identity. Hence $\bar{a}c = -1$, since $z' - w = -(z - w)$. But from $\bar{a}c = -1$ we find that $(\text{am } a)/2 - (\text{am } c)/2 = \pm \pi/2$, and therefore l is perpendicular to m.

The above theorem can certainly be proved by pure geometry, but the next, also used in the Spiegelungsbegriff geometry, is rather harder without algebra.

51.3   The incidence of a point and a line

Theorem   *The product of the reflexions in a point w and in a line l is involutory if and only if w lies on l.*

Proof. The necessity of the theorem is clear geometrically (see Fig. 51.2). If w lies on l the order of the operations: reflexion in w, reflexion in l does not affect the product of the operations. Since we shall need the algebra for the converse of the theorem, which is not as clear by pure geometry, we prove the necessity algebraically also. Let

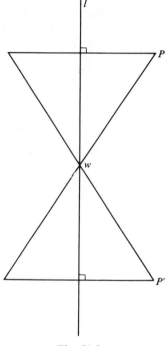

Fig. 51.2

the reflexion in l be $M_l:z' = a\bar{z} + b$, where $a\bar{b} + b = 0$, and the reflexion in w be $M_w:z' = -z + 2w$. Then the product $M_w M_l$ is M, say, where M is given by

$$z' = a\overline{(-z + 2w)} + b = -a\bar{z} + 2a\bar{w} + b.$$

Assume now that w lies on l. Then w is invariant under M_l: that is, $w = a\bar{w} + b$. Hence we may write for M^2: $z' = -a\overline{(-a\bar{z} + 2w - b)} + 2w - b$, that is

$$z' = z - 2a\bar{w} + a\bar{b} + 2w - b$$
$$= z - (2a\bar{w} + 2b) + (b + a\bar{b}) + 2w$$
$$= z - 2w + 0 + 2w = z,$$

and therefore M is involutory.

Conversely, suppose that $M^2 = $ Identity. Since M is an opposite isometry, and is involutory, Thm 50.1 tells us that M must be *a line reflexion*. The condition for $z' = p\bar{z} + q$ to have a fixed point, which is sufficient to make it a line reflexion, is $p\bar{q} + q = 0$ (§49.1), which in our case, where $z' = -a\bar{z} + 2a\bar{w} + b$ becomes

$$-a(\overline{2a\bar{w} + b}) + 2a\bar{w} + b = 0,$$

or

$$-2w - a\bar{b} + 2a\bar{w} + b = 0 \quad \text{(since } |a| = 1\text{)}.$$

Writing this equation in the form

$$-2w - (a\bar{b} + b) + 2a\bar{w} + 2b = 0,$$

and using the condition $a\bar{b} + b = 0$, we end up with

$$-2w + 2a\bar{w} + 2b = 0,$$

or

$$w = a\bar{w} + b,$$

which shows that w is invariant under the mapping M_l: that is, w lies on l.

51.4   Isometries as reflexions

Theorem *Every isometry is the product of at most three reflexions. If the isometry has a fixed point, at most two line reflexions produce the isometry.*

Remark. A translation has no fixed points, and can be expressed as the product of two line reflexions (in parallel lines). The second part of the theorem does not therefore have a converse.

Proof. This can be geometric–algebraic, or purely geometric. If the isometry is direct, it is either a translation or a rotation (Thm 48.4). We have seen that a translation can be expressed as the product of reflexions in two parallel lines, perpendicular to the vector which gives the direction of the translation (Exercise 14.2), and that a rotation of angle $2A$ about a point O can be effected by reflexion in two lines intersecting at O at an angle A (Exercise 14.1). Hence a direct isometry can be expressed as the product of two reflexions.

If the isometry is of the opposite type, it is either a line reflexion, when it has fixed points, or if it has no fixed points it is a glide reflexion, which is a reflexion followed by a translation parallel to the line of reflexion (§50.1 and §50.3). Since a translation is equivalent to two line reflexions, a glide reflexion is equivalent to three line reflexions. We have therefore proved the first part of the theorem. If the isometry does have a fixed point, it must be a rotation or a reflexion in a line, and so not more than two reflexions are necessary.

We complete the proof, and this chapter, by giving the algebraic proofs that a translation is expressible as the product of reflexions in parallel lines, and that a rotation is expressible as the product of reflexions in two lines through the center of rotation.

If the translation is $z' = z + b$, we assume that $b \neq 0$, since the identity mapping $z' = z$ can be effected by reflexions in the same line, taken twice. Hence we can solve the equation $a\bar{b} + b = 0$ for a. Having done this, consider the mapping $z' = a\bar{z} + b$, which is a reflexion, since $a\bar{b} + b = 0$. We know that it can be written in the form $T_{b/2}^{-1} M_a T_{b/2}$ (§49.2), since $w = b/2$ is always a fixed point of the mapping, that is a point on the line making an angle (am a)/2 with the X-axis in which reflexion takes place. Then M_a itself is given by $z' = a\bar{z}$. If we form the composition of $z' = a\bar{z} + b$ with $z' = a\bar{z}$, we have $z' = a(\overline{a\bar{z}}) + b = z + b$, since $|a| = 1$. Hence we have found two parallel lines, at a distance $|b|/2$ apart, such that reflexion in one followed by reflexion in the other is equivalent to the translation $z' = z + b$. The lines, of course, are perpendicular to the direction of the vector b.

In the rotational case, the isometry is $T_w^{-1} R_a T_w$, where w is the center of rotation, and R_a is the rotation $z' = az$. We may write this in the form

$$T_w^{-1} R_a T_w = (T_w^{-1} M_1 T_w)(T_w^{-1} M_a T_w),$$

where M_1 is the reflexion $z' = \bar{z}$, and M_a is the reflexion $z' = a\bar{z}$, and we do have $R_a = M_1 M_a$ since $a\bar{\bar{z}} = az$. Hence the rotation about a point w has been expressed as the product of reflexions in two lines through w.

Exercises

51.1 Show that the dilatation and reflexion that give their names to the dilative reflexion $z' = c\bar{z}$ ($|c| \neq 1$), are commutative.

51.2 If the lines p, q, r all pass through a point, show that the product $M_p M_q M_r$ of line-reflexions in them is a line-reflexion in a line which also passes through the point.

51.3 Show that if the lines p, q and r are all parallel, the product $M_p M_q M_r$ of line-reflexions in them is a line-reflexion in a parallel line.

51.4 Show that the product of a line-reflexion $z' = a\bar{z} + b$, where $|a| = 1$ and $a\bar{b} + b = 0$, and a half-turn about the point w, $z'' = -z' + 2w$, where w is not a fixed point of the first mapping, produces an indirect isometry without fixed points, that is, a glide-reflexion.

51.5 If ABC is a triangle, and M_{BC}, M_{CA} and M_{AB} denote reflexions in the sides, show that the product $M_{BC} M_{CA} M_{AB}$ is equivalent to the product $M_1 M_2 M_{AB}$, where M_1 and M_2 are both line-reflexions in lines which pass through C, and M_2 is the reflexion in a line perpendicular to AB. Replacing $M_2 M_{AB}$ by a half-turn about the point F in which the perpendicular from C to AB intersects AB, deduce, from the previous Exercise, that the product of reflexions in lines which form a triangle is equivalent to a glide-reflexion.

51.6 Show that the product of line-reflexions in three distinct lines is never equivalent to the identity.

51.7 ABC is an acute-angled triangle, and D, E, F are the feet of the perpendiculars from A, B and C onto the opposite sides. The transformation $M_{BC} M_{CA} M_{AB}$ has been shown (Exercise 51.5 above) to be a glide-reflexion. Show that the *axis* (the line in which reflexion takes place) is the line DF, and that the translation is in the direction $\overrightarrow{DF}$ and equal in magnitude to the perimeter of triangle DEF. (The axis is the only line mapped onto itself by the transformation. Remembering that DF and DE make equal angles, in opposite senses, with BC, it follows that the line DF, after reflexion in BC, becomes the line DE. Continue thus. To obtain the magnitude of the translation, and its direction, consider the map of the point D on the axis. If this be D', the vector $\overrightarrow{DD'}$ gives the translation. Now, the image of D in BC is D. The image of D in CA is D_1, where D_1 lies on EF, and $ED_1 = ED$, while $|D_1F| = |DE| + |EF|$. The final image point D' lies on DF, and its distance from F, on the side opposite to D, is $|DE| + |EF|$.)

51.8 Show that Exercise 48.5 is equivalent to the following result: $H_A H_B H_C H_D = $ Identity, if and only if the points A, B, C, D about which the half-turns operate form the vertices of a parallelogram.

51.9 Show that an opposite isometry can only be involutory, (so that its square is the identity), if it is a line reflexion.

51.10 From the result of Exercise 51.5, show that $(M_p M_q M_r)^2 = $ Identity if and only if the lines p, q, r form a pencil, which means that either they all pass through a point, or they are all parallel.

51.11 From the result of Exercise 48.3, which says that the product of a half-turn about the point w and a half-turn about the point w' is a translation T_d, where $d = 2(w' - w)$, show that $H_C H_B H_C H_A = $ Identity if and only if C is the midpoint of AB.

51.12 If $P' = (P)H_A$, and $Q = (P)M_f$, while $Q' = (P')M_f$, show that $Q' = (Q)M_f^{-1}H_A M_f$, and that the mapping which produces Q' from Q is also a half-turn. Deduce that $M_f H_B M_f H_A = $ Identity if and only if the line f is the perpendicular bisector of the segment AB.

51.13 Find a product of line-reflexions equivalent to the Identity which gives a necessary and sufficient condition for a line f to be the bisector of the angle between two lines p and q, if they intersect, or to be half-way between them if they are parallel.

51.14 ABC, $A'B'C'$ are two congruent triangles. Show geometrically that a composition of at most three line-reflexions suffices to map ABC onto $A'B'C'$. Deduce Theorem 51.4.

VI

MAPPINGS OF THE INVERSIVE PLANE

We now consider a larger group of mappings of the extended Gauss plane, named after Moebius, and to do so we extend the plane of the complex numbers by adjoining one additional point, which we call *the point at infinity*, and denote by the symbol ∞.

We stipulate that $1/z = \infty$, when $z = 0$, and calculations with the symbol ∞ will be performed according to the following rules:

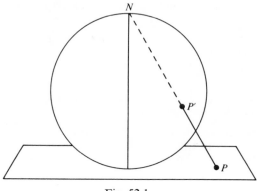

Fig. 52.1

For any finite complex number c,

$$c \pm \infty = \infty, \qquad c/\infty = 0,$$

and for non-zero c,

$$c \cdot \infty = \infty, \qquad c/0 = \infty.$$

The expressions $\infty - \infty$, ∞/∞, $0 \cdot \infty$ and $0/0$ are considered to be undefined.

Definition. The Gauss plane, with the point ∞ adjoined to it, is called *the inversive plane*.

Remark. If we recall that the complex number plane, the Gauss plane, is in one-to-one correspondence, by means of stereographic projection, with the surface of a punctured sphere, a sphere from which the point N of projection has been removed, all that we are doing now is to take the whole surface of the sphere as a map of the plane of complex numbers, adding the point N, to which we attach the symbol ∞ (Fig. 52.1).

52.1   Moebius transformations

These are transformations of the inversive plane given by

$$z' = \frac{az + b}{cz + d},$$

where a, b, c and d are complex numbers with the restriction that

$$\triangle = ad - bc \neq 0.$$

The number $\triangle$ is called the *determinant* of the mapping.

    We do not intend to give a detailed account of these mappings. Our purpose in this chapter is to develop the theory sufficiently to discuss the Poincaré model of a hyperbolic non-Euclidean geometry. To do this we shall have to use many of the ideas we have introduced in preceding chapters.

52.2   The determinant of a Moebius mapping

Let us assume that in the expression $(az + b)/(cz + d)$, the number $c \neq 0$. Then we may write

$$\frac{az + b}{cz + d} = \frac{a}{c} - \frac{(ad - bc)}{c(cz + d)}.$$

If $\triangle = ad - bc = 0$, our mapping would be $z' = a/c$, a mapping of the whole inversive plane onto one point. If $c = 0$, the mapping is $z' = (az + b)/d$, which is a dilative rotation if $d \neq 0$. But if $\triangle = 0$ also, $ad = 0$. If $d = 0$, all points of the inversive plane are mapped onto the one point $z' = \infty$, and if $a = 0$, $z' = b/d$. If both $a = 0$ and $d = 0$, $z' = \infty$. Hence we see that there is a very sound reason for insisting that $\triangle \neq 0$, since we wish to consider non-degenerate mappings. Degenerate mappings, in which all points of the inversive plane are mapped onto one point, are of little interest.

    We now see that the condition $\triangle \neq 0$ is sufficient for a Moebius transformation to be a one-to-one onto mapping of the inversive plane, a bijective mapping. In the first place $(az + b)/(cz + d)$ defines a unique z' unless $cz + d = 0$. But if $cz + d = 0$, so that $z = -(d/c)$, we cannot also have $az + b = 0$ for this value of z, since this would imply that $a(-d/c) + b = 0$, or $\triangle = ad - bc = 0$. Hence $cz + d = 0$ gives $z' = $ (non-zero constant)/0, and we have agreed that this gives $z' = \infty$.

    The point $z = \infty$ has a unique image-point, since

$$\frac{az + b}{cz + d} = \frac{a + b/z}{c + d/z} = \frac{a}{c}$$

when $z = \infty$, by our rules. Hence every point of the inversive plane maps onto a uniquely defined point of the inversive plane.

The inverse mapping is given by $z = (-dz' + b)/(cz' - a)$, and this is again a Moebius transformation, since the determinant of this mapping is

$$(-d)(-a) - bc = ad - bc = \triangle \neq 0.$$

Hence the object-point z of an image-point z' is uniquely determined, and since every point z of the inversive plane has an image z', and every point z' of the inversive plane arises from a uniquely defined point z of the inversive plane, the mapping given by our Moebius transformation $z' = (az + b)/(cz + d)$ is one-to-one and onto, in other words, *a bijective mapping*.

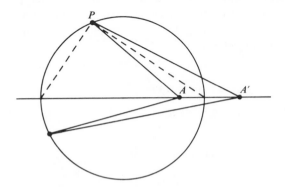

Fig. 52.2

By choosing special values of the constants a, b, c and d, we obtain a number of mappings which we have already encountered. But we can also *dissect* the given mapping into a product of familiar mappings. Before we do this, we examine the effect of a Moebius mapping on lines and circles.

52.3   The map of a circle or line under a Moebius transformation

Theorem   *A Moebius transformation is a circular transformation, that is, it maps the set of circles and lines into the set of circles and lines.*

Proof. We can prove this theorem and an additional one by taking the equation of a circle or line in the z-plane in the form $|z - p| = k|z - q|$, where k is a real number (Fig. 52.2). This is an expression of the theorem that any circle is the locus of a point which moves so that the ratio of its distances from two fixed points is a constant, these fixed points being inverse points for the circle. When the locus is a line, $k = 1$, and the points are mirror images in the line. The circle is called an Apollonius circle with respect to the two points (see §18.3). If we now transform this equation by the Moebius mapping $Z = (az + b)/(cz + d)$, we must write $z = (-dZ + b)/(cZ - a)$, and we obtain the equation

$$\left| \frac{-dZ + b}{cZ - a} - p \right| = k \left| \frac{-dZ + b}{cZ - a} - q \right|$$

and, after a little reduction this leads to

$$|Z(-d - pc) + b + pa| = k|Z(-d - qc) + b + qa|$$

or

$$\left| Z - \frac{ap + b}{cp + d} \right| = k \left| \frac{d + qc}{d + pc} \right| \left| Z - \frac{aq + b}{cq + d} \right|,$$

which is of the form

$$|Z - p'| = k'|Z - q'|,$$

where p', q' are the maps of the points p and q under the Moebius transformation, and $k' = k|d + qc|/|d + pc|$.

The form of the equation for Z shows that the transform is either a line or a circle, and we also have:

Corollary. The map of a circle and a pair of inverse points under a Moebius transformation is a circle and a pair of inverse points, where if the circle is a line, the points are mirror images in the line.

Other proofs of this theorem and Corollary are indicated in Exercises 52.1, 52.3.

52.4   Dissection of a Moebius transformation

We may write

$$z' = \frac{az + b}{cz + d} = \frac{a}{c} + \frac{(bc - ad)}{c(cz + d)} = \frac{a}{c} + \frac{(bc - ad)}{c^2} \left(\frac{1}{z + d/c} \right).$$

We now write

(i) $T_{d/c} : z_1 = z + d/c,$

(ii) $N : z_2 = 1/z_1,$

(iii) $D : z_3 = \frac{(bc - ad)}{c^2} z_2,$

(iv) $T_{a/c} : z' = a/c + z_3.$

The mapping (i) is a translation, (ii) is an inversion in the unit circle, center at the origin, followed by a reflexion in the real axis, (iii) is a dilative rotation, and (iv) is also a translation. If we give names to these familiar mappings as indicated, then

$$z' = \frac{az + b}{cz + d} = zT_{d/c}NDT_{a/c}.$$

In other words, a general Moebius transformation is the composition of a translation, an inversion in the unit circle followed by a reflexion in the X-axis, a dilative rotation and a translation. Many important consequences follow from this dissection, as the exercises following indicate.

Of course, a Moebius transformation is not, in general, a direct isometry, nor a direct similitude, although these mappings are included amongst Moebius transformations as special cases.

52.5   The group of Moebius transformations

Theorem   *Moebius transformations form a group $\mathscr{B}$ under composition of mappings. If B and C are two Moebius mappings, Δ_B and Δ_C their determinants, then $\Delta_{BC} = \Delta_B\Delta_C$ is the determinant of the Moebius mapping BC.*

Proof. We have seen already that if B is a Moebius transformation, so is B^{-1}. The identity mapping $z' = z$ is a Moebius transformation. If $B:z' = (az + b)/(cz + d)$, $\Delta_B = ad - bc$, $C:z' = (a'z + b')/(c'z + d')$, $\Delta_C = a'd' - b'c'$ are two Moebius transformations, then the composition is

$$BC:z' = \frac{a'(az + b) + b'(cz + d)}{c'(az + b) + d'(cz + d)} = \frac{(aa' + b'c)z + a'b + b'd}{(ac' + cd')z + bc' + dd'},$$

and $\Delta_{BC} = (aa' + b'c)(bc' + dd') - (a'b + b'd)(ac' + cd') = (ad - bc)(a'd' - b'c')$, after some cancellation, so that $\Delta_{BC} = \Delta_B\Delta_C$, as we wished to prove, and since $\Delta_B \neq 0$, $\Delta_C \neq 0$, $\Delta_{BC} \neq 0$, and so the composition is also a Moebius transformation. The associative law holds for composition of bijective mappings, and therefore all the group axioms are satisfied, and the theorem is proved.

Exercises

52.1   By looking at the dissection of a Moebius transformation show that a Moebius transformation maps lines onto lines or circles, and circles onto lines or circles.

52.2   Show that the mapping $z' = 1/z$ maps the points $2, 1 + i, 1 - i$ of the circle $|z - 1| = 1$ onto the collinear points $1/2, 1/2 - i/2, 1/2 + i/2$. Can you explain why this is so?

52.3   Give another proof of the Corollary to Thm 52.3, by examining the effect of the component parts of a Moebius transformation.

52.4   In the proof of Thm 52.3, the condition that the map of our circle or line be a line is $k' = k|d + qc|/|d + pc| = 1$. Show that an equivalent condition is that the point $z = -d/c$ which maps onto the point $Z = \infty$ lie on the circle $|z - p| = k|z - q|$. Why is this equivalence to be expected?

52.5   Show that the condition that $Z = (az + b)/(cz + d)$ should map the circle $|Z| = 1$ onto a line in the z-plane is $|a| = |c|$.

52.6   If a given Moebius transformation maps a circle $\mathscr{C}$ onto a circle $\mathscr{C}'$, show that a point P moving continuously on $\mathscr{C}$ is mapped onto a point P' which moves continuously on $\mathscr{C}'$, but that clockwise motion on $\mathscr{C}$ may be changed into anti-clockwise motion on $\mathscr{C}'$.

52.7 By Thm 52.5, the determinant of the mapping inverse to $z' = (az + b)/(cz + d)$ should be $(ad - bc)^{-1}$, whereas the mapping $z = (-dz' + b)/(cz' - a)$, which is the inverse mapping, has determinant $ad - bc$. Can you explain this?

53.1 Fixed points of a Moebius transformation

Theorem *If a Moebius transformation has more than two distinct fixed points, it is the identity mapping, $z' = z$.*

Proof. If w is a fixed point, then $w = (aw + b)/(cw + d)$. This leads to the quadratic equation $cw^2 + (d - a)w - b = 0$. If this equation has more than two distinct roots, it must be the identical equation $0 \cdot w^2 + 0 \cdot w + 0 = 0$, so that $c = d - a = b = 0$. Then the mapping is $z' = z$, since $\Delta = ad \neq 0$, and so neither a nor d is zero, and $z' = az/d = z$.

It is sometimes possible to say that a given mapping is uniquely determined when a certain number of distinct points and their images under the mapping are given. We came across theorems of this nature with isometries (Thm 43.5). A Moebius transformation depends on the ratios of four constants, a, b, c, d, subject to the conditions $ad - bc \neq 0$, and we expect such a transformation to be given if three distinct points and their images are assigned. But a stronger theorem is true.

53.2 The determination of a Moebius transformation

Theorem *A Moebius transformation is uniquely determined by the assignment of three distinct points z_j and their three distinct image points z_j' $(j = 1, 2, 3)$.*

Proof. Consider the mapping $B: z' = (z - z_2)(z_1 - z_3)/(z - z_1)(z_2 - z_3)$. This is a Moebius transformation, since its determinant is

$$\Delta_B = (z_1 - z_3)(z_2 - z_3)(-z_1 + z_2) \neq 0.$$

Under the mapping B, the image of $z = z_1$ is $z' = \infty$, the image of $z = z_2$ is $z' = 0$, and the image of $z = z_3$ is $z' = 1$.

Now define a mapping $C: z' = (z - z_2')(z_1' - z_3')/(z - z_1')(z_2' - z_3')$. Here the determinant is

$$\Delta_C = (z_1' - z_3')(z_2' - z_3')(-z_1' + z_2') \neq 0,$$

so that this is also a Moebius transformation, and under this mapping the points z_1', z_2' and z_3' are mapped, respectively, onto the points ∞, 0 and 1. The composition of transformations BC^{-1} is again a Moebius mapping, and maps z_1 onto ∞ and then onto z_1', z_2 onto 0 and then onto z_2', and finally z_3 onto 1 and then onto z_3'.

Hence BC^{-1} is a Moebius transformation which maps z_1, z_2, z_3 onto the points z_1', z_2', z_3'. Is it unique? If A were a distinct Moebius mapping with the same properties, the Moebius mapping $(BC^{-1})A^{-1}$ would have three distinct fixed points, z_1, z_2 and z_3, and by Thm 53.1, the transformation $(BC^{-1})A^{-1} =$ Identity, which means that $BC^{-1} = A$.

We are on ground familiar to those who know something about projective geometry (§69.1), and without saying so, we have been using the concept of *cross-ratio*.

53.3   The cross-ratio of four complex numbers

Definition. The cross-ratio of the four points (or complex numbers) z_j ($j = 1, 2, 3, 4$) is the complex number

$$(z_1, z_2; z_3, z_4) = \frac{(z_1 - z_3)}{(z_2 - z_3)} \bigg/ \frac{(z_1 - z_4)}{(z_2 - z_4)}$$

$$= \frac{(z_1 - z_3)(z_2 - z_4)}{(z_2 - z_3)(z_1 - z_4)}.$$

It will be noticed that the order of the points is significant, and in fact there are, on the face of it, $4! = 24$ different values of the cross-ratio of four given points, if the order of the points is not assigned. But these fall into six sets, since there are relations between the values which arise after a permutation of the points. The full details will be given later (§69.1). Here we need one property of the permutations:

Theorem $(z_1, z_2; z_3, z_4) = (z_2, z_1; z_4, z_3)$.

Proof. This is simple verification:

$$(z_2, z_1; z_4, z_3) = \frac{(z_2 - z_4)}{(z_2 - z_3)} \bigg/ \frac{(z_1 - z_4)}{(z_1 - z_3)} = \frac{(z_1 - z_3)(z_2 - z_4)}{(z_2 - z_3)(z_1 - z_4)}$$

$$= (z_1, z_2; z_3, z_4).$$

Remark. The above is a special case of the theorem that *interchanging one pair* of the four numbers *and also the other pair* leaves the cross-ratio unchanged (compare Exercise 69.1).

One reason for the importance of the cross-ratio of four numbers is its *invariance* under any Moebius transformation. We prove this in the next section, but remark here that we proved in §23.2 that the cross-ratio of four collinear points is unchanged if we *invert* with respect to a center of inversion on the line. Now, inversion by itself is not a Moebius mapping, but *inversion followed by reflexion* in the line on which the points lie is a Moebius mapping, given by $z' = 1/z$, and the points are unchanged by the reflexion in the line. Hence we can regard the theorem of §23.2 as a very special case of the one we shall now prove.

53.4   An invariant under Moebius transformations

Theorem   *The cross-ratio of four points is an invariant under Moebius transformations.*

Proof. This can be shown by direct verification. If the four points are $z_j (j = 1, 2, 3, 4)$, and these are mapped onto the points z_j', where $z_j' = (az_j + b)/(cz_j + d)$, then straightforward calculation shows that

$$\frac{(z_1 - z_3)(z_2 - z_4)}{(z_2 - z_3)(z_1 - z_4)} = \frac{(z_1' - z_3')(z_2' - z_4')}{(z_2' - z_3')(z_1' - z_4')}.$$

Alternatively, we may use the method of Thm 53.2. With the notations of that theorem, we have $z_j BC^{-1} = z_j'$, which means that $z_j B = z_j' C$. If we write out what this means in algebraic terms, we have

$$\frac{(z_j - z_2)(z_1 - z_3)}{(z_j - z_1)(z_2 - z_3)} = \frac{(z_j' - z_2')(z_1' - z_3')}{(z_j' - z_1')(z_2' - z_3')},$$

and writing $z_j = z_4$, and $z_j' = z_4'$, we have the desired result. Of course, the proof of Thm 53.2 was devised with a previous knowledge of the cross-ratio function in mind.

We now prove a very important geometric property of the cross-ratio.

53.5   The reality of the cross-ratio

Theorem *The cross-ratio of four distinct points is real if and only if the four points lie on a straight line or a circle.*

Remark. This theorem can be proved geometrically, more or less, by a method indicated in Exercise 53.3. The proof we give here is purely algebraic.

Proof. Let the four distinct points be $z_j (j = 1, 2, 3, 4)$, and suppose in the first instance that the cross-ratio $(z_1, z_2; z_3, z_4) = k$, where k is real. Let T be the unique Moebius transformation which maps z_1, z_2 and z_3 onto the three distinct points z_1', z_2' and z_3' of the real axis (Thm 53.2). Let $z_4 T = z_4'$. By the invariance of the cross-ratio (Thm 53.4), we have

$$\frac{(z_4 - z_2)(z_1 - z_3)}{(z_4 - z_1)(z_2 - z_3)} = \frac{(z_4' - z_2')(z_1' - z_3')}{(z_4' - z_1')(z_2' - z_3')} = k.$$

Hence

$$(z_4' - z_2')/(z_4' - z_1') = k(z_2' - z_3')/(z_1' - z_3')$$

$$= \text{a real number.}$$

Since z_1', z_2', z_3' are real, if we solve this equation for z_4', we find that $z_4' = $ a real number, since the real numbers form a field. Hence $z_4 T = z_4'$ also lies on the real axis. The four points z_1, z_2, z_3 and z_4 therefore lie on the image of (real line)T^{-1}. But we know that a Moebius transformation maps a line onto a circle or a line. Since T^{-1} is a Moebius transformation, and the real line is a line, it follows that the four points $z_j (j = 1, 2, 3, 4)$ lie on a circle or a line when the cross-ratio of the four points is real.

Now let us assume that the four given points lie on a line or a circle. There is a unique Moebius transformation T which maps z_1, z_2 and z_3 into three given distinct points z_1', z_2', z_3' of the real axis, and since T maps circles and lines onto circles or lines the map of the line or circle which contains the four points z_j must be a line or circle through the three points z_1', z_2', z_3'. Since these points lie on a line, the real axis, the map must be this line. Hence the map of z_4 is also a point z_4' on the real axis. The cross-ratio of the four points z_j' on the real axis is a real number, and since cross-ratios are invariant under a Moebius transformation, it follows that the cross-ratio $(z_1, z_2; z_3, z_4)$ of the four points on a line or a circle is the same real number.

53.6   Lines in the inversive plane and the point ∞

The point ∞ was added to the Euclidean plane, considered as a representative of the complex line, as soon as we introduced inversion in §20.1. We then found that inversion is a one-to-one onto transformation of the completed plane, which we call the inversive plane. In §21.1 it was seen to be reasonable to consider all lines of the inversive plane to pass through this point ∞. The theorem that the inverse of a line which does not pass through the center of inversion is a curve (a circle, in fact) which does pass through the center of inversion is then an immediate deduction. When we consider Moebius transformations of the Gauss plane, inversion is the one component of the dissected Moebius transformation which is not one-to-one onto, and by adding the point ∞ to the Gauss plane, obtaining the inversive plane, we ensure that a Moebius transformation is one-to-one onto for the inversive plane.

We have already mentioned stereographic projection, which gives a one-to-one correspondence between the points of the Gauss plane and the points of a sphere from which one point has been removed. Adding the point ∞ to the Gauss plane is equivalent to patching the hole in the punctured sphere. We shall see later (Exercise 88.1) that circles in the Gauss plane project into circles on the sphere, if we project from the point N (Fig. 52.1), the Gauss plane being regarded as the tangent plane at the point on the sphere diametrically opposite to N. If we join N to a line in the Gauss plane, the intersection of the resulting plane with the sphere is a circle which passes through N. The inversive plane can be regarded as the surface of a sphere, the point N being the point ∞. This was already remarked at the beginning of the Chapter. The further point made here is that lines in the inversive plane all pass through the point ∞.

Exercises

53.1 Show that any Moebius transformation with the two distinct fixed points p and q may be written in the form

$$\frac{z' - p}{z' - q} = k\frac{(z - p)}{(z - q)},$$

where k is a complex number.

53.2 Deduce from Thm 53.5 that a Moebius transformation maps the set of lines and circles into the set of lines and circles.

53.3 If the complex numbers z_1, z_2, z_3, z_4 are attached to the points A, B, C and D respectively of the real Euclidean plane, prove that the cross-ratio

$$(z_1, z_2; z_3, z_4) = k(\cos \theta + i \sin \theta),$$

where $k = |AC| \cdot |BD|/|AD| \cdot |BC|$, and $\theta = \alpha - \beta$, where α and β are the measures of the angles CAD and CBD respectively. Deduce Thm 53.5.

53.4 Show that the points of any given circle in the inversive plane may be represented thus:

$$z = (at + b)/(ct + d) \qquad (ad - bc \neq 0),$$

where a, b, c, and d are complex numbers, and t is real.

53.5 Prove that if z_1, z_2, p, q and r are distinct complex numbers, the following relation between cross-ratios is true:

$$(z_1, z_2; p, r) = (z_1, z_2; p, q) \cdot (z_1, z_2; q, r).$$

53.6 Prove that

$$(z_1, z_2; z_4, z_3) = (z_2, z_1; z_3, z_4) = 1/(z_1, z_2; z_3, z_4).$$

53.7 Show that if the points z_1, z_2, z_3 and z_4 which lie on a circle are mapped by a Moebius transformation onto the points A, B, C and ∞ of a line, then

$$(z_1, z_2; z_3, z_4) = (A, B; C, \infty) = \overline{AC}/\overline{BC}.$$

53.8 Prove the following theorem for cross-ratios:

$$\frac{(z_1, w_2; z_2, w_1)}{(z_2, w_3; z_3, w_2)} \cdot \frac{(z_3, w_4; z_4, w_3)}{(z_4, w_1; z_1, w_4)}$$

$$= (z_1, z_3; z_2, z_4) \cdot (w_1, w_3; w_2, w_4).$$

53.9 Prove that if $\mathscr{C}_1$, $\mathscr{C}_2$, $\mathscr{C}_3$ and $\mathscr{C}_4$ are four circles (or lines), and $\mathscr{C}_1$, $\mathscr{C}_2$ intersect in P_1, Q_1, the circles $\mathscr{C}_2$, $\mathscr{C}_3$ intersect in P_2, Q_2, the circles $\mathscr{C}_3$, $\mathscr{C}_4$ intersect in P_3, Q_3, and finally the circles $\mathscr{C}_4$, $\mathscr{C}_1$ intersect in P_4, Q_4, then if the four points P_i lie on a line or circle, so do the four points Q_i. (Use the preceding Exercise. This result appears again as Thm III of §94.1, with applications. Note that one point of the intersection of two circles, when the circles are lines, is always the point ∞.)

53.10 In the stereographic projection of the surface of a sphere onto the inversive plane (Fig. 52.1), the lines in the plane arise from circles through the point N. What can you say about the two circles through N which arise from two parallel lines in the plane?

53.11 If $z_i = a + r(\cos \alpha_i + i \sin \alpha_i)$ $(i = 1, 2, 3, 4)$, so that the four points z_i are points on a circle, center a, radius r, angular coordinates α_i, show that

$$(z_1, z_2; z_3, z_4) = \frac{\sin \frac{1}{2}(\alpha_1 - \alpha_3) \sin \frac{1}{2}(\alpha_2 - \alpha_4)}{\sin \frac{1}{2}(\alpha_1 - \alpha_4) \sin \frac{1}{2}(\alpha_2 - \alpha_3)}.$$

53.12 If the four points A, B, C, D represented by the complex numbers z_1, z_2, z_3, z_4 form a set of *orthocyclic points* (see Exercise 21.9), show that $(z_1, z_2; z_3, z_4)$ has modulus 1.

53.13 Prove that the triangles ABC and $A'B'C'$ are directly similar if and only if $(a, b; c, \infty) = (a', b'; c', \infty)$, where a, b, c are the complex numbers representing the points A, B, C, and a', b', c' the complex numbers representing the points A', B', C'.

54.1　The map of the inside of a circle

Theorem *If a circle $\mathscr{C}$ be mapped onto a circle $\mathscr{C}'$ by a Moebius mapping, then the interior of $\mathscr{C}$ is mapped onto the interior of $\mathscr{C}'$, or onto the exterior of $\mathscr{C}'$.*

Proof. We saw that a given Moebius transformation may be written as the composition of a translation, an inversion in the unit circle followed by a reflexion in the real axis, a dilative rotation and a translation. All these mappings, with the exception of the inversion, map the inside of a circle onto the inside of the mapped (image) circle. Inversion may map the inside of the object circle onto the inside of the image circle, or it may map the inside of the object circle onto the outside of the image circle. A simple example is the effect of inversion on the circle of inversion itself. Points on the circle of inversion map onto themselves, but points *inside* the circle are mapped onto points *outside* the circle. On the other hand, if the circle of inversion is *orthogonal* to the circle $\mathscr{C}$, points on $\mathscr{C}$ are merely interchanged by the inversion, and it is clear that the interior of $\mathscr{C}$ maps into the interior of $\mathscr{C}' = \mathscr{C}$.

Algebraic proofs of the theorem in inversion: The interior of a circle $\mathscr{C}$ inverts into the interior or the exterior of the image circle $\mathscr{C}'$, and this depends on whether the center of inversion is outside or inside $\mathscr{C}$, are indicated in the following Exercise (Exercise 54.3).

54.2　Map of a circle into itself

We are especially interested in those Moebius transformations which map a given circle $\mathscr{C}$ onto itself, and the inside of $\mathscr{C}$ onto the inside of $\mathscr{C}$. We take the unit circle, center at the origin, $|z| = 1$, as the circle $\mathscr{C}$.

There are, of course, Moebius transformations which map $|z| = 1$ onto $|z| = 1$, and the inside onto the inside. The identity mapping $Z = z$ is one such mapping, so the set of transformations is not empty. Another Moebius mapping which will serve is the rotation $Z = az$, where $|a| = 1$. This merely rotates the unit circle about the origin. If we take the composition of two such mappings, we have a mapping of the unit circle onto itself, inside mapping onto inside, and more generally we see that *under composition the set of Moebius transformations which map a circle and its interior onto themselves form a group.* This is merely a special case of the theorem discussed in §0.12: 'One-to-one onto mappings of a set form a group under composition.'

It is interesting to determine what form a Moebius transformation takes when it maps $|z| = 1$ onto $|Z| = 1$ and the interior onto the interior. If the transformation is $Z = (az + b)/(cz + d)$, we want $z\bar{z} = 1$ to map onto $Z\bar{Z} = 1$, in the first instance. The map of $Z\bar{Z} = 1$ arises from

$$\frac{(az + b)}{(cz + d)} \frac{\overline{(az + b)}}{\overline{(cz + d)}} - 1 = 0,$$

which simplifies to

$$z\bar{z}(a\bar{a} - c\bar{c}) + z(a\bar{b} - c\bar{d}) + \bar{z}(\bar{a}b - \bar{c}d) + b\bar{b} - d\bar{d} = 0.$$

This is the same as the equation $z\bar{z} - 1 = 0$ if

$$a\bar{b} - c\bar{d} = \bar{a}b - \bar{c}d = 0,$$

and

$$a\bar{a} - c\bar{c} + b\bar{b} - d\bar{d} = 0.$$

We therefore have only two conditions:

$$a\bar{b} - c\bar{d} = 0,$$

and

$$|a|^2 + |b|^2 = |c|^2 + |d|^2.$$

These are sufficient conditions. Write $c = k\bar{b}$. Then $\bar{c}d - \bar{a}b = 0$ leads to $\bar{k}bd - \bar{a}b = 0$, so that $d = \bar{a}/\bar{k}$. If we now substitute in the condition $|c|^2 + |d|^2 = |a|^2 + |b|^2$, this leads to $|k|^2|b|^2 + |a|^2/|k|^2 = |a|^2 + |b|^2$, and so we can take $|k| = 1$. Then $1/\bar{k} = k$, and so the form of the Moebius transformation is

$$Z = \frac{az + b}{k(\bar{b}z + \bar{a})}, \text{ where } |k| = 1,$$

and, of course, we may also write this as $Z = k(az + b)/(\bar{b}z + \bar{a})$, $|k| = 1$.

This mapping transforms $|z| = 1$ into $|Z| = 1$. The determinant of the mapping is $\Delta = k^2(a\bar{a} - b\bar{b}) = k^2(|a|^2 - |b|^2)$. If we choose $k = 1$, and $\Delta = 1$, then $|a|^2 - |b|^2 = 1$, and so $|b| < |a|$. The point 0 is then mapped onto the point $b/\bar{a}$, and since $|b| < |a|$, this point is *inside* the circle $|Z| = 1$. This is sufficient for us to be able to assert that the mapping $Z = (az + b)/(\bar{b}z + \bar{a})$, where $|a|^2 - |b|^2 = 1$, is a Moebius transformation which maps the circle $|z| = 1$ onto the circle $|Z| = 1$, and the inside of $|z| = 1$ onto the inside of $|Z| = 1$.

54.3   Conformal property of Moebius transformations

Theorem   *The circles $\mathscr{C}$ and $\mathscr{D}$ intersect at the point z_0, and u is a unit tangent vector to $\mathscr{C}$ at z_0, v is a unit tangent vector to $\mathscr{D}$ at z_0. If $\mathscr{C}$ and $\mathscr{D}$ are mapped by a Moebius transformation onto the circles $\mathscr{C}'$ and $\mathscr{D}'$ intersecting at Z_0, and u is mapped on u', v is mapped on v', then u' and v' are tangent circles at Z_0 to $\mathscr{C}'$ and $\mathscr{D}'$ respectively, and the angle between u' and v' is equal to that between u and v both in measure and in sense. (See Fig. 54.1.)*

Proof. We assume that $cz_0 + d \neq 0$ in the Moebius mapping $Z = (az + b)/(cz + d)$.

Let z_1 be a point on $\mathscr{C}$ distinct from z_0, and z_2 a point on $\mathscr{D}$ distinct from z_0. Then

$$Z_0 - Z_1 = \frac{az_0 + b}{cz_0 + d} - \frac{az_1 + b}{cz_1 + d} = \frac{(ad - bc)(z_0 - z_1)}{(cz_0 + d)(cz_1 + d)},$$

$$Z_0 - Z_2 = \frac{(ad - bc)(z_0 - z_2)}{(cz_0 + d)(cz_2 + d)}.$$

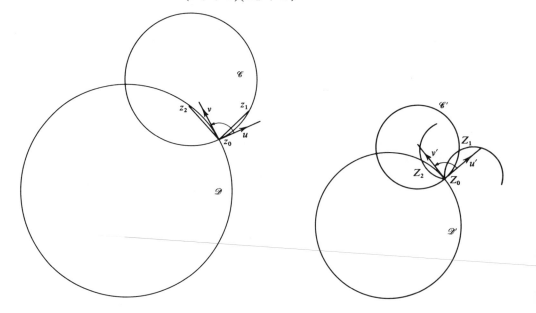

Fig. 54.1

Hence, since $ad - bc \neq 0$,

$$\frac{Z_0 - Z_1}{Z_0 - Z_2} = \frac{(z_0 - z_1)}{(z_0 - z_2)} \cdot \frac{(cz_2 + d)}{(cz_1 + d)}.$$

As z_1 approaches z_0 on $\mathscr{C}$, Z_1 approaches Z_0 on $\mathscr{C}'$, and since the *line* joining z_0 to z_1 is mapped onto a circle through Z_0 and Z_1 (which may be a line), the unit tangent vector u to $\mathscr{C}$ at z_0 is mapped onto a circle tangent to $\mathscr{C}'$ at Z_0. Similarly, the unit tangent vector v to $\mathscr{D}$ at z_0 is mapped onto a circle tangent to $\mathscr{D}'$ at Z_0. As z_2 and z_1 both approach z_0 on the circles $\mathscr{C}$ and $\mathscr{D}$ respectively, we have, in the limit

$$\frac{Z_0 - Z_1}{Z_0 - Z_2} = \frac{z_0 - z_1}{z_0 - z_2},$$

and comparison with Thm 47.9 shows us that the triangles $Z_0Z_1Z_2$ and $z_0z_1z_2$ are *directly similar*, and therefore the angle between u' and v' is equal to that between u and v both in measure and sense.

Another proof of this theorem is suggested in Exercise 54.1.

Exercises

54.1 From the dissection of a Moebius transformation (§52.4) prove Thm 54.3, by examining the effect of the component transformations on the measure and the sense of the angle between two circles.

54.2 Prove directly that the mappings $Z = (az + b)/(\bar{b}z + \bar{a})$, where $a\bar{a} - b\bar{b} = 1$, map the circle $|z| = 1$ onto the circle $|Z| = 1$, and the interior onto the interior, and that the set of mappings form a group under composition.

54.3 The *inside* of the circle $\mathscr{C}$, given by the equation

$$C(X, Y) \equiv X^2 + Y^2 - 2pX - 2qY + r = 0$$

may be characterized as the set of points (x, y) for which the *power* $C(x, y)$ of (x, y) with respect to the circle $\mathscr{C}$ is not positive. Using the formula of §21.1, show that, after inversion in the circle $X^2 + Y^2 = k^2$,

$$C(x, y) \equiv r\left\{X^2 + Y^2 - \frac{2k^2}{r}(pX + qY) + \frac{k^4}{r}\right\}\bigg/(X^2 + Y^2),$$

where the point (x, y) is mapped onto the point (X, Y) by the inversion, so that if the circle $\mathscr{C}$ is mapped onto the circle $\mathscr{C}'$, we have

Power of (x, y) with respect to $\mathscr{C} = r(\text{Power of } (X, Y) \text{ with respect to } \mathscr{C}')/(X^2 + Y^2)$.

Deduce that if the center of inversion $(0, 0)$ is *inside* $\mathscr{C}$, the inside of $\mathscr{C}$ inverts into the *outside* of $\mathscr{C}'$, and that if the center of inversion is *outside* $\mathscr{C}$, the inside of $\mathscr{C}$ inverts into the *inside* of $\mathscr{C}'$.

55.1   Special Moebius transformations

We now concentrate on the subgroup of Moebius transformations which map the unit circle onto itself, and the inside onto the inside. We know that these transformations do form a subgroup of the group of all Moebius transformations (§54.2), and we shall call these special Moebius mappings M-transformations.

We also give the boundary of the unit circle $|z| = 1$ a special name, ω, and we call the inside of the unit circle Ω. The points of ω are not included among the points of Ω, which is regarded as an open disk.

Our aim is to set up a geometry in Ω, defining points and lines and incidence properties in the first instance. Our points are ordinary Euclidean points in Ω, but our lines are circles orthogonal to ω, and since we are only considering the space Ω, our lines are the arcs of circles orthogonal to ω which lie in Ω. We shall show that we can obtain a geometry in Ω which marches step by step alongside Euclidean geometry, until there is a dramatic parting of ways when we arrive at the concept of parallel lines in the two geometries. In Euclidean geometry there is a unique parallel to a given line l through a given point P. We shall find in our space Ω that there are *two* parallels. This geometry is called *hyperbolic geometry*, and is an example of a *non-Euclidean* geometry.

We shall show that the group of M-transformations corresponds to the group of direct isometries of the Euclidean plane, in the sense that similar theorems are true for each group in the respective geometries. We begin our study of Ω by proving some theorems on M-transformations.

55.2   Theorems on M-transformations

Theorem   *Let X and A be two points of Ω, ξ and α directions through X and A respectively. Then there is a unique M-transformation which maps X on A, and the direction ξ onto the direction α. (See Fig. 55.1.)*

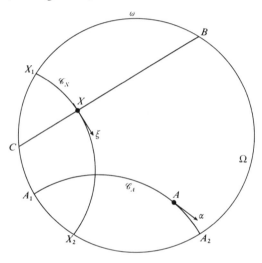

Fig. 55.1

Proof. We note in the first place that through any point X, and touching any given line through X, there passes exactly one circle orthogonal to ω. (See Exercise 22.3, and Exercise 55.1, following, for proofs of this theorem.) Call this circle $\mathscr{C}_X$, and let X_1, X_2 be its intersections with ω. In the same way, let $\mathscr{C}_A$ be the unique circle through A which has the direction α at A and is orthogonal to ω. Let A_1, A_2 be its intersections with ω. Since Moebius transformations conserve angles of intersection, and our M-transformations map ω onto itself, circles orthogonal to ω are mapped onto circles orthogonal to ω.

Hence any M-transformation which maps (X, ξ) onto (A, α) must necessarily map $\mathscr{C}_X$ onto $\mathscr{C}_A$, since $\mathscr{C}_X$ is mapped onto a circle through A, with direction α, orthogonal to ω, and there is only one such circle, and it must be $\mathscr{C}_A$.

Let us now suppose that the points of intersection of $\mathscr{C}_X$ and $\mathscr{C}_A$ with ω are so numbered that the direction ξ is that of X_1 towards X_2 in Ω, and the direction α is that of A_1 towards A_2 in Ω. The M-transformation we are seeking must map X_1 on

A_1, X_2 on A_2, and X on A. But three pairs of corresponding points determine a *unique* Moebius transformation (Thm 53.2). Our proof will therefore be complete if we can show that this unique Moebius transformation is an M-transformation, mapping Ω onto itself.

This is clear, since the unique Moebius transformation maps the circle through X_1 and X_2 orthogonal to $\mathscr{C}_X$, which is ω, on the circle through A_1 and A_2 orthogonal to $\mathscr{C}_A$, which is again ω. Hence ω is mapped onto ω. Since the interior point X of Ω is mapped on the interior point A of Ω, the interior of ω is mapped on the interior of ω; that is, Ω is mapped on Ω. The unique Moebius transformation we have been discussing is therefore an M-transformation.

Corollary. Let $\mathscr{C}_A$ be a circle through a point A of Ω orthogonal to ω, and let $\mathscr{C}_B$ be a circle through a point B of Ω orthogonal to ω. Then there exist just two M-transformations which map A on B and $\mathscr{C}_A$ on $\mathscr{C}_B$ (Fig. 55.2).

Proof. There are two directions tangent to $\mathscr{C}_A$ at A, and each is mapped by a unique M-transformation on a given direction tangent to $\mathscr{C}_B$ at B.

Remark. The theorem on direct isometries which compares with the above is Thm 43.2. If X and A are two points of the real Euclidean plane, there is a unique direct isometry which maps X on A and a given direction ξ through X on a given direction α through A.

55.3   Interchanging two points

Theorem   *There exists a unique M-transformation which interchanges two given points A and B of Ω.*

Proof. Let $\mathscr{C}$ be the unique circle through A and B orthogonal to ω (Fig. 55.2). Any M-transformation which maps A on B and B on A must map the circle $\mathscr{C}$ onto itself. By the corollary to the preceding theorem we know, taking $\mathscr{C}_A = \mathscr{C}_B = \mathscr{C}$, that there are only two M-transformations which map A on B and $\mathscr{C}$ on itself. By the proof to the Corollary, one preserves orientation on $\mathscr{C}$, and the other reverses it. The one which preserves orientation on $\mathscr{C}$ cannot map B on A, since the sense of A towards B in Ω would become the sense of B towards A in Ω. Let M' be the M-transformation which maps A on B and *reverses* the orientation of $\mathscr{C}$. We prove that M', besides mapping A on B, also maps B on A.

Let X_1 and X_2 be the intersections of $\mathscr{C}$ and ω. Let the transform of B under M' be A'. We want to prove that $A' = A$. Since $\mathscr{C}$ and ω are mapped on themselves, the set of two intersections X_1 and X_2 of $\mathscr{C}$ and ω is mapped onto itself. Hence either X_1 is mapped on X_1 and X_2 on X_2, or X_1 is mapped on X_2 and X_2 is mapped on X_1. In the first case, since A is mapped on B, the orientation $\mathscr{C}$ would be preserved under M', contrary to hypothesis, so that we must have X_1 mapped on X_2 and X_2 on X_1. By the cross-ratio property (Thm 53.4),

$$(X_1, X_2; A, B) = (X_2, X_1; B, A'),$$

but since (Thm 53.3) for any cross-ratio

$$(X_1, X_2; A, B) = (X_2, X_1; B, A),$$

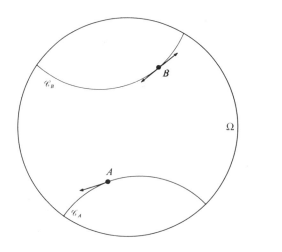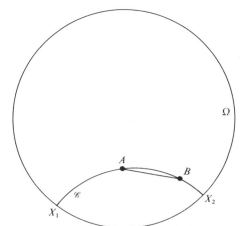

Fig. 55.2

we have

$$(X_2, X_1; B, A) = (X_2, X_1; B, A'),$$

and therefore

$$A' = A.$$

We have now established the correspondence between the direct isometries of the Euclidean plane and the M-transformations of the inversive plane which will enable us to develop a geometry in Ω which bears a strong resemblance to Euclidean geometry as far as, but not including, the parallel axiom. We must point out that the parallel axiom in the form: *through a given point P which does not lie on a given line l there is a unique parallel to the line l*, does not appear in Appendix I, which lists Euclid's Definitions, Postulates and the First Thirty Propositions of Book I. It was Playfair who showed that this version is equivalent to Euclid's, and in this form it is sometimes called Playfair's Axiom. This is the form we shall use.

Exercises

55.1 X is a point inside a circle ω and ξ is a given direction through X. The perpendicular to ξ at X intersects ω at B and C. Show that the circle through X which touches ξ and is orthogonal to ω is the unique member of the coaxal system which has B and C as limiting points and passes through X.

55.2 In the previous exercise show that the circle under consideration can also be defined as the Apollonius circle with ratio $|XB|/|XC|$, with regard to the points B and C (that is, the locus of a point P which moves so that $|PB|/|PC| = |XB|/|XC|$).

55.3 Show that the cases which may arise in Exercise 22.3, when there is no circle, or an infinity, through a point X, with a given direction ξ at X, and orthogonal to ω, do not arise in Exercise 55.1 above.

56.1   The Poincaré model of a hyperbolic non-Euclidean geometry

We now discuss Poincaré's model of a hyperbolic geometry. The term *non-Euclidean* refers to the fact that in this geometry Euclid's parallel axiom is not valid.

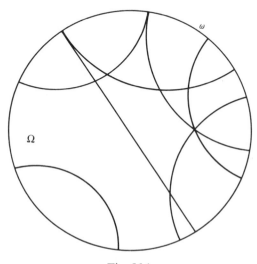

Fig. 56.1

Our space is *the interior* Ω of a circle ω. We shall call this *the p-plane*. A point of Ω is to be a point, or *p-point*, in our new geometry. But our *p-lines* are the arcs of circles (or straight lines) in Ω, limited by ω and orthogonal to ω (Fig. 56.1). These were the arcs considered in the theorems of §55.2.

Incidence in our new geometry is defined in a natural way. If a point in Ω lies on an arc of a circle or line orthogonal to ω, we say that the corresponding *p-point lies on* the corresponding *p-line*.

The ordinary incidence theorems of real Euclidean plane geometry hold in the new geometry. Since there passes through two distinct points of Ω a unique circle or line orthogonal to ω, we can say:

Through two distinct *p-points* there passes a unique *p-line*.

It follows immediately that:

Two distinct *p-lines* intersect in at most one *p-point*.

We shall consider pairs of *p-lines* which do *not* intersect in §57.2.

We define the *angle* between two *p*-lines to be the ordinary Euclidean angle, measured in the ordinary way at a point of intersection in Ω of the lines or arcs of circles which make up *p*-lines. The relation of congruence of the angles between *p*-lines is to be the conventional Euclidean one.

56.2   The congruence of *p*-segments

We naturally define *a p-segment* as the set of *p*-points on a *p*-line *between* two *p*-points. The relation of *being between* is to be the same as in Euclidean geometry. In ordinary Euclidean geometry two segments are equal (or *congruent*) if one can be *applied* to the other. This involves a rigid motion of the plane, which preserves distances: in other words, *an isometry*. Lines are mapped on lines, and angles are preserved. If the isometry is *direct*, the *sense* of an angle is also preserved.

In our *p*-geometry the group of *M*-transformations is the equivalent of the group of the *direct* isometries. The *M*-transformations map *p*-lines on *p*-lines, and conserve angles, together with the sense of the angles. This is clear since an *M*-transformation maps ω onto itself, and maps a *p*-line into a line or circle which intersects the map of ω, which is again ω, orthogonally. Since an *M*-transformation also maps the inside of ω into the inside of ω, *p*-lines are mapped into *p*-lines. Again, any Moebius transformation is a conformal transformation, and preserves the measure and the sense of angles (Thm 54.3). Our *M*-transformations are Moebius transformations.

With these preliminaries we may say that *two p-segments AB and CD are p-congruent*, written

$$AB \overset{p}{=} CD,$$

if and only if there is an M-transformation which maps A on C and B on D.

Since *M*-transformations form a group, the relation of *p*-congruence satisfies the usual axioms of equivalence:

(i) $AB \overset{p}{=} AB$ (Identity)
(ii) If $AB \overset{p}{=} CD$, then $CD \overset{p}{=} AB$ (Symmetry)
(iii) If $AB \overset{p}{=} CD$, and $CD \overset{p}{=} EF$, then $AB \overset{p}{=} EF$ (Transitivity).

The proof of these statements is left as an Exercise for the reader. We also have Thm 55.3, which tells us that there is an *M*-transformation which maps *A* on *B* and *B* on *A*. where *A* and *B* are any two distinct *p*-points, so that

$$AB \overset{p}{=} BA.$$

Definition. A *p*-triangle is a triad of points which do not lie on a *p*-line. As in Euclidean geometry, there are important theorems about the congruence of *p*-triangles. Two *p*-triangles *ABC*, *A'B'C'* are *p*-congruent if there is an *M*-transformation which maps *A* on *A'*, *B* on *B'* and *C* on *C'*.

Exercises

56.1 Circles orthogonal to ω intersect in a pair of points which have a special relationship to ω. What is the relationship? Deduce that two distinct *p*-lines which intersect do so in only one point which lies inside ω.

56.2 If $AB \stackrel{p}{=} CD$, show that the *M*-transformation which maps *A* on *C* and *B* on *D* is unique, and that it also maps *p*-points on the *p*-segment *AB* onto *p*-points on the *p*-segment *CD*.

56.3 Prove that the relation of *p*-congruence is an equivalence relation (§56.2).

57.1 The SAS theorem

Theorem I *Given two p-triangles ABC and A′B′C′ with equal orientation, the congruences $AB \stackrel{p}{=} A′B′$, $AC \stackrel{p}{=} A′C′$, and $\angle CAB \stackrel{p}{=} \angle C′A′B′$ imply the p-congruence of the triangles, so that $\angle ABC \stackrel{p}{=} \angle A′B′C′$, $\angle ACB \stackrel{p}{=} \angle A′C′B′$ and $BC \stackrel{p}{=} B′C′$.*

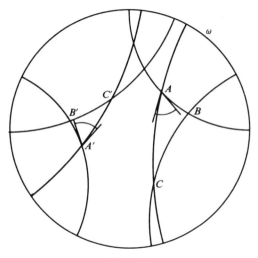

Fig. 57.1

Proof. We show that there is an *M*-transformation which maps *A* on *A′* (Fig. 57.1), *B* on *B′* and *C* on *C′*. By hypothesis there is an *M*-transformation which maps *A* on *A′* and *B* on *B′*. By Exercise 56.2 we know that this *M*-transformation is unique. We consider its effect on the *p*-line *AC*. Since *M*-transformations conserve both angles and sense (Thm 54.3) the *direction* of the map of the *p*-line *AC* at *A′* coincides with the *direction* of the *p*-line *A′C′*. Since there is only one *p*-line through *A′* in the direction of the *p*-line *A′C′*, and by hypothesis there is an *M*-transformation which maps *A* on *A′* and *C* on *C′*, and this *M*-transformation is unique, it must be the one we started with, which maps *A* on *A′* and *B* on *B′*. This *M*-transformation maps *B* on *B′* and *C* on *C′*, and therefore $BC \stackrel{p}{=} B′C′$, and the equality of the angles *ABC* and *A′B′C′*, and the angles ACB and $A'C'B'$ is a consequence of M-transformations conserving both angles and sense.

This proof should be compared with the Euclidean Side-Angle-Side proof. If the terms "there is an M-transformation" and "there is a direct isometry of the plane" are interchanged, the proofs are the same except for automatic interchanges between "p-line" and "line". We shall take advantage of this correspondence later when we

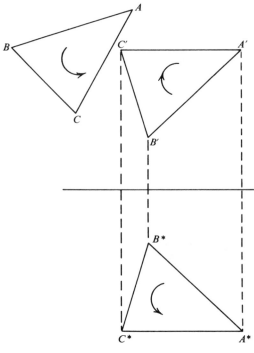

Fig. 57.2

shall prove that in our p-geometry two triangles are congruent if the three angles of the one are equal respectively to the three angles of the other (Thm 59.1).

It will be noticed that the formulation of Thm 57.1 above postulates triangles with equal orientation, whereas this concept is omitted in Euclidean geometry. If two triangles ABC, $A'B'C'$ (Fig. 57.2) in the Euclidean plane are not equally oriented, we cannot slide the plane until one falls on the other, which is the essence of the proof by direct isometries. We lift one triangle out of the plane, *turn it over*, and then replace it with its orientation changed. The usual proof then proceeds. Equivalently, we operate on one triangle with an *indirect* isometry, and change its orientation.

Our M-transformations preserve orientation, and to change the orientation of a triangle we would have to use a transformation outside the group of M-transformations, such as a reflexion in the real axis.

It is clear that if we reflect a p-triangle in the real axis, it has an image which is a p-triangle, but with opposite orientation, and this p-triangle is congruent, in the Euclidean sense, to the original triangle. Hence, if we have two p-triangles ABC and $A'B'C'$ with opposite orientation, and $AB \stackrel{p}{=} A'B'$, $AC \stackrel{p}{=} A'C'$, and $\sphericalangle CAB \stackrel{p}{=} \sphericalangle C'A'B'$ (we remember that with angles, p-congruence is the same as Euclidean congruence, and orientation is not involved), we can reflect p-triangle ABC in the real axis, and obtain a p-triangle $A*B*C*$ which is congruent in the Euclidean sense to the p-triangle ABC, and this new triangle has the same orientation as the p-triangle $A'B'C'$. But now we cannot assert that $A*B* \stackrel{p}{=} A'B'$, since the reflexion in the real axis, which is given by $Z = \bar{z}$, is not a Moebius transformation.

The dilemma is exactly similar to the one encountered in Euclidean geometry if we wish to restrict ourselves to direct isometries. The SAS theorem cannot be proved when ABC and $A'B'C'$ are not equally oriented, if only direct isometries are allowed. This dilemma is overcome in some axiomatic treatments of Euclidean geometry by *postulating* the SAS theorem, without any mention of orientation.

We must extend our group of M-transformations to include those transformations of the form (*conjugate* Moebius transformations)

$$Z = \frac{a\bar{z} + b}{c\bar{z} + d} \qquad (ad - bc \neq 0)$$

which map the unit circle $|z| = 1$ onto itself, and the inside into the inside. Let us call these transformations M_--transformations, and the former M-transformations M_+-transformations, since they correspond to the two kinds of isometry in the Euclidean plane. The M_--transformations by themselves do not form a group, but the totality of M_+ and M_--transformations form a group, with composition as group multiplication. The M_+-transformations preserve orientation, and the M_--transformations reverse orientation. We list those properties of M_--transformations which differ from M_+-transformations as Exercises (Exercises 57.7, 57.9).

We must now redefine p-congruence. If we call the elements of our extended group M^*-*transformations*, we say that $AB \stackrel{p}{=} A'B'$ if there is an M^*-transformation which maps A on A' and B on B'. Angles, if congruent in the Euclidean sense are congruent under M^*-transformations, and we now have *the general SAS theorem*, which is independent of the orientation of the triangles ABC and $A'B'C'$: the formulation is as in Thm 57.1, but the phrase 'with equal orientation' is omitted.

Theorem II *Given two p-triangles ABC and $A'B'C'$, the congruences $AB \stackrel{p}{=} A'B'$, $AC \stackrel{p}{=} A'C'$ and $\sphericalangle CAB \stackrel{p}{=} \sphericalangle C'A'B'$ imply the p-congruence of the triangles, so that $\sphericalangle ABC \stackrel{p}{=} \sphericalangle A'B'C'$, $\sphericalangle ACB \stackrel{p}{=} \sphericalangle A'C'B'$ and $BC \stackrel{p}{=} B'C'$.*

57.2 The parallel axiom

As far as we have gone, there seems to be a complete correspondence between our *p*-geometry and that part of Euclidean geometry which is developed before the introduction of the parallel axiom. This axiom, in one form, (shown by Playfair to be equivalent to Euclid's formulation), says:

Through a given point *P* which does not lie on a given line *l* there passes a unique line which is parallel to *l*.

In Euclidean geometry, of course, two coplanar lines are said to be *parallel* if they do not intersect.

We could say that two *p*-lines in our *p*-geometry are *p*-parallel if they do not intersect.

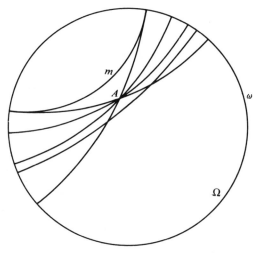

Fig. 57.3

Let *A* be a point in Ω (Fig. 57.3), and let *m* be a given *p*-line which does not contain *A*. Then we see that through *A* there is an infinity of *p*-lines which do not intersect *m* in Ω. This property is independent of whether *m* is convex or concave to *A*. We could say, therefore, that there is *an infinity* of *p*-lines through *A* which are *p*-parallel to the given *p*-line *m*.

But we notice (see Exercise 57.1) that there are exactly *two p*-lines through *A* which touch the given *p*-line *m*, the points of contact lying on ω. These two *p*-lines through *A* separate the *p*-lines through *A* into two classes: the *p*-lines which cut the *p*-line *m* are in one class, and those which do not intersect *m* are in the second class.

We reserve the term *p-parallel* for *the two p-lines through A which touch the given p-line m at points of* ω. We now have, in our geometry, instead of the Euclidean parallel axiom, the theorem:

Theorem   *Through any given p-point A which does not lie on a given p-line m there pass exactly two p-lines p-parallel to the given p-line.*

Hence our p-geometry is not Euclidean geometry, and it is naturally called *non-Euclidean geometry*. Our p-plane is a *model* for such a geometry. If Euclidean geometry is an internally consistent system, without contradictions, then so is the model, and its existence shows that the parallel axiom in Euclidean geometry is *independent* of the other axioms, that is *it cannot be deduced from them*.

Exercises

57.1 Prove that through any given point A in Ω which does not lie on a given p-line m there pass just two p-lines which touch m, and these do so on ω.

57.2 Show that through any given p-point A there exists a unique p-line l' perpendicular to a given p-line l, and that if m, n be the p-lines through A which are p-parallel to l (it is assumed that A does not lie on l), then m and n make equal angles with l'.

57.3 Prove that two distinct p-lines which are perpendicular to the same p-line have no intersection.

57.4 If we allow the vertices A, B and C of a p-triangle to lie on ω, what can we say about the angles of this triangle?

57.5 Prove that two distinct p-lines which have no intersection have a unique p-line perpendicular to both of them.

57.6 Prove that the totality of M_+ and M_- transformations, with composition as group multiplication, form a group, which is a subgroup of the group of all Moebius and conjugate Moebius transformations.

57.7 If z_1, z_2, z_3 and z_4 are mapped by a conjugate Moebius transformation into z_1', z_2', z_3' and z_4', prove that the cross-ratio of the object set is equal to the conjugate of the cross-ratio of the image set:

$$(z_1, z_2; z_3, z_4) = \overline{(z_1', z_2'; z_3', z_4')}$$

57.8 Show that the composition of the two mappings

$$Z' = (\bar{a}\bar{z} + \bar{b})/(\bar{c}\bar{z} + d), \quad Z = \bar{Z}'(\bar{a}d - \bar{b}\bar{c} \neq 0)$$

produces the mapping

$$Z = (az + b)/(cz + d), \quad (ad - bc \neq 0).$$

Deduce that Thm 52.3 and its Corollary are true for conjugate Moebius transformations.

57.9 Show that Thm 54.3 continues to hold for conjugate Moebius transformations, except that the sense of an angle is changed by the transformation.

57.10 Show that if z_1, z_2, z_3, z_4 are four points on a p-line, and z_1', z_2', z_3', z_4' are their images under a Moebius *or* a conjugate Moebius transformation, then

$$(z_1, z_2; z_3, z_4) = (z_1', z_2'; z_3', z_4').$$

57.11 Show that inversion in the circle $cz\bar{z} - z\bar{p} - \bar{z}p - b = 0$, where b and c are real, the center is the complex number p/c, the square of the radius is given by $r^2 = (p\bar{p} + bc)/c^2$, can be obtained by writing

$$(Z - p/c)\overline{(z - p/c)} = r^2 = (p\bar{p} + bc)/c^2,$$

which leads to the conjugate Moebius transformation

$$Z = \frac{p\bar{z} + b}{c\bar{z} - \bar{p}} \quad (p\bar{p} + bc > 0).$$

57.12 Show that the circles $cz\bar{z} - z\bar{p} - \bar{z}p - b = 0$, $ez\bar{z} - z\bar{q} - \bar{z}q - d = 0$ are orthogonal if and only if $be + cd + q\bar{p} + \bar{q}p = 0$.

57.13 Show that the composition of inversion in two circles is equivalent to a Moebius transformation.

57.14 Show that if the Moebius transformation of the preceding Exercise is of the form $Z = (az + b)/(cz + d)$, then if the circles are orthogonal we find that $a + d = 0$.

57.15 Verify that a Moebius transformation $Z = (az + b)/(cz + d)$, where $a + d = 0$ is an involutory transformation, that is, if the map of any point z is Z, then the map of Z is z. (Compare Exercise 25.2. Since $(T_\mathscr{C})^{-1} = T_\mathscr{C}$, and $(T_\mathscr{D})^{-1} = T_\mathscr{D}$, $T_\mathscr{C}T_\mathscr{D} = T_\mathscr{D}T_\mathscr{C}$ leads to $(T_\mathscr{C}T_\mathscr{D})^2 = $ Identity. See also Thm I, §76.2, and Appendix III, on *the complex line*.)

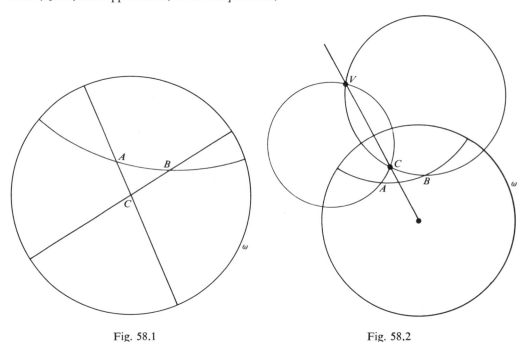

Fig. 58.1   Fig. 58.2

58.1   The angle-sum of a *p*-triangle

Theorem   *The sum of the angles of a p-triangle ABC is less than π.*

Proof. If C is the center of ω (Fig. 58.1), then the *p*-lines CA and CB are diameters of ω, and the *p*-line AB is an arc of a circle which is orthogonal to ω. If two circles are orthogonal, their centers do not both lie inside one of the circles. The center of the *p*-line AB therefore lies outside ω, and the center C of ω lies outside the circle which determines the *p*-line AB. Hence the *p*-line AB is *convex* to C, as shown in Fig. 58.1, and the angles of the curvilinear triangle ABC at A and at B are less than the angles of the Euclidean triangle ABC at A and at B. Since the angle-sum of the Euclidean triangle ABC is equal to π, the angle-sum of the *p*-triangle ABC is less than π.

If no vertex of the *p*-triangle ABC is the center of ω (Fig. 58.2), let V be the other intersection of the circles which determine the p-lines AC and BC. Then V and C are inverse points in ω, being the intersection of two circles orthogonal to ω. Choose as circle of inversion the circle $\mathscr{C}$ which has V as center and which is orthogonal to ω.

If we invert in this circle, ω inverts into itself, circles through V invert into lines, and since C is the intersection of two circles orthogonal to ω, C inverts into the center of ω, and the circles through V invert into diameters of ω. Hence the p-triangle ABC inverts into a p-triangle $A'B'C'$, where C' is the center of ω, and so we are back in the case we examined first. Since angles are unchanged by inversion, the angle-sum of p-triangle $ABC =$ the angle-sum of triangle $A'B'C'$, and this is less than π.

58.2 The exterior angle is greater than either of the interior and opposite angles

Theorem *In any p-triangle, if one of the p-sides be produced, the exterior angle is greater than either of the interior and opposite angles.*

Proof. This follows directly from the proof of Thm 58.1, since angles of p-triangle ABC are unchanged by inversion, and if we are considering p-line CB produced, say, the exterior angle at B of p-triangle ABC is equal to the exterior angle at B' of p-triangle $A'B'C'$. Since the p-line $B'A'$ is convex to C', the exterior angle at B' of p-triangle $A'B'C'$ is greater than the same angle for the Euclidean triangle $A'B'C'$, and the angle at A' for the p-triangle $A'B'C'$ is less than the same angle for the Euclidean triangle $A'B'C'$. Since, for the Euclidean triangle $A'B'C'$ the exterior angle at B' exceeds the angle A', this is even more the case for the p-triangle $A'B'C'$.

58.3 Non-Euclidean distance

We have shown how to compare two p-segments for p-congruence. We now obtain a *distance-function* for any two p-points A and B. This function will satisfy the usual conditions:

$$d(A, B) = d(B, A) \geqslant 0:$$
$$d(A, B) = 0 \text{ if and only if } B = A:$$
$$d(A, B) + d(B, C) \geqslant d(A, C),$$

with equality if and only if B lies in the p-segment connecting A and C. This last inequality is called *the triangle inequality*.

Our M^*-transformations are equivalent to the isometries of the Euclidean plane, and these, of course, are distance-preserving, and we hope that our distance-function will be invariant under M^*-transformations. We shall see that this is the case.

Our distance-function is to be defined in terms of the cross-ratio $(\alpha, \beta; A, B)$, where the p-line cuts ω in the points α and β (Fig. 58.3). We therefore begin by obtaining some further properties of this cross-ratio.

We know that there is a unique Moebius transformation which maps the points

α, A and β onto three given distinct points α', A' and β' of the real line, where we choose A' to lie between α' and β'. A point moving continuously on the p-segment $\alpha\beta$ is mapped by this Moebius transformation onto a point which moves continuously on the segment of the real line defined by α' and β'. (Compare Exercise 52.6.) Since we are considering a pair of points A, B on the p-line whose terminal points on ω are α and β, all points A', B' will lie on the segment of the real line $\alpha'\beta'$. We also know that cross-ratios are invariant under a Moebius transformation, and so

$$(\alpha, \beta; A, B) = (\alpha', \beta'; A', B'),$$

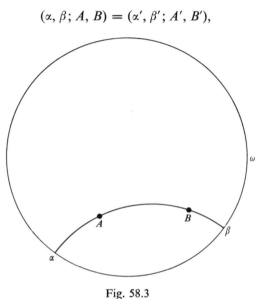

Fig. 58.3

which is a real number. We found this result in Thm 53.5. But we now obtain a stronger result. The order α, A, B, β on the p-line is the same as the order α', A', B', β' on the real line, and the cross-ratio of the real numbers is equal to $\dfrac{\alpha'A'}{A'\beta'} \Big/ \dfrac{\alpha'B'}{B'\beta'}$. As A' moves on the real line from α' towards β', the ratio $\alpha'A'/A'\beta'$ increases from 0 to $+\infty$, in a continuous manner. It therefore follows, that if B' is *between* A' and β', the ratio $\alpha'B'/B'\beta'$ is *greater* than the ratio $\alpha'A'/A'\beta'$. We have therefore proved that *if the points α, A, B, β are in this order on a p-line, the cross-ratio $(\alpha, \beta; A, B)$ is a positive real number less than 1.*

If α, β be interchanged, the cross-ratio is the reciprocal of this (Ex. 53.6), and similarly if A, B be interchanged: so that interchanging either α, β or A, B produces a positive real number greater than 1.

Definition. Let the p-line ab cut ω in the points α, β. Then

$$d(A, B) = |\log (\alpha, \beta; A, B)|.$$

We have just seen that the cross-ratio in the definition is a positive real number, so that the logarithm exists, and the logarithm is positive if the cross-ratio exceeds 1, and negative if the cross-ratio is less than 1. Since we define $d(A, B)$ to be the modulus of the cross-ratio, we have

$$d(A, B) \geqslant 0.$$

Again,

$$d(B, A) = |\log(\alpha, \beta; B, A)| = |\log 1/(\alpha, \beta; A, B)|$$
$$= |-\log(\alpha, \beta; A, B)| = d(A, B).$$

Furthermore, if $d(A, B) = 0$, then we must have $(\alpha, \beta; A, B) = 1$, and since α, β are distinct points, this leads to $A = B$. Hence $d(A, B) = 0$ if and only if $B = A$.

We note also that interchanging α and β leaves $d(A, B)$ unchanged, the proof being similar to the proof that $d(A, B) = d(B, A)$.

Theorem I   *If A, B and C are p-points, in this order, on a p-line then*

$$d(A, C) = d(A, B) + d(B, C).$$

Proof. Suppose that α, A, B, C, β lie, in this order, on a *p*-line, α, β being the intersections of the *p*-line with ω. Then we know that the cross-ratios $(\alpha, \beta; A, B)$, $(\alpha, \beta; B, C)$ and $(\alpha, \beta; A, C)$ are all positive real numbers less than 1. Their logarithms are therefore all negative real numbers. From Exercise 53.5 we may write

$$(\alpha, \beta; A, C) = (\alpha, \beta; A, B) . (\alpha, \beta; B, C)$$

(this result is independent of the order of A, B and C). Taking logarithms,

$$\log(\alpha, \beta; A, C) = \log(\alpha, \beta; A, B) + \log(\alpha, \beta; B, C),$$

and since these three logarithms are all negative numbers, we now have

$$|\log(\alpha, \beta; A, C)| = |\log(\alpha, \beta; A, B)| + |\log(\alpha, \beta; B, C)|,$$

and so

$$d(A, C) = d(A, B) + d(B, C).$$

Before proving the general triangle inequality, we remark that our distance-function $d(A, B)$ is, in fact, invariant under an M^*-transformation. We know that an M_+-transformation leaves cross-ratios unchanged (Thm 53.4), and an M_--transformation maps a cross-ratio onto its conjugate complex value (Exercise 57.7). But when the four points under consideration are on a *p*-line, the cross-ratio is a real number (Thm 53.5), and so both M_+ and M_--transformations leave the cross-ratio invariant. Hence, if the points α, β, A and B on a *p*-line are mapped onto the points α', β', A' and B' by an M^*-transformation, these points are again on a *p*-line, α' and β' lie on ω, and since

$$(\alpha, \beta; A, B) = (\alpha', \beta'; A', B'),$$

we have

$$d(A, B) = d(A', B').$$

Conversely, if we are given that $d(A, B) = d(A', B')$, we can prove that there is an M_+-transformation which maps A on A' and B on B'. It will be recalled that our original definition of p-congruence between p-segments (§56.2) involved the existence of an M_+-transformation which mapped A on A' and B on B'. The result we prove now ties the concepts of p-congruence and p-distance together.

We assume that the p-line AB cuts ω in the points α and β, and that the p-line $A'B'$ cuts ω in the points α' and β'. We name the points of intersection of the p-lines with ω so that the order α, A, B, β on the one p-line corresponds to the order α', A', B', β' on the other p-line. We know there is a unique M_+-transformation which maps α on α', A on A' and β on β'; the proof is exactly as in that of Thm 55.2. We assert that this M_+-transformation also maps B on B'.

We are given that $|\log(\alpha, \beta; A, B)| = |\log(\alpha', \beta'; A', B')|$, and because of our method of choosing α' and β' we know that each cross-ratio is a positive real number less than 1. Hence we can say that the two logarithms are equal, dropping the modulus signs, and that therefore the cross-ratios are equal. Since α is mapped on α', A on A' and β on β', this means that B is mapped on B'.

The proof that the cross-ratio $(\alpha, \beta; A, B)$ is a positive real number less than 1 also shows that if C is a p-point in the p-segment $B\beta$, the cross-ratio $(\alpha, \beta; A, C)$ is also positive, real and less than 1 but *less than the value of the cross-ratio* $(\alpha, \beta; A, B)$. Since the logarithms of a sequence of steadily decreasing positive real numbers less than 1 tend steadily towards $-\infty$, and the distance-function is the modulus of the logarithm of this cross-ratio, we see that:

The distance-function $d(A, B)$ for fixed A is a monotone increasing function of the position of the p-point B on the p-line $A\beta$. As B moves from A towards β, $d(A, B)$ increases steadily from 0 to $+\infty$.

Of course, a similar result is true if A moves steadily from B towards α.

It follows that if we have four p-points A, B, C and D (Fig. 58.4), and we know that

$$d(A, B) > d(C, D)$$

and we find the unique M-transformation which maps A on C, and the direction along the p-line AB at A onto the direction along the p-line CD at C (Thm 55.2), the point B will be mapped onto a p-point on the p-segment $D\beta'$; that is, it will fall on a point *outside* the p-segment CD.

Hence, there is an exact equivalent in our p-geometry to the Euclidean construction: 'Mark off from the segment CD, or CD produced, a length equal to a given length AB.'

Suppose now that A, B and C are p-points (Fig. 58.5), and we consider the p-lines through B which intersect the p-line AC in p-points which lie between A and C. The p-line BA intersects the p-line AC at A, and if B' be the p-point of intersection of a p-line through B and the p-line AC, we see that as B' moves continuously towards C from A, the p-angle ABB' increases continuously. We also note that if B' be any p-point in the p-segment AC, the p-line BB' lies inside the p-angle ABC and the p-angle ABB' is less than the p-angle ABC.

With these preparatory remarks *we may now prove theorems in our p-geometry with the terminology and the diagrams of Euclidean geometry.*

We shall not, of course, use the Euclidean parallel axiom!

Theorem II   *In a p-triangle ABC, if d(A, B) = d(A, C), then the p-angles ABC and ACB are congruent.*

Proof. By the General SAS Theorem of §57.1, if we consider the two triangles *ABC*

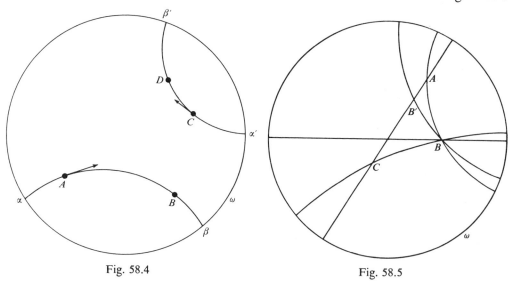

Fig. 58.4   Fig. 58.5

and *ACB*, we have $AB \overset{p}{=} AC$, $AC \overset{p}{=} AB$, and $\angle BAC \overset{p}{=} \angle CAB$. The two triangles are therefore *p*-congruent, and so $\angle ABC \overset{p}{=} \angle ACB$.

Note that triangles *ABC* and *ACB* have opposite orientation, so that we must use the General SAS Theorem. A direct proof is indicated in Exercise 58.4.

Theorem III   *In any p-triangle ABC, the greatest side is opposite the greatest angle*

Proof. Suppose that $d(A, C) > d(A, B)$ (see Fig. 58.6), and mark off a *p*-segment AB' on *p*-line AC equal to AB, so that $d(A, B') = d(A, B)$. Then B' lies in the segment AC, and the *p*-line BB' lies inside the angle ABC. Since $d(A, B') = d(A, B)$, the angles ABB' and $AB'B$ are equal. Therefore angle ABC is greater than angle $AB'B$, since angle ABC is greater than angle ABB'. But angle $AB'B$ is the exterior angle to angle $BB'C$ in *p*-triangle $BB'C$, and therefore exceeds the interior angle $B'CB$ which is the angle ACB (Thm 58.2). Hence angle ABC is greater than angle ACB.

Corollary. In a right-angled *p*-triangle, the longest side is that opposite the right angle.

Proof. Since the angle-sum of a *p*-triangle is less than two right angles, (Thm 58.1), the right angle is the largest angle.

We can now prove the triangle inequality for *p*-triangles and *p*-distances.

Theorem IV   *In any p-triangle ABC, d(A, B) + d(B, C) > d(A, C).*

Proof. If $d(A, C) \geqslant d(A, B)$ and $d(A, C) \geqslant d(B, C)$, then

$$d(A, C) + d(C, B) > d(A, B) \quad \text{and} \quad d(A, C) + d(A, B) > d(B, C),$$

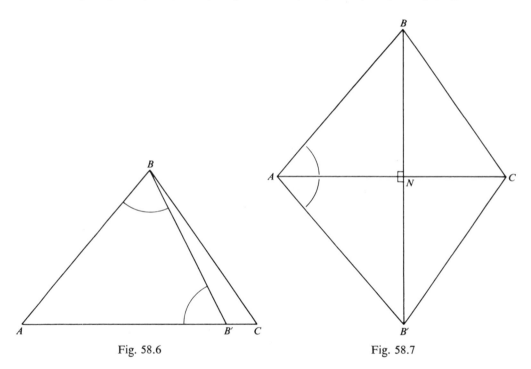

Fig. 58.6   Fig. 58.7

so that if *AC* is the longest side of the *p*-triangle *ABC* (Fig. 58.7), the proof of the theorem is only necessary with this assumption for the general inequality to be true. On the opposite side of the *p*-line *AC* to *B* mark off angle $B'AC = $ angle BAC, where $d(A, B') = d(A, B)$. Let the intersection of the *p*-lines BB' and AC be N. The *p*-triangles ANB and ANB' are *p*-congruent, by the S A S theorem (Thm 57.1, II), and since the sum of the angles BNA, $B'NA$ is two right angles, both BNA and $B'NA$ are right angles. Therefore $d(A, B) > d(A, N)$, and $d(B, C) > d(N, C)$. Hence, *if N lies in the segment AC,*

$$d(A, B) + d(B, C) > d(A, N) + d(N, C) = d(A, C)$$

(by Thm I, above).

If N lay outside the p-segment AC, we still have $d(A, B) > d(A, N)$, and since in this case $d(A, N) > d(A, C)$, we should have $d(A, B) > d(A, C)$, which is a contradiction to our initial hypothesis.

Exercises

58.1 Show that the stronger form of the triangle inequality is: If A, B and C are three p-points, then $d(A, B) + d(B, C) \geqslant d(A, C)$, with equality if and only if A, B and C lie, in this order, on a p-line.

58.2 Does Thm III of §58.3 imply Thm II of the same Section? (Examine the proofs of both Theorems carefully before answering.)

58.3 If A is at the center of ω, and B, C are p-points such that $d(A, B) = d(A, C)$, show that the line-segments AB and AC are equal in length from the Euclidean point of view, and deduce that $\measuredangle ABC = \measuredangle ACB$.

58.4 Using the preceding Exercise, prove that if in a p-triangle ABC we have $d(A, B) = d(A, C)$, then the p-angles ABC and ACB are equal.

58.5 Prove the ASA Theorem: (If for two triangles ABC, $A'B'C'$ (p-triangles, of course) $AB \overset{p}{=} A'B'$, $\measuredangle BAC \overset{p}{=} \measuredangle B'A'C'$ and $\measuredangle ABC \overset{p}{=} \measuredangle A'B'C'$, then the triangles are p-congruent).

58.6 Prove the SSS Theorem: (If for two p-triangles ABC, $A'B'C'$ we have $AB \overset{p}{=} A'B'$, $BC \overset{p}{=} B'C'$ and $CA \overset{p}{=} C'A'$, then the corresponding angles of the two triangles are also p-congruent). (Hint: Copy triangle $A'B'C'$ onto triangle ACQ, where $AQ \overset{p}{=} A'B'$, $\measuredangle CAQ \overset{p}{=} \measuredangle C'A'B'$, and Q is on the opposite side of AC to B. Prove triangles ABC, AQC are p-congruent, and deduce the theorem.)

58.7 Prove the SAA Theorem: (If for two p-triangles ABC, $A'B'C'$ we have $AB \overset{p}{=} A'B'$, $\measuredangle CAB \overset{p}{=} \measuredangle C'A'B'$ and $\measuredangle ACB \overset{p}{=} \measuredangle A'C'B'$, then the two triangles are p-congruent.)

59.1 Similar p-triangles are p-congruent

We have already seen (Thm 58.1) that in our p-geometry the sum of the angles of a p-triangle is always less than π, whereas in Euclidean geometry the angle-sum is always equal to π. Perhaps an even more surprising difference is the fact that a p-triangle is completely determined by its angles.

Theorem *Two p-triangles are p-congruent if the three angles of the one are respectively equal to the three angles of the other.*

Remark. We have established above that we may prove theorems in our p-geometry with the terminology and diagrams of Euclidean geometry, and have taken advantage of this freedom in the proofs of Theorems II and III and IV in §58.3. We continue in the same vein in the following proof.

Proof. In Fig. 59.1 suppose that $\measuredangle A = \measuredangle A'$, $\measuredangle B = \measuredangle B'$, $\measuredangle C = \measuredangle C'$ but that the p-triangles ABC and $A'B'C'$ are not p-congruent. Then we may assume that $d(A, B) \neq d(A', B')$, and that in fact $d(A, B) > d(A', B')$. On AB and AC mark off lengths equal to

$A'B'$ and $A'C'$ respectively, calling the endpoints on AB and AC the points D and E. Then D certainly lies in the segment AB, but E may lie on C, or on AC produced, or in the segment AC. If E falls on C, then by the SAS Theorem (§57.1) the triangles ADC and $A'B'C'$ are congruent, so that the angle ADC = angle $A'B'C'$ = angle ABC. But by Thm 58.2, angle ADC > angle ABC, Hence E cannot fall on C. If E falls on AC produced then again triangles ADE and $A'B'C'$ are congruent, and angle AED = angle $A'C'B'$ = angle ACB, and by Thm 58.2 we should have angle ACB > angle AED. Hence E can only fall between A and C. If we now consider the quadrilateral $BCED$, the sum of the angles is 2π. But this contradicts Thm 58.1, since a quadrilateral can

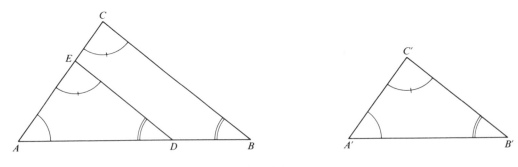

Fig. 59.1

be divided into two triangles, and the angle-sum of one of the triangles would be $\geqslant \pi$. Hence the original triangles ABC, $A'B'C'$ are p-congruent.

59.2 p-circles in the p-plane

There are some obvious constructs in our p-geometry which merit investigation. The definition of a p-circle is evidently as follows:

Definition. The locus of a p-point which moves so that its p-distance from a given p-point A is a constant r is a p-circle, center (p-center) A and p-radius r.

Theorem I   *A p-circle, center A is a Euclidean circle orthogonal to the family of p-lines which pass through A.*

Proof. The p-lines through A all pass through the fixed point A' (Fig. 59.2), where A' is the inverse of A in ω. Let P be the point whose locus we are considering, and $\mathscr{C}$ the p-line through P and A. This passes through A'. We wish to find the locus of P, subject to the condition $|\log (\alpha, \beta : A, P)| = r$, for a constant r, where α, β are the points in which $\mathscr{C}$ intersects ω. If we invert with respect to a circle, center A', orthogonal to ω, ω inverts into itself, the inside of ω inverts into the inside of ω, and A inverts into the center O of ω. The circle $\mathscr{C}$ inverts into the line OP', where P' is the inverse of P. Let the diameter OP' meet ω in the points α', β', which are the inverses of α, β. We know that although the inversion is a conjugate Moebius transformation, the cross-ratios $(\alpha, \beta; A, P)$ and $(\alpha', \beta'; O, P')$ are equal (Exercise 57.10). Hence

$$|\log (\alpha', \beta'; O, P')| = |\log (\alpha, \beta; A, P)| = r.$$

This is readily seen to imply that P' is at a given distance from O, on either side of O, and therefore the locus of P' is a circle in the Euclidean sense, center O. This circle is

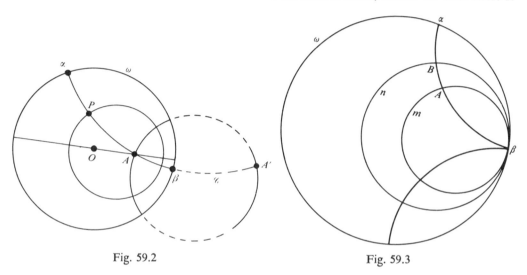

Fig. 59.2   Fig. 59.3

orthogonal to the pencil of lines through O, and therefore arises from the Euclidean circle which is orthogonal to the family of circles through A and A' in the original figure.

We note that the Euclidean center of this circle only coincides with the *p*-center A in one case. We leave the elucidation of this happening as an Exercise. (Exercise 59.3.)

Again, as the center A moves towards ω, and we keep P fixed, the *p*-distance $d(AP)$ tends towards infinity, and the *p*-circle center A tends towards a Euclidean circle which touches ω internally. Such a circle is called a *horocycle*.

Theorem II *Two horocycles tangent to ω at the same point β cut off equal p-distances on the p-lines through β.*

Proof. Let m and n be two horocycles touching ω at the point β (Fig. 59.3), and suppose that a *p*-line through β meets ω again in the point α, and intersects m in the *p*-point A and n in the *p*-point B. We know (Exercise 57.11) that inversion is a conjugate Moebius transformation. We also know (Exercise 57.10) that *p*-lengths are unaffected by a conjugate Moebius transformation. Let us therefore invert the given configuration in a circle center β (Fig. 59.4). The circles ω, m and n invert into parallel lines ω', m' and n', the p-line AB inverts into a line $\alpha'B'A'$, where α' is the point where it meets ω', B' is the point where it meets n', and A' is the point where it meets m'. The point β itself has inverted into the point ∞. What has happened to the cross-ratio $(\alpha, \beta; A, B)$? It has become the cross-ratio $(\alpha', \infty; A', B')$, and this is simply the ratio $\overline{\alpha'A'}/\overline{\alpha'B'}$. But for fixed parallel lines we know that this is constant. Hence $d(A, B)$ is independent of the particular p-line through β which cuts A and B on the given horocycles.

An alternative proof is indicated in Exercise 59.1.

It has been made clear that to the possible two-dimensional inhabitants of our

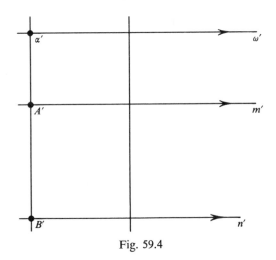

Fig. 59.4

space Ω, the boundary circle ω is unattainable, and corresponds to the line at infinity in the real Euclidean plane. Poincaré suggested physical conditions which might obtain in Ω which would make ω unattainable. He imagined a variation of temperature in Ω which would make the dimensions of the two-dimensional inhabitants shrink to zero as they approach ω, the Great Divide. As they shrink, their steps also shrink, and the law of shrinkage can be so chosen that the geodesic, or shortest path, in the Euclidean sense, between two points in Ω is precisely the p-line through the two points. The Poincaré model is an illustration of Hamlet's: 'I could be bounded in a nutshell and count myself a king of infinite space, were it not that I have bad dreams'.

We have seen that in our hyperbolic non-Euclidean geometry there are two p-parallels through a given p-point to a given p-line. It is possible to have a non-Euclidean geometry in which there are *no* parallels through a given point to a line.

We shall not investigate other types of non-Euclidean geometry. The Poincaré model, which we now leave, is a model of *hyperbolic* non-Euclidean geometry, and a non-Euclidean geometry with no parallels through a point to a given line is called an *elliptic* non-Euclidean geometry. In Exercise 59.2 a model for such a geometry is introduced.

The invention of non-Euclidean geometries in the nineteenth century was one of the great leaps forward of the human mind. We can well understand the excitement of János Bolyai, one of the inventors of non-Euclidean geometry, when he wrote to his father, who first taught him mathematics: 'Out of nothing I have created a new, different, world!' For further explorations of this different world we refer the interested reader to other books (**4, 19, 20, 23**).

We have listed the axioms of Euclid in an Appendix (p. 432) so that the reader may be able to refer to them, and compare the Euclidean theorems deduced from them with those we have obtained. It is hoped that the impact of a non-Euclidean geometry may serve to interest the reader in the foundations of geometry.

In our next chapter we discuss *projective geometry*. In this geometry two lines in a plane always intersect, and so this geometry also ranks as a non-Euclidean geometry.

Exercises

59.1   Let m and n be two horocycles touching ω at the point β, and suppose that one p-line through β meets m again in A, and n again in B, and that another p-line through β meets m again in A', and n again in B'. Show that, by an inversion in a suitable circle of the coaxal system defined by the two p-lines, the p-line βAB can be inverted into the p-line $\beta A'B'$, and that this inversion maps ω and the horocycles onto themselves, ω onto ω, m onto m and n onto n, so that A is mapped onto A' and B onto B'. Deduce, that since an inversion is a conjugate Moebius transformation, $d(A, B) = d(A', B')$.

59.2 S is the surface of a sphere in real three-dimensional Euclidean space, and O is the center. Show that we have a model of an elliptic non-Euclidean geometry if we take S as our Plane, and identify a *pair* of points on S which are diametrically opposed (that is, their join passes through the center) as a Point in the Plane. The Lines in our Plane are to be given by the intersections with S of Euclidean planes through O. Show that two distinct Points determine a unique Line, and that two distinct Lines always intersect in a unique Point, so that there are no parallels to a given Line through a given Point which does not lie on it.

59.3   The points A and A' are inverse in a circle ω, A lying inside ω, and $\mathscr{C}$ is a circle of radius > 0 which lies inside ω and is a member of the coaxal system polar to the coaxal system defined by circles which pass through A and A'. Show that if A' is a finite point, the center of $\mathscr{C}$ is never at A. When is the center of $\mathscr{C}$ at A?

59.4   Consider a Euclidean half-plane Ω bounded by the line ω, and all (semi-) circles in Ω which have their centers on ω, together with the set of all lines in Ω perpendicular to ω. Show how to set up a model of a hyperbolic geometry with this arrangement. What is the connexion with the model discussed in § 56.1?

VII

THE PROJECTIVE PLANE AND PROJECTIVE SPACE

We shall now investigate a geometry which is more fundamental than Euclidean geometry, in the sense that there will be no mention of metrical properties, distances or angle-measure, and the only concept we shall make use of at first is that of *incidence*. This geometry is called *projective geometry*, since its properties are invariant under *projection*, which is a geometrical transformation we shall investigate later. We shall enquire into the properties of projective space of n dimensions, after a special study of the cases $n = 2$ and $n = 3$, and since our treatment will be algebraic (non-algebraic treatments are also possible), we must say what kind of algebraic construct our numbers are to be taken from.

We shall work in a *commutative field*, although some of our theorems are also valid in a *skew* (non-commutative) field. For most of our purposes it will suffice to assume that our field is that of *the complex numbers*.

The outstanding property of this field which we may want to use is sometimes called *the fundamental theorem of algebra*: any polynomial with coefficients in the field of complex numbers splits into linear factors with coefficients which are complex numbers.

60.1 The complex projective plane

We must define the points of our geometry. A *point* is defined as an ordered triad of complex numbers (y_0, y_1, y_2) *not all zero*, with the understanding that the ordered triad $(y_0 k, y_1 k, y_2 k)$, where k is *any non-zero complex number*, represents the same point.

We must show that this relation is *an equivalence relation*, which is easily done, but this becomes evident if we remark that the ordered triad (y_0, y_1, y_2) represents a point distinct from the origin in three-dimensional complex affine space, and the set of points $(y_0 k, y_1 k, y_2 k)$ as k varies is the *ray*, or line through the origin joining the origin to the point. We are therefore defining a *line* or *ray* through the origin to be a *point* in our geometry, which is that of the complex projective plane.

The coordinates (y_0, y_1, y_2) we are using are usually called *homogeneous coordinates*, since only the ratios $y_0 : y_1 : y_2$ of the coordinates are significant, these being the same for the point $(y_0 k, y_1 k, y_2 k)$.

A *line* in our complex projective plane is defined to be the subset of points in the plane whose coordinates satisfy a linear homogeneous equation

$$u^0 X_0 + u^1 X_1 + u^2 X_2 = 0,$$

where not all the u^i ($i = 0, 1, 2$) are zero. Since, multiplying on the right by $k \neq 0$ we also have

$$u^0(X_0 k) + u^1(X_1 k) + u^2(X_2 k) = 0,$$

it follows that if one member of the equivalence class representing a point of our projective plane satisfies the equation, so does every member.

We note that in three-dimensional affine space the equation above is that of *a plane through the origin*. Hence rays through the origin are to be points of the projective plane, and *planes* through the origin are to be *lines* of the projective plane. A *point* of the projective plane is *incident* with, or *lies on* a *line* of the projective plane if and only if the *ray* corresponding to the point lies in the *plane* corresponding to the line.

We now prove two incidence theorems which explain why our algebraic formulation is chosen in what may appear, at first sight, to be a curious manner.

Theorem I   *In the projective plane, two distinct points determine a unique line with which they are incident.*

Theorem II   *In the projective plane two distinct lines intersect in a unique point.*

Remark. It is the second theorem which distinguishes this geometry from ordinary Euclidean geometry, where lines which are distinct may not intersect. In this geometry distinct lines *always* intersect.

Proof. We prove the second theorem first. Let

$$u^0 X_0 + u^1 X_1 + u^2 X_2 = 0 = v^0 X_0 + v^1 X_1 + v^2 X_2$$

be the equations of the two lines. Since they are distinct, we do *not* have

$$u^0 : u^1 : u^2 = v^0 : v^1 : v^2.$$

Solving the two homogeneous equations for the point of intersection of the two lines gives us as point of intersection:

$$(u^1 v^2 - u^2 v^1, \ u^2 v^0 - u^0 v^2, \ u^0 v^1 - u^1 v^0).$$

This triad could only fail to be a point if *each* of the elements of the triad displayed above were zero. This could only be the case if

$$[u^0, u^1, u^2] = k[v^0, v^1, v^2] \qquad (k \neq 0),$$

which would mean that the two lines are identical, and this possibility is contrary to hypothesis.

To prove Thm I we remark that if the two distinct points are

$$(x_0, x_1, x_2) \quad \text{and} \quad (y_0, y_1, y_2),$$

then we do *not* have

$$(x_0, x_1, x_2) = (y_0 k, y_1 k, y_2 k) \qquad (k \neq 0),$$

and the line

$$u^0 X_0 + u^1 X_1 + u^2 X_2 = 0$$

contains both points, where

$$u^0 = x_1 y_2 - x_2 y_1, \quad u^1 = x_2 y_0 - x_0 y_2, \quad \text{and} \quad u^2 = x_0 y_1 - x_1 y_0,$$

as can be immediately verified. This line exists, since we do not have $u^0 = u^1 = u^2 = 0$, the two points (x_0, x_1, x_2) and (y_0, y_1, y_2) being distinct points. Hence there is *a* line through the two points. There cannot be more than one, since we should then have two distinct lines intersecting in more than one point, contrary to Thm II.

To Theorems I and II we add a third, designed to ensure that the geometry we are considering is not trivial.

Theorem III The projective plane contains at least four distinct points, no three of which are collinear.

Proof. The points $(1, 0, 0)$, $(0, 1, 0)$, $(0, 0, 1)$ and $(1, 1, 1)$ are easily seen to satisfy this condition over the field of the complex numbers, since the first three points lie, in pairs, on the lines $X_2 = 0$, $X_0 = 0$ and $X_1 = 0$ respectively, and the fourth point does not lie on any of these lines.

It will be noticed that we use capital letters for the variables in an equation, lower-case letters for actual coordinates of points, and also for what we may evidently call *the coordinates of a line*: $[u^0, u^1, u^2]$, the line being

$$u^0 X_0 + u^1 X_1 + u^2 X_2 = 0.$$

We note, that just as with points, the ordered triad $[k u^0, k u^1, k u^2]$ $(k \neq 0)$ represents the same line as does the ordered triad $[u^0, u^1, u^2]$.

60.2   The principle of duality in the projective plane

The symmetrical nature of the incidence relation

$$u^0 y_0 + u^1 y_1 + u^2 y_2 = 0,$$

which expresses the fact that the line $[u^0, u^1, u^2]$ *contains* the point (y_0, y_1, y_2), or equivalently, that the point (y_0, y_1, y_2) *lies on* the line $[u^0, u^1, u^2]$ gives rise to the *Principle of Duality* in the projective plane. This asserts that by an automatic interchange of the terms *point* and *line*, *lying on* and *passing through*, *join* and *intersection*, *collinear* and *concurrent*, and so on, any theorem in the projective plane which involves only incidence properties of points and lines becomes, on transliteration with the help of the dictionary of interchanges we have just listed, a theorem involving lines and points.

To give an immediate example, the theorem of Desargues, which we have already encountered (§5.2) and which is essentially a theorem of the projective plane, says that if ABC, $A'B'C'$ are two triangles in the projective plane which are such that the lines AA', BB' and CC' are concurrent (that is, pass through the same point), then the three points of intersection of corresponding sides, $BC \cap B'C'$, $CA \cap C'A'$ and $AB \cap A'B'$ are collinear (lie on a line).

The theorem obtained by applying the Principle of Duality says that if abc, $a'b'c'$ are triangles (we should change the term *triangle*, which refers to a configuration formed by three *points*, to *triline*, to be precise, since we now consider two configurations formed by triads of lines, a, b and c and a', b' and c', these not being concurrent: but we shall not do this) in the projective plane which are such that the points aa', bb' and cc' are collinear (lie on a line) then the lines joining corresponding vertices, $bc \cup b'c'$, $ca \cup c'a'$ and $ab \cup a'b'$ are concurrent (pass through a point).

We shall see, in fact, that exactly the same algebra which we shall use to establish the Desargues Theorem will establish the dual theorem, which is the converse of the Desargues Theorem (§62.2).

It will be noted that we are using the symbol AB to denote the *line* joining two distinct points A and B, and dually the symbol ab to denote the *point* of intersection of the two distinct lines a and b. Since we use the symbol $\cap$ for the intersection of two point-sets, the point of intersection of two lines AB and $A'B'$ is written as $AB \cap A'B'$, and therefore dually we write $ab \cup a'b'$ for the line joining the points ab and $a'b'$, although the $\cup$ symbol here does *not* stand for the join of two point-sets. Another symbol, if used, would also create difficulties. Some authors use the symbol $A + B$ for the set of points on the line joining A to B. We shall not do this, and we do not think that there will be any misunderstanding in interpreting our symbolism.

The reader may wonder why we talk of the Principle of Duality, when our statement describing it amounts to a proof. Historically the statement was put in the form of a Principle. For us it is almost a self-evident theorem.

Exercises

60.1 Show that the equation of the line joining the two points (x_0, x_1, x_2) and (y_0, y_1, y_2) is the result of eliminating U^0, U^1, U^2 from the three equations $U^0X_0 + U^1X_1 + U^2X_2 = 0$, $U^0x_0 + U^1x_1 + U^2x_2 = 0$, $U^0y_0 + U^1y_1 + U^2y_2 = 0$, and is therefore the determinantal equation

$$\begin{vmatrix} X_0 & X_1 & X_2 \\ x_0 & x_1 & x_2 \\ y_0 & y_1 & y_2 \end{vmatrix} = 0.$$

60.2 Write out the dual of the Exercise above, changing the phrase 'equation of the line' to 'equation of the point'. (Note that the equation of a *line* is the condition that a point (X_0, X_1, X_2) lie on a line $[u^0, u^1, u^2]$. The equation of a *point* is the condition that a line $[U^0, U^1, U^2]$ pass through a point (x_0, x_1, x_2).)

60.3   Four distinct points A, B, C and D, no three of which are collinear, are said to form a complete quadrangle (or four-point), with *sides* formed by joining the points in pairs, and with *diagonal triangle* formed by the *diagonal points*, which are the intersections of the sides, other than A, B, C and D. Write down the dual of this definition, drawing a diagram in each case.

60.4   The Pappus Theorem says that if A, B, C are three distinct points on a line l, and A', B', C' three distinct points on a distinct line m, then the three points $BC' \cap B'C$, $CA' \cap C'A$ and $AB' \cap A'B$ are collinear. Write down the dual of this theorem, drawing a diagram in each case.

60.5   A set of points $(P, Q, R, \ldots,)$ on a line l is said to be *in perspective* with a set of points $(P', Q', R', \ldots)$ on a line l', if the joins PP', QQ', $RR', \ldots$ all pass through a point V. Formulate a dual concept for lines $(p, q, r, \ldots)$ on a point L, and lines $(p', q', r', \ldots)$ on a point L'.

60.6   The points on the surface of a given sphere which are diametrically opposed are identified as Points, and great circles cut on the surface of the sphere by planes through the center are identified as Lines. Show that these Points and Lines satisfy Theorems I, II and III of §60.1, the ground field being that of the real numbers.

60.7   Show that the points $A = (1, 0, 0)$, $B = (0, 1, 0)$, $C = (0, 0, 1)$, $D = (0, 1, 1)$, $E = (1, 0, 1)$, $F = (1, 1, 0)$ and $G = (1, 1, 1)$, where the field is that of the integers, mod 2 (that is $1 + 1 = 0$) form a finite projective geometry, with three points on every line and three lines through every point (See Pedoe, **12**, p. 85).

60.8   If all the elements involved, with the exception of the symbol ∞, are real numbers, show that the set of points and lines described satisfy Thms. I, II and III of §60.1. The points are: (*i*) ordinary points with coordinates (a, b); (*ii*) infinite points (m), together with the point ∞. The lines are given by the equations $X = c$, and $Y = mX + b$, and there is also the line at infinity ω. The line at infinity contains the point ∞ and all points (m), the line $X = c$ contains the point ∞ and all points (c, d), and the line $Y = mX + b$ contains the point (m) and all the points $(a, ma + b)$.

61.1   A model for the projective plane

We have already associated a *ray* through the origin of a three-dimensional space with a *point* of our projective plane. We pursue this further. Let V_3 be the complex vector space of vectors $y = (y_0, y_1, y_2)$, where the y_i are complex numbers. If $y \neq (0, 0, 0)$, the zero vector, then we may define a *ray* of V_3 as the set of vectors $yk = (y_0k, y_1k, y_2k)$ $(k \neq 0)$. We say that a *point* of our complex projective plane is a *ray* of V_3.

   Two vectors $y = (y_0, y_1, y_2)$ and $z = (z_0, z_1, z_2)$ of V_3 are said to be *linearly dependent* if there exist complex numbers λ and μ, not both zero, so that

$$y\lambda + z\mu = 0,$$

where 0 is the zero vector. If such a relation necessarily implies that $\lambda = \mu = 0$, we say that y and z are *linearly independent*. Since a point of the complex projective plane is defined to be a ray of V_3, we see that if two non-zero vectors y and z are linearly dependent, so that they define the same ray of V_3, then they also define the same point of our complex projective plane.

   If y and z are *independent*, and therefore neither is the zero vector, each defines a ray of V_3, and the set of vectors linearly dependent on y and z form a subspace of V_3.

In fact, if $x = y\lambda + z\mu$, and we write down the three equations for the coordinates of the vectors on each side of this equation, we have

$$y_0\lambda + z_0\mu - x_0 = 0,$$
$$y_1\lambda + z_1\mu - x_1 = 0,$$
$$y_2\lambda + z_2\mu - x_2 = 0,$$

and on eliminating λ and μ from these equations we have the determinantal equation

$$\begin{vmatrix} X_0 & X_1 & X_2 \\ y_0 & y_1 & y_2 \\ z_0 & z_1 & z_2 \end{vmatrix} = 0.$$

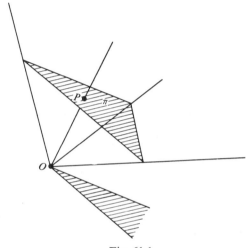

Fig. 61.1

This is a linear homogeneous equation in (X_0, X_1, X_2), and represents both the *plane* in V_3 defined by two distinct rays $y\lambda$ and $z\mu$ and the *line* in our projective plane which joins the two distinct points y and z.

It is clear that we may use a single symbol such as y or z for a point in our projective plane, with the understanding that $y\lambda$ represents the same point as y. We have seen that the set of points linearly dependent on the points y and z is the set $y\lambda + z\mu$. This gives the set of points on the *line* joining y and z.

If we assume that we are perfectly familiar with vector spaces V_3 over the complex number field, we have set up a *model* for the complex projective plane, since all the mathematical properties of the projective plane can be described or verified for V_3, and we assume that the mathematical properties of V_3 are free from contradiction. If we wish to *see* something, and this is what a model is for, let us consider how to set up homogeneous coordinates in a plane in real affine space. Call the plane π, and choose a system of three-dimensional coordinates for which the origin $O = (0, 0, 0)$ does not lie in π (Fig. 61.1). Then a ray through O which is not parallel to π intersects

π in a unique point P, and if the ray is the set of points (y_0k, y_1k, y_2k), where k varies over the real number field, the point P can be assigned the coordinates (y_0, y_1, y_2), where (y_0k, y_1k, y_2k) are also the coordinates of the same point P. This is precisely the development already given. Here there is a difference, since rays through O may be parallel to π. If such a ray is $(z_0, z_1, z_2)k$, we say that the point (z_0, z_1, z_2) is *an ideal point* in π, or *a point at infinity*. If the equation of the plane π is $u^0X_0 + u^1X_1 + u^2X_2 =$ constant, in our system of three-dimensional coordinates, the equation of the plane through O which is *parallel* to π is simply the equation $u^0X_0 + u^1X_1 + u^2X_2 = 0$, and we may take this same equation as *the equation of the line at infinity in the plane π*.

By introducing ideal points into the plane π, we have made it into a projective plane, in which all points have equal status, and two distinct lines always intersect.

Returning to our complex projective plane, which we shall also label π, we know that the vector space V_3 contains sets of 3 linearly independent vectors. One such set is $E_0 = (1, 0, 0)$, $E_1 = (0, 1, 0)$, $E_2 = (0, 0, 1)$. Vectors are, of course, linearly independent if there is no non-trivial linear homogeneous relation between them; that is, if the vectors be x, y and z, they are linearly independent when any linear relation $x\lambda + y\mu + z\nu = 0$ necessarily implies $\lambda = \mu = \nu = 0$. The vectors E_0, E_1 and E_2 are clearly independent. Since we may write any vector $y = (y_0, y_1, y_2)$ in V_3 as

$$y = E_0y_0 + E_1y_1 + E_2y_2,$$

and this same expression holds for points in π, we have shown that:

The system of homogeneous coordinates in π arises from the expression of a point P being linearly dependent on three fixed points which form a proper triangle in π.

We know that any three independent vectors in V_3 form *a basis* for the vectors of V_3: that is, any vector of V_3 can be expressed linearly in terms of three independent vectors. If the three independent vectors be named $E_0{}^*$, $E_1{}^*$, $E_2{}^*$ then any vector y of V_3 has a unique expression in the form

$$y = E_0{}^*\lambda + E_1{}^*\mu + E_2{}^*\nu.$$

Independent vectors in V_3 correspond, if there are two vectors, to two distinct points of π, and if there are three vectors, to three non-collinear points of π, forming a proper triangle. The theory of the basis for vectors in V_3 shows us that if any three non-collinear points be chosen in π, with coordinates $E_0{}^*$, $E_1{}^*$ and $E_2{}^*$, then any point in π may be expressed in the form

$$P = E_0{}^*\lambda + E_1{}^*\mu + E_2{}^*\nu,$$

where the ordered triad (λ, μ, ν) is any one of an equivalence class $(\lambda k, \mu k, \nu k)$ $(k \neq 0)$ (Fig. 61.2.)

If we call the triangle we have chosen *the triangle of reference*, we may call (λ, μ, ν) *the coordinates of the point P with respect to the given triangle of reference*.

We shall investigate, in §64.1 the effect of a change of triangle of reference on the coordinates of a given point. But for a given triangle of reference we now have enough apparatus to prove theorems on points, lines, their joins and intersections in the projective plane π.

Since we shall be investigating the projective geometry of n dimensions in §66.1, we have introduced a suffix notation (y_0, y_1, y_2) for our coordinates, but we may now revert to a simpler notation, calling the triangle of reference ABC, and the coordinates of a point in π: (x, y, z).

We note that the coordinates of A, B and C are respectively $(1, 0, 0)$, $(0, 1, 0)$ and $(0, 0, 1)$ referred to ABC as triangle of reference. Any point on BC is $(0, 1, 0)y + (0, 0, 1)z = (0, y, z)$, and in fact the equation of the line BC is $X = 0$.

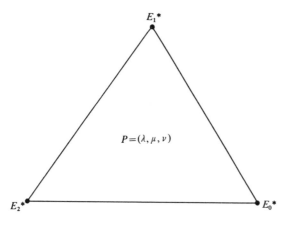

Fig. 61.2

If A' is the point $(0, q, r)$ on BC, any point on the line AA' is given by $(1, 0, 0)p + (0, q, r) = (p, q, r)$. Conversely, if the point (p, q, r) be joined to the vertex A, the intersection of this line with the side BC is the point $(0, q, r)$, and the equation of the line joining A to the point (p, q, r) is simply $rY - qZ = 0$.

The reader will find it interesting to compare the treatment of points and lines given here with that of points and lines in the real Euclidean plane given in Chapter I. Some of the results appear to be similar, but we do not deal with metrical results in the projective plane. We have a greater freedom, since our coordinates are not normalized (for instance, if A, B and C are three non-collinear points, any point P may be expressed as $P = Ax + By + Cz$, without having $x + y + z = 1$, as in Thm 4.2), but we do not have a metrical interpretation.

On the other hand, some theorems, such as the Theorem of Ceva, do have a corresponding theorem in the projective plane, since the metrical interpretation is not fundamental to the theorem.

Example. The lines AP, BP and CP cut the opposite sides of triangle ABC in the points L, M and N respectively. If $L = xB + x'C$, $M = yC + y'A$, $N = zA + z'B$, then

$$xyz = x'y'z'.$$

If $P = (p, q, r)$ with respect to ABC as triangle of reference, then $L = (0, q, r)$, $M = (p, 0, r)$ and $N = (p, q, 0)$, Since we are given that $L = (0, x, x')$, $M = (y', 0, y)$ and $N = (z, z', 0)$, we have

$$x : x' = q : r, \quad y : y' = r : p, \quad z : z' = p : q,$$

and therefore

$$(x : x')(y : y')(z : z') = (q : r)(r : p)(p : q) = 1.$$

Example. Theorem of Menelaus (compare Thm 4.1). A line cuts the sides BC, CA and AB of a triangle ABC in the points L, M and N respectively. If $L = Bx + Cx'$, $M = Cy + Ay'$ and $N = Az + Bz'$, then $xyz = -x'y'z'$.

Let the equation of the line be $eX + fY + gZ = 0$. Then L is given by the equation $fY + gZ = 0$, so that $L = (0, -g, f) = B(-g) + C(f)$. Hence $x : x' = -g : f$. Similarly $y : y' = -e : g$, and $z : z' = -f : e$, and the theorem follows immediately.

Exercises

61.1 Prove that if $A' = B + Ck$ is a point on BC, and if M be taken on AA' distinct from A or A', then if BM intersects CA in Q and CM intersects AB in R, the line QR will intersect BC in the point $B - Ck$.

61.2 The sides BC, CA and AB of the triangle of reference have coordinates $[1, 0, 0]$, $[0, 1, 0]$ and $[0, 0, 1]$ respectively. If we write

$$a = [1, 0, 0], \quad b = [0, 1, 0], \quad c = [0, 0, 1],$$

show that $a' = b + kc$ is a line through the point bc. Write down the theorem dual to that given in Exercise 61.1, and prove it, using essentially the same algebra.

61.3 If ABC be the triangle of reference, show that the equation $lX + nZ = 0$ represents a line through the point B, and the equation $pX + qY = 0$ represents a line through the point C. If the first line intersects the side AC in the point E, and the second line intersects the side AB in the point F, show that the equation of the line EF is $lpX + lqY + npZ = 0$.

61.4 If the lines $uX + vY + wZ = 0$, $u'X + v'Y + w'Z = 0$ intersect at the point P, show that the equation

$$(uX + vY + wZ)\lambda + (u'X + v'Y + w'Z)\mu = 0$$

represents a line which always passes through P, for all values of λ and μ except $\lambda = \mu = 0$. Find the equation of the line which joins P to the vertex A of the triangle of reference.

62.1 The normalization theorem for points on a line

Theorem *If P, Q and R are three distinct collinear points in π, then a coordinate system may be set up in which the homogeneous coordinates of P, Q and R, written as vectors y, z and t, satisfy the equation*

$$t = y + z.$$

Remark. We use a single symbol to represent a vector. When we use a single symbol to represent a point in π, this means that we have chosen a specific representative from the equivalence class of ordered triads which represent the point.

Proof. In any system of homogeneous coordinates where P, Q and R are represented by the vector symbols y', z' and t' respectively, we have a relation

$$y'\lambda + z'\mu + t'\nu = 0,$$

since the three points are linearly dependent. The λ, μ, v are not all zero. In fact, none of the multipliers is zero, since this would involve two of the points under consideration being identical, and this is excluded. Hence $v \neq 0$, and we may write

$$t' = y'\rho + z'\sigma,$$

where neither ρ nor σ is zero. Now write

$$y'\rho = y, \quad z'\sigma = z \quad \text{and} \quad t' = t,$$

so that y is a specific representative of the point P, z a specific representative of the point Q and t a specific representative of the point R, and we have

$$t = y + z.$$

Remark. If we set up a coordinate system on the line for which P and Q are base points, any point on the line being $X = Px_0 + Qx_1$, our procedure assigns the unit point, with coordinates $(1, 1)$, to R.

This tidying-up process enables us to prove the Desargues Theorem very simply.

62.2   The Desargues theorem

Theorem   *If two triangles ABC, $A'B'C'$ in π are such that the lines AA', BB' and CC' pass through a point V, then the three points $BC \cap B'C'$, $CA \cap C'A'$ and $AB \cap A'B'$ all lie on a line.*

Proof. We assume that the seven points V, A, B, C, A', B', C' are distinct (Fig. 62.1). In a system of homogeneous coordinates let ε denote the point V, and let $A = \alpha$, $B = \beta$, $C = \gamma$, $A' = \alpha'$, $B' = \beta'$, and $C' = \gamma'$.

By the Normalization Theorem of §62.1, we may write

$$\varepsilon = \alpha + \alpha' = \beta + \beta' = \gamma + \gamma'$$

simultaneously for the three pairs of points with which V is collinear. Let us write:

$$\rho = \beta - \gamma = \gamma' - \beta', \quad \sigma = \gamma - \alpha = \alpha' - \gamma', \quad \tau = \alpha - \beta = \beta' - \alpha'.$$

Then the point represented by ρ is linearly dependent on B and C, and therefore lies on the line BC, and it is also linearly dependent on B' and C', and therefore also lies on the line $B'C'$. If these lines are distinct, ρ is therefore the symbol for the point of intersection $BC \cap B'C'$. Similarly,

$$\sigma = CA \cap C'A' \quad \text{and} \quad \tau = AB \cap A'B'.$$

But

$$\rho + \sigma + \tau = \beta - \gamma + \gamma - \alpha + \alpha - \beta = 0.$$

Therefore ρ, σ and τ are linearly dependent, and represent three collinear points. Hence the points

$$BC \cap B'C', \quad CA \cap C'A' \quad \text{and} \quad AB \cap A'B'$$

are collinear.

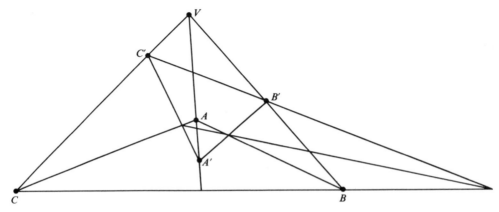

Fig. 62.1

Remark. With a similar definition of homogeneous coordinates to that we have given, and with lines defined to be sets of points linearly dependent on pairs of given distinct points, the projective plane can be defined over a skew field. The Normalization Theorem holds for such a field, and so does the Theorem of Desargues. It will be observed that commutativity under multiplication has not been used in the two proofs above.

When projective geometry is built up from incidence axioms, the Desargues Theorem for the plane is often invoked as an axiom. We can prove the Theorem here because we have assumed a system of homogeneous coordinates for our projective plane. It can be shown that the assumption of the Desargues Theorem leads to the possibility of the construction of a system of homogeneous coordinates over a skew field.

Finally, we note that when two triangles ABC, $A'B'C'$ are such that the lines AA', BB' and CC' pass through a point V, they are said to be *in perspective* from the point V, which is called *the center of perspective*. The line which contains the points $BC \cap B'C'$, $CA \cap C'A'$ and $AB \cap A'B'$ is called *the axis of perspective*. We may also say that the two triangles are *in perspective* from the axis of perspective, and the Desargues Theorem says that two triangles in perspective from a point are also in perspective from a line. The dual, or converse, reverses the order of these statements.

Exercises

62.1 State and prove the dual to the Normalization Theorem (§62.1).

62.2 State and prove the dual of the Desargues Theorem.

62.3 Show that the dual of the Desargues Theorem is implied by the Desargues Theorem, without the intervention of the Principle of Duality. (Hint: we wish to show that three certain lines pass through a point. This can be shown to be equivalent to three certain points lying on a line. Choosing two triangles which satisfy the Desargues Theorem, this follows.)

62.4 If the sides of a variable triangle pass through three fixed collinear points, while two vertices move along fixed lines, prove that the third vertex must move along a third fixed line concurrent with the other two.

62.5 Three triangles $A_iB_iC_i$ ($i = 1, 2, 3$) are such that any two of them are in perspective, the center of perspective of $A_iB_iC_i$ and $A_jB_jC_j$ being V_{ij}. If V_{12}, V_{23} and V_{31} are collinear, prove that the three triangles have a common axis of perspective l, and that any two of the triangles $A_1A_2A_3$, $B_1B_2B_3$ and $C_1C_2C_3$ are in perspective, the three centers of perspective lying on l.

62.6 A, B, C, D is a quadrangle. P is a point on BD, Q, R are points on AB, AD, and S, T are points on CB, CD. Prove that if P, Q, R are collinear, and also P, S, T, then the lines QS, RT and AC are concurrent.

62.7 D is a fixed point on the side AB of a given triangle ABC, H and K are fixed points which do not lie on any side of the triangle. P is a variable point on CD, and AP intersects DH in Q, and BP intersects DK in R. Prove that QR meets AB in a fixed point. (If P' be another position of P, and Q', R' the corresponding constructed points, we wish to show that QR, AB, and $Q'R'$ are concurrent. Application of the converse of Desargues' Theorem to two suitable triangles produces the result.)

62.8 D is a point on the side BC of triangle ABC, and P, Q are points on AB. The line PD meets AC in H, the line QD meets AC in K, and CP meets AD in M, while CQ meets AD in N. Prove that KM and HN meet on AB. (For a geometrical proof, not using coordinates, see Exercise 63.6.)

63.1 The normalization theorem for points in a projective plane

Theorem I *If A, B, C and P are four distinct points in a projective plane π, and no three are collinear, then a system of homogeneous coordinates can be set up so that the symbols x, y, z and t for the respective points satisfy the relation $t = x + y + z$.*

Proof. Since four points in π are linearly dependent, if x', y', z' and t' are the vector symbols for A, B, C and P in our system of homogeneous coordinates, we have a relation $x'\lambda + y'\mu + z'\nu + t'\tau = 0$, where not all of the multipliers λ, μ, ν, τ are zero. In fact, none of them can be zero since if, say, $\lambda = 0$, this would imply the linear dependence, and therefore collinearity of the points B, C and P. Hence we may write

$$t' = x'\lambda' + y'\mu' + z'\nu',$$

where none of the multipliers λ', μ', ν' is zero. We may now *absorb* the multipliers λ', μ' and ν', and write

$$x = x'\lambda', \quad y = y'\mu', \quad z = z'\nu',$$

where x is a definite symbol for A, y for B and z for C, and writing $t = t'$ we have

$$t = x + y + z.$$

Remark. If A, B and C are chosen as vertices of the triangle of reference in π, our theorem asserts that any point P *not on the sides* of ABC can be taken to have the coordinates $(1, 1, 1)$. This is called *fixing the unit point* of the coordinate system.

Care has to be taken in applying this theorem, so that we do not assign the same coordinates to two distinct points in the plane. Students tend to do this at first, and are therefore able to prove all true and untrue theorems they set their minds to!

The dual of the above theorem is also important, and we state it without proof.

Theorem II *If a, b, c and p are four distinct lines in a projective plane π, and no three are concurrent, then a system of homogeneous coordinates may be set up so that the symbols l, m, n and q for the respective lines satisfy the relation*

$$q = l + m + n.$$

Remark. If a, b and c are chosen as sides of the triangle of reference, the fourth line p, which does not pass through any vertex, can be written as

$$X + Y + Z = 0.$$

Example. ABC is a triangle, and O is a point in the plane of the triangle which does not lie on any side. The line AO meets BC in A', the line BO meets CA in B', and the line CO meets AB in C'. We prove that the points

$$L = B'C' \cap BC, \quad M = C'A' \cap CA, \quad N = A'B' \cap AB$$

are collinear, and that a coordinate system may be chosen so that the equation of the line LMN is $X + Y + Z = 0$.

Remark. The line LMN is called *the polar line* of the point O with respect to the triangle ABC.

We choose ABC as triangle of reference (Fig. 63.1), and O as unit point, as we may do, by Thm I above. Then $A' = (0, 1, 1)$, $B' = (1, 0, 1)$ and $C' = (1, 1, 0)$. To find L, we consider the set of points $(1, 0, 1)\lambda + (1, 1, 0)\mu$, which is the set of points on $B'C'$, and choose λ, μ so that the resulting point lies on BC, which means that its first coordinate is zero. This is evidently done by taking $\lambda = 1$ and $\mu = -1$, so that $L = (1, 0, 1) - (1, 1, 0) = (0, -1, 1)$.

Similarly $M = (1, 0, -1)$ and $N = (-1, 1, 0)$, and these three points satisfy the equation $X + Y + Z = 0$.

Theorem III *The Pappus Theorem. If D, A, E are distinct points on a line l, and C, F, B are distinct points on another line m, coplanar with l, then the three points $DF \cap AC = V$, $DB \cap CE = U$ and $EF \cap AB = W$ are collinear. (See Fig. 63.2.)*

Remark. The Pappus Theorem appeared in §9.5, when the line of collinearity UVW is the line at infinity. Our naming of the points here is not the standard one, since we

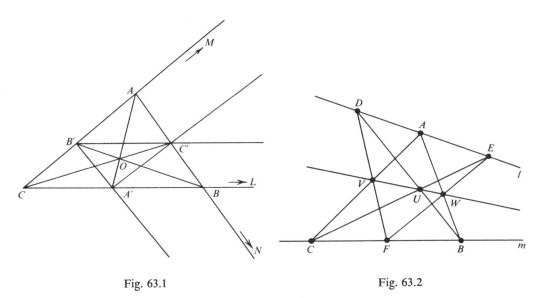

Fig. 63.1   Fig. 63.2

wish to name the triangle of reference we intend to use to prove the theorem as ABC, and the unit point as U.

Proof. We choose ABC as triangle of reference, and U as the unit point, so that $U = A + B + C$. We then take E, which is on UC as $E = U + Ck$. We now find $D = AE \cap BU$. Since $E = A + B + C(1 + k)$, we see that

$$E + Ak = A(1 + k) + B(1 + k) + C(1 + k) - Bk$$
$$= U(1 + k) - Bk,$$

and therefore the point represented, lying on AE and BU, must be D. Hence

$$D = E + Ak = A(1 + k) + B + C(1 + k).$$

The point F is not yet defined, and we take

$$F = B + Ck'.$$

V lies on AC and FD, and we find that

$$V = -F + D = A(1 + k) + C(1 + k - k').$$

Finally, W lies on AB and EF, and to find W we have to eliminate C from

$$E = A + B + C(1 + k), \quad \text{and} \quad F = B + Ck',$$

and

$$Ek' - F(1 + k) = Ak' + B(k' - 1 - k) = W.$$

We note that in obtaining this result we assumed that $k'(1 + k) = (1 + k)k'$, that is, we assumed the commutativity of multiplication for the symbols of our ground field. Since we began by assuming that it is the field of complex numbers, we may do this.

We now have

$$U = A + B + C$$
$$V = A(1 + k) + B \cdot 0 + C(1 + k - k')$$
$$W = Ak' + B(k' - 1 - k) + C \cdot 0,$$

and evidently

$$U(1 + k - k') - V + W = 0,$$

so that the points U, V and W are collinear.

The proof is not as elegant as that of the Desargues Theorem, and algebraically we expect this, since it can be shown that the Pappus Theorem is equivalent to the ground field being commutative, and so at some stage of the proof this assumption enters. No such assumption, as we saw, entered into the proof of the Desargues Theorem. Both theorems will appear again later, when we shall have the means to prove them without using algebra (Thm V, §71.2, Exercise 71.5).

Exercises

63.1 In the proof of the Pappus Theorem (Thm III above) we tacitly assumed, in choosing U as unit point with respect to triangle ABC, that U does not lie on any side of ABC. Show that if U does lie on a side of ABC the initial assumptions of the theorem are invalidated, or the theorem becomes trivial.

63.2 The line LMN is the polar line of the point O with respect to the triangle ABC, and $A' = AO \cap BC$, $B' = BO \cap CA$, $C' = CO \cap AB$. The point P does not lie on a side of the triangle ABC, and AP, BP and CP intersect the polar line in the points F, G and H respectively. Prove that the lines FA', GB' and HC' are concurrent. (With the notation of the preceding Example, p. 254, take P as (p, q, r), and show that the equation of the line FA' is $(q - r)(X + Y + Z) + 2rq(Y/q - Z/r) = 0$. If the lines FA', GB' and HC' are $u = 0$, $v = 0$ and $w = 0$ respectively, we have

$$pu + qv + rw \equiv 0,$$

which proves that the lines are linearly dependent, that is, concurrent.)

63.3 ABC is a given triangle, and D, E and F are fixed non-collinear points on BC, CA and AB respectively. A variable line through A meets DE and DF in P and Q respectively. Prove that the intersection R of BP and CQ lies on a fixed line.

63.4 *L*, *M* and *N* are collinear points on the sides *BC*, *CA* and *AB* of a triangle *ABC*, and *A'*, *B'* and *C'* are chosen on *AL*, *BM* and *CN* respectively. *B'C'*, *C'A'* and *A'B'* meet *BC*, *CA* and *AB* respectively in the points *P*, *Q* and *R*. Prove that *A'P*, *B'Q* and *C'R* are concurrent.

63.5 If *ABC* and *DEF* are two coplanar triangles such that *AD*, *BE*, *CF* are concurrent, and also *AE*, *BF*, *CD* are concurrent, prove that *AF*, *BD* and *CE* are concurrent. (With *ABC* as triangle of reference, and the points *D*, *E*, *F* as $(1, p, p')$, $(q', 1, q)$ and $(r, r', 1)$, the first condition gives $pqr = p'q'r'$. Proceed from this result. The problem can also be solved by applying the Pappus Theorem.)

63.6 *D* is a point on the side *BC* of a triangle *ABC*, and *P*, *Q* are points on *AB*. The line *PD* meets *AC* in *H*, the line *QD* meets *AC* in *K*, and *CP* meets *AD* in *M*, while *CQ* meets *AD* in *N*. Prove that *KM* and *HN* meet on *AB*. (This can be proved by using the Pappus Theorem, noting that the line *AB* is also the line *PQ*. For a proof involving only the Desargues Theorem, see Exercise 69.8.)

64.1 Change of triangle of reference

Since we wish to introduce matrix notation, we now switch back to the suffix notation for coordinates. Let *ABC* and *A'B'C'* be two triangles in the projective plane π, and suppose that a point *P* has coordinates (y_0, y_1, y_2) with respect to *ABC* as triangle of reference, and coordinates (y_0', y_1', y_2') with respect to *A'B'C'* as triangle of reference. This means that we may write

$$P = Ay_0 + By_1 + Cy_2 \quad \text{and} \quad P = A'y_0' + B'y_1' + C'y_2'.$$

Now suppose that *A* has coordinates (a_{00}, a_{10}, a_{20}) with respect to triangle *A'B'C'*, so that

$$A = A'a_{00} + B'a_{10} + C'a_{20},$$

and similarly suppose that

$$B = A'a_{01} + B'a_{11} + C'a_{21},$$
$$C = A'a_{02} + B'a_{12} + C'a_{22}.$$

Then

$$\begin{aligned} P &= (A'a_{00} + B'a_{10} + C'a_{20})y_0 + (A'a_{01} + B'a_{11} + C'a_{21})y_1 \\ &\qquad + (A'a_{02} + B'a_{12} + C'a_{22})y_2 \\ &= A'(a_{00}y_0 + a_{01}y_1 + a_{02}y_2) + B'(a_{10}y_0 + a_{11}y_1 + a_{12}y_2) \\ &\qquad + C'(a_{20}y_0 + a_{21}y_1 + a_{22}y_2) \\ &= A'y_0' + B'y_1' + C'y_2'. \end{aligned}$$

Since *A'*, *B'* and *C'* are linearly independent, any linear relation between them must be trivial, with zero multipliers. Hence we may equate the multipliers of *A'*, *B'* and *C'* on both sides of our last equation, and we have

$$y_0' = a_{00}y_0 + a_{01}y_1 + a_{02}y_2,$$
$$y_1' = a_{10}y_0 + a_{11}y_1 + a_{12}y_2,$$
$$y_2' = a_{20}y_0 + a_{21}y_1 + a_{22}y_2,$$

so that, in matrix form

$$y' = (y_0', y_1', y_2')^T = \mathbf{A}y,$$

where $y = (y_0, y_1, y_2)^T$, and $\mathbf{A} = [a_{ij}]$ $(i, j = 0, 1, 2)$.

Of course, an equivalent equation is

$$y'k = \mathbf{A}y \qquad (k \neq 0),$$

since we are dealing with homogeneous coordinates.

We can make an immediate statement about the square matrix $\mathbf{A}$. Since we may also express y in terms of y', *the matrix $\mathbf{A}$ has an inverse*, and the expression of y in terms of y' must be

$$y = \mathbf{A}^{-1}y'.$$

We have therefore proved

Theorem I   *A change of triangle of reference in the projective plane is equivalent to a non-singular linear transformation of coordinates.*

We consider the effect of a change of triangle of reference on the equation of a line. If the equation of a line is $u^0 Y_0 + u^1 Y_1 + u^2 Y_2 = 0$, we saw that we may talk of *line coordinates* $[u^0, u^1, u^2]$, and designate line coordinates by a single symbol $u = [u^0, u^1, u^2]$. Now suppose that the equation of a line is $v^0 Y_0' + v^1 Y_1' + v^2 Y_2' = 0$ in the primed system of coordinates. Then its original equation in the unprimed system was

$$v^0(a_{00} Y_0 + a_{01} Y_1 + a_{02} Y_2) + v^1(a_{10} Y_0 + a_{11} Y_1 + a_{12} Y_2)$$
$$+ v^2(a_{20} Y_0 + a_{21} Y_1 + a_{22} Y_2) = 0,$$

that is

$$(v^0 a_{00} + v^1 a_{10} + v^2 a_{20}) Y_0 + (v^0 a_{01} + v^1 a_{11} + v^2 a_{21}) Y_1$$
$$+ (v^0 a_{02} + v^1 a_{12} + v^2 a_{22}) Y_2 = 0.$$

Hence, if $y' = \mathbf{A}y$ is the matrix transformation of coordinates,

$$[u^0, u^1, u^2] = [v^0, v^1, v^2] \begin{bmatrix} a_{00} & a_{01} & a_{02} \\ a_{10} & a_{11} & a_{12} \\ a_{20} & a_{21} & a_{22} \end{bmatrix}$$

is the matrix transformation between the original and the new line coordinates. Once more, if $y' = \mathbf{A}y$ gives the new point-coordinates y' in terms of the original point-coordinates y, then $u' = u\mathbf{A}^{-1}$ gives the new line-coordinates u' in terms of the original line-coordinates u.

We note that a point is written as a *column*-vector, and a line as a *row*-vector in these computations. Of course, the calculations may be considerably simplified by using matrix notation. The equation of a line is $uy = 0$, and in the new coordinates it is $vy' = 0$. If we substitute $y' = \mathbf{A}y$, the equation becomes $v\mathbf{A}y = 0$, so that $u = v\mathbf{A}$, or if we call the new line coordinates u', then $u = u'\mathbf{A}$, which is the result obtained above, with the working given in detail.

We therefore have

Theorem II   *If we change to a new triangle of reference by the non-singular transformation of point coordinates* $y' = Ay$, *then the corresponding transformation in line-coordinates is given by* $u' = uA^{-1}$.

Example. Let O be a point not on the sides of a triangle ABC. Let $AO \cap BC = A'$, $BO \cap CA = B'$ and $CO \cap AB = C'$. We have proved in the Example of §63.1 that we may choose the coordinates so that the equation of the line on which the points $L = B'C' \cap BC$, $M = C'A' \cap CA$ and $N = A'B' \cap AB$ lie is given by the equation $X_0 + X_1 + X_2 = 0$. We now show that with respect to $A'B'C'$ as triangle of reference,

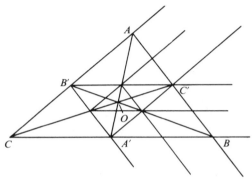

Fig. 64.1

the equation of this line can also be taken as $X_0' + X_1' + X_2' = 0$, and we shall deduce *that O has the same polar line with respect to the triangles ABC and $A'B'C'$.*

Once again we take O as unit point with respect to triangle ABC (Fig. 64.1), so that $A' = (0, 1, 1)$, $B' = (1, 0, 1)$ and $C' = (1, 1, 0)$.

Since A', B' and C' form a proper triangle, we may take $A'B'C'$ as a new triangle of reference, and express the points A, B and C linearly in terms of A', B' and C'. Clearly

$$-2A = (-2, 0, 0) = \quad (0, 1, 1) - (1, 0, 1) - (1, 1, 0),$$
$$-2B = (0, -2, 0) = -(0, 1, 1) + (1, 0, 1) - (1, 1, 0),$$
$$-2C = (0, 0, -2) = -(0, 1, 1) - (1, 0, 1) + (1, 1, 0),$$

and the above theory tells us that it is the transpose of this matrix, which gives A, B and C in terms of A', B' and C' which is the matrix $\mathbf{A}$ we are looking for. The matrix is symmetric, and it is

$$\mathbf{A} = \begin{bmatrix} 1 & -1 & -1 \\ -1 & 1 & -1 \\ -1 & -1 & 1 \end{bmatrix}.$$

The point $O = (1, 1, 1) = (A' + B' + C')/2$, and so its coordinates are still $(1, 1, 1)$ with respect to the triangle $A'B'C'$. Now let us see what happens to the equation $X_0 + X_1 + X_2 = 0$, which has line coordinates $[1, 1, 1]$. The new line coordinates u' are connected with the original ones by the transformation $u' = [1, 1, 1]\mathbf{A}^{-1}$.

Since

$$\begin{bmatrix} 0 & -1 & -1 \\ -1 & 0 & -1 \\ -1 & -1 & 0 \end{bmatrix}\begin{bmatrix} 1 & -1 & -1 \\ -1 & 1 & -1 \\ -1 & -1 & 1 \end{bmatrix} = 2\begin{bmatrix} 1 & 0 & 0 \\ 0 & 1 & 0 \\ 0 & 0 & 1 \end{bmatrix},$$

$$u' = [1, 1, 1]\begin{bmatrix} 0 & -1 & -1 \\ -1 & 0 & -1 \\ -1 & -1 & 0 \end{bmatrix} = [-2, -2, -2]$$

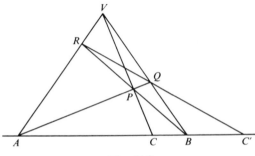

Fig. 64.2

so that the new line-coordinates of the polar line may also be taken as $[1, 1, 1]$. But these are the line coordinates of the polar line of O with respect to the triangle $A'B'C'$, since O has coordinates $(1, 1, 1)$ with respect to this triangle. Hence O has the same polar line with respect to triangles ABC and $A'B'C'$.

The geometric meaning of this result is left as an exercise for the reader, and the explanation is simple in terms of harmonic ranges and pencils, with which we end this discussion of the projective geometry of the plane, leaving many other important matters to be dealt with later.

64.2 Harmonic ranges

A, B and C are three distinct collinear points, and V is any point not on the line AB. We take any point P on VC distinct from V and from C, and call the intersection of BP and VA the point R and let Q be the intersection of AP and VB. Then if the line QR intersects AB in C', we show that C' depends only on A, B and C, and not on the choice of V and P (Fig. 64.2).

In fact, we show that if $C = A + B\lambda$, then $C' = A - B\lambda$.

Since P is on VC, we may write

$$P = C + V\mu = A + B\lambda + V\mu.$$

Therefore

$$P - A = B\lambda + V\mu = Q,$$

since the point represented by each term of this equation lies on AP and also on BV. Again

$$P - B\lambda = A + V\mu = R.$$

Also, C' is on RQ and also on AB, and since

$$R - Q = P - B\lambda - (P - A) = A - B\lambda,$$

this must be the point C', so that we have obtained

$$C' = A - B\lambda.$$

We call C' *the harmonic conjugate* of C with respect to A and B. Our algebraic result shows that if we start with C', A and B, and perform the above construction, we arrive at C, since $-(-\lambda) = \lambda$. Hence C and C' are paired, just as A and B are, to start with. But the pairing is in the same sense, since

$$2B\lambda = B = (A + B\lambda) - (A - B\lambda) = C - C'$$

shows that if we begin with C and C' instead of A and B, and B instead of C, the harmonic construction leads to

$$C + C' = (A + B\lambda) + (A - B\lambda) = 2A = A.$$

Hence, *if we say that C and C' are harmonic conjugates with respect to A and B, we have also proved that A and B are harmonic conjugates with respect to C and C'.* The points A, B and C, C' are said to form *a harmonic range*.

Metrically, we encountered harmonic ranges in §19.1. Points which divide a segment AB internally and externally in the same numerical ratio are harmonic conjugates with respect to A and B. This is indicated by the symbols $A + B\lambda$ and $A - B\lambda$ for C and C'.

There is a dual theory for harmonic pencils, and an important connecting link which asserts that *the range of points given by the intersections of a line with the four rays of a harmonic pencil is a harmonic range,* but we leave the proof of this until §69.1.

Here we remark that the point of intersection, in the construction above, of VC and RQ can be written as $R + Q = 2V\mu + (A + B\lambda)$, and since $R - Q = C'$, this shows that *the harmonic conjugate of C' with respect to R and Q is the point $VC \cap RQ$.* Hence, if $BR \cap QC = R'$ and $AQ \cap CR = Q'$, the line $Q'R'$ passes through C'.

We now see why the polar line of O, in the Example of §64.1, with respect to both triangles, ABC and $A'B'C'$, is the same. Another method of proof is indicated in Exercise 64.1.

64.3   Abstract projective planes

As we have already indicated in Exercises 60.7 and 60.8, the concept of *projective plane* can be generalized. Suppose that we have a finite or infinite set S of elements P_1, P_2, . . . together with a finite or infinite collection of *special subsets* l_1, l_2, . . . of S, and the concept of *inclusion*. Thus, if P is an element of S which is in a special subset l of S, we shall write $P \in l$. Then an *abstract projective plane* is a set S, which with its special subsets l satisfies the following three properties:

I. If P and Q are any two distinct elements of S, there is a unique special subset l of S such that $P \in l$ and $Q \in l$.

II. If l_1 and l_2 are any two distinct special subsets of S, there is a unique element P of S such that $P \in l_1$ and $P \in l_2$.

III. There are at least four distinct elements of S such that no three of them are elements of a special subset l of S.

In §60.1, these properties were written as Theorems I, II and III, our complex projective plane being a special case of an abstract projective plane. In Exercise 60.7 a finite projective plane with seven points was introduced, the ground field being that of the integers modulo 2, and in Exercise 60.8 it was indicated that once coordinates have been introduced into the real affine plane, further 'ideal' points and lines can be added so that we obtain a projective plane.

The subject of finite projective planes is a most important one. It will not be discussed here. Further discussion can be found in Pedoe (**12**), Chapter V. But we give some Exercises which indicate some consequences of the three axioms which are to be satisfied by the elements (points) and special subsets (lines) of an abstract projective plane.

Exercises

64.1   Prove the result of the Example in §64.1 by considering the triangles ABC and $A'B'C'$ as perspective triangles, and then the triangles $A'B'C'$ and $A''B''C''$ as perspective triangles, where $A'' = B'C' \cap AA'$, $B'' = C'A' \cap BB'$ and $C'' = A'B' \cap CC'$.

64.2   Show that if we have a harmonic range A, B, $A + B\lambda$, $A - B\lambda$ referred to one system of coordinates, and with respect to another triangle of reference A becomes A' and B becomes B', then $A + B\lambda$ becomes $A' + B'\lambda$ and $A - B\lambda$ becomes $A' - B'\lambda$.

64.3   If A, B, C, D are four points in a plane, no three of which are collinear, prove that their coordinates a, b, c, d may be so chosen that $a + b + c + d = 0$. If $P = AB \cap CD$, $Q = AC \cap BD$ and $R = AD \cap BC$, so that P, Q and R are the diagonal points of the quadrangle formed by A, B, C and D, show that the coordinates p, q and r of P, Q and R satisfy the equations

$$p = a + b = -c - d,$$
$$q = a + c = -b - d,$$
$$r = a + d = -b - c.$$

64.4   Deduce from the preceding Exercise that any two diagonal points of a complete quadrangle are harmonic conjugates with respect to the points in which their join meets the two sides of the quadrangle which do not contain the two diagonal points.

64.5   State and prove the dual theorems corresponding to the two preceding Exercises.

64.6 OA, OB, OC are given lines. Variable lines MPQ, $MP'Q'$ meet OA in M, OB in P, P' and OC in Q, Q'. Finally PQ' meets $P'Q$ in R. Prove that R lies on a fixed line.

64.7   Let the diagonal triangle of a complete quadrangle be LMN, and let P be a given point in the plane. Prove that the harmonic conjugates of LP, MP and NP relative to the pairs of opposite sides of the quadrangle which meet in L, M and N respectively are concurrent.

64.8 ABC is a triangle, and P, Q are given points. A', B' and C' are the harmonic conjugates relative to P, Q of the points in which PQ meets BC, CA and AB respectively. Prove that AA', BB' and CC' are concurrent.

64.9   Consider the non-singular linear mapping given by the equations

$$(X_0', X_1', X_2') = (3X_0 - X_1 + 2X_2, -2X_0 + 2X_1 - 2X_2, 2X_0 - X_1 + 3X_2),$$

and let the transforms of P and Q be P' and Q'. Prove that PP' passes through the point $(1, -1, 1)$, that the line PP' is mapped onto itself, and that the lines PQ and $P'Q'$ intersect on the line $2X_0 - X_1 + 2X_2 = 0$.

64.10   Prove that in an abstract projective plane at least three points lie on every line, and at least three lines pass through every point.

64.11   A tennis match is played between two teams, each player playing one or more members of the other team. We are told that:

(i) any two members of the same team have exactly one opponent in common,

(ii) no two members of the same team play all the members of the other team between them.
Prove that two players who do not play each other have the same number of opponents. Deduce that any two players, whether in the same or different teams, have the same number of opponents.

64.12   Let $n \geqslant 2$ be any integer, and suppose that we are considering an abstract projective plane with a finite number of points. Show that the following properties are equivalent:

(1) One line contains exactly $n + 1$ points.
(2) One point is on exactly $n + 1$ lines.
(3) Every line contains exactly $n + 1$ points.
(4) Every point is on exactly $n + 1$ lines.
(5) There are exactly $n^2 + n + 1$ points.
(6) There are exactly $n^2 + n + 1$ lines.

65.1   Projective space of three dimensions

We now introduce homogeneous coordinates into what we call a *three-dimensional projective space*. We consider a four-dimensional vector space V_4 over the complex field, and define *a point of three-dimensional projective space*, which we shall call S_3, to be *a ray*, which arises from *a vector of V_4*. That is, a point of S_3 is to be the equivalence class (y_0k, y_1k, y_2k, y_3k) $(k \neq 0)$, where $y = (y_0, y_1, y_2, y_3)$ is *a non-zero vector of V_4*.

Once again, linear dependence of points in S_3 is defined in the obvious way, and the points in S_3 linearly dependent on two distinct points of S_3 determine a line. If y and z are any representatives of the two points, the set of points $y\lambda + z\mu$ determine a line which contains the two points. But now that we obtain S_3 from a V_4, we may consider the set of points linearly dependent on three points, not already linearly dependent, which means that they are not collinear. If the points be denoted by the symbols y, z and t, the set is $y\lambda + z\mu + t\nu$, and we say this set of points determines *a plane* in S_3, the plane (yzt).

In V_4 there are sets of four independent vectors, but any five vectors are dependent. Hence in S_3 there are sets of four independent points, but any set of five points is a dependent set. Suppose that we now let A, B, C, D and E be the symbols for five points in S_3. Then we have a relation

$$Aa + Bb + Cc + Dd + Ee = 0,$$

where not all of the multipliers are zero. If we now write

$$Aa + Bb + Cc = -Dd - Ee,$$

the left-hand side represents a point in the plane ABC, and the right-hand side a point on the line DE. Hence we have one of the Propositions of Incidence in S_3:

A plane either contains a given line, or intersects it in one point.

Suppose now that we have two distinct planes, the one defined by the points A, B, C and the other by the points D, E, and F. Then the line DE either lies in the plane ABC or intersects it in a point, and the line EF either lies in the plane ABC or intersects it in a point. Since the planes are distinct, we do not have both lines, DE and EF in the plane ABC. If one of these lines lies in the plane ABC, the two planes ABC and DEF already intersect in a line. If neither lies in the plane ABC, we have two points in the plane DEF which lie in the plane ABC, and therefore the line joining the two points lies in ABC. Hence

Two planes which are distinct always intersect in the points of a line.

We note that there are no exceptions to the Propositions of Incidence in S_3, since planes are never parallel.

We have not written all the Propositions of Incidence in a form which would indicate that the Principle of Duality (§60.2) also applies to S_3. It does, of course, but in S_3 the fundamental interchange is between point and plane. Algebraically this is seen as soon as we show that a plane can also be given by homogeneous coordinates. If

$$(X_0,\ X_1,\ X_2,\ X_3) = (a_0,\ a_1,\ a_2,\ a_3)\lambda + (b_0,\ b_1,\ b_2,\ b_3)\mu + (c_0,\ c_1,\ c_2,\ c_3)\nu,$$

the elimination of the parameters λ, μ, and ν leads to the determinantal equation

$$\begin{vmatrix} X_0 & X_1 & X_2 & X_2 \\ a_0 & a_1 & a_2 & a_3 \\ b_0 & b_1 & b_2 & b_3 \\ c_0 & c_1 & c_2 & c_3 \end{vmatrix} = 0,$$

which is of the form

$$u^0 X_0 + u^1 X_1 + u^2 X_2 + u^3 X_3 = 0,$$

a homogeneous linear equation. Hence a plane can be assigned homogeneous coordinates $[u^0, u^1, u^2, u^3]$, and belongs to an equivalence class $k[u^0, u^1, u^2, u^3]$ $(k \neq 0)$. Two distinct planes intersect in a unique line, and the equation of a line is often given by the assignment of two distinct planes which contain it. We may use matrix notation, and write the equation of a plane as

$$uX = 0,$$

where $u = [u^0, u^1, u^2, u^3]$ and $X = (X_0, X_1, X_2, X_3)^T$, so that once again we write *plane-coordinates* as *row-vectors* and *point-coordinates* as *column-vectors*. Then the equation of a line may be written

$$uX = 0 = vX,$$

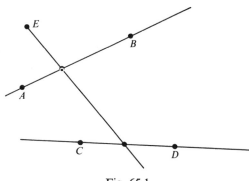

Fig. 65.1

where u and v are two distinct planes through the line.

The condition that a given plane u should contain a given point x is

$$ux = 0,$$

and this is a symmetrical relation between the plane-coordinates u and the point-coordinates x. Hence we have a Principle of Duality for S_3, in which *point* and *plane* and *plane* and *point* are interchanged. Since a line is determined by two points, and also by two planes, *lines remain unchanged* in the transliteration which produces a dual theorem from a given theorem.

We may prove easily, or regard the Proposition as an application of the Principle of Duality, that since three distinct points determine a unique plane, provided the points are linearly independent, so *three distinct planes determine a unique point*, provided they are linearly independent, *which lies in each of them*.

We consider some applications of the Propositions of Incidence.

(*a*) From a point which does not lie on either of two skew (non-intersecting) lines in S_3 one and only one line can be drawn to intersect each line. (Such a line is called a *transversal* of the two lines).

Take points A, B on one line, and C, D on the other (Fig. 65.1), and let E be the symbol for the given point. Since five points are always linearly dependent, we have a relation

$$Aa + Bb + Cc + Dd + Ee = 0$$

between the symbols for the five points, where not all of a, b, c, d and e are zero. Writing this as

$$-Ee = (Aa + Bb) + (Cc + Dd),$$

we see that the point E is collinear with the point $Aa + Bb$ on the one line and the point $Cc + Dd$ on the other.

If there were two distinct transversals to the two lines, through E, the two lines

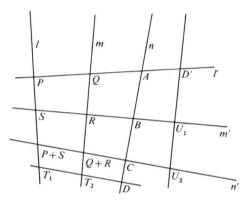

Fig. 65.2

would lie in the plane determined by the two transversals, which intersect at E, and therefore the given lines, being coplanar, would intersect, whereas we are told that they are skew.

Another way of obtaining this result is suggested in Exercise 65.2.

(b) We investigate the set of lines which are transversal to *three* given mutually skew lines in S_3 (Fig. 65.2). Let l, m and n be mutually skew lines, and draw the transversals l', m' and n' from three distinct points A, B and C lying on n to l and m. From the point D on n, distinct from A, B and C, we draw a fourth transversal to l and m, and from D', any point on l' which is not on l, m or n we draw the unique transversal to m' and n'. We prove that, subject to certain conditions, the transversal from D to l and m intersects the transversal from D' to m' and n'.

We take the four points shown in the diagram as base-points, P, Q, R and S. They are independent, since dependence would involve l and m being coplanar, and we have assumed that they are skew. By the Normalization Theorem, §62.1, we may choose $A = P + Q$, where $A = n \cap l'$, and $B = R + S$, where $B = n \cap m'$. With A and B determined except for a multiplying constant, we may also choose $C = A + B$, where $C = n \cap n'$. (This possibility should be examined carefully.) We then take

$D = A + B\lambda$, for some λ, and $D' = P + Q\mu$, for some μ, where D is the point chosen on n distinct from A, B and C, and D' is the point chosen on l' distinct from A, P and Q.

Since $C = A + B = P + Q + R + S = (P + S) + (Q + R)$, the transversal from C to l and m intersects these lines at the points $P + S$ and $Q + R$ respectively. Similarly $D = A + B\lambda = P + Q + (R + S)\lambda = (P + S\lambda) + (Q + R\lambda)$, and therefore the transversal from D to l and m intersects these lines in the points $T_1 = P + S\lambda$ and $T_2 = Q + R\lambda$.

Finally, we note that $D' = P + Q\mu = -(S + R\mu) + (P + S) + (Q + R)\mu$, and therefore the transversal from D' to m' and n' intersects these lines in the points

$$U_1 = -(S + R\mu) \quad \text{and} \quad U_2 = (P + S) + (Q + R)\mu.$$

We have now obtained the coordinates of the points we are considering. We wish to show that there is a point on the line U_1U_2 which is the same point as one on the line T_1T_2: that is, we require that for some values of v, v' and k,

$$(T_1 + T_2v)k = [(P + S\lambda) + (Q + R\lambda)v]k = U_2 - U_1v'$$
$$= (P + S) + (Q + R)\mu + (S + R\mu)v'.$$

Since P, Q, R and S are independent, the multipliers of P, Q, R and S on each side of this equation are proportional, and therefore

$$1 : v : \lambda v : \lambda = 1 : \mu : \mu + \mu v' : 1 + v',$$

are equations to be solved, if possible, for v and v'. Taking $v = \mu$, then we want both

$$\lambda\mu = \mu + \mu v',$$

and

$$\lambda = 1 + v'.$$

If we assume commutativity for our multipliers, we have

$$\lambda\mu = \mu\lambda = \mu + \mu v', \qquad \mu[\lambda - (1 + v')] = 0,$$

and since $\mu \neq 0$, we may take

$$v' = \lambda - 1,$$

and the final equation is also satisfied.

In this case the point of intersection of the transversals through D and D' is

$$T_1 + T_2v = (P + S\lambda) + (Q + R\lambda)v$$
$$= P + Qv + Rv\lambda + S\lambda$$
$$= P + Q\mu + R\mu\lambda + S\lambda,$$

so that the coordinates of the point may be taken as

$$(y_0, y_1, y_2, y_3) = (1, \mu, \mu\lambda, \lambda)$$

and this always lies on the surface

$$X_1X_3 - X_0X_2 = 0.$$

Such a surface is called *a quadric surface*, and it is one of the fundamental constructs in S_3. We shall investigate quadric surfaces in more detail in §81.1. We remark that the surface contains two systems of *generators*, or lines which lie in the surface, one system containing the lines l, m and n, and the other containing the lines l', m' and n'. The lines in a system are mutually skew, but any line in one system intersects any line in the other system. The reader is asked to prove this result in Exercise 65.5.

We have already mentioned that the assumption of commutativity of our coordinates is equivalent to the assumption of the Pappus Theorem, and in some texts the intersection of the lines U_1U_2 and T_1T_2 is derived directly from that theorem [Baker, (2), Vol. 1]. Another proof is suggested in Exercises 71.16, 17, 18, using the Fundamental Theorem of Projective Geometry, which is also equivalent to the Pappus Theorem. This theorem is proved in §71.1.

65.2   Moebius tetrahedra

Theorem   *If two tetrahedra $ABCD$, $A'B'C'D'$ are such that $A' \in (BCD)$, $B' \in (CDA)$, $C' \in (DAB)$, $D' \in (ABC)$ (so that the first tetrahedron is* escribed *to the second), and also $A \in (B'C'D')$, $B \in (C'D'A')$ and $C \in (D'A'B')$, then necessarily $D \in (A'B'C')$.*

Such mutually inscribed and escribed tetrahedra are called *Moebius* tetrahedra, after their discoverer. They are a surprising phenomenon, since it is not possible in S_2 to have two triangles which are mutually inscribed and escribed (Exercise 65.7). We leave the proof of the theorem as stated for an Exercise (65.6), and demonstrate the existence of Moebius tetrads by remarking that if we take $ABCD$ as tetrad of reference, and

$$
\begin{aligned}
A' &= (0, & q_1, & r_1, s_1) \\
B' &= (-q_1, & 0, & r_2, s_2) \\
C' &= (-r_1, -r_2, & 0, s_3) \\
D' &= (-s_1, -s_2, -s_3, & 0)
\end{aligned}
$$

then we have a tetrad which is inscribed in $ABCD$. To show that it is also escribed, we have to show that the plane $A'B'C'$ contains the point $D = (0, 0, 0, 1)$, the plane $B'C'D'$ contains the point $A = (1, 0, 0, 0)$, the plane $C'D'A'$ contains the point $B = (0, 1, 0, 0)$, and the plane $D'A'B'$ the point $C = (0, 0, 1, 0)$. If we prove that the plane $A'B'C'$ contains D, this will indicate the method. Since $(0, q_1, r_1, s_1)r_2 - (-q_1, 0, r_2, s_2)r_1 + (-r_1, -r_2, 0, s_3)q_1 = (0, 0, 0, k)$, where $k = s_1r_2 - s_2r_1 + s_3q_1$, this is so. We also remark that the matrix of entries for A', B', C', D' is skew-symmetric. Another proof of our last result arises from the theorem that the determinant of a skew-symmetric matrix of odd order is zero. We shall see in §85.4 that Moebius tetrads and skew-symmetric matrices are linked.

This section on S_3 has been kept short since we shall obtain results in S_n, projective space of n dimensions, which naturally apply to the case $n = 3$, and the methods are not essentially different. The Exercises which follow indicate what a fuller treatment of the case $n = 3$ would imply. We must stress a point which is often not clear to students when the properties of S_3 are first studied: a line in S_3 is not different from a line in S_2, and a plane in S_3 is a projective plane, so that properties in S_2 continue to hold in any plane in S_3. For example, if we have any three distinct collinear points in S_3, which we denote by A, B and C, and if $C = A + B\lambda$, then the harmonic conjugate of C with respect to A and B can be constructed, as in §64.2, in any plane which contains the line ABC, and we find that the harmonic conjugate $C' = A - B\lambda$, and therefore it is independent of the particular plane chosen through the line.

Exercises

65.1 Show that in S_3 a plane is determined by the assignment of a point and a line which does not contain the point, if the plane is to contain both the point and the line. Show also that two intersecting lines determine a plane, which is to contain them.

65.2 Give another proof that from a point which does not lie on either of two skew lines in S_3 there is a unique transversal to the two lines, by considering the intersection of the one line with the plane defined by the point and the other line.

65.3 State and prove a Normalization Theorem for five points A, B, C, D and E in S_3, where E does not lie in any face of the tetrahedron formed by A, B, C and D.

65.4 Show that in Fig. 65.2 the point C does not lie in any face of the tetrahedron formed by P, Q, R and S, so that, by the Normalization Theorem, we could have taken $C = P + Q + R + S$ to start with, and then deduced that $A = P + Q$ and $B = R + S$, since there is a unique transversal through C to the lines l' and m'.

65.5 Show that the quadric surface $X_1 X_3 - X_0 X_2 = 0$ contains the line of intersection of the two planes $\lambda X_1 - X_2 = 0 = \lambda X_0 - X_3$ for any value of λ, and the line of intersection of the two planes $\mu X_0 - X_1 = 0 = X_2 - \mu X_3$ for any value of μ. If we call the first set of lines the λ-system, and the second set the μ-system, show that the λ-system consists of mutually skew lines, as does the μ-system, but that every line of the λ-system intersects every line of the μ-system. Finally, show that the point of intersection of a line of the λ-system and a line of the μ-system is the point $(1, \mu, \mu\lambda, \lambda)$ of the surface. Do any more lines in the surface pass through this point?

65.6 Prove the theorem on Moebius tetrads of §65.2. (Begin with A' as given, and show that if we take $B' = (-q_1, 0, r_2, s_2)$, $C' = (-r_1, q_3, 0, s_3)$, and $D' = (-s_1, q_4, r_4, 0)$, as we may, then we obtain $q_3 = -r_2$, $r_4 = -s_3$ and $q_4 = -s_2$, so that $(A'B'C')$ contains D.)

65.7 Show that it is impossible to find two proper triangles in S_2 which are mutually inscribed and escribed.

65.8 AB, CD are given skew lines in S_3. The transversal of AB, CD through a variable point P meets AB, CD in R, S. The point P' is the harmonic conjugate of P with respect to R, S. Find the locus of P' if P moves on (i) a given plane, (ii) a given line.

65.9 If ABC, $A'B'C'$ are two triangles in S_3, which are not coplanar, and the lines AA', BB' and CC' are concurrent at a point V (so that the triangles are perspective from a point), show that the points $BC \cap B'C'$, $CA \cap C'A'$ and $AB \cap A'B'$ exist and are collinear, using only the Propositions of Incidence.

65.10 Prove the converse of the Desargues Theorem in S_3, that if two non-coplanar triangles are perspective from a line, then they are also perspective from a point.

65.11 Given five mutually skew lines p, a, b, c, d in S_3, a transversal is drawn to a and c from a point P on p, intersecting a in A and c in C, and another transversal is drawn to b and d, intersecting b in B and d in D. Prove that the intersection $Q = AB \cap CD$ of AB and CD describes a line as P describes p.

65.12 Show that a change of tetrad of reference in S_3 is equivalent to a non-singular transformation $y' = Ay$ between coordinates. Show that plane-coordinates change in the manner indicated by the matrix equation $u = u'A$, where $y' = Ay$ is the relation between point-coordinates.

65.13 Restate the Normalization Theorem (Exercise 65.3) as a theorem on non-singular linear transformations in S_3.

65.14 Let A_i ($i = 1, 2, 3, 4$) be a tetrad of points in S_3 which are not coplanar. If P be any given point, and PA_1 meet the plane $A_2A_3A_4$ in the point P_1, let P_1' be the harmonic conjugate of P with respect to the points A_1 and P_1. Let us write $PT_1 = P_1'$, where the transformation T_1 is called *harmonic homology* (see §39.1). Show that T_1 is represented by a non-singular matrix. Prove further that if T_2, T_3 and T_4 be defined in a similar way,

$$(P)T_1T_2T_3T_4 = P.$$

(Compare Exercise 25.3, and the link to this Exercise provided by §39.1.)

65.15 If A, B, C and D be four non-coplanar points in S_3, and

$$P = Ax_0 + Bx_1 + Cx_2 + Dx_3,$$

show that the harmonic conjugate of P with respect to the pair of points in which the transversal from P to AB and CD intersects them is

$$P' = Ax_0 + Bx_1 - Cx_2 - Dx_3,$$

and that the relation between P and P' is represented by a non-singular matrix. What can we say about the square of this matrix?

65.16 $ABCD$ is a tetrahedron, A' an arbitrary point not lying in a face. The transversal is drawn from A' to AB, CD, and B' is the harmonic conjugate of A' with respect to the intersections of this transversal with AB and CD. Points C' and D' are defined similarly, with regard to AC and BD and AD and BC. Prove that the two tetrads $ABCD$, $A'B'C'D'$ are such that each edge of either meets a pair of opposite edges of the other, and is divided harmonically by them.

Prove also that AA', BB', CC' and DD' meet in a point A'', that AB', BA', CD' and DC' meet in a point B'', that AC', BD', CA', DB' meet in a point C'', and that AD', BC', CB', DA' meet in a point D''. Finally, show that the three tetrads $ABCD$, $A'B'C'D'$ and $A''B''C''D''$ form a completely symmetrical system, in the sense that the edge of any one of the tetrads meets a pair of opposite edges of each of the others and is divided harmonically by them. (Such a set of tetrahedra is called a *desmic system*. Begin with $A' = A + B + C + D$, and then $B' = (A + B) - (C + D)$, etc. The solution is shorter than the enunciation of the problem! For a real, and simple model of a set of desmic tetrads, see Exercise 88.6.)

65.17 π and π' are distinct planes in S_3, and V is a point not in either plane, W a distinct point not in either plane. The points P in π' are projected into points Q in π from V, and into points Q' in π from W, $(VP \cap \pi = Q, WP \cap \pi = Q'.)$ Show that the mapping $Q \to Q'$ in π is a collineation, that the joins of corresponding points, QQ', pass through a fixed point, and that the intersection of corresponding lines, $m \cap m'$, where m' is the map of m, always lies on a fixed line. After doing this geometrically (without the use of coordinates) set up the equations for the collineation $Q \to Q'$ in π, and verify your results.

VIII

THE PROJECTIVE GEOMETRY OF n DIMENSIONS

We have not dwelt too long on the projective geometry of two and three dimensions because the projective geometry of space of n dimensions needs very little more linear algebra for its study, and after obtaining general theorems we shall return to the cases $n = 2$ and $n = 3$.

Once again, our definition of S_n, projective space of n dimensions, is bound up with the concept of V_{n+1}, an $(n + 1)$-dimensional vector space over the field of complex numbers. It will have been noticed that the work done in the previous chapter holds in any field. *We have not, as yet, made use of any special properties of the field of complex numbers.* When we do so, we shall draw attention to the fact. We denote our ground field by the symbol K, and assume that it is commutative.

Our space V_{n+1} is a vector space. Its elements consist of ordered $(n + 1)$-tuples $(y_0, y_1, \ldots, y_n)$, where the y_i are elements of K. We denote such ordered sets by the symbol y, and say that V_{n+1} consists of the *vectors* y where $y = (y_0, y_1, \ldots, y_n)$. The addition of two vectors y and $z = (z_0, z_1, \ldots, z_n)$ is defined as

$$y + z = (y_0 + z_0, y_1 + z_1, \ldots, y_n + z_n),$$

and it is clear that under addition our vectors form an Abelian group, with zero the zero-vector $0 = (0, 0, \ldots, 0)$. We also define multiplication of a vector y by a scalar λ, an element of K, thus:

$$y\lambda = \lambda y = (y_0\lambda, y_1\lambda, \ldots, y_n\lambda).$$

Naturally, our vector space satisfies the well-known axioms for a vector space which deal with multiplication by a scalar:

$$\lambda(\mu y) = (\lambda\mu)y,$$
$$(y + z)\lambda = y\lambda + z\lambda,$$
$$\lambda(y + z) = \lambda y + \lambda z,$$

and
$$1y = y.$$

The vectors $y^1, y^2, \ldots, y^m$ are said to be *linearly independent* if the relation

$$y^1\lambda_1 + y^2\lambda_2 + \ldots + y^m\lambda_m = 0$$

implies that $\lambda_1 = \lambda_2 = \ldots = \lambda_m = 0$. If this is not the case, they are said to be *dependent*.

[271]

Our space V_{n+1} contains $n + 1$ independent vectors. These are the vectors

$$e_0 = (1, 0, 0, \ldots, 0), \; e_1 = (0, 1, 0, \ldots, 0), \ldots, e_n = (0, 0, \ldots, 1).$$

We say that these vectors *span* V_{n+1}, since we can write any vector y thus:

$$y = (y_0, y_1, \ldots, y_n) = e_0 y_0 + e_1 y_1 + \ldots + e_n y_n.$$

This equation also shows that the set y, e_0, e_1, $\ldots$, e_n is a *linearly dependent* set. There are no more than $n + 1$ linearly independent vectors in V_{n+1}, which is the reason for saying this vector space has *dimension* $n + 1$, and since any set of $n + 2$ vectors is linearly dependent, we see that if we have any set $v^1, v^2, \ldots, v^{n+1}$ of *independent* vectors, then since the addition of any vector y to the set makes it a dependent set, we must have

$$y = v^1 \lambda_1 + v^2 \lambda_2 + \ldots + v^{n+1} \lambda_{n+1},$$

so that *any set of $n + 1$ linearly independent vectors spans the vector space.*

The set of linear combinations of m ($\leqslant n + 1$) linearly independent vectors v^1, $v^2, \ldots, v^m$ given by

$$v = v^1 \lambda_1 + v^2 \lambda_2 + \ldots + v^m \lambda_m$$

is called *an m-dimensional linear subspace V_m* of the vector space V_{n+1}. It is also a vector space, and sometimes denoted by $L\{v^1, v^2, \ldots, v^m\}$. The dimension m is independent of the choice of the independent vectors $v^1, \ldots, v^m$. Any set of m independent vectors in the subspace serve equally well for defining the subspace. Any such set is called *a set of base vectors.*

In particular, a *one-dimensional subspace* consists of the set of vectors $v\lambda$, where $v = (v_0, v_1, \ldots, v_n)$ is a fixed non-zero vector. We associate the dimension zero with the ray $v\lambda$ if v is the zero vector.

We are now ready to define projective space of n dimensions over the ground field K.

66.1 Projective space of n dimensions

Definition. A point of projective space S_n is a one-dimensional subspace, or *ray* of V_{n+1}. Projective space S_n is the set of rays of a vector space V_{n+1} defined over a given ground field K.

A subspace S_m of S_n can now be defined as the set of rays of a subspace V_{m+1} of V_{n+1}. The subspace S_m therefore consists of the set of points of S_n which depend linearly on $m + 1$ fixed linearly independent points. If these be V^0, V^1, $\ldots$, V^m, the points V of S_m can be written

$$V = V^0 \lambda_0 + V^1 \lambda_1 + \ldots + V^m \lambda_m,$$

so that if $V^i = (y_0^i, y_1^i, \ldots, y_n^i)$ and $V = (y_0, y_1, \ldots, y_n)$, we have

$$y_k = y_k^0 \lambda_0 + y_k^1 \lambda_1 + \ldots + y_k^m \lambda_m \qquad (k = 0, 1, \ldots, n).$$

Since the points V^0, V^1, $\ldots$, V^m, which we call the *base-points* of S_m, are linearly independent, the expression of a given point in S_m as a linear combination of this set of base-points is unique, except for a non-zero multiplicative factor. If we have two combinations for a point V, say

$$V = V^0\lambda_0 + V^1\lambda_1 + \ldots + V^m\lambda_m$$

and

$$V = V^0\mu_0 + V^1\mu_1 + \ldots + V^m\mu_m,$$

then since the coordinates are homogeneous coordinates, we can say that

$$(V^0\lambda_0 + V^1\lambda_1 + \ldots + V^m\lambda_m) = (V^0\mu_0 + V^1\mu_1 + \ldots + V^m\mu_m)k,$$

where $k \neq 0$. Hence

$$V^0(\lambda_0 - \mu_0 k) + V^1(\lambda_1 - \mu_1 k) + \ldots + V^m(\lambda_m - \mu_m k) = 0,$$

and since $V^0, V^1, \ldots, V^m$ are linearly independent, every multiplier in the above equation is zero, and therefore

$$(\lambda_0, \lambda_1, \ldots, \lambda_m) = (\mu_0, \mu_1, \ldots, \mu_m)k.$$

The elements $(\lambda_0, \lambda_1, \ldots, \lambda_m)$ may therefore be regarded as *a system of homogeneous coordinates* in the subspace S_m.

If we take $m = n$, the equations

$$y_k = y_k{}^0\lambda_0 + y_k{}^1\lambda_1 + \ldots + y_k{}^n\lambda_n \quad (k = 0, 1, \ldots, n)$$

may be written as a matrix equation

$$(y_0, y_1, \ldots, y_n)^T = \mathbf{A}(\lambda_0, \lambda_1, \ldots, \lambda_n)^T,$$

where $\mathbf{A}$ is the matrix $[y_j{}^i]$ $(i, j = 0, 1, \ldots, n)$. This gives the expression of our original coordinates $(y_0, y_1, \ldots, y_n)$ in terms of the new coordinates $(\lambda_0, \lambda_1, \ldots, \lambda_n)$, corresponding to a change from the simplex of reference given by the $n + 1$ points

$$E_0 = (1, 0, \ldots, 0), E_1 = (0, 1, \ldots, 0), \ldots, E_n = (0, 0, \ldots, 1)$$

to the simplex given by the $n + 1$ points

$$V^0, V^1, \ldots, V^n.$$

Since this latter set of points is linearly independent, we may express $E_0, E_1, \ldots, E_n$ in terms of $V^0, V^1, \ldots, V^n$, and we shall then obtain a matrix equation

$$(\lambda_0, \lambda_1, \ldots, \lambda_n)^T = \mathbf{B}(y_0, y_1, \ldots, y_n)^T.$$

This shows that *the matrix A has an inverse*, and it is therefore a *regular* or *non-singular* matrix (both terms are in use). We therefore have:

Theorem I   *A change of simplex of reference in S_n is equivalent to a linear mapping $y = \mathbf{A}x$,* *where A is a non-singular matrix, and a point has coordinates x with respect to one simplex, and y with respect to the other.*

The proof of this theorem given above is essentially the same as that given in §64.1 for the case $n = 2$ of the projective plane.

The one-dimensional subspaces in S_n are called *lines*. These arise as the set of points linearly dependent on two given distinct points. The two-dimensional subspaces are called *planes*. These arise as the set of points which are linearly dependent on three independent (that is non-collinear) points. The $(n-1)$-dimensional subspaces deserve a special name. They are called *hyperplanes*. They arise as the set of points linearly dependent on n independent points of S_n. The only subspace of S_n of higher dimension is S_n itself.

It will be noticed that m points define an S_{m-1}, that is, they are linearly independent, unless they already lie in a space of lower dimension. We investigate the Propositions of Incidence in S_n in §67.1. Here we look at hyperplanes first of all.

If we take n independent points $V^0, V^1, \ldots, V^{n-1}$, our work above shows us that we obtain $n + 1$ equations

$$y_k = y_k{}^0\lambda_0 + y_k{}^1\lambda_1 + \ldots + y_k{}^{n-1}\lambda_{n-1} \ (k = 0, 1, \ldots, n)$$

for the coordinates $(y_0, y_1, \ldots, y_n)$ of a point Y in the hyperplane spanned by the n independent points. If we eliminate the parameters $\lambda_0, \lambda_1, \ldots, \lambda_{n-1}$, we obtain the determinantal equation

$$\begin{vmatrix} y_0 & y_1 & \cdots & y_n \\ y_0{}^0 & y_1{}^0 & \cdots & y_n{}^0 \\ \cdot & \cdot & \cdots & \cdot \\ y_0{}^{n-1} & y_1{}^{n-1} & \cdots & y_n{}^{n-1} \end{vmatrix} = 0,$$

which shows that $(y_0, y_1, \ldots, y_n)$ satisfy an equation of the form:

$$u^0 Y_0 + u^1 Y_1 + \ldots + u^n Y_n = 0.$$

This is *the equation of a hyperplane* in S_n, a homogeneous linear equation in the coordinates. It is of interest to prove that conversely, *a non-trivial equation of this form represents a subspace of dimension $n - 1$.*

Theorem II   *The solutions of the non-trivial equation*

$$u^0 Y_0 + u^1 Y_1 + \ldots + u^n Y_n = 0$$

depend linearly on n independent points.

Proof. Since not all the u^i are zero we may assume, by renumbering the coordinates if necessary, that $u^0 \neq 0$. Let $(y_0, y_1, \ldots, y_n)$ be a solution of the equation. We may write (if $k = (u^0)^{-1}$),

$$y_0 = -ku^1 y_1 - ku^2 y_2 - \ldots - ku^n y_n,$$

and we may now consider the $y_1, y_2, \ldots, y_n$ as variables, and take the solution of our equation as

$$(-ku^1y_1 - ku^2y_2 \ldots - ku^ny_n, y_1, y_2, \ldots, y_n)$$
$$= y_1(-ku^1, 1, 0, \ldots, 0) + y_2(-ku^2, 0, 1, \ldots, 0) + \ldots + y_n(-ku^n, 0, 0, \ldots, 1).$$

The set of n points

$$(-ku^1, 1, 0, \ldots, 0) = V^1,$$
$$(-ku^2, 0, 1, \ldots, 0) = V^2,$$
$$\cdots \cdots \cdots$$
$$(-ku^n, 0, 0, \ldots, 1) = V^n,$$

is easily seen to be linearly independent, since any linear relation

$$V^1k_1 + V^2k_2 + \ldots + V^nk_n = 0$$

leads to $k_1 = 0$, if we look at the second coordinate on the left-hand side of the equation, to $k_2 = 0$ if we look at the third coordinate, and so on. Hence we have found a set of n linearly independent points which form a base for the solutions of our equation, which therefore fill a hyperplane.

Since the equation of a hyperplane is unchanged if we multiply by a non-zero constant, we may refer to *the homogeneous coordinates* $[u^0, u^1, \ldots, u^n]$ *of a hyperplane* with equation

$$u^0X_0 + u^1X_1 + \ldots + u^nX_n = 0.$$

If we write these coordinates as a row-vector: $U = [u^0, u^1, \ldots, u^n]$, then since we write a point $X = (X_0, X_1, \ldots, X_n)^T$ as a column-vector, we may write the equation of a hyperplane in the simple form

$$UX = 0.$$

It is reasonable to speak of *a set of linearly independent linear homogeneous equations,* since each equation gives, and is given by, the row-vector u of the coordinates of the corresponding hyperplane. We recall the theorem in linear algebra which asserts that *d linearly independent linear homogeneous equations in $n + 1$ unknowns have solutions which depend linearly on $n + 1 - d$ linearly independent solutions.* (Pedoe, p. 190 **(11)**.)

The case treated above is $d = 1$. This theorem states, in the language of S_n, that d *linearly independent hyperplanes intersect in an S_{n-d}.*

We assume that it is clear that *the common solutions* of a set of linear equations correspond to *the intersection* of the subspaces represented by these linear equations.

We conclude this section by proving a converse to the theorem that $n - d$ linearly independent hyperplanes intersect in an S_d.

Theorem III   *A linear space S_d in S_n can be defined by $n - d$ linearly independent homogeneous linear equations.*

Proof. Let S_d be defined by the points $V^0, V^1, \ldots, V^d$, which are linearly independent. Consider the set of linear equations in the U's as unknowns:

$$(UV^0) = 0, (UV^1) = 0, \ldots, (UV^d) = 0.$$

The matrix of coefficients of this set of equations is a $(d + 1) \times (n + 1)$ matrix, and the rows are merely the coordinates of the V^i $(i = 0, 1, \ldots, d)$. Since these rows, considered as vectors, are independent by hypothesis, the matrix we are considering has rank $d + 1$. Hence the set of equations, by the result quoted above, has exactly $n - d$ linearly independent solutions in the U's. Each of these solutions defines a hyperplane in S_n, and each of these hyperplanes contains the points $V^0, V^1, \ldots, V^d$. Hence the intersection of these $n - d$ independent hyperplanes, which we know is an S_d, contains the points $V^0, V^1, \ldots, V^d$. But these points define an S_d. The two S_d therefore coincide.

66.2 The principle of duality in S_n

The Principle of Duality in S_n becomes clear as soon as we have found that a point $(y_0, y_1, \ldots, y_n)$ lies in a hyperplane $[u^0, u^1, \ldots, u^n]$ if the following relation is satisfied:

$$UY = u^0 y_0 + u^1 y_1 + \ldots + u^n y_n = 0.$$

This relation involves the homogeneous coordinates of Y and the homogeneous coordinates of U symmetrically, and is still valid if these coordinates be interchanged. Hence we have *The Principle of Duality for S_n*:

In any theorem which involves the incidence of points and hyperplanes, these terms can be interchanged, and the result is also a theorem.

More formally, we set up a mapping in which a point $(y_0, y_1, \ldots, y_n)$ is mapped on a hyperplane $[y_0, y_1, \ldots, y_n]$, and a hyperplane $[u^0, u^1, \ldots, u^n]$ is mapped on a point $(u^0, u^1, \ldots, u^n)$. The incidence relation $UY = 0$ is preserved under this mapping. We then obtain from any construct which consists of points and hyperplanes a construct which consists of hyperplanes and points, and this has the same incidence relations with respect to hyperplanes and points that the original construct has with respect to points and hyperplanes.

Such a mapping of S_n is called *a correlation*. The space of the points $(u^0, u^1, \ldots, u^n)$, also a projective space of n dimensions, is called the space *dual* to S_n.

Since not all constructs involve only points and hyperplanes, it is expedient to consider what an S_m is mapped onto by this correlation. If we regard the S_m as being given by $n - m$ linearly independent equations in the point-coordinates y, the mapping gives $n - m$ linearly independent points in the dual space. The mapping of the points which *lie in* the S_m produce hyperplanes which *contain* the $n - m$ independent points. These $n - m$ points define an S_{n-m-1}, and therefore the points of the S_m are mapped on the hyperplanes which contain an S_{n-m-1}.

Now suppose that $S_p \subseteq S_q$, that is, all points of S_p are also points of S_q. In the dual mapping, S_p is mapped on the hyperplanes which contain an S_{n-p-1}, and S_q on the hyperplanes which contain an S_{n-q-1}, and a point in S_p also being in S_q shows that all hyperplanes through S_{n-p-1} also pass through S_{n-q-1}. Hence $S_{n-q-1} \subseteq S_{n-p-1}$. The relation $\subseteq$ becomes $\supseteq$ under the duality transformation.

We now have a more flexible statement of The Principle of Duality in S_n: An S_p is mapped onto an S_{n-p-1}, and if $S_p \subseteq S_q$ then for the corresponding S_{n-p-1} and S_{n-q-1} we have $S_{n-p-1} \supseteq S_{n-q-1}$.

We note that the relation $S_0 \subseteq S_{n-1}$ becomes $S_{n-1} \supseteq S_0$, so that incidence between points and hyperplanes is preserved, as we expect.

66.3   The intersection and join of linear subspaces of S_n

The *intersection* of two subspaces S_p and S_q of S_n is defined to be the set of points common to the two subspaces, and is denoted thus: $S_p \cap S_q$. *This set is also a subspace of S_n.* If the set of common points is empty, or if it consists of only one point, we are finished. If the set contains at least two distinct points, P and Q, it also contains all points of the line $P\lambda + Q\mu$, since both S_p and S_q contain all these points. If the set contains at least three linearly independent points P, Q and R, it contains all points of the plane $P\lambda + Q\mu + R\nu$, since both S_p and S_q contain these points. There is a maximum number of linearly independent points which can lie in the set of common points, namely $n + 1$, so the proof is complete, since we can go on as we have been doing.

If we consider the dual process, we think of S_p and S_q as being given by the *intersection of hyperplanes* instead of by the *join of points*. If u is a hyperplane which contains both S_p and S_q, and there is no other, we say that this hyperplane is the *join* of S_p and S_q, written $S_p \cup S_q$. Note that this does *not* agree with the point-set convention, which would mean either the points in S_p or the set of points in S_q or both. Some authors write $S_p + S_q$. We could write $[S_p S_q]$, and shall sometimes do so.

If there is no hyperplane which contains both S_p and S_q, we say that the join is S_n, *the whole space.* If u, v are independent hyperplanes which contain both S_p and S_q, any hyperplane $\lambda u + \mu v$ also contains both S_p and S_q, and so on. There is a finite maximum number of linearly independent hyperplanes which contain both S_p and S_q. *The linear subspace which is the intersection of these hyperplanes* is called *the join of S_p and S_q.* The join evidently contains both S_p and S_q, and can also be defined in a number of alternative and equivalent ways, as we suggest in the following Exercises.

Once again, we shall denote the *intersection* of S_p, S_q by the symbol $S_p \cap S_q$, and the *join* by $S_p \cup S_q$ or $[S_p S_q]$.

Exercises

66.1   A hyperplane U is determined by the n linearly independent points $V^1, V^2, \ldots, V^n$. Show that the coordinates u^i of this hyperplane are proportional to the $n \times n$ subdeterminants of the matrix $[v_j{}^i]$ ($i = 1, \ldots, n$, $j = 0, 1, \ldots, n$).

66.2   A point Y is determined by the n linearly independent hyperplanes $U_1, U_2, \ldots, U_n$. Show that the coordinates y_i of this point are proportional to the $n \times n$ subdeterminants of the matrix $[u_i{}^j]$ $(i = 1, \ldots, n, j = 0, 1, \ldots, n)$.

66.3   Show that in S_n a given line always intersects a given hyperplane in one point, or lies in the hyperplane.

66.4   Show that in S_n a given plane always intersects a given hyperplane in a line or lies in the hyperplane.

66.5   Prove The Normalization Theorem For S_n: If $Y^0, Y^1, \ldots, Y^n$ are $n + 1$ linearly independent points in S_n, and P is a point which does not lie in any *face* of the simplex formed by these $n + 1$ points (*any hyperplane formed by sets of n of the $n + 1$ given points*), then the coordinates of these $n + 1$ points may be so modified by the absorption of multiplicative constants that we may write

$$P = Y^0 + Y^1 + \ldots + Y^n.$$

66.6   Show that the *intersection* of two subspaces of S_n corresponds to the *join* of the dual spaces, when we apply the dual mapping, and conversely.

66.7   What is the dual in S_3 for the Desargues Theorem for two triangles in S_3?

66.8   Show that the intersection of S_p and S_q is the *largest* subspace which lies in both of them, and the join is the *smallest* subspace which contains both of them.

66.9   Show that the join of S_p and S_q contains the set of lines which join any point of S_p to any point of S_q. Does it contain any other lines?

66.10   If l, m are two skew (non-intersecting) lines in S_3, show that the join of the two lines is S_3. Deduce that there is a transversal, cutting both lines, through any point of S_3.

67.1   The propositions of incidence in S_n

From our definitions we have the following two theorems:

Theorem I   *Any $m + 1$ points in S_n which do not lie in an S_q, where $q < m$, determine an S_m.*

Theorem II   *Any d hyperplanes in S_n which do not have in common an S_q, where $q > n - d$, intersect in an S_{n-d}.*
These are dual theorems. We now prove

Theorem III   *An S_p and an S_q in S_n have as intersection, if $p + q \geqslant n$, a linear space S_d of dimension $d \geqslant p + q - n$.*

Proof. S_p can be given by $n - p$ independent homogeneous linear equations, and S_q by $n - q$ independent equations. Their intersection satisfies, and is determined by the whole set of $2n - p - q$ equations. If these equations are all independent, they define a linear space of dimension $n - (2n - p - q) = p + q - n$. If they are not independent, some can be removed from the set without affecting the intersection, which is then of dimension $> p + q - n$, since the rank of the remaining set of equations is then $< 2n - p - q$.

Theorem IV   *An S_p and an S_q which have an S_d in common lie in an S_m, where*

$$m \leqslant p + q - d.$$

Proof. The intersection S_d of the two spaces is determined by $d + 1$ independent points (Fig. 67.1). To determine S_p we add $p - d$ points to this set, making up a set of $p + 1$ independent points. In a similar fashion we add $q - d$ points to the $d + 1$ points in the intersection and obtain $q + 1$ points which determine S_q. All these

$$(d + 1) + (p - d) + (q - d) = p + q - d + 1$$

points determine, if they are independent, an S_{p+q-d}, or else an S_m with $m < p + q - d$. This S_m with $m \leqslant p + q - d$ contains a set of base-points for S_p, and therefore

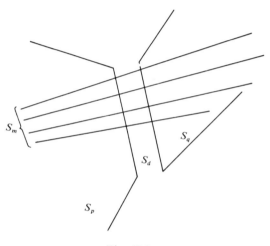

Fig. 67.1

contains S_p, and a set of base-points for S_q, and it therefore also contains S_q. This completes the proof.

If there is no intersection S_d, the same argument gives:

Theorem V   *An S_p and an S_q with no intersection always lie in an S_m, where*

$$m \leqslant p + q + 1.$$

We now strengthen these theorems to:

Theorem VI   *An S_p and an S_q which intersect in an S_d lie in a uniquely determined S_{p+q-d}. If their intersection is empty, they lie in a uniquely determined S_{p+q+1}.*

Proof. Let the intersection of S_p and S_q be S_d. By Thm IV, S_p and S_q lie in an S_m, with $m \leqslant p + q - d$. By Thm III, on the other hand, applied to S_m as the ambient space for S_p and S_q,

$$d \geqslant p + q - m,$$

so that we must have $m = p + q - d$. If S_p and S_q were both contained in S_m* distinct from S_m, the intersection of S_m and S_m* would also contain both spaces, and would necessarily be of dimension less than m, which we have shown is not the case. Hence the S_m we have found is unique.

If the intersection of S_p and S_q is empty, Thm V shows that S_p and S_q lie in an S_m, with $m \leqslant p + q + 1$. If however $m \leqslant p + q$, then by Thm III the spaces S_p and S_q would have a non-empty intersection. Hence we must have $m = p + q + 1$, and, as above, the S_m is unique.

This uniquely defined space S_{p+q-d} (or S_{p+q+1}) which we have found in Thm VI is, of course, the *join* of S_p and S_q, already introduced in §66.3. Here we have obtained more information about it. It is the smallest linear space (space with the lowest dimension) which contains both S_p and S_q, just as the *intersection* of S_p and S_q is the largest linear space which lies in both spaces.

An easy way to remember the main dimension formula is:

dimension of join + dimension of intersection = $p + q$.

Our final theorem in this section is:

Theorem VII   (The Complementation Theorem). *Given any subspace S_p in S_n, there exists an S_{n-p-1} which is such that $S_p \cap S_{n-p-1}$ is empty, and the join $[S_p S_{n-p-1}] = S_n$.*

Proof. Let the points $P_0, P_1, \ldots, P_p$ be a set of base-points for S_p, and choose additional points $Q_{p+1}, \ldots, Q_n$ so that the complete set of points forms a base for S_n. Then the points $Q_{p+1}, \ldots, Q_n$ determine a space of $n - p - 1$ dimensions. This is the S_{n-p-1} we seek. It is clear that the join of S_p and S_{n-p-1} is S_n, and if the intersection $S_p \cap S_{n-p-1}$ contained a point X, this point would be linearly dependent on the P_i and also linearly dependent on the Q_j. Hence the set of points P_i, Q_j would be a dependent set, and this is not the case. Hence S_p does not intersect S_{n-p-1}.

Remark. If $x \in S_n$, we may write

$$x = \lambda_0 P_0 + \lambda_1 P_1 + \ldots + \lambda_p P_p + \lambda_{p+1} Q_{p+1} + \ldots + \lambda_n Q_n,$$

that is

$$x = y + z$$

where

$$y \in S_p \text{ and } z \in S_{n-p-1}.$$

67.2   Some examples

We apply the formula we have just found to S_3. If $p = q = 2$, then dimension of join + dimension of intersection = 4. If our two planes are distinct, then the dimension of their join >2, and is therefore $= 3$, so that, by the formula, the dimension of their intersection $= 1$. Hence, once again, two distinct planes in S_3 always intersect in a line.

If $p = 1$, $q = 1$, then dimension of join $+$ dimension of intersection $= 2$. If the intersection of the two lines is a point, of dimension $= 0$, the dimension of the join $= 2$, so that two intersecting lines determine a plane. If the intersection is empty, the dimension of the join $= 3$, which is the whole space S_3. We note that we may take the dimension of the empty set as $= -1$.

If $p = 1$, $q = 2$, we see that the line either lies in the plane, or meets it in a point, and the join is then S_3.

We consider S_4. In this space we have points, lines, planes and subspaces of dimension three, which we shall call *solids* (they are, of course, hyperplanes). We take a line, $p = 1$, and a plane, $q = 2$. Then:

$$\text{dimension of join} + \text{dimension of intersection} = 1 + 2 = 3.$$

If the dimension of the intersection $= -1$, the dimension of the join $= 4$, and this is the general situation, when the line does not intersect the plane. If it does, the join is of dimension three, and the line and the plane it intersects determine a solid, in which they lie.

If $p = 1$, $q = 3$, then dimension of join $+$ dimension of intersection $= 4$, and since dimension of join $\leqslant 4$, dimension of intersection $\geqslant 0$, so that a line always meets a solid in a point, or lies in the solid.

Consider two planes, with $p = q = 2$, then dimension of join $+$ dimension of intersection $= 4$, and again, since dimension of join $\leqslant 4$, dimension of intersection $\geqslant 0$. In the general case two planes intersect in a point. If they intersect in a line, they determine a solid, in which they lie.

67.3   Projection

Let l and l' be two distinct lines in S_2, and V a point in S_2 not on either line (Fig. 67.2). If we have a point $P \in l$, then VP is a definite line, and this intersects l' in a uniquely determined point P'. It is clear that there is a one-to-one correspondence between the points of l and those of l', P being mapped on P' if the points P, P' and V are collinear. This correspondence is called *a perspectivity*, and is sometimes written thus:

$$P \stackrel{v}{\barwedge} P'.$$

We also say that P is *projected* into P' from V.

In S_3 let π and π' be two distinct planes, and V a point in neither plane (Fig. 67.3). Then again, if $P \in \pi$, the line VP is uniquely defined, and intersects π' in a uniquely defined point P'. There is a one-to-one correspondence between the points of π and those of π'. We note that if the points P in π lie on a line l, the corresponding points P' in π' lie on a line l', where $l' = \pi' \cap [Vl]$. Again, the points P' are said to be the *projections* of the points P.

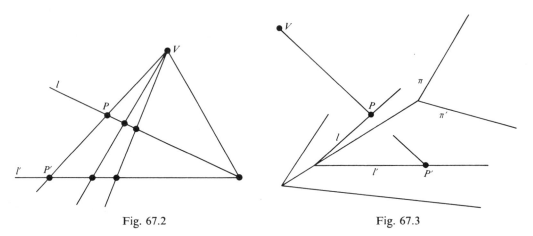

Fig. 67.2   Fig. 67.3

There is another kind of projection in S_3. Let l, m, m' be three mutually skew (non-intersecting) lines in S_3, and let P be a point on m (Fig. 67.4). Then the plane $[lP]$ is uniquely determined, and never contains m'. It therefore intersects m' in a uniquely determined point P', and there is a one-to-one correspondence between the points of m and those of m'. We can regard this correspondence as being effected by *projection*

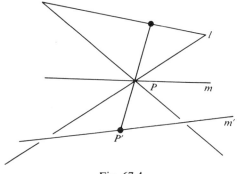

Fig. 67.4

from a line. It can be shown that this projection is equivalent to a product of two perspectivities, between m and m^*, and then between m^* and m', where m^* is any line chosen to intersect both m and m'. (See Exercise 67.4.)

In S_4, since two planes intersect in at least a point, we may project the points of a plane π onto the points of a plane π' by means of a line l. We assume that l does not intersect either plane, and join l to any point P in π. Then the plane $[lP]$ intersects π' in a point P'. If it intersected π' in a line, this line would intersect l, lying in the plane $[lP]$, and so l would intersect π', contrary to hypothesis. Hence once again we have a one-to-one correspondence between the points of π and π', effected by projection.

The process evidently has meaning in S_n for any value of n. Take two distinct spaces

S_m and S_m', and let S_{n-m-1} be a space which has no points in common with either of the two spaces of m dimensions. If P be any point in S_m, the join of P and S_{n-m-1} produces an S_{n-m}, and this intersects S_m' in a unique point P'. If it intersected S' in a line, this would intersect S_{n-m-1}, and so S_{n-m-1} would intersect S_m', contrary to hypothesis. Hence the join of S_{n-m-1} to a point P in S_m does not intersect S_m any further, and intersects S_m' in a unique point P', and again we have a one-to-one correspondence between S_m and S_m'.

67.4   General subspaces in S_n

The adjective 'general', when used of subspaces in S_n, refers to the mutual relationship between the subspaces. If S_p and S_q are two given subspaces, with $p \leqslant q$, and $p + q - n \geqslant 0$, the intersection can vary in dimension from p, when S_q contains S_p, down to $p + q - n$. When the dimension of the intersection is the *least* possible number, and therefore the dimension of the join is the *greatest* possible number, namely n, we say that S_p and S_q are *general spaces* in S_n, or *in general position*. Any *one* subspace, considered by itself, is as good as any other, from our point of view, and the term 'general' does not apply. If we have an S_p and an S_q, and $p + q < n$, then they are said to be general subspaces when their intersection is empty.

If we have three subspaces S_p, S_q and S_r, they are said to be *in general position*, or *general subspaces* when they are general subspaces, taken in pairs, and each subspace is in general position with respect to the intersection and join of the other two subspaces. For example, if we have three general lines in S_4, they are skew to each other, but they could satisfy this condition and yet all three lie in an S_3, and so we must have the condition that one of the lines does not lie in the S_3 which joins the other two. Then, of course, none of the lines will lie in the S_3 which joins the other two lines. The conditions we have given are not the minimum set of conditions, but they are sufficient for our purpose. The extension of the notion to more than three subspaces is clear.

Exercises

67.1   From the theorem that the join of two lines l, m in S_4 which do not meet is an S_3, and that a line n intersects an S_3 in a point, or lies in it, deduce that in S_4 three lines in general position have a unique transversal.

67.2   Show that the Normalization Theorem (Exercise 66.5) for S_5 leads to the following theorem: in S_5 a plane can be drawn through an arbitrary point to intersect three general lines.

67.3   Prove Exercise 67.2 from the consideration that the plane joining the given point to one line intersects the solid joining the two other lines in a point.

67.4   The lines l, m and m' in S_3 are mutually skew. Let m^* be any line which intersects m and m'. Let P be any point on m, and let the plane $[lP]$ intersect the line m^* in P^* and the line m' in P'. Show that the line PP^* passes through $V = l \cap [mm^*]$ and that the line P^*P' passes through $V' = l \cap [m^*m']$. Deduce that

$$P \overset{V}{\barwedge} P^*, \qquad P^* \overset{V'}{\barwedge} P',$$

so that the projection of P onto P' from l can be effected by a product of two perspectivities. (§67.3, Fig. 67.4.)

67.5   A simplex of reference in S_n is given by the points $P_0, P_1, \ldots, P_n$. Show that the projection of the point $P = (y_0, y_1, \ldots, y_n)$ from P_0 onto the hyperplane determined by the points $P_1, P_2, \ldots, P_n$ is the point $P' = (0, y_1, \ldots, y_n)$. Show also that the projection of P from the line P_0P_1 onto the S_{n-2} determined by the points $P_2, \ldots, P_n$ is the point $(0, 0, y_2, \ldots, y_n)$. Generalize to the projection of P from the S_{n-m-1} formed by $P_0, P_1, \ldots, P_{n-m-1}$ onto the S_m formed by $P_{n-m}, \ldots, P_n$.

67.6   In Exercise 67.5 show that the projection of P from the line P_0P_1 onto the S_{n-2} determined by the points $P_2, \ldots, P_n$ can be effected by a product of two perspectivities, the first from P_0 onto the hyperplane determined by the points $P_1, P_2, \ldots, P_n$, and the second projecting the resulting point from P_1 onto the hyperplane determined by the points $P_0, P_2, \ldots, P_n$. Generalize this result to the projection of P from an S_{n-m-1} onto an S_m. (Show that the result is the same as a product of projections from $n-m$ independent points of S_{n-m-1} onto $n-m$ suitably chosen independent hyperplanes containing S_m.)

68.1   Dedekind's law for linear subspaces in S_n

A subspace S_p of S_n may, or may not be contained in another subspace S_q of S_n. If it is, we write

$$S_p \subseteq S_q,$$

and the possibility that $S_p = S_q$ is not excluded. It is clear from the definition of linear subspaces that if $S_p \subseteq S_q$ and also $S_q \subseteq S_r$, then $S_p \subseteq S_r$. We also have the condition for identity of sets of points: if $S_p \subseteq S_q$ and also $S_q \subseteq S_p$, then $S_p = S_q$. The subspaces of S_n therefore satisfy the following conditions with regard to the inclusion relation $\subseteq$:

  I   For all S_p, $S_p \subseteq S_p$.
  II   If $S_p \subseteq S_q$ and also $S_q \subseteq S_p$, then $S_p = S_q$.
  III   If $S_p \subseteq S_q$ and $S_q \subseteq S_r$ then $S_p \subseteq S_r$.

We say that the subspaces of S_n are *partially ordered* with respect to the inclusion relation $\subseteq$.

As soon as we have a partial ordering relation, the possibility of the existence of a least upper bound and greatest lower bound for any pair of elements of our partially ordered set suggests itself. If we have two subspaces S_p, S_q, it is clear that the join of the two subspaces is a least upper bound, and the intersection of the two subspaces is a greatest lower bound. Our subspaces therefore form a *lattice*. Since it is possible in a variety of ways to show that the join of a given set of subspaces is a least upper bound of the set, and the intersection is a greatest lower bound, the set of linear subspaces of S_n is what is known as *a complete lattice*. There are many types of lattice, and ours is even more specialized, since it is *a complemented lattice* (§67.1, Thm VII). It is also *a modular lattice*, since it obeys the Dedekind Law, and this is what we are investigating in this section. This is a distributive law, of the $a \cdot (b + c) = a \cdot b + a \cdot c$ type, but in our case the symbols represent subspaces, the $\cdot$ symbol stands for *intersection*, and the $+$ symbol for *join*.

Theorem Dedekind's Law *If R, S and T are linear subspaces of S_n and $R \subseteq S$, then*

$$S \cap (R \cup T) = R \cup (S \cap T).$$

Remark. This distributive law, as we shall see, is valid when $R \subseteq S$, but it may be untrue if $R \nsubseteq S$. In any case, the right-hand side of the expression above is only equal to $(S \cap R) \cup (S \cap T)$ because $R \subseteq S$ implies that $S \cap R = R$. The dimensions of the subspaces play no part in this theorem, which is our reason for using different symbols to denote subspaces from those we used previously.

Proof. If $R = T$, then $R \cup T = R$, and $S \cap (R \cup T) = S \cap R = R$, so that

$$R \cup (S \cap T) = R \cup (S \cap R) = R \cup R = R,$$

and the theorem is true. We therefore now assume that $R \neq T$.

If $S = R$, we have to prove that $S \cap (S \cup T) = S \cup (S \cap T)$. But each side of this equation is equal to S. We therefore also assume that $S \neq R$.

From our definitions of intersection and join we see that $S \cap T \subseteq S$, and since also $R \subseteq S$, we see that $R \cup (S \cap T) \subseteq S$.

Again, since $S \cap T \subseteq T$, it follows that $R \cup (S \cap T) \subseteq R \cup T$. Therefore

$$R \cup (S \cap T) \subseteq S \cap (R \cup T).$$

If we now prove that

$$S \cap (R \cup T) \subseteq R \cup (S \cap T),$$

we shall have shown the identity of both sides of our inequality, and completed the proof.

Let s be a point in $S \cap (R \cup T)$. Since $R \subseteq S$, such a point exists. Since $s \, \varepsilon \, R \cup T$, we can find a point $r \, \varepsilon \, R$ and a point $t \, \varepsilon \, T$ such that $s = r + t$, so that the three points are collinear. We have $S \neq R$ and $R \neq T$, and therefore the points r and t can be chosen to be distinct from s and from each other.

Since $R \subseteq S$, both r and s are in S, and therefore t is linearly dependent on points in S. Hence t must lie in S. But t also lies in T. Therefore t lies in $S \cap T$. The point s is linearly dependent on r and t, and therefore lies in $R \cup (S \cap T)$. Since s is any point in $S \cap (R \cup T)$, this proves that

$$S \cap (R \cup T) \subseteq R \cup (S \cap T),$$

and completes the proof.

Verb. sap. We have proved that $t \in S \cap T$. But suppose that the intersection $S \cap T$ is empty? It is true that we can find a point s in $S \cap (R \cup T)$, but our next assumption is based on the unproved assertion that s lies *outside* R, and in S. If s lies in R, we have $s = r$, and we cannot write $s = r + t$. Hence care has to be taken to deal with the case in which S and T do not intersect. The contradiction we arrive at, that t lies in $S \cap T$, shows that the only points in $S \cap (R \cup T)$ necessarily lie in R, and so $S \cap (R \cup T) \subseteq R \cup (S \cap T) = R$, and the theorem is proved. On the other hand, if S and T do intersect, and s lies in R, then it certainly lies in $R \cup (S \cap T)$. If s, chosen to lie in $S \cap (R \cup T)$, lies in S but outside R, the proof given initially applies.

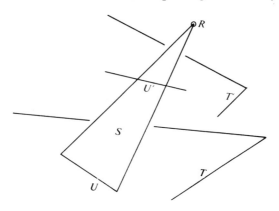

Fig. 68.1

68.2   Dedekind's law and projections

Let R be a given linear subspace of S_n (Fig. 68.1), and let T and T' be distinct subspaces which are both complementary to R (§67.1), so that

$$S_n = R \cup T = R \cup T',$$

$$\emptyset = R \cap T = R \cap T',$$

where $\emptyset$ stands for the empty set. If now S is a subspace which contains R, Dedekind's Law gives us

$$S \cap (R \cup T) = S \cap S_n = S = R \cup (S \cap T) = R \cup U,$$

where we write $U = S \cap T$. Hence $S = R \cup U$. For subspaces U of T and spaces S which contain R the two relations

$$U = S \cap T \quad \text{and} \quad S = R \cup U$$

are equivalent, one being derivable from the other.

We now begin with any $U \subseteq T$, form the join $S = U \cup R$, and intersect S with T', any complement to R in S_n which is distinct from T. Define $U' = S \cap T'$. Then this gives us a one-to-one correspondence between the subspaces U of T and the subspaces U' of T', and this correspondence is completely described by the one equation

$$U \cup R = U' \cup R.$$

Since $U_1 \subseteq U_2$ for two U_1, U_2 in T implies $U_1' \subseteq U_2'$ in T', this one-to-one correspondence preserves the property of points being collinear, and is called *a collineation*. If U_1 is a point, and U_2 a line, and U_1 lies in (or on) U_2, then the image point U_1' lies in the image line U_2'. We shall see later what the equations of such a collineation look like. Of course, the collineation is called *the projection of T onto T' from the center R*, and this is our second encounter with projections. (See §67.3. The equations of a projection are given in §71.2.)

Exercises

68.1   Show that if in S_3 the spaces R, S and T are, respectively, a line, a point lying on the line (so that $R \supseteq S$), and a line skew to the line R, then

$$S \cap (R \cup T) = S, \quad \text{and} \quad R \cup (S \cap T) = R,$$

but $(S \cap R) \cup (S \cap T) = S$, so that the Dedekind Law can be said to hold. On the other hand, if S does not lie on R or T, then

$$S \cap (R \cup T) = S, \quad \text{and} \quad (S \cap R) \cup (S \cap T) = \emptyset,$$

so that the Dedekind Law does not necessarily hold if $S \not\subseteq R$ or $R \not\subseteq S$.

68.2   If in S_3 the space R is a point lying on a line S which is skew to the line T, so that S and T do not intersect, show that $S \cap (R \cup T) = R$, and therefore

$$S \cap (R \cup T) = R \cup (S \cap T) = R$$

in this case.

69.1   Cross-ratio

By Thm I of §66.1 we know that a change of simplex of reference in S_n is equivalent to a linear mapping $y = \mathbf{A}x$, where $\mathbf{A}$ is a non-singular matrix and a point has co-ordinates x with respect to one simplex, and y with respect to the other. We are interested in functions of sets of points which remain unchanged under a change of coordinate system. Distinct points, that is points $(x_0, x_1, \ldots, x_n)$ and $(x_0', x_1', \ldots, x_n')$, where there is no $k \neq 0$ such that $x_i = x_i'k$ $(i = 0, 1, \ldots, n)$, map into distinct points. Collinear points map into points with coordinates of the form Y, Y', $Y + Y'\lambda$, but since the simplex of reference may be chosen so that these three points become Y, Y', $Y + Y'$, where originally they were the points X, X', and $X + X'\lambda$, there is no function of three collinear points which remains unchanged. The situation changes with *four* collinear points, and this is where the *cross-ratio* enters. It is a simple function of four collinear points which is independent of the coordinate system.

Let X_1^*, X_2^*, X_3^* and X_4^* be four collinear points in S_n, all distinct, and suppose that $X_3^* = X_1^*\lambda_1 + X_2^*\lambda_2$, and $X_4^* = X_1^*\mu_1 + X_2^*\mu_2$. Then the number

$$\frac{\lambda_2}{\lambda_1} \Big/ \frac{\mu_2}{\mu_1} = \frac{\lambda_2\mu_1}{\lambda_1\mu_2}$$

is said to be *the cross-ratio* of the four points X_1^*, X_2^*, X_3^*, X_4^*, in this order, and is written

$$[X_1^*, X_2^*; X_3^*, X_4^*] = \frac{\lambda_2\mu_1}{\lambda_1\mu_2}.$$

We show that this number is independent of the coordinates assigned to the four points. Suppose that $Y_1^* = \mathbf{A}X_1^*$, and $Y_2^* = \mathbf{A}X_2^*$. Then

$$\begin{aligned}
Y_3^* = \mathbf{A}(X_1^*\lambda_1 + X_2^*\lambda_2) &= \mathbf{A}(X_1^*\lambda_1) + \mathbf{A}(X_2^*\lambda_2) \\
&= (\mathbf{A}X_1^*)\lambda_1 + (\mathbf{A}X_2^*)\lambda_2 \\
&= Y_1^*\lambda_1 + Y_2^*\lambda_2,
\end{aligned}$$

and similarly

$$Y_4^* = \mathbf{A}X^* = Y_1^*\mu_1 + Y_2^*\mu_2.$$

Hence

$$[X_1^*, X_2^*; X_3^*, X_4^*] = [Y_1^*, Y_2^*; Y_3^*, Y_4^*].$$

To a set of points on a line, sometimes called a *range* of points, there corresponds dually a *pencil* of hyperplanes through an S_{n-2}. If U^*, U^{**} denote two distinct hyperplanes of the pencil, then any two hyperplanes of the pencil can be written in the form $V^* = \lambda^1 U^* + \lambda^2 U^{**}$, and $V^{**} = \mu^1 U^* + \mu^2 U^{**}$. We now prove that if two lines l and m cut the four hyperplanes of the pencil in the points X_1^*, X_2^*, X_3^*, X_4^* and $X_1^{*\prime}$, $X_2^{*\prime}$, $X_3^{*\prime}$, $X_4^{*\prime}$, then the two cross-ratios $[X_1^*, X_2^*; X_3^*, X_4^*]$ and $[X_1^{*\prime}, X_2^{*\prime}; X_3^{*\prime}, X_4^{*\prime}]$ are equal, and each is equal to the cross-ratio $[U^*, U^{**}; V^*, V^{**}]$ of the four hyperplanes. This is a powerful theorem, and may be stated thus:

Theorem I   *If four points on a line l are projected from an S_{n-2} which is skew to l onto four points on a line m, also skew to the S_{n-2}, then the cross-ratio of the first set of points is equal to the cross-ratio of the second set of points.*

Proof. We write the equation of a hyperplane U in the form $(U, X) = 0$, where

$$(U, X) = u^0 X_0 + u^1 X_1 + \ldots + u^n X_n.$$

Then we have the equations:

$$(U^*, X_1^*) = (U^{**}, X_2^*) = (V^*, X_3^*) = (V^{**}, X_4^*) = 0,$$

where the last two equations can also be written in the form:

$$(\lambda^1 U^* + \lambda^2 U^{**}, X_1^*\lambda_1 + X_2^*\lambda_2) = 0,$$
$$(\mu^1 U^* + \mu^2 U^{**}, X_1^*\mu_1 + X_2^*\mu_2) = 0.$$

These equations are bilinear, and on expansion they become

$$\lambda^1(U^*, X_1^*)\lambda_1 + \lambda^1(U^*, X_2^*)\lambda_2 + \lambda^2(U^{**}, X_1^*)\lambda_1 + \lambda^2(U^{**}, X_2^*)\lambda_2 = 0,$$

which reduces to

$$\lambda^1(U^*, X_2^*)\lambda_2 + \lambda^2(U^{**}, X_1^*)\lambda_1 = 0,$$

and

$$\mu^1(U^*, X_2^*)\mu_2 + \mu^2(U^{**}, X_1^*)\mu_1 = 0.$$

If we write $c = (U^*, X_2^*)$, then $c = 0$ would imply that the hyperplane U^* contains not only X_1^*, but also X_2^*, and therefore the l on which X_1^*, X_2^* lie. The line l would

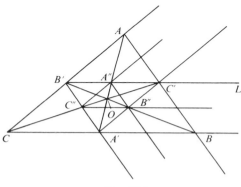

Fig. 69.1

therefore intersect the S_{n-2} to which it is supposed to be skew. Hence $c \neq 0$, and our two equations give us

$$(U^{**}, X_1^*) = -(\lambda^1\lambda_2)c/(\lambda^2\lambda_1) = -(\mu^1\mu_2)c/(\mu^2\mu_1),$$

from which

$$\frac{\lambda^2\mu^1}{\lambda^1\mu^2} = \frac{\lambda_2\mu_1}{\lambda_1\mu_2},$$

and therefore

$$[X_1^*, X_2^*; X_3^*, X_4^*] = [U^*, U^{**}; V^*, V^{**}]$$
$$= [X_1^{*\prime}, X_2^{*\prime}; X_3^{*\prime}, X_4^{*\prime}].$$

In §64.2 we discussed sets of four collinear points with the coordinates A, B, $A + B\lambda$, $A - B\lambda$. We called such a set *a harmonic range*, and now see that $[A, B; A + B\lambda, A - B\lambda] = -1$. We also have the result that *if we join these four points to a point V, obtaining four lines of a pencil, then any line which does not pass through V cuts the four lines in the four points of a harmonic range*. Hence in Fig. 64.1, which we reproduce again as Fig. 69.1, if $AA' \cap B'C' = A''$, then since the joins of A to the points C, B, A', L are joins to the points of a harmonic range, the intersections of the line $B'C'$ with these lines gives a harmonic range. Hence the points B', C'. $4''$ and L are points of a harmonic range. If now $BB' \cap C'A' = B''$, and $CC' \cap A'B' = C''$, the harmonic construction of §64.2 shows that *the line $B''C''$ intersects $B'C'$ at the harmonic conjugate of A'' with respect to B' and C'*. But this is the point L. Hence $B''C''$ passes through L, and with corresponding results for $C''A''$ (which passes through M) and $A''B''$ (which passes through N), we see that once again O has the same polar line with respect to the triangles ABC and $A'B'C'$. This was proved in §64.1 as an Example on changing the triangle of reference.

By definition, the value of the cross-ratio of four collinear points depends on the order in which they are taken. If we permute the four points, we obtain the following results:

$$[X_1, X_2; X_3, X_4] = [X_3, X_4; X_1, X_2] = [X_2, X_1; X_4, X_3] = [X_4, X_3; X_2, X_1],$$

which shows that the interchange of one pair of points, together with the interchange of the other pair, does not alter the value of the cross-ratio. Hence, of the possible $4! = 24$ possible distinct values of the cross-ratio, we only obtain $24/4 = 6$, at most.

Again we have

$$[X_1, X_2; X_3, X_4][X_1, X_2; X_4, X_3] = 1,$$

and

$$[X_1, X_2; \ldots, X_4] + [X_4, X_2; X_3, X_1] = 1.$$

We shall prove this final formula, which is the most difficult, and leave the other results to be done as Exercises. We have

$$X_3 = X_1\lambda_1 + X_2\lambda_2,$$
$$X_4 = X_1\mu_1 + X_2\mu_2,$$

with $[X_1, X_2; X_3, X_4] = \lambda_2\mu_1/\lambda_1\mu_2$. To find $[X_4, X_2; X_3, X_1]$, we must express X_3 and X_1 in terms of X_4 and X_2. We find that

$$X_3 = X_4\mu_1^{-1}\lambda_1 + X_2(\lambda_2 - \mu_2\mu_1^{-1}\lambda_1),$$
$$X_1 = X_4\mu_1^{-1} + X_2(-\mu_2\mu_1^{-1}).$$

Therefore

$$[X_4, X_2; X_3, X_1] = \frac{\lambda_2 - \mu_2\mu_1^{-1}\lambda_1}{\mu_1^{-1}\lambda_1} \times \frac{\mu_1^{-1}}{(-\mu_2\mu_1^{-1})}$$

$$= 1 - \frac{\lambda_2\mu_1}{\lambda_1\mu_2} = 1 - [X_1, X_2; X_3, X_4].$$

Exercises

69.1 Show that $[X_1, X_2; X_3, X_4] \cdot [X_1, X_2; X_4, X_3] = 1$, and that
$$[X_1, X_2; X_3, X_4] = [X_3, X_4; X_1, X_2] = [X_2, X_1; X_4, X_3] = [X_4, X_3; X_2, X_1].$$

69.2 Show that if k is one value of the cross-ratio of four given collinear points, then the other possible values are k^{-1}, $1 - k$, $(1 - k)^{-1}$, $-k(1 - k)^{-1}$, and $(k - 1)k^{-1}$. If $k = -1$, show that the only other possible values are 2 and $1/2$.

69.3 Write down in a 6×4 matrix the 24 permutations of four points, with rows corresponding to permutations with the same cross-ratio.

69.4 Prove the theorem which corresponds to Exercise 53.5, that if X_1, X_2, P, Q and R are distinct collinear points, then

$$[X_1, X_2; P, R] = [X_1, X_2; P, Q] . [X_1, X_2; Q, R].$$

69.5 Two lines PQR, $P'Q'R'$ meet the sides BC, CA, AB of a triangle ABC in the points P, P' and Q, Q' and R, R'. Prove that

$$[P, P'; B, C] . [Q, Q'; C, A] . [R, R'; A, B] = 1.$$

69.6 A line meets the sides BC, CA, AB of a triangle ABC in P, Q, R. The lines AO, BO, CO meet BC, CA and AB in P', Q' and R'. Prove that

$$[P, P'; B, C] . [Q, Q'; C, A] . [R, R'; A, B] = -1.$$

69.7 Show that the line-pairs

$$aX_0{}^2 + 2hX_0X_1 + bX_1{}^2 = 0, \quad a'X_0{}^2 + 2h'X_0X_1 + b'X_1{}^2 = 0$$

through the point $(0, 0, 1)$ form *a harmonic pencil* (intersect any line not through their common vertex in the points of a harmonic range) if and only if

$$ab' + a'b - 2hh' = 0.$$

69.8 D is a point on the side BC of a triangle ABC, and P, Q are points on AB. The line PD meets AC in H, the line QD meets AC in K, and CP meets AD in M, while CQ meets AD in N. Prove that KM and HN meet on AB. (See Exercise 63.6. We can now prove that HM and KN meet BC at the same point, using harmonic properties, and can then apply the Desargues Theorem to triangles HNC and MKD. Since HM, NK and $CD = BC$ are concurrent, the Theorem applies, and gives the required result. There is a great temptation to obtain the result with one application of Desargues, choosing pairs of triangles of which at least one is a set of three collinear points. The Desargues Theorem does not then apply, of course.)

70.1 Projective transformations

By Thm I of §66.1 we have seen that a change of simplex of reference in S_n is equivalent to a linear transformation $Y = AX$ between the coordinates X and Y in the two systems, where A is a non-singular matrix. Thus a given point may have different names, according to the simplex of reference to which it is referred. We now change our point of view and consider the one-to-one onto mapping of S_n given by the equation $Y = AX$, where again A is a non-singular matrix. Essentially the same mapping is given by $Yk = AX$, where $k \neq 0$. Such a mapping is called *a projective transformation of S_n into itself*.

The existence of the inverse mapping, $X = A^{-1}Y$ shows that the mapping is one-to-one. The only possibility of a map of a point X in S_n not being a point would arise if $Y = (0, 0, . . ., 0)$, and then $X = A^{-1}Y$ would be of the form $(0, 0, . . ., 0)$ also, and not a point, contrary to hypothesis. Hence the mapping is one-to-one and onto. If A

is not a regular matrix, the map of a point X may well not be a point, as we shall see later (§73.1).

Our mapping $Y = \mathbf{A}X$ is a collineation, as we expect from the interpretation we have had already when it referred to a change of simplex of reference, but we prove this independently. The points X_1, X_2, $X_1 + X_2k$ map into Y_1, Y_2 and $\mathbf{A}(X_1 + X_2k) = \mathbf{A}X_1 + \mathbf{A}(X_2k) = \mathbf{A}X_1 + (\mathbf{A}X_2)k = Y_1 + Y_2k$. Hence *collinear points map into collinear points*. For this reason the mapping is often called *a projective collineation of S_n*. If P is a point in the set of points which constitute a line l, and $P' = \mathbf{A}(P)$, then P' is among the set of points which constitute the line l', the map of l. Similarly, if $S_p \subseteq S_q$, then the maps S'_p and S'_q satisfy the relation $S'_p \subseteq S'_q$. We are assuming that an S_p maps into an S'_p, that is, *the maps of linearly independent points P_1, P_2, . . ., P_m are linearly independent points P'_1, P'_2, . . ., P'_m*. The proof is immediate. If the mapped points were dependent, then for some k_1, k_2, . . ., k_m, not all zero, we should have

$$P'_1k_1 + P'_2k_2 + . . . + P'_mk_m = 0.$$

Hence

$$\mathbf{A}^{-1}(P'_1k_1 + P'_2k_2 + . . . + P'_mk_m) = \mathbf{A}^{-1}(0) = 0,$$

which leads to

$$P_1k_1 + P_2k_2 + . . . + P_mk_m = 0,$$

from which we deduce that $k_1 = k_2 = . . . = k_m = 0$, a contradiction.

The mapping $Y = \mathbf{A}X$ therefore preserves dimensionality and incidence. Hence points map onto points, lines onto lines, planes onto planes, etc. and any configuration of points, lines, planes, etc. with incidence properties maps onto a similar configuration. For example, the configuration formed by two triangles in perspective, with which we are already familiar in the Desargues Theorem (§62.2), maps into a configuration which also consists of two triangles in perspective, and the center and axis of perspective map into the corresponding center and axis of perspective. Since the only configurations we have considered in this chapter are those which involve incidence between linear subspaces of S_n, it is evident that a projective mapping of these configurations reveals nothing new!

This does not mean that we have exhausted our study of projective geometry. Following Felix Klein we can say that: *Projective geometry in S_n is a study of those properties of constructs in S_n which are left invariant under projective transformations.*

If $Y = \mathbf{A}X$ and $Z = \mathbf{B}Y$ are two projective transformations of S_n, the *composition*, or *product* of the two transformations is $Z = \mathbf{B}(\mathbf{A}X) = \mathbf{B}\mathbf{A}(X)$, a projective transformation with matrix $\mathbf{B}\mathbf{A}$, the product of the two matrices. We know that if $\mathbf{A}$ and $\mathbf{B}$ are both non-singular, so is the product $\mathbf{B}\mathbf{A}$. The *inverse* of the projective transformation $Y = \mathbf{A}X$ exists, and is given by $X = \mathbf{A}^{-1}Y$ if $\mathbf{A}$ is non-singular. This is also a projective transformation. The associative law always holds for mappings of a space, so that the projective transformations of S_n form *a group of transformations*. This is usually called *PGL(n, K)*, *the projective general linear group of S_n over the field K*.

We have not made any special use of the field K as yet. We began by saying that K is the field of complex numbers, but have not used any special properties of this field. We now use such a property, to give a counterexample. It might be thought that every one-to-one onto point transformation which preserves incidence relations in S_n is a projective transformation, but this is not so. The anti-collineation

$$(Y_0, Y_1, \ldots, Y_n) = (\bar{X}_0, \bar{X}_1, \ldots, \bar{X}_n),$$

where $\bar{X}_i$ is the *conjugate* complex number to X_i, is a counterexample. In fact, if $X_i{}^\phi$ denotes the map of X_i under the automorphism ϕ of the field K from which our co-ordinates are taken, then if $\mathbf{A} = [\alpha_{ik}]$,

$$Y_i = \sum_0^n \alpha_{ik} X_k{}^\phi \; (i = 0, 1, \ldots, n)$$

is a one-to-one point transformation of S_n onto itself which maps lines on lines, etc., preserving incidence relations, but it is only a projective transformation if ϕ is the identity mapping. Such a mapping is called a *semi-linear* mapping.

70.2   Change of coordinate system

To simplify our notation suppose that we are considering a projective mapping $Y' = \mathbf{A}Y$, and we change the coordinate system in S_n, so that $Y = \mathbf{B}Z$, and $Y' = \mathbf{B}Z'$, where $\mathbf{B}$ is also non-singular. Then we have

$$Y' = \mathbf{B}Z' = \mathbf{A}(\mathbf{B}Z) = (\mathbf{A}\mathbf{B})Z,$$

so that

$$Z' = (\mathbf{B}^{-1}\mathbf{A}\mathbf{B})Z$$

is the projective transformation *in the new coordinates*.

The matrix $\mathbf{A}$ is said to be *similar* to the matrix $\mathbf{B}^{-1}\mathbf{A}\mathbf{B}$. Since $\mathbf{A} = \mathbf{I}^{-1}\mathbf{A}\mathbf{I}$, where $\mathbf{I}$ is the unity matrix, $\mathbf{A}$ is similar to itself. If

$$\mathbf{C} = \mathbf{B}^{-1}\mathbf{A}\mathbf{B}, \text{ then } \mathbf{A} = \mathbf{B}\mathbf{C}\mathbf{B}^{-1} = (\mathbf{B}^{-1})^{-1}\mathbf{C}(\mathbf{B}^{-1}),$$

so that if $\mathbf{A}$ is similar to $\mathbf{C}$, then $\mathbf{C}$ is similar to $\mathbf{A}$. Finally, if

$$\mathbf{C} = \mathbf{B}^{-1}\mathbf{A}\mathbf{B}, \text{ and } \mathbf{D} = \mathbf{F}^{-1}\mathbf{C}\mathbf{F},$$

so that $\mathbf{C}$ is similar to $\mathbf{A}$, and $\mathbf{D}$ is similar to $\mathbf{C}$, then

$$\mathbf{D} = \mathbf{F}^{-1}\mathbf{B}^{-1}\mathbf{A}\mathbf{B}\mathbf{F} = (\mathbf{B}\mathbf{F})^{-1}\mathbf{A}(\mathbf{B}\mathbf{F}),$$

and $\mathbf{D}$ is similar to $\mathbf{A}$. Hence *the notion of similarity between matrices is an equivalence relation*. This enables us to assemble all $(n + 1) \times (n + 1)$ non-singular matrices over the field K into disjoint classes, all the matrices in any one class being similar to each other. We expect similar matrices to share geometric properties, and we shall return to this subject later (§75.1).

For some problems it is convenient to have different simplices of reference for a point and its image under a projective transformation. If the original transformation is $Y' = AY$, let $Y = BZ$, and $Y' = CZ'$ be coordinate transformations. Then in the new coordinate systems the transformation between Z and Z' is $CZ' = A(BZ)$, or $Z' = C^{-1}ABZ$.

We make use of this freedom in the next section.

Exercises

70.1 If we are investigating the linear mapping $Y' = AY$ in the plane, where

$$A = \begin{bmatrix} 3 & -1 & 2 \\ -2 & 2 & -2 \\ 2 & -1 & 3 \end{bmatrix},$$

the point $(1, -1, 1)$ and the line $2X_0 - X_1 + 2X_2 = 0$ have a special significance (see Exercise 64.9). We take the points $(-2, 2, 3)$ and $(3, 2, -2)$ on the line, and change the coordinate system by the transformation $Y = BZ$, where

$$B = \begin{bmatrix} 1 & -2 & 3 \\ -1 & 2 & 2 \\ 1 & 3 & -2 \end{bmatrix}.$$

Show that in the new coordinate system the mapping is $Z' = (B^{-1}AB)Z$, where

$$B^{-1}AB = \begin{bmatrix} 6 & 0 & 0 \\ 0 & 1 & 0 \\ 0 & 0 & 1 \end{bmatrix},$$

and verify again the results of Exercise 64.9 in the new coordinate system.

70.2 Show that the linear mapping $Y' = AY$, where

$$A = \begin{bmatrix} 2 & 2 & 3 \\ -2 & -3 & -6 \\ 1 & 2 & 4 \end{bmatrix},$$

when subjected to the coordinate transformation $Y = BZ$, where

$$B = \begin{bmatrix} 1 & 2 & 0 \\ -2 & -1 & 0 \\ 1 & 0 & 1 \end{bmatrix}$$

becomes $Z' = (B^{-1}AB)Z$, where

$$B^{-1}AB = \begin{bmatrix} 1 & 0 & 3 \\ 0 & 1 & 0 \\ 0 & 0 & 1 \end{bmatrix}.$$

Show that if P' is the map of P, the lines PP' pass through a fixed point, and that if Q' is the map of Q, the lines PQ and $P'Q'$ meet on a fixed line which contains the fixed point.

70.3 If the points $(Y_0^i, Y_1^i, \ldots, Y_n^i)$ $(i = 0, 1, \ldots, n)$ are taken as the vertices of a simplex of reference, the coordinates with reference to the simplex being called Y', show that the columns of the matrix A in the coordinate transformation $Y = AY'$ are $(Y_0^i, Y_1^i, \ldots, Y_n^i)^T$ $(i = 0, 1, \ldots, n)$.

71.1   The fundamental theorem for projective transformations

Theorem   *A projective transformation T of S_n is uniquely determined if we are given $n + 2$ points Y^0, Y^1, . . ., Y^n, Y^*, and their image points under the mapping: TY^0, TY^1, . . ., TY^*, where no $n + 1$ of the points Y or of the points TY lie in a hyperplane.*

Remark. We note that for $n = 1$ this is the fundamental theorem for projective correspondences on a line, which says that such a correspondence is uniquely determined by the assignment of three distinct points A, B and C and their distinct image points A', B' and C'. This theorem has far-reaching consequences, as we shall see.

Proof. We choose individual coordinate systems for both the points Y and their image points TY. We take the points Y^0, Y^1, . . ., Y^n as simplex of reference for the points Y, which we may do by the given conditions, and we also choose the points TY^0, TY^1, . . ., TY^n as simplex of reference for the image points. Then in the new coordinate systems the projective transformation we seek is given by

$$Z' = \mathbf{C}^{-1}\mathbf{A}\mathbf{B}Z$$

(§70.2), and in this mapping the point $(1, 0, . . ., 0)$ is mapped on $(1, 0, . . ., 0)$, the point $(0, 1, . . ., 0)$ is mapped on $(0, 1, . . ., 0)$, . . ., and the point $(0, 0, . . ., 1)$ is mapped on $(0, 0, . . ., 1)$. Hence the matrix $\mathbf{D} = \mathbf{C}^{-1}\mathbf{A}\mathbf{B}$ of the transformation must be a diagonal matrix:

$$\mathbf{D} = \begin{bmatrix} \delta_0 & & & & \\ & \delta_1 & & & \\ & & \cdot & & \\ & & & \cdot & \\ & & & & \delta_n \end{bmatrix}$$

with zeros everywhere except on the principal diagonal. We also have the condition that T maps the point $Y^* = (z_0, . . ., z_n)$ into the given point $TY^* = (z'_0, . . ., z'_n)$. Hence, for some $k \neq 0$,

$$z'_i k = \delta_i z_i \quad (i = 0, . . ., n).$$

By the hypotheses of the theorem, no $z_i = 0$, and no $z'_i = 0$. Hence the δ_i are uniquely determined by this set of equations, except for a common multiplicative factor $k \neq 0$. This implies that the matrix $\mathbf{C}^{-1}\mathbf{A}\mathbf{B}$ is uniquely determined, except for a non-zero multiplicative factor. Since the transformation we are seeking is determined by the matrix $\mathbf{C}^{-1}\mathbf{A}\mathbf{B}$, the theorem is proved.

If we have two identical projective transformations they will produce the same points TY from given points Y. Hence we have:

Corollary. Two projective transformations are identical if and only if their respective matrices $\mathbf{A}$ and $\mathbf{A}'$ are such that $\mathbf{A} = \mathbf{A}'k$ ($k \neq 0$).

71.2   Projections as projective transformations

The definitions and reasoning of the preceding section are also valid if we consider a projective transformation of S_n not necessarily into itself, but into a distinct S'_n. In particular we consider projective transformations from an S_m into an S'_m, where we suppose that both spaces are subspaces of an S_n. Now, of course, the coordinates $(\gamma_0, \gamma_1, \ldots, \gamma_m)$ in S_m and $(\gamma'_0, \gamma'_1, \ldots, \gamma'_m)$ in S'_m refer to simplices of reference in the respective spaces, but a projective transformation between the spaces is still given by equations of the form

$$\gamma'_i = \sum_0^m \alpha_{ik}\gamma_k$$

or

$$\gamma' = A\gamma,$$

in matrix notation.

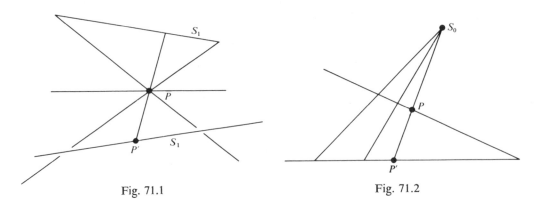

Fig. 71.1   Fig. 71.2

We have already discussed a *projection* from an S_{n-m-1} (as vertex of projection) of points lying in an S_m onto points of an S'_m, where both S_m and S'_m lie in an S_n. (§67.3.) We wish to prove that such a projection effects *a projective transformation* between the points of S_m and those of S'_m.

For example, in S_3 if we project the points P of a line S_1 ($m = 1$) from a skew S_1 ($n - m - 1 = 1$) onto a third skew line S_1 (Fig. 71.1), so that the planes $[S_1P]$ intersect the third S_1 in points P', the mapping from P to P' is a projective transformation. Or even more simply, if in S_2 we have a point S_0 ($m = 0$), and we join this point to the points P of a line $S_1(n - m - 1 = 1)$, intersecting another line S_1 in the points P' (Fig. 71.2), the mapping from P to P' is a projective transformation. We have seen that the apparently more sophisticated mapping in S_3 can be expressed as the product of two of the second type of mappings, called *perspectivities* (§67.3).

Theorem I   *The spaces S_m and S'_m are subspaces of equal dimension lying in S_n. A third subspace S_{n-m-1} is skew both to S_m and to S'_m (has no points in common with* *either). If the points Y of S_m are projected into points Y' of S'_m by joining Y to S_{n-m-1} by an S_{n-m} which intersects S'_m in the corresponding point Y', then the relation between Y and Y' is that of projective transformation.*

Proof. Let the S_{n-m-1} have the equations

$$(U^{(0)}, X) = (U^{(1)}, X) = \ldots = (U^{(m)}, X) = 0. \tag{1}$$

All the points of the space which joins this S_{n-m-1} to a point Y of S_m are linear combinations of Y and $n - m$ points $Z^1, Z^2, \ldots, Z^{n-m}$ which form a basis for S_{n-m-1}. These $n - m$ points Z^i satisfy the equations (1). In particular the point Y' of intersection of the joining S_{n-m} and S'_m is a linear combination of the point Y and the points Z^i. Hence

$$Y' = Y\lambda + Z^1\lambda_1 + Z^2\lambda_2 + \ldots + Z^{n-m}\lambda_{n-m}. \tag{2}$$

Since Y' does not lie in S_{n-m-1}, we must have $\lambda \neq 0$, and we can therefore normalize it, and write $\lambda = 1$, by absorbing it into the coordinate Y. Using the equations (1), which are linear, and the fact that the Z^i satisfy these equations, we have

$$(U^{(0)}, Y') = (U^{(0)}, Y) = \beta_0,$$

say,

$$(U^{(1)}, Y') = (U^{(1)}, Y) = \beta_1,$$

$$\cdots \cdots \cdots$$

$$(U^{(m)}, Y') = (U^{(m)}, Y) = \beta_m,$$

where the β_i are merely the common values of the forms $(U^{(i)}, Y')$ and $(U^{(i)}, Y)$. Now, the point Y lies in S_m, and we may therefore write

$$Y = Y^0\gamma_0 + Y^1\gamma_1 + \ldots + Y^m\gamma_m,$$

where the Y^i are fixed base-points for the space S_m, and $(\gamma_0, \gamma_1, \ldots, \gamma_m)$ are the coordinates of Y in S_m with respect to these base-points. Hence

$$(U^{(i)}, Y) = (U^{(i)}, Y^0)\gamma_0 + (U^{(i)}, Y^1)\gamma_1 + \ldots + (U^{(i)}, Y^m)\gamma_m,$$

which we write in the form

$$\beta_i = (U^{(i)}, Y) = \sum_0^m \delta_{ik}\gamma_k, \tag{3}$$

Similarly,

$$\beta_i = (U^{(i)}, Y') = \sum_0^m \varepsilon_{ik}\gamma'_k, \tag{4}$$

where the $(\gamma'_0, \gamma'_1, \ldots, \gamma'_m)$ are the coordinates of Y' with reference to some set of base-points in S'_m.

Since S_m and S'_m have no point in common with our S_{n-m-1}, the expressions β_i are never simultaneously zero, whatever the point Y and its image Y' are. Let us consider the matrix $\mathbf{D} = (\delta_{ik})$. It is $(m + 1) \times (m + 1)$, and if its rank were less than $m + 1$, the equations

$$\sum_0^m \delta_{ik}\gamma_k = 0 \qquad (i = 0, 1, \ldots, m)$$

would have non-trivial solutions (that is, not of the form $(0, 0, \ldots, 0)$) for $\gamma_0, \gamma_1, \ldots,$ γ_m. Hence the matrix $\mathbf{D}$ has the maximum rank, that is, it is non-singular, and possesses an inverse. This means that the γ_i can be expressed as linear functions of the β_i, and since the β_i are linear functions of the γ'_i, we find that the γ_i can be expressed as linear functions of the γ'_i. The argument also holds for the γ'_i and the γ_i, and we have found that

$$\gamma_i = \sum_0^m p_{ik}\gamma'_k \qquad\qquad \gamma'_i = \sum_0^m q_{ik}\gamma_k,$$

where the matrices $[p_{ik}]$ and $[q_{ik}]$ are both non-singular. Hence the projection of Y into Y' is a projective transformation, and our theorem is proved.

Definition. A projective transformation of the above type, between spaces of equal dimension m, from a vertex of dimension $n - m - 1$, is called *a perspectivity*. We have used this term before only when $n = 2$, and $m = 1$, when we were projecting from a point, and the object point Y and the image point Y' were both on lines.

As a Corollary to the above theorem, we have:

Theorem II   *A product of perspectivities is a projective transformation.*

We now have all the main tools we need for obtaining theorems in S_n. We recollect that we proved in §69.1 that the cross-ratio of four given collinear points is independent of the simplex of reference to which they are referred. This is equivalent to saying that *cross-ratio is an invariant under projective transformation.* We also showed, in Thm I of the same section, that *cross-ratio is invariant under a perspectivity.* Now that we know, by the theorem we have just proved, that a perspectivity is equivalent to a projective transformation, we are not surprised that this is so.

Projection and section are the main procedures in obtaining theorems in projective geometry, and many theorems in a higher-dimensional space are proved by projection onto a space of lower dimension. Thus the next theorem, although it refers to the projective line, turns out to be a very powerful theorem for proving theorems in S_2 and in spaces of higher dimension.

Theorem III   *A projective mapping of a line onto itself, in which three distinct points are left fixed, must be the identity mapping.*

Proof. A fixed point Z under a projective transformation $Y = \mathbf{A}X$ is a point such that $Z = \mathbf{A}Z$, or more generally, $Zk = \mathbf{A}Z$ $(k \neq 0)$. Invoking the Fundamental Theorem of §71.1, we know that a projective transformation is uniquely determined, in the case $n = 1$, if three distinct points A, B, C are to be mapped onto three distinct points A', B' and C'. The theorem still holds if $A' = A$, $B' = B$ and $C' = C$. But then we know that the identity mapping, $Y = X$, will map A on A, B on B and C on C, and since the projective mapping is uniquely determined, and the identity mapping is a projective mapping which satisfies the requirements of the theorem, the projective mapping of the theorem must be the identity mapping.

Since a product of perspective mappings is a projective transformation, we have a

Corollary. If, by a product of perspective transformations, three distinct points of a line are mapped onto themselves, then every point of the line is mapped onto itself.

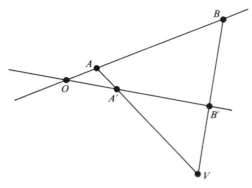

Fig. 71.3

As a first application of this theorem, we prove the very useful theorem:

Theorem IV   *A projective mapping between two distinct lines which intersect in a point O which is such that O is mapped onto itself must be a perspectivity.*

Proof. We work in the projective plane S_2 which contains the two lines (Fig. 71.3). Suppose that the points A, B on the one line are mapped onto the points A', B' of the other by the projective mapping. Let $V = AA' \cap BB'$. Call the perspective mapping which has vertex V the mapping T^*, and the given projective mapping T. Then

$$(O, A, B)T = (O, A', B'),$$

and

$$(O, A, B)T^* = (O, A', B'),$$

so that

$$(O, A, B)T^*T^{-1} = (O, A, B).$$

Hence the mapping T^*T^{-1} is the identity mapping (Thm III, above), and therefore $T = T^*$, which is a perspectivity.

Another useful way of regarding this theorem is the following: if in a projective mapping between two intersecting lines the point of intersection is mapped on itself, then the joins of corresponding points on the two lines all pass through a point.

We complete this section by applying the methods we have developed above to the proof of the Desargues Theorem, already encountered in §62.2, where we gave an algebraic proof.

Theorem V *If two triangles ABC, A'B'C' in a projective plane are such that the lines AA', BB', and CC' pass through a point V, then the three points $L = BC \cap B'C'$, $M = CA \cap C'A'$ and $N = AB \cap A'B'$ are collinear.*

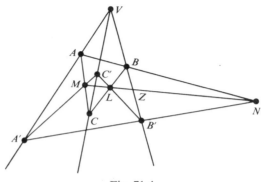

Fig. 71.4

Proof. Project points on the line VBB' (Fig. 71.4) from the vertex L onto the line VCC'. Then we have

$$(V, B, B') \overset{L}{\barwedge} (V, C, C').$$

Project points on the line VCC' from the vertex M onto the line VAA'. We have

$$(V, C, C') \overset{M}{\barwedge} (V, A, A').$$

Finally, project the points on the line VAA' from the vertex N onto the line VBB'. We have

$$(V, A, A') \overset{N}{\barwedge} (V, B, B').$$

Hence the product of these three perspectivities gives a projective mapping of the points of the line VBB' which maps V on V, B on B and B' on B'. The mapping is therefore the identity (Thm III, above). Let

$$Z = LM \cap VB,$$

and trace the image of Z under the three perspectivities. Under the first, vertex L, the point Z is mapped onto the point $LM \cap VC$, and under the second perspectivity, vertex M, the point $LM \cap VC$ is mapped onto the point $LM \cap VA$. The point $LM \cap VA$ is to be mapped by the third perspectivity, vertex N, onto the same point $Z = LM \cap VB$ which we started with, since the product of the three mappings is the identity mapping on the line VBB'. The join of N to $LM \cap VA$ must intersect VB in the point in which LM intersects VB. Since the line joining N to $LM \cap VA$ has two points in common with the line LM, namely $LM \cap VA$ and $Z = LM \cap VB$, the lines coincide. Hence the points L, M and N are collinear.

Remark. There are certain special cases which have to be examined, to see that the above proof still holds. We suggest these in the Exercises.

If a projective transformation exists between ranges or pencils, we say they are *projective.* A perspectivity is a special projectivity, and denoted by the symbol $\overset{o}{\overline{\wedge}}$, where O is the vertex of perspective. When ranges of points or pencils of lines are projective, we use a modified symbol $\overline{\wedge}$, where, of course, the vertex is not indicated, since not all projectivities are simple perspectivities.

Exercises

71.1 Carry out the proof of Thm I when our space is S_2, and the vertex of projection is a point V, and the S_m and S'_m are both lines. (The point V will be given by two linear equations, $(U^{(0)}, X) = (U^{(1)}, X) = 0$. After carrying out the proof as in the text, you may wish to begin again, and to represent V by its point-coordinates.)

71.2 If A, B, C and D are four collinear points, prove that the range (A, B, C, D) is projective with the range (B, A, D, C), and deduce the equality of the cross-ratios $[A, B; C, D]$, $[B, A; D, C]$. (Take S, P collinear with B, and not on AB and let $DS \cap AP = U$, $AS \cap CP = V$. Let $DU \cap CP = T$, and project A, B, C, D from P onto DU, then project these points from A onto PC, and finally project this last set from S back onto AB.)

71.3 ABC is a triangle in S_2, O a point not on any side, and a line through O intersects AC in B', AB in C' and CB in A'. Prove that

$$[O, A'; B', C'] = [O(A', A; B, C)],$$

where the expression on the right refers to the cross-ratio of the pencil of lines vertex O joining O to the designated points. (Take sections of the pencil of lines by BC, and project the range of points onto BC.)

71.4 If $ABCD$ be a tetrahedron in S_3, and a line l meets the face BCD in A', the face CDA in B', the face DAB in C' and the face ABC in D', prove that the range (A', B', C', D') is projective with the pencil of planes $l(A, B, C, D)$. (Project from D onto the face ABC, and show that this theorem reduces to the one given in the preceding Exercise.)

71.5 Show that a correlation $U^T = AX$ (between points and planes) can be set up in the preceding Exercise which maps A on face BCD, B on face CDA, C on face DAB, D on face ABC, and the line l onto itself. Deduce the theorem of Exercise 71.4.

71.6 Write out the dual in S_2 of Thm IV of §71.2, and prove it by using the theorems provided in this section.

71.7 Prove the Desargues Theorem by proving that the pencils

$$B'(B, L, M, N), \qquad B(B', L, M, N)$$

are projective, by taking sections on $A'C'$ and AC respectively, and then using the result of Exercise 71.6. (The notation is as in Thm V of §71.2.)

71.8 A, B, C, D and A', B', C', D' are sections of distinct pencils in S_2, vertices V and V' respectively, and $[A, B; C, D] = [A', B'; C', D']$. Let U be a point on VC which does not lie on AB and is distinct from V, and let U' be a point on $V'C'$ which does not lie on $A'B'$ and is distinct from V'. Show that the unique projective transformation which maps V on V', A on A', B on B' and U on U' also maps C on C' and D on D', so that we can say the pencils $V(A, B, C, D)$ and $V'(A', B', C', D')$ are projective. Generalize to pencils in S_n, with vertices distinct S_{n-2} and S'_{n-2}.

71.9   Show that the projective mapping $Y' = \mathbf{A}Y$ between points Y and Y' of S_n induces the projective mapping $U = U'\mathbf{A}$ between hyperplane coordinates U and U'. (Compare Thm II of §64.1.) Deduce that a projective mapping of S_n maps four hyperplanes of a pencil into four hyperplanes of a pencil with the same cross-ratio as the object set. What is the relation of this theorem to the generalization of the preceding Exercise?

71.10   If ABC, $A'B'C'$ are two triangles in perspective in S_2 from a point V, we know that the points $L = BC \cap B'C'$, $M = CA \cap C'A'$ and $N = AB \cap A'B'$ are collinear (Thm V, §71.2). The ten points described lie in sets of three on ten lines, and three lines of the ten pass through every point. Show that this *configuration* (as it is known technically) can be obtained by taking a plane section of the ${}^5C_2 = 10$ joins of five distinct points in space of three dimensions.

71.11   If the five points in S_3 be numbered 1, 2, 3, 4, 5, show that the points and lines of the Desargues configuration described in the preceding Exercise may be given a binary notation, $(12) = (21)$, etc. for points and $[12] = [21]$, etc. for lines, where the points on the line $[12]$ are the points (34), (45) and (53), etc. Describe the notation more precisely.

71.12   Show that any point of the ten in the Desargues configuration can be chosen as the point V of the original formulation in Exercise 71.10, and two appropriate triangles in perspective from V can then be found.

71.13   Show by drawing two appropriate triangles ABC, $A'B'C'$ that in the Desargues configuration it is possible to have M lying on BB'. By renaming the points, show that this is equivalent to B' lying on CA.

71.14   Show that the proof of Thm V (The Desargues Theorem) given in §71.2 continues to hold for either of the specializations described in the preceding Exercise.

71.15   Prove the Pappus Theorem. A, B, C are distinct points on a line l, and A', B', C' distinct points on a coplanar line l'. Then the points $L = BC' \cap B'C$, $M = CA' \cap C'A$ and $N = AB' \cap A'B$ are to be proved collinear. If $U = AC' \cap A'B$, $X = A'C \cap C'B$, and $T = LM \cap A'B$, prove that

$$(A', N, U, B) \overset{A}{\doublewedge} (A', B', C', Z) \overset{C}{\doublewedge} (X, L, C', B) \overset{M}{\doublewedge} (A', T, U, B),$$

where $Z = AB \cap A'B'$ is assumed to be distinct from any of the points A, B, C, A', B', C'. Deduce from the above that $T = N$, so that L, M, N are collinear. (See Thm III, §63.1.)

71.16 l, m and m' are mutually skew lines in S_3, and with l as vertex of perspective points P on m are projected into points P' on m'. (See Fig. 67.4.) Prove that P and P' are corresponding points on m and m' if and only if the line PP' intersects l.

71.17   Prove that a projective mapping between two skew lines m and m' in S_3 is always a perspectivity. (If A, B, C on m correspond to A', B' and C' on m' under the projective correspondence, let l be any line which intersects the three lines AA', BB' and CC'. Then the perspective mapping with vertex l maps A on A', B on B' and C on C'. By the Fundamental Theorem (§71.1) this perspective mapping coincides with the given projective mapping.)

71.18   From the result of the two preceding Exercises prove Application (*b*) of the Propositions of Incidence of §65.1, that if l, m, n are mutually skew lines in S_3, and l', m', n' are transversals of these three lines, then any further transversals of l, m, n and l', m', n' must intersect each other.

71.19 Give another proof of Exercise 71.4 by projecting A' from B onto the point P' on CD, and B' from A onto the point Q' on CD, so that the line CD intersects the pencil of planes l (A, B, C, D) in the points (Q', P', C, D). Note that the planes through the line AB which pass through Q', P', C, D intersect l in the points B', A', D', C', and proceed from this point.

71.20 If all the points designated lie on the same line, show that the two conditions $(A, B, C, D) \barwedge (A', B', C', D')$ and $[A, B; C, D] = [A', B'; C', D']$ are equivalent.

72.1   Derivation of the tetrahedral complex

We recall (Exercise 71.4) that the range in which a line l meets the four faces of a tetrahedron in S_3 is projective with the pencil of planes obtained by joining the line to the four corresponding opposite vertices of the tetrahedron. If the line l moves so

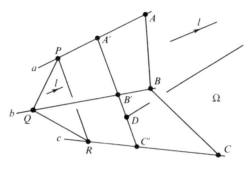

Fig. 72.1

that the range of points in which it meets the faces is always projective with a fixed range of points, we say that l generates *a tetrahedral complex*. We show that such a complex of lines arises naturally from a construct in S_4.

First we note that if π be a plane in S_4, and we use π as vertex of projection, projecting points P_1, P_2, P_3, P_4 on a line l, then the intersections of a line m with the hyperplanes $[\pi P_1], [\pi P_2], [\pi P_3]$ and $[\pi P_4]$ will give points P_1', P_2', P_3' and P_4' which form a set projective with the set P_1, P_2, P_3, P_4 (Thm I, §71.2).

Let a, b, c be three general lines in S_4 (Fig. 72.1), and let Ω be an S_3 which does not contain any of a, b, c, or the unique transversal of a, b, c. We recall that this unique transversal is the unique transversal to b and c in the hyperplane $[bc]$ which passes through the point in which a intersects this solid. Our S_3, which we have denoted by Ω, is to be the S_3 in which the tetrahedral complex will appear, and we can choose it so that it intersects the lines a, b, c in points A, B and C respectively which are not on the unique transversal of a, b and c. We suppose that this unique transversal intersects a, b and c in the points A', B' and C' respectively. Furthermore, let the unique transversal meet Ω in the point D. This is, of course, the only point of the unique transversal in Ω. We now assert that any plane π which meets a, b and c in the points P, Q and R and does not lie in Ω will intersect Ω in the lines of a tetrahedral complex related to the tetrad of points A, B, C and D. (We note that a general plane in S_4 does not intersect any of the lines a, b, c, so that our planes π are planes of a special set.)

The hyperplane $[\pi A]$ contains A and P, and therefore the line a, and hence also the point A'. Arguing similarly for the hyperplanes $[\pi B]$ and $[\pi C]$, and remembering that D lies on $A'B'C'$, we can say that the four hyperplanes $[\pi A]$, $[\pi B]$, $[\pi C]$ and $[\pi D]$ intersect the transversal of a, b and c in a set of fixed points A', B', C' and D. This set is independent of the plane PQR under consideration.

Now let l be the line in which the plane $\pi = PQR$ intersects Ω. Then the hyperplane $[\pi A]$ intersects Ω in the plane $[lA]$, and any other line of Ω intersects the plane $[lA]$ in the point in which this line intersects the hyperplane $[\pi A]$ in S_4. Hence the planes $[lA]$, $[lB]$, $[lC]$ and $[lD]$ are intersected by any line of Ω in the range in which this same line intersects the four hyperplanes $[\pi A]$, $[\pi B]$, $[\pi C]$ and $[\pi D]$. By our initial remark on projection, this range of points is projective with the fixed range A', B', C', D.

Hence the lines l in which the planes π intersect Ω form a tetrahedral complex with respect to the tetrad A, B, C and D.

73.1   Singular projectivities

It is sometimes necessary to consider a mapping between points Y, Y' of S_n, given by the matrix equation $Y' = AY$, where the matrix $\mathbf{A}$ is singular, having rank less than $n + 1$. We call such a projectivity a *singular* projectivity.

There are now points in S_n which have no image, since all the y_i' of the image point Y' are zero. The points Y which have no image satisfy the equations

$$\sum_0^n \alpha_{ik} y_k = 0 \quad (i = 0, 1, \ldots, n),$$

and therefore fill a subspace S_{n-r}, the matrix $\mathbf{A}$ having rank r (§66.1).

On the other hand, the image points Y' which do exist fill a space S_{r-1}. To see this we note that the matrix $\mathbf{A}$ being of rank r, it has r independent columns, and the equations of the mapping

$$y_0' = \alpha_{00} y_0 + \alpha_{01} y_1 + \ldots + \alpha_{0n} y_n,$$
$$y_1' = \alpha_{10} y_0 + \alpha_{11} y_1 + \ldots + \alpha_{1n} y_n,$$
$$\ldots \quad \ldots \quad \ldots \quad \ldots$$
$$y_n' = \alpha_{n0} y_0 + \alpha_{n1} y_1 + \ldots + \alpha_{nn} y_n,$$

show that the coordinates $(y_0', \ldots, y_n')$ are linear multiples of the r independent column vectors, each of which represents a point. Hence the points Y' lie in an S_{r-1}.

We call the S_{n-r} of points which have no map *the null space* of the mapping, and the S_{r-1} onto which all points with maps are mapped is called *the range* of the mapping. We now prove:

Theorem I   *The map of an S_{n-r+1} which contains the null space S_{n-r} is a point of the range.*

Proof. Let $Z_{(0)}, Z_{(1)}, \ldots, Z_{(n-r)}$ be a basis for the null-space. Then we know that $AZ_{(i)} = 0$ $(i = 0, 1, \ldots, n - r)$. If Y is any point not in the null-space, any point in the S_{n-r+1} which joins Y to the null-space is

$$P = Y\lambda + Z_{(0)}\mu_0 + \ldots + Z_{(n-r)}\mu_{n-r} \quad (\lambda \neq 0).$$

The map of P is

$$Y' = AP = A(Y\lambda) + (AZ_{(0)})\mu_0 + \ldots + (AZ_{(n-r)})\mu_{n-r} = (AY)\lambda,$$

and we therefore obtain the same image point Y' for all $\lambda, \mu_0, \ldots, \mu_{n-r}$. This point Y' lies, of course, in the range, S_{r-1}.

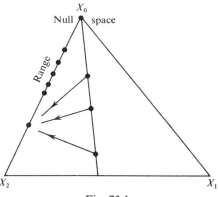

Fig. 73.1

Remark. It may be thought that the null-space and the range have no points in common. But we may have a point Y for which $Y' = AY$ is a point in the range, but $AY' = A^2Y = 0$, so that Y' is a point in the null-space also.

Example. Let the matrix be

$$A = \begin{bmatrix} 0 & 1 & 0 \\ 0 & 0 & 0 \\ 0 & 0 & k \end{bmatrix},$$

so that $Y' = AY = (y_1, 0, ky_2)$. All the points of S_2 are mapped into points of $X_1 = 0$. Here the rank $r = 2$, and $S_{n-r} = S_0$, and $S_{r-1} = S_1$. The null-space is a point, and in fact the point $(1, 0, 0)$ has no map. The null-space lies in the range (Fig. 73.1). If we take $Y = (\alpha, \beta, 0)$, then $Y' = A(\alpha, \beta, 0)^T = (\beta, 0, 0)^T$ is a point in the null-space, and also in the range, and $A^2(\alpha, \beta, 0)^T = (0, 0, 0)^T$. Any point on the line joining $(1, 0, 0)$ to the point (α, β, γ) is $(\lambda + \alpha, \beta, \gamma)$ and this is mapped on the point $(\beta, 0, k\gamma)$, which is independent of the point chosen. Hence lines through the null-space are mapped onto points in the range.

Remark. We proved in Thm I, §71.2, that *projection* of a point Y lying in an S_m onto a point Y' lying in an S'_m from a vertex S_{n-m-1} which intersected neither subspace is a non-singular projectivity as regards the mapping between the S_m and the S'_m. But if the point Y is regarded as a point in S_n, and although Y' is in S'_m it is also regarded as a point in S_n, the map is a singular projectivity with the null-space S_{n-m-1}, and the range S'_m. The rank of the mapping is $r = m + 1$.

For example, the projection of the point $Y = (y_0, y_1, \ldots, y_n)$ from the vertex $(1, 0, \ldots, 0)$ of the simplex of reference onto the hyperplane $X_0 = 0$ is the point $Y' = (0, y_1, y_2, \ldots, y_n)$, and the matrix of the mapping is

$$A = \begin{bmatrix} 0 & 0 & \ldots & 0 \\ 0 & 1 & \ldots & 0 \\ 0 & 0 & 1 & .. & 0 \\ . & . & .... & . \\ 0 & 0 & \ldots & 1 \end{bmatrix}$$

which has rank $r = n$. The projection of the point Y from the line which joins $(1, 0, \ldots, 0)$ to $(0, 1, 0, \ldots, 0)$ onto the S_{n-2} with equations $X_0 = X_1 = 0$ is $Y' = (0, 0, y_2, \ldots, y_n)$, and the corresponding matrix A is the unity matrix with *two* zeros replacing the unities, and is of rank $n - 1$, etc.

It is to be expected, geometrically, that the vertex of projection has no map, and is therefore the null-space, and that the join of any point not in the vertex of projection to the vertex of projection maps into a point, since this is the geometrical process we are investigating. All points which lie in the same join naturally have the same projection.

Exercise

73.1 Show that the projectivity $Y' = A Y$, where

$$A = \begin{bmatrix} 1 & 3 & 5 \\ 2 & -1 & 3 \\ 3 & 2 & 8 \end{bmatrix}$$

is a singular projectivity in S_2, and that:

(i) the point $P = (-2, -1, 1)$ has no image;
(ii) every other point of the plane is mapped onto the line $X_0 + X_1 - X_2 = 0$.
(iii) the points of any line through P, excluding P, map onto the same point on the line $X_0 + X_1 - X_2 = 0$, there being a one-to-one correspondence between lines through P and image points of the line.
(iv) Any line in the plane distinct from $X_0 + X_1 - X_2 = 0$ maps onto this line.

74.1 Fixed points of projectivities

In earlier chapters (I, II, V) we discussed one-to-one onto mappings of the real Euclidean plane. A non-singular projectivity is a one-to-one onto mapping of S_n, and the reader may have wondered whether the concepts are connected in the case $n = 2$. There is a connection, which we shall discuss, via the notion of *fixed points* of a non-singular projectivity.

Definition. A fixed point P of a projective mapping $Y' = \mathbf{A}Y$ of S_n onto itself is a point such that $P = \mathbf{A}(P)$.

Note that the geometrical *point* is invariant under the mapping, but since we are using homogeneous coordinates, the point P may reappear as $P\lambda$.

Hence, if $P = Y = (y_0, \ldots, y_n)$ is a fixed point under the mapping,

$$Y\lambda = \mathbf{A}Y,$$

which may be written

$$(\mathbf{A} - \lambda\mathbf{I}_{n+1})Y = 0,$$

where $\mathbf{I}_{n+1}$ is the $(n+1) \times (n+1)$ unity matrix. If the matrix $\mathbf{A} - \lambda\mathbf{I}_{n+1}$ were not singular, the equation above would lead to $Y = (0, 0, \ldots, 0)$, which is not the case. Hence the matrix $\mathbf{A} - \lambda\mathbf{I}_{n+1}$ is singular, and therefore

$$\det(\mathbf{A} - \lambda\mathbf{I}_{n+1}) = 0.$$

Now this is an equation for λ, and its expansion begins thus:

$$(-1)^{n+1}\lambda^{n+1} + \ldots = 0,$$

so that λ is a root of a polynomial equation of degree $n + 1$. *For the first time we use the fact that our ground field K, the field of complex numbers, is algebraically closed,* so that the equation in λ has at least one root in the field K. Let λ^* be a root. Then λ^* is called an *eigenvalue*, or *characteristic root of the matrix* $\mathbf{A}$. To every λ^* there correspond one or more fixed points under the mapping $Y' = \mathbf{A}Y$. For we have the equations

$$(\mathbf{A} - \lambda^*\mathbf{I}_{n+1})Y = 0$$

for fixed points, and we now know that the matrix $\mathbf{A} - \lambda^*\mathbf{I}_{n+1}$ has rank r which is less than $n + 1$. The solutions y of this set of equations fill a linear subspace of S_n of dimension $n - r$, and all the points in this space are fixed points which correspond to the characteristic root λ^*.

We note that although one characteristic root can give rise to an infinity of fixed points, each fixed point belongs to a uniquely defined characteristic root. Hence the subspaces of fixed points corresponding to distinct characteristic roots do not intersect each other. These subspaces of fixed points are called *fundamental* subspaces under the collineation.

It is important to distinguish between subspaces which are mapped onto themselves by the projective mapping, but not necessarily so that every point is fixed, and those which are invariant point-wise, consisting entirely of fixed points, like those we have just considered. If we consider what the mapping $Y' = \mathbf{A}Y$ does to hyperplanes, we see that every such mapping has at least one invariant hyperplane, that is an S_{n-1} every point of which is mapped into the same S_{n-1}, but not every point of which is necessarily a fixed point. The associated mapping for hyperplanes U and U' is given by $U = U'A$ (Exercise 71.9). If U is an invariant hyperplane under this mapping, and $U' = \lambda^{-1}U$, then $U = U'A$ leads again to $\lambda U = UA$ and $U(A - \lambda I_{n+1}) = 0$, so that

$$\det (A - \lambda I_{n+1}) = 0,$$

which is the same equation as the one considered above. Hence, *any projectivity has at least one invariant hyperplane.* If λ^* is a root of the characteristic equation which makes the matrix $A - \lambda^* I_{n+1}$ of rank r, we have seen that there is an S_{n-r} of fixed points which arises from this eigenvalue λ^*. It is also clear that the intersection of $n - r + 1$ linearly independent hyperplanes, each one of which is mapped into itself by the projectivity, gives a subspace of dimension $n - (n - r + 1) = r - 1$ which is mapped into itself by the projectivity. Let us consider some examples, taking $n = 2$.

If

$$A = \begin{bmatrix} \lambda_1 & 0 & 0 \\ 0 & \lambda_2 & 0 \\ 0 & 0 & \lambda_3 \end{bmatrix},$$

the characteristic equation has the roots $\lambda_1, \lambda_2, \lambda_3$ and taking $\lambda = \lambda_1$, we have to solve the equations

$$0 . y_0 = 0,$$
$$(\lambda_1 - \lambda_2)y_1 = 0, \quad (\lambda_1 \neq \lambda_2),$$
$$(\lambda_1 - \lambda_3)y_2 = 0, \quad (\lambda_1 \neq \lambda_3),$$

which are satisfied by $(1, 0, 0)$ and no other point. This is a fixed point, and similarly $(0, 1, 0)$ and $(0, 0, 1)$ are fixed points. The line joining two fixed points is evidently a fixed line, and if we consider the fixed lines of the collineation we obtain $[1, 0, 0]$, $[0, 1, 0]$ and $[0, 0, 1]$, the sides of the triangle of reference, and no other lines. In this case $r = 2$, and $n - r = 0$, giving a fixed point for S_{n-r}, and $r - 1 = 1$, giving a line which is mapped onto itself.

Now consider the case when

$$A = \begin{bmatrix} \lambda_1 & 0 & 0 \\ 0 & \lambda_1 & 0 \\ 0 & 0 & \lambda_2 \end{bmatrix} \quad (\lambda_1 \neq \lambda_2).$$

The characteristic equation is $(\lambda_1 - \lambda)^2(\lambda_2 - \lambda) = 0$. If we take $\lambda^* = \lambda_1$, the fixed points are given by

$$(\lambda_2 - \lambda_1)y_2 = 0,$$

and therefore $X_2 = 0$ is *a line of fixed points.* If we take $\lambda^* = \lambda_2$, we have the equations

$$y_0 = y_1 = 0,$$

and therefore $(0, 0, 1)$ is a fixed point, and there are no others. For invariant lines, taking $\lambda^* = \lambda_1$, we have the equation $U^2 = 0$, and this represents any line through the point $(0, 0, 1)$. Therefore *any line through this point is invariant*. If we take $\lambda^* = \lambda_2$, we find that $U^0 = U^1 = 0$, which means that $[0, 0, 1]$, that is $X_2 = 0$, is invariant, which we already know.

This mapping is called *a central collineation* (see Fig. 74.1). The point $P = (y_0, y_1, y_2)$ is mapped onto the point $P' = (\lambda_1 y_0, \lambda_1 y_1, \lambda_2 y_2)$, so that

$$P' = \lambda_1(y_0, y_1, y_2) + (\lambda_2 - \lambda_1)y_2(0, 0, 1).$$

Hence *the join of P to P' always passes through the point* $(0, 0, 1)$. This point is called *the center of the collineation.*

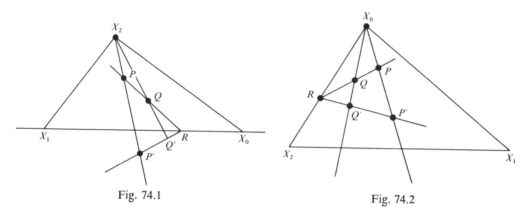

Fig. 74.1   Fig. 74.2

If the line PQ intersects the line of fixed points in the point R, the line PQR maps onto the line $P'Q'R$, where P maps on P', and Q on Q'. Hence *a line and its map always intersect on the line* $[0, 0, 1]$. This line is called *the axis of the central collineation*. We shall see that this is the mapping which corresponds to the central similarity transformation in the real Euclidean plane (§41.5).

It is possible for the center of the collineation we are examining to lie on the axis. Consider the mapping $y' = Ay$, where now

$$A = \begin{bmatrix} \lambda_1 & 1 & 0 \\ 0 & \lambda_1 & 0 \\ 0 & 0 & \lambda_1 \end{bmatrix}.$$

The characteristic equation is $(\lambda_1 - \lambda)^3 = 0$, and $\lambda^* = \lambda_1$ gives $y_1 = 0$ for fixed points, so that again we have *a line of fixed points*, given by $X_1 = 0$.

When we look for invariant lines (Fig. 74.2), we solve the equations

$$[U^0, U^1, U^2]\begin{bmatrix} 0 & 1 & 0 \\ 0 & 0 & 0 \\ 0 & 0 & 0 \end{bmatrix} = [0, 0, 0]$$

and this gives us $U^0 = 0$, which represents the pencil of lines through the point $(1, 0, 0)$. This point, which we shall see is *the center of the collineation*, lies on the line of fixed points $X_1 = 0$, which is the axis. The mapping gives

$$P' = (\lambda_1 y_0 + y_1, \lambda_1 y_1, \lambda_1 y_2)$$
$$= \lambda_1(y_0, y_1, y_2) + y_1(1, 0, 0),$$

and this shows that the join of P to its map P' always passes through the point $(1, 0, 0)$.

A central collineation in which the vertex lies on the axis is called *an elation*. This is the projectivity of the plane which, as we shall see, corresponds to translations in the real Euclidean plane.

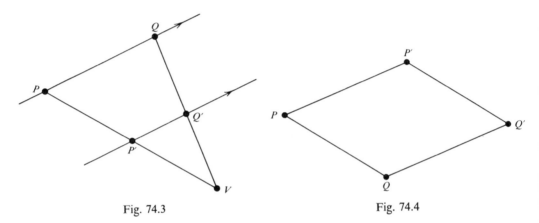

Fig. 74.3   Fig. 74.4

That the general central collineation, for which the vertex does not lie on the axis, gives rise to central similitudes or similarities is immediate if we regard the axis as *the line at infinity*, which means that we consider lines which intersect on the axis to be *parallel*. If V is the center of the collineation (Fig. 74.3), and the points P, Q are mapped onto P', Q' respectively, the points V, P, P' are collinear, as are the points V, Q, Q' and since PQ intersects $P'Q'$ on the axis, the lines PQ and $P'Q'$ are parallel. This is sufficient to show that the mapping of P on P' is a central similarity.

If we consider an elation, and take the axis to be the line at infinity, we now have PP' parallel to QQ' (Fig. 74.4), since the vertex is on the line at infinity, and also PQ parallel to $P'Q'$. Hence the mapping of P on P' is a translation, given by the fixed vector $\overrightarrow{QQ'}$.

If in a central collineation which is not an elation we consider the vertex to be on the line at infinity, we have a mapping of the real Euclidean plane in which the line PP' is always in a fixed direction (Fig. 74.5), and if PP' intersects the axis in R, the ratio of lengths $\overline{PR} : \overline{RP'}$ is fixed, since if QQ' meets the axis in S, and PQ intersects $P'Q'$ on the axis at T, $\overline{PR} : \overline{QS} = \overline{TR} : \overline{TS} = \overline{RP'} : \overline{SQ'}$, so that $\overline{PR} : \overline{RP'} = \overline{QS} : \overline{SQ'}$. If we take the case when $\overline{PR} = \overline{RP'}$, and the fixed direction of PP' is perpendicular to the axis, we obtain geometrical reflexion in the axis.

Exercises

74.1   Show that the non-singular projectivity given by the equations

$$(X_0', X_1', X_2') = (3X_0 - X_1 + 2X_2, -2X_0 + 2X_1 - 2X_2, 2X_0 - X_1 + 3X_2)$$

has a fixed point $(1, -1, 1)$ and a line of fixed points $2X_0 - X_1 + 2X_2 = 0$. (Compare Exercise 64.9.) Investigate the fixed lines of the mapping.

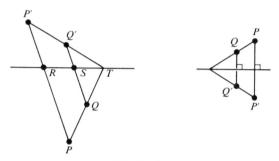

Fig. 74.5

74.2   Show that in the projectivity given by the equations

$$(X_0', X_1', X_2') = (2X_0 + 2X_1 + 3X_2, -2X_0 - 3X_1 - 6X_2, X_0 + 2X_1 + 4X_2)$$

there is a line of fixed points corresponding to the one characteristic root. Investigate the fixed lines, and show that these form a pencil, with vertex on the line of fixed points.

75.1   The Jordan normal form for a collineation

If we consider collineations (projective transformations) of an S_n onto itself, $Y' = AY$, a change of coordinate simplex $Y = BZ$, $Y' = BZ'$, leads to the mapping

$$BZ' = A(BZ),$$

that is

$$Z' = (B^{-1}AB)Z.$$

By a suitable choice of the matrix **B**, the matrix $B^{-1}AB$ may be written in a *canonical*, or *normal* form, the *Jordan normal form*. For proofs of this result we refer to other books (see Hodge and Pedoe (8), Chapter VIII in Vol. I). Here we investigate the geometrical meaning of the normal form. We have already seen that the matrices $B^{-1}AB$, for non-singular **A** and **B**, form disjoint classes of similar matrices (§70.2), the relation of similarity *an equivalence relation*. The matrices in an equivalence class are essentially the same matrices, connected with a projectivity, viewed from different coordinate systems, and the normal form exposes the geometric properties common to the class. If two normal forms differ, the collineations they represent differ, and every non-singular collineation has a representative Jordan normal form. The geometric description of a Jordan normal form is bound up with the fixed points and invariant hyperplanes of the equivalence class of collineations represented by the appropriate Jordan normal form. The three examples of the preceding section illustrated essentially distinct types of plane collineation by means of their Jordan normal forms.

The characteristic roots of the matrix we are discussing figure prominently in the Jordan form. We have a number of submatrices, all square, ranged along the principal diagonal of the matrix, and non-overlapping, each submatrix being of the form given below:

$$
\begin{bmatrix}
\lambda & 1 & 0 & \cdots & 0 \\
0 & \lambda & 1 & \cdots & 0 \\
\cdot & \cdot & & \cdots & \cdot \\
0 & \cdot & & \cdots & \lambda
\end{bmatrix}
\tag{1}
$$

where λ is a characteristic root, and 1's lie in the diagonal immediately above the principal diagonal. If the submatrix is a 1×1 matrix, it merely has the entry λ. Of course, the total order of all these submatrices adds to $n + 1$, and there is at least one submatrix for every distinct eigenvalue of $\mathbf{A}$. Corrado Segre characterized Jordan normal forms by a set of integers which give the number of rows of each submatrix, the matrices corresponding to the same characteristic root having their schematic integers enclosed in parentheses. Thus, for $n = 2$, the Segre symbols are:

$$[1 \quad 1 \quad 1], \, [(1 \quad 1) \quad 1], \, [(1 \quad 1 \quad 1)], \, [2 \quad 1], \, [(2 \quad 1)], \, [3],$$

and these represent the following matrices, each in normal form:

$$
\begin{bmatrix} \lambda_1 & 0 & 0 \\ 0 & \lambda_2 & 0 \\ 0 & 0 & \lambda_3 \end{bmatrix},
\begin{bmatrix} \lambda_1 & 0 & 0 \\ 0 & \lambda_1 & 0 \\ 0 & 0 & \lambda_2 \end{bmatrix},
\begin{bmatrix} \lambda_1 & 0 & 0 \\ 0 & \lambda_1 & 0 \\ 0 & 0 & \lambda_1 \end{bmatrix},
\begin{bmatrix} \lambda_1 & 1 & 0 \\ 0 & \lambda_1 & 0 \\ 0 & 0 & \lambda_2 \end{bmatrix},
\begin{bmatrix} \lambda_1 & 1 & 0 \\ 0 & \lambda_1 & 0 \\ 0 & 0 & \lambda_1 \end{bmatrix},
\begin{bmatrix} \lambda_1 & 1 & 0 \\ 0 & \lambda_1 & 1 \\ 0 & 0 & \lambda_1 \end{bmatrix}
$$

We have examined the disposition of fixed points and invariant lines for the first, second and fifth of the above matrices. We note that if $\mathbf{A}$ is non-singular, we do not have $\lambda = 0$ for any eigenvalue, since this would involve det $\mathbf{A} = 0$, which is not the case. It is possible to characterize singular projective transformations, and then $\lambda = 0$ must be allowed for, but we shall examine only the non-singular case.

Let us see what fixed points appear if we have the submatrix (1), which we suppose is an $e \times e$ submatrix, in the first position on the diagonal of the Jordan matrix. The point $V_1 = (1, 0, \ldots, 0)$ is fixed, its map being $(\lambda, 0, \ldots, 0)$. The point $V_2 = (0, 1, 0, \ldots, 0)$ is mapped onto the point $(1, \lambda, 0, \ldots, 0) = V_1 + V_2\lambda$. Hence the line joining the points V_1 and V_2 is an invariant line. The point $V_3 = (0, 0, 1, 0, \ldots, 0)$ is mapped onto the point $(0, 1, \lambda, 0, \ldots, 0) = V_2 + V_3\lambda$, and therefore there is an invariant plane, the plane $V_1 V_2 V_3$, through the invariant line and the fixed point. Eventually we find an invariant S_{e-1}.

This argument does not depend on the relative position of our submatrix (1) on the principal diagonal of the Jordan form for the matrix **A**. We naturally put all sub-matrices corresponding to the same value of a characteristic root together. If we use the equations for fixed points, it is clear that we only obtain the one fixed point, the one we have described, corresponding to any submatrix, and no fixed points for any characteristic root which does not correspond to the root which appears in the sub-matrix. That is, the only fixed points which the collineation possesses appear as we carry out the above procedure for each submatrix, and move down the principal diagonal.

Suppose that there are g Jordan submatrices grouped together on the principal diagonal and corresponding to the same characteristic root λ. There are then g fixed points corresponding to the characteristic root λ, and it is easily verified (Exercise 75.5) that linear combinations of fixed points which correspond to the same value of a characteristic root are also fixed points. Hence the g fixed points determine a space S_{g-1} which is point-wise invariant. There are no other fixed points under the collineation, so that a rapid survey of all the submatrices gives us information about all the point-wise invariant subspaces.

If we survey the six distinct types of collineation in S_2 whose Jordan forms are displayed on p. 312, our reasoning above tells us that: in case

(1) There are three distinct fixed points.

(2) There are two distinct fixed points corresponding to the same eigenvalue, so that the line joining them is fixed, and a third fixed point, not on the fixed line.

(3) There are three fixed points, each corresponding to the same eigenvalue, so that the whole plane joining them is fixed. In fact, this is the identity mapping.

(4) There is a fixed point, an invariant line through the fixed point, and another fixed point, not on the fixed line.

(5) There is a fixed point, a fixed line through the fixed point, and another fixed point corresponding to the same eigenvalue, so that there is a line of fixed points. The fixed line through the fixed point indicates that we have an elation in this case (Exercise 75.4).

(6) There is a fixed point, an invariant line through the fixed point, and an invariant plane through the fixed line. Since we are in a plane, the last item of information adds nothing to our knowledge.

This examination of the possible non-singular collineations in S_2 indicates that if there is a line of fixed points, we necessarily have a central collineation, which may be an elation. Of course, an examination of the effect of the collineation on the lines of the plane indicates this, since we have a pencil of invariant lines, but, as we indicate in the Exercises, non-algebraic proofs are also possible. We consider the algebraic treatment again in the next section.

Exercises

75.1   If the matrix of the Jordan normal form be called $\mathbf{J}$, and the points $V_1, V_2, \ldots, V_e$ are as described in the preceding section, so that $\mathbf{J}V_1 = V_1\lambda$, $\mathbf{J}V_2 = V_1 + V_2\lambda$, $\mathbf{J}V_3 = V_2 + V_3\lambda, \ldots$, show that

$$(\mathbf{J} - \lambda)^e V_e = (\mathbf{J} - \lambda)^{e-1} V_{e-1} = \ldots = (\mathbf{J} - \lambda)V_1 = 0.$$

75.2   The matrix of a projective mapping of S_n onto itself is $\mathbf{B} = [b_{ij}]$, and the simplex of reference consists of the points $B^0, B^1, \ldots, B^n$. We know that B^0 is fixed under the mapping. Show that this implies that $b_{i0} = 0$ $(i = 1, \ldots, n)$. We also know that the line $B^0 B^1$ is invariant under the mapping. Show that this implies that $b_{i1} = 0$ $(i = 2, \ldots, n)$. What is the form of the matrix if we also know that the plane $B^0 B^1 B^2$ is invariant under the mapping, $\ldots$, the hyperplane $B^0 B^1 \ldots B^{n-1}$ is invariant under the mapping?

75.3   We know that over the complex field every non-singular projectivity of S_n onto itself has at least one fixed point (§74.1). Choose the simplex of reference so that B^0 is a fixed point. Consider the line which joins B^0 to the variable point $(0, y_1, y_2, \ldots, y_n)$. Show that this latter point may be chosen in the S_{n-1} formed by $B^1, B^2, \ldots, B^n$ so that the line is invariant under the given projectivity, and that this involves the existence of at least one fixed point for a non-singular projectivity of an S_{n-1} onto itself. Change the simplex of reference to $C^0, C^1, \ldots, C^n$, where $C^0 = B^0$, and the line $C^0 C^1$ is fixed. Proceed thus, showing that we arrive at the situation of the preceding Exercise.

75.4   Examine the effect of the six distinct types of collineation in S_2, whose Jordan forms are displayed on p. 312, on the lines in S_2, and reconcile your results with those obtained on p. 313 for the effect on the points in S_2. In particular, show that in case (5) there is a pencil of invariant lines with vertex on the line of fixed points.

75.5   If $\mathbf{A}V_1 = \lambda V_1$, $\mathbf{A}V_2 = \lambda V_2, \ldots, \mathbf{A}V_e = \lambda V_e$, show that

$$\mathbf{A}(k_1 V_1 + k_2 V_2 + \ldots + k_e V_e) = \lambda(k_1 V_1 + \ldots + k_e V_e).$$

Interpret this result in relation to the space spanned by fixed points which correspond to the same characteristic root.

75.6   In this and the following Exercises we consider collineations of the projective plane, where these are *defined* as one-to-one onto mappings of the plane which preserve collinearity. Fixed points and fixed lines are defined as usual, a fixed line not being necessarily point-wise fixed. Prove that

(*a*) A collineation in the projective plane which fixes every point on each of two distinct lines is the identical collineation.
(*b*) A collineation in the projective plane which fixes every point on one line and also fixes two points not on the line is the identical collineation (see Pedoe, (**12**)).

75.7   Prove that a collineation of the projective plane which fixes every point on a given line l also fixes a point V, which may or may not be on l, and fixes every line through V.

75.8   State and prove the dual of the result of the preceding Exercise.

76.1   Special Collineations in S_n

The Jordan canonical form for non-singular collineations enables us to examine certain special collineations with ease.

Over the complex field we may call the [1   1   . . .   1] case, in which the Jordan form is a diagonal matrix with distinct eigenvalues $\lambda_1, \lambda_2, \ldots, \lambda_{n+1}$ on the diagonal, the "general case". The fixed points are the vertices of the coordinate simplex, in the new coordinate system, of course, and the invariant subspaces form the faces of this simplex.

The *central collineations* are characterized by the fact that they possess *a hyperplane* of fixed points. In the Segre symbol, corresponding to some value of λ, there must be n submatrices. Since the total length of the principal diagonal is $n + 1$, either each submatrix is of length 1 along the diagonal, and there is one submatrix left over, and the symbol is $[(1 \quad 1 \quad \ldots \quad 1) \quad 1]$, or one of the submatrices can have length 2 along the principal diagonal, and the symbol is $[(2 \quad 1 \quad 1 \quad \ldots \quad 1)]$. In the first case the normal form of the mapping is

$$y'_0 = \lambda y_0,$$
$$. \qquad . \qquad .$$
$$y'_{n-1} = \lambda y_{n-1},$$
$$y'_n = \mu y_n.$$

The hyperplane of fixed points is $X_n = 0$, and a hyperplane through the point $(0, 0, \ldots, 0, 1)$ is mapped onto itself. This point is the vertex of the central collineation. The point $P = (y_0, \ldots, y_n)$ is mapped onto the point

$$P' = \lambda P + (\mu - \lambda)(0, 0, \ldots, 1)y_n,$$

and so the line joining any point P to its map P' passes through a fixed point which does not lie in the hyperplane of fixed points.

In the other case, $[(2 \quad 1 \quad \ldots \quad 1)]$, the equations are

$$y'_0 = \lambda y_0 + y_1,$$
$$y'_1 = \lambda y_1,$$
$$. \qquad . \qquad . \qquad .$$
$$y'_n = \lambda y_n.$$

The hyperplane of fixed points is spanned by the vertices $Y^0, Y^2, Y^3, \ldots, Y^n$ of the simplex of reference. That is, the hyperplane of fixed points is given by $X_1 = 0$. The point $P = (y_0, \ldots, y_n)$ is mapped onto the point

$$P' = \lambda P + (1, 0, \ldots, 0)y_1,$$

and once again the line PP' passes through a fixed point, but this time the point, which is *the vertex* of the collineation, lies in the hyperplane of fixed points $X_1 = 0$. This collineation is called an *elation*.

76.2　Cyclic Collineations in S_n

We know that the non-singular collineations of S_n onto itself form a group (§70.1). We now consider which elements of this group are of *finite order*. That is, we seek the collineations $Y' = AY$ which are such that for some integer m,

$$\mathbf{A}^m = k\mathbf{I}_{n+1},$$

where $k \neq 0$, and $\mathbf{I}_{n+1}$ is the unity matrix. Such collineations are called *cyclic* collineations of order m. If we set $P_1 = \mathbf{A}(P)$, $P_2 = \mathbf{A}(P_1)$, $P_3 = \mathbf{A}(P_2)$, . . ., then since $P_r = \mathbf{A}^r(P)$, the points P, P_1, P_2, . . . form a finite set with at most m distinct elements, since $P = \mathbf{A}^m(P)$. The points of S_n are divided by such a cyclic collineation into disjoint sets (Exercise 76.7).

When $m = 2$, so that the points of S_n are packaged into disjoint pairs, we say that the collineation is *an involution*. These are especially important in projective geometry.

In order that we may be able to investigate cyclic collineations, we observe that if $\mathbf{A}^m = k\mathbf{I}_{n+1}$, then since

$$(\mathbf{T}^{-1}\mathbf{A}\mathbf{T})(\mathbf{T}^{-1}\mathbf{A}\mathbf{T}) = \mathbf{T}^{-1}\mathbf{A}^2\mathbf{T},$$

and inductively

$$(\mathbf{T}^{-1}\mathbf{A}\mathbf{T})^m = \mathbf{T}^{-1}\mathbf{A}^m\mathbf{T},$$

we have

$$(\mathbf{T}^{-1}\mathbf{A}\mathbf{T})^m = \mathbf{T}^{-1}k\mathbf{I}_{n+1}\mathbf{T} = k\mathbf{I}_{n+1},$$

so that we may examine the Jordan normal form of $\mathbf{A}$ to discuss cyclic collineations.

If we begin to find the powers of a matrix in the normal form, we see that the square is of the form

$$\begin{bmatrix} \lambda^2 & 2\lambda & 1 & 0 & \ldots & 0 \\ 0 & \lambda^2 & 2\lambda & 1 & \ldots & 0 \\ \cdot & \cdot & \cdot & \cdot & \cdot \cdot \cdot & \cdot \\ 0 & \cdot & \cdot & \cdot & \ldots & \lambda^2 \end{bmatrix},$$

where we have indicated the square of a typical submatrix, and these are strung out along the principal diagonal. This shows us immediately that the only possible way in which the Jordan form can be such that its mth power is the unity matrix, or proportional to the unity matrix, is for each submatrix to be a 1×1 matrix, and each eigenvalue to be an mth root of unity, or some non-zero number μ.

In particular we examine the case $m = 2$, that of an *involution*. We now require $\lambda^2 = \mu$, for all values of the eigenvalues and the same μ, so that $\lambda = \pm \sqrt{\mu}$. Since the matrix $\mathbf{A}$ may be multiplied by a non-zero constant, we may take $\lambda = +1$ or -1, and therefore in the involutory case the Jordan form of $\mathbf{A}$ is a diagonal matrix with only $+1$ or -1 as entries on the principal diagonal, all other entries being zero.

We gather the $+1$ and the -1 together, by renaming the coordinates. It now follows, if there are $r + 1$ entries equal to $+1$, and $n - r$ equal to -1, that there is an S_r of fixed points, and a complementary S_{n-r-1} of fixed points under the involutory collineation. If the image of P is P', then

$$P' = (y_0, y_1, \ldots, y_r, -y_{r+1}, \ldots, -y_n),$$

so that the point $P + P'$ lies in the S_r of fixed points, and the point $P - P'$ lies in the S_{n-r-1} of fixed points, and the four points P, P', $P + P'$, $P - P'$ form a harmonic range (§64.2).

A geometrical description of the mapping is simple. The point P is joined to the S_r, forming an S_{r+1}, since we assume that P does not lie in either of the spaces of fixed points. This S_{r+1} intersects the S_{n-r-1} in a unique point Q, say, and the join PQ, lying in an S_{r+1} which contains S_r intersects the S_r in a unique point R. The image of P is defined to be the harmonic conjugate of P with respect to the points Q and R. It is clear that the map of P' is P, so that the mapping is involutory.

For the case $r = 0$, we have encountered *harmonic homologies*, as they are sometimes called, in E_3 (§39.1), when we represented *inversion* in a circle as a process in E_3, and

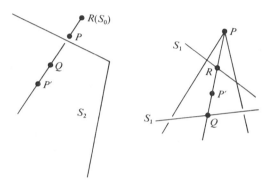

Fig. 76.1

the case $r = 1$ in S_3 is familiar (Fig. 76.1), when we have two skew lines as the spaces of fixed points (Exercise 65.15).

We note that the S_r of fixed points is the fundamental subspace of S_n under the collineation of S_n onto itself which corresponds to the characteristic root $\lambda = 1$, and the S_{n-r-1} of fixed points is the fundamental subspace corresponding to the characteristic root $\lambda = -1$. The S_r and S_{n-r-1} naturally have no points in common since they correspond to different values of the characteristic root. Being complementary subspaces of S_n, any point x in S_n may be written in the form $x = y + z$, where $y \in S_r$ and $z \in S_{n-r-1}$ (see *Remark*, §67.1).

In this form we can obtain the above results for involutions of S_n very rapidly, without making use of canonical forms of matrices. Let $y' = \mathbf{A}y$ be a linear map of S_n into itself which is such that $\mathbf{A}^2 = \mathbf{I}_{n+1}$. If $x \in S_n$, let $y = (x + \mathbf{A}(x))/2$, and let $z = (x - \mathbf{A}(x))/2$. Then $x = y + z$, and

$$\mathbf{A}(y) = (\mathbf{A}(x) + \mathbf{A}^2(x))/2 = (\mathbf{A}(x) + x)/2 = y,$$

whereas

$$\mathbf{A}(z) = (\mathbf{A}(x) - \mathbf{A}^2(x))/2 = (\mathbf{A}(x) - x)/2 = -z.$$

Hence y belongs to the fundamental space which corresponds to the characteristic root $\lambda = 1$, and z belongs to the fundamental subspace corresponding to the characteristic root $\lambda = -1$. Since we can write $x = y + z$ for any $x \in S_n$, the join of these two subspaces is S_n, and since they have no points in common they are complementary subspaces, so that if the one is of dimension r, the other is of dimension $n - r - 1$.

It may seem surprising to conclude this section with a theorem for the case $n = 1$, but the following theorem is fundamental for consideration of involutions on a line, and, by duality, for pencils of lines in the case $n = 2$.

Theorem I   *A projectivity T between the points of a line which is such that $T(A) = A'$ and $T(A') = A$, for any one pair of distinct points A, A' on the line is such that $T^2 = I_2$, that is, T is an involution.*

Proof. Let X be any point on the line, distinct from A, let $T(X) = X'$, and $T(X') = X''$. We wish to prove that $X'' = X$. The points A, A', X, X' are projective with the points A', A, X', X'', by hypothesis. By Exercise 71.2, the range (A, A', X, X') is projective with (A', A, X', X). We therefore find that the ranges

$$(A', A, X', X'') \text{ and } (A', A, X', X)$$

are projective. But this projectivity must be the identity (Thm III, §71.2) and therefore $X'' = X$.

Corollary. An involution on a line is determined by the assignment of two point-pairs (P, P') and (X, X'), say, for it is the unique projectivity determined by the mapping

$$(X, P, P') \to (X', P', P).$$

Remark. The discussion given above for involutions in S_n applies to the case $n = 1$. We have two *fixed points*, taking $r = 0$, there being no other possibility, and if these be called E and F, every point-pair P, P' of the involution forms a harmonic range with E and F. Conversely, if we begin with two distinct points E and F, and consider all point-pairs P, P' which form a harmonic range with E and F, the point-pairs P, P' are in involution, with fixed points E and F.

Since an involution is uniquely determined by the assignment of two point-pairs (X, X') and (P, P'), if we find two points E, F such that

$$[E, F; P, P'] = [E, F; X, X'] = -1,$$

we can say that E and F are the fixed points of the involution determined by (X, X') and (P, P').

Corresponding to the theory of involution ranges on a line there is a dual theory, in S_2, say, of involution pencils. One of the cases which occurs with frequency in metrical geometry in the real Euclidean plane will concern us later (§80.1). If we have a triangle SXX', and we consider the internal and external bisectors of the angle at S, these will be orthogonal lines. Suppose they intersect the base XX' at E and F respectively. We know that $|XE| : |EX'| = |XS| : |X'S| = |XF| : |FX'|$, where the lengths involved are not taken algebraically. This is one of the simplest examples of a harmonic range on a line, using Euclidean rather than projective geometry. If we keep SE and SF fixed (Fig. 76.2), and consider any other point-pair (Y, Y') which is such that Y, Y' are harmonic conjugates with respect to E and F, we can say that SE and SF are *bisectors of the angle YSY'.* For if we suppose that Y is given, its harmonic conjugate

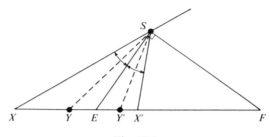

Fig. 76.2

with respect to E and F is uniquely determined. If we take Y' so that SE bisects the angle YSY', and therefore so does SF, the points Y, Y' are harmonic conjugates with respect to E and F. Our statement follows.

Exercises

76.1 Prove that a necessary and sufficient condition for three pairs of points A, A'; B, B' and C, C' on a line to belong to an involution is

$$(A, B, C, C') \,\overline{\wedge}\, (A', B', C', C).$$

76.2 Prove that the three pairs of opposite sides of a quadrangle meet any line in the same plane which does not pass through a vertex in three pairs of points in involution.

76.3 State and prove the dual of the theorem of the preceding Exercise.

76.4 The four given points A, B, Q, V lie on a line, and in a projective plane through the line a given line through V meets a given line through Q at O. The point X is variable on OV, and AX meets QO at P, and BP meets QX at J. Prove that J moves on a fixed line through O.

76.5 Prove that if ABC is a triangle the sides of which intersect a line in the points P, Q and R respectively, and P', Q' and R' are mated with P, Q, and R respectively in an involution on this line, then AP', BQ' and CR' are concurrent.

76.6 Two projectivities T_1, T_2 on a line are said to commute if

$$(P)T_1T_2 = (P)T_2T_1$$

for all points P of the line. Prove that if T_1 and T_2 are both involutions, they commute if and only if the product T_1T_2 is also an involution. If U and V are the fixed points of the involution T_1, and E, F are the fixed points of the involution T_2, prove that a necessary and sufficient condition that T_1 and T_2 commute is that U, V and E, F are pairs in a harmonic range.

76.7 If $\mathbf{A}$ is an $(n + 1) \times (n + 1)$ matrix which is such that $\mathbf{A}^m = k\mathbf{I}_{n+1}$, where $k \neq 0$, and $\mathbf{I}_{n+1}$ is the unity matrix, show that the relation

$$\mathbf{A}^r(P) = Q \qquad (r \leqslant m)$$

between points P and Q of S_n is an equivalence relation, so that the points of S_n can be divided into disjoint sets with the help of a cyclic collineation.

76.8   Representing a harmonic homology with vertex (or center) O and plane of fixed points π by the symbol $\{O, \pi\}$, show that the product $\{O, \pi\}\{O', \pi'\}$ is the harmonic homology with fixed lines $l = OO'$ and $l' = \pi \cap \pi'$.

IX

THE PROJECTIVE GENERATION OF CONICS AND QUADRICS

We have already encountered a quadric surface in S_3 (§65.1, (b)) as the locus of lines which intersect three mutually skew lines. We now discuss this surface once more, and a corresponding construct in S_2, the conic, from another point of view, that of *projective generation*. This leads to a much deeper insight into the nature of conics and quadrics, which appear in some books on mathematics as unadorned loci given by second-degree equations, to be studied because two is the next higher degree after one.

77.1 The conic

Let A and B be two distinct points of the projective plane S_2, and let us suppose that a projectivity has been established between the *lines* of the pencils with vertex A and

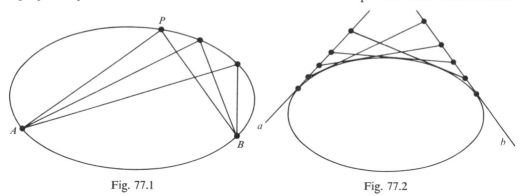

Fig. 77.1   Fig. 77.2

B respectively. If l be a line vertex A, and l' the corresponding line, vertex B, the locus of the point $P = l \cap l'$, the intersection of corresponding rays in the projectivity, is called a *conic* (Fig. 77.1).

Since we shall also be considering the dual construct, we sometimes refer to the locus just defined as a *point-conic*. Dually, if we have two distinct lines a and b in S_2 (Fig. 77.2), between the points of which a projectivity has been established, then the *join* of corresponding points on the two lines gives a set of lines which is called a *line-conic*.

We shall see that every point-conic has a line-conic associated with it, and conversely. There is a special case to be considered in the first definition, and a corresponding one in the second.

If, in the projectivity between the pencils vertices A and B, the line AB corresponds to the line BA, *the locus of intersection of corresponding rays is a line* (Fig. 77.3). This is the dual of Thm IV of §71.2, and we run through the proof rapidly. Let P and Q be two points on the locus, and let the line PQ intersect AB in the point R. Then the rays $A(R, P, Q)$ correspond in the projectivity to the rays $B(R, P, Q)$. But the line PQR

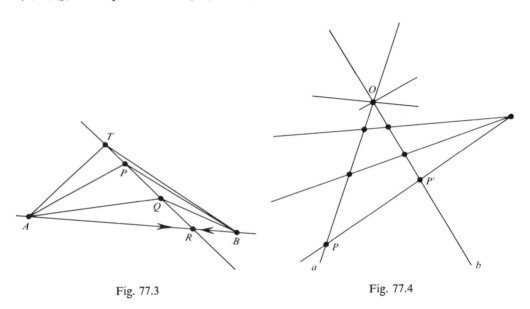

Fig. 77.3   Fig. 77.4

intersects the pencils with vertex A and B in projective ranges. Since the points P, Q and R are fixed points of these projective ranges, the projectivity between them is the identity (Thm III, §71.2). Hence if T is any point on the line PQR, the ray AT corresponds to the ray BT. The locus of intersection of corresponding rays is therefore the line PQR.

To this locus we may add the line AB, since the intersection of AB and BA is the line AB.

The dual of this theorem says that if we have a projectivity between the points of two distinct lines which is such that the point of intersection O of the lines, (which are coplanar), corresponds to itself, then the joins of corresponding points on the two lines always pass through a point (Fig. 77.4). This is Thm IV of §71.2. To the set of lines which pass through the point we have just mentioned we may add the set of lines which pass through O, the intersection of the two given distinct lines, since any line through O satisfies the conditions of joining corresponding points on the two lines.

Each of the above theorems has an evident converse. In a perspectivity between two lines, the point of intersection of the lines is mapped onto itself, and in a perspectivity between two pencils of lines, the line joining the vertices of the pencils is mapped onto itself. Hence the point-locus defined to be a conic is *not* a line if the line AB joining the vertices of the pencils does *not* correspond to the line BA, and the line-locus defined to be a line-conic is *not* a pencil of lines if the point of intersection O of the two ranges of points does *not* correspond to itself.

We now return to the generation of our point-conic. It is possible to proceed without the use of algebra, but we develop an algebraic treatment which carries over to the generation of a quadric in S_3.

Let $U_0 = 0$, $U_1 = 0$ be two distinct lines intersecting in the point A. Then any line of the pencil vertex A may be written in the form $\lambda_0 U_0 + \lambda_1 U_1 = 0$, where we may

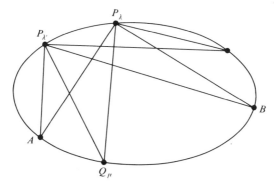

Fig. 77.5

take (λ_0, λ_1) as a system of homogeneous coordinates for the lines of the pencil vertex A. Each line of the pencil determines a unique equivalence class of ordered pairs (λ_0, λ_1), and conversely each (λ_0, λ_1) determines a unique line of the pencil.

Similarly, any line of the pencil vertex B may be written in the form $\lambda_0^* V_0 + \lambda_1^* V_1 = 0$, where $V_0 = 0$, $V_1 = 0$ are two distinct lines which intersect in B, and $(\lambda_0^*, \lambda_1^*)$ may be taken as a system of homogeneous coordinates for the lines of the pencil vertex B.

The projective correspondence between the lines of the pencil vertex A, and the lines of the pencil vertex B may be written in the form

$$\begin{bmatrix} \lambda_0^* \\ \lambda_1^* \end{bmatrix} = \mathbf{T} \begin{bmatrix} \lambda_0 \\ \lambda_1 \end{bmatrix},$$

where $\mathbf{T}$ is a non-singular 2×2 matrix over the ground field K. If we change the coordinate system in either pencil, we may arrange that $\mathbf{T}$ is the unity matrix, and then our equations are $\lambda_0^* = \lambda_0$, $\lambda_1^* = \lambda_1$.

We now have two projective pencils of lines, vertices A and B respectively:

$$A_\lambda \equiv \lambda_0 U_0 + \lambda_1 U_1 = 0, \text{ representing the line } a_\lambda,$$
$$B_\lambda \equiv \lambda_0 V_0 + \lambda_1 V_1 = 0, \text{ representing the line } b_\lambda.$$

We denote by P_λ the point of intersection of the lines a_λ and b_λ (Fig. 77.5), and note that by definition our conic consists of the set of points $\{P_\lambda\}$. Since, by hypothesis, the line AB does not correspond to itself, no two of the lines a_λ, b_λ can coincide for the same value of λ.

Hence, for any pair of numbers μ_0, μ_1, not both zero, the equation $\mu_0 A_\lambda + \mu_1 B_\lambda = 0$ represents a line, which clearly passes through the point P_λ. (If for some value of λ the lines a_λ, b_λ were the same, it would be possible for the equation $\mu_0 A_\lambda + \mu_1 B_\lambda = 0$ to be of the form $0 = 0$.) The equation of this line is

$$\mu_0 A_\lambda + \mu_1 B_\lambda \equiv \lambda_0(\mu_0 U_0 + \mu_1 V_0) + \lambda_1(\mu_0 U_1 + \mu_1 V_1) = 0.$$

This, for *fixed* μ_0, μ_1 and *variable* λ_0, λ_1, represents a line of the pencil whose vertex is the point Q_μ common to the lines

$$\mu_0 U_0 + \mu_1 V_0 = 0 = \mu_0 U_1 + \mu_1 V_1.$$

Now this pencil has an equation of the form

$$\lambda_0 C_0 + \lambda_1 C_1 = 0,$$

where $C_0 = 0$ and $C_1 = 0$ are fixed lines, and it is therefore projective with the pencils (a_λ) and (b_λ). But corresponding lines of this pencil and of the pencils (a_λ) and (b_λ) intersect in the point P_λ. Hence the point-conic we are considering, locus of points P_λ, is such that its points are the intersections of corresponding lines of an infinity of projective pencils, whose vertices Q_μ are the points determined by the lines

$$\mu_0 U_0 + \mu_1 V_0 = 0 = \mu_0 U_1 + \mu_1 V_1,$$

as μ_0 and μ_1 vary.

The points A and B belong to the set of points $\{Q_\mu\}$, and correspond, respectively, to the values $(1, 0)$ and $(0, 1)$ of (μ_0, μ_1).

We have now shown that the same point-locus, the set of points $\{P_\lambda\}$, can be derived by replacing A and B by any two distinct points of the set $\{Q_\mu\}$.

We may now start again, and write

$$\mu_0 A_\lambda + \mu_1 B_\lambda \equiv \mu_0(\lambda_0 U_0 + \lambda_1 U_1) + \mu_1(\lambda_0 V_0 + \lambda_1 V_1) = 0.$$

For *fixed* λ_0, λ_1 and *varying* μ_0, μ_1 this line describes a pencil with vertex P_λ. All the pencils so obtained, corresponding to different (λ_0, λ_1) are projective with each other, corresponding lines arising from the same values for (μ_0, μ_1) and intersecting in the points Q_μ. Thus:

The points Q_μ are the points of intersection of corresponding lines of projective pencils whose vertices are the points P_λ.

We are thus led to consider two sets of points, the set $\{P_\lambda\}$ and the set $\{Q_\mu\}$, between which there exists a symmetrical relationship, namely: the points of either set are the intersection of corresponding lines of an infinite set of projective pencils, whose vertices constitute the other set.

This reciprocity, as we shall see, is characteristic of projectively generated loci.

In the special case of a point-conic both sets $\{P_\lambda\}$ and $\{Q_\mu\}$ coincide, for by eliminating λ_0, λ_1 and μ_0, μ_1, we see that a point belongs to either set if and only if its coordinates satisfy the equation

$$\begin{vmatrix} U_0 & U_1 \\ V_0 & V_1 \end{vmatrix} = 0.$$

This is called *the equation* of the point-conic described by the set $\{P_\lambda\}$ or $\{Q_\mu\}$, and it is of the second degree in our homogeneous coordinates. We have proved:

Theorem I *The pencils of lines joining the points of a conic to any two fixed points of the conic are projective.*

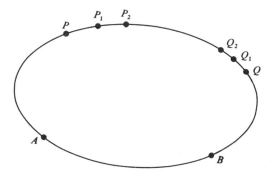

Fig. 77.6

This is a theorem with far-reaching consequences. It enables us to talk of *projective ranges on a point-conic*. For suppose that we have a set of points $\{P, P_1, P_2, \ldots\}$ on a point-conic (Fig. 77.6), and a set of points $\{Q, Q_1, Q_2, \ldots\}$ such that the *pencils* $A\{P, P_1, P_2, \ldots\}$ and $A\{Q, Q_1, Q_2, \ldots\}$ are projective. Then by the Theorem above, if B is any other point of the conic

$$A\{Q, Q_1, Q_2, \ldots\} \barwedge B\{Q, Q_1, Q_2, \ldots\},$$

where we are using the same symbol $\barwedge$ for 'is projective with' as we did in §71.2. Since the relation of projectivity between pencils is evidently transitive whether the pencils have the same or distinct vertices, we deduce from

$$A\{P, P_1\ P_2, \ldots\} \barwedge A\{Q, Q_1, Q_2, \ldots\}$$

and

$$A\{Q, Q_1, Q_2, \ldots\} \barwedge B\{Q, Q_1, Q_2, \ldots\}$$

that

$$A\{P, P_1, P_2, \ldots\} \barwedge B\{Q, Q_1, Q_2, \ldots\}.$$

Hence we may speak of *projective ranges on the conic*, it being understood that this means that if *any* point on the conic be joined to one set of points and any other point on the conic (it may be the same point) be joined to the other set of points, we obtain projective pencils.

77.2   Quadratic forms

We may now proceed to develop the theory of conics in a synthetic (that is, non-algebraic) manner, and we shall indicate later how some of the main theorems can be obtained this way. But first we look at the equation of a conic. This was seen to be $U_0V_1 - U_1V_0 = 0$, where the U_i and V_i are linear, homogeneous forms in X_0, X_1, X_2, the homogeneous coordinates in S_2. Hence the equation of a conic is given by equating a quadratic form to zero, that is, by

$$a_{00}X_0{}^2 + a_{11}X_1{}^2 + a_{22}X_2{}^2 + 2a_{01}X_0X_1 + 2a_{12}X_1X_2 + 2a_{02}X_0X_2 = 0.$$

If we adopt the convention that

$$a_{ij} = a_{ji} \quad (i \neq j),$$

the quadratic form may be written in matrix notation as

$$[X_0, \ X_1, \ X_2] \begin{bmatrix} a_{00} & a_{01} & a_{02} \\ a_{10} & a_{11} & a_{12} \\ a_{20} & a_{21} & a_{22} \end{bmatrix} \begin{bmatrix} X_0 \\ X_1 \\ X_2 \end{bmatrix},$$

that is as X^TAX, where $\mathbf{A} = [a_{ij}]$, and X^T is the matrix $[X_0, X_1, X_2]$. We note that $\mathbf{A}$ is now a *symmetric* matrix, so that $\mathbf{A}^T = \mathbf{A}$.

The matrix representation makes it easy to see what happens if we change the co-ordinates. If $X = \mathbf{P}Y$, where $\mathbf{P}$ is a non-singular matrix, then the form becomes

$$(\mathbf{P}Y)^T\mathbf{A}(\mathbf{P}Y) = Y^T(\mathbf{P}^T\mathbf{A}\mathbf{P})Y,$$

so that in the new coordinates the matrix of the form is $\mathbf{P}^T\mathbf{A}\mathbf{P}$. The matrices $\mathbf{A}$ and $\mathbf{P}^T\mathbf{A}\mathbf{P}$ are said to be *congruent*, and the relation is *an equivalence relation. Over the complex field* the one invariant of a quadratic form under coordinate transformations is its rank, and if rank $(\mathbf{A}) = r$, the matrix $\mathbf{A}$ is congruent to the matrix

$$\begin{bmatrix} \mathbf{I}_r & 0 \\ 0 & 0 \end{bmatrix},$$

where $\mathbf{I}_r$ is the $r \times r$ unity matrix. This is shown by *completing the square* and choosing a suitable coordinate transformation (see Hodge and Pedoe, **8**, Vol. I, Chapter IX).

The above discussion holds for a quadratic form in any number of variables. What concerns us at the moment is that for 3×3 symmetric matrices we can say:

Any two symmetric 3×3 matrices of the same rank are congruent.

Hence, in dealing with quadratic forms in three homogeneous variables we know that there are essentially only three distinct types, those congruent to $Y_0{}^2$, to $Y_0{}^2 + Y_1{}^2$, and to $Y_0{}^2 + Y_1{}^2 + Y_2{}^2$.

The equaton $Y_0^2 = 0$ represents two *coincident* lines, the term 'coincident' being introduced for reasons similar to those given when we say that the quadratic equation $X^2 = 0$ *has two coincident* roots equal to zero. We wish to 'conserve the number' in the case of quadratic equations, and we wish to preserve the degree of the algebraic equation representing a conic, even in special cases.

The equation $Y_0^2 + Y_1^2 = 0$, over the complex field, represents two distinct lines $Y_0 + iY_1 = 0$, and $Y_0 - iY_1 = 0$. In this case, and also the above, we say that the conic represented by the quadratic equation is *reducible*.

The equation $Y_0^2 + Y_1^2 + Y_2^2 = 0$ can be written as

$$\begin{vmatrix} Y_0 + iY_2 & Y_1 \\ -Y_1 & Y_0 - iY_2 \end{vmatrix} = 0,$$

and this shows that we now have a projectively generated conic, since this is of the form we obtained previously. We note that this conic, having rank 3, is *irreducible*, since it is simple to show that if a quadratic form can be written as the product of two linear factors, it necessarily has rank less than three. If the product is UV, where U and V are both linear forms, we have

$$UV = (U + V)^2/4 - (U - V)^2/4,$$

and a simple change of coordinates shows that this quadratic form has rank 2 at most.

The case when our point-conic is irreducible is the one we are interested in, and the coordinate transformation

$$X_0 = Y_0 \qquad\qquad +iY_2,$$
$$X_1 = \qquad iY_1,$$
$$X_2 = Y_0 \qquad\qquad -iY_2,$$

is non-singular, and gives us the equation of an irreducible conic in the standard, normal, or canonical form over the complex field as

$$X_0 X_2 - X_1^2 = 0.$$

Hence we have our main theorem of this section:

Theorem I   *Over the complex field the coordinates may be chosen so that the equation of any irreducible point-conic is $X_0 X_2 - X_1^2 = 0$.*

We may repeat all the preceding algebra, and interpret it dually in order to investigate *line-conics*. The definition is as follows: let a, b be distinct lines of S_2, and let there be a projectivity between the points of a and b which is not a perspectivity. If L be a point on a, and L' the corresponding point on b, the locus of the line $p = LL'$ is called a *line-conic*.

We clearly arrive at

Theorem II   *Over the complex field the coordinates may be so chosen that the equation* *of any irreducible line-conic is* $U^{(0)}U^{(2)} - (U^{(1)})^2 = 0$, *the coordinates of a line being* $[U^{(0)}, U^{(1)}, U^{(2)}]$.

Exercises

77.1 ABC is a variable triangle which is such that the vertices B and C respectively move on two given lines l and m, and the sides BC, CA and AB respectively pass through given points U, V and W. Prove that the pencils $W\{A\}$ and $V\{A\}$ are projective, so that A describes a conic through V and W. (This is Maclaurin's construction for a conic.)

   Show that if the given points U, V and W are collinear, the locus of the vertex A is a line which is concurrent with the lines l and m.

77.2   In Exercise 77.1 let $l \cap UV = W'$, $m \cap UW = V'$ and $l \cap m = A'$. Show, by taking special positions of the triangle ABC, that the points W', V' and A' are on the conic locus of the point A.

77.3   Show that the conic

$$X^2 + 5Y^2 + 2XY + 2XZ + 14YZ = 0$$

can be written in the form

$$(X + Y + Z)^2 + 4Y^2 - Z^2 + 12YZ = 0$$

(completing the squares in the X-terms), and then in the form

$$(X + Y + Z)^2 + (2Y + 3Z)^2 - 10Z^2 = 0,$$

so that the coordinate transformation given by

$$X' : Y' : Z' = X + Y + Z : 2Y + 3Z : \sqrt{10}iZ$$

transforms the conic into the conic $(X')^2 + (Y')^2 + (Z')^2 = 0$.

77.4   Show that the conic

$$XY + YZ + ZX = 0$$

may be written in the form

$$(X + Y)^2 - (X - Y)^2 + 4(X + Y)Z = 0,$$

so that the coordinate transformation

$$X' : Y' : Z' = X + Y : X - Y : Z$$

transforms this equation into the equation

$$(X')^2 - (Y')^2 + 4X'Z' = 0.$$

By completing the square on the X'-terms, transform the equation into

$$X_0^2 + X_1^2 + X_2^2 = 0.$$

78.1   Point-conics and line-conics

The point-conic $X_0X_2 - X_1^2 = 0$ is the locus of points of intersection of the lines

$$\lambda X_0 - \mu X_1 = 0,$$
$$\mu X_2 - \lambda X_1 = 0.$$

These are projective pencils, vertices $(0, 0, 1)$ and $(1, 0, 0)$ respectively (Fig. 78.1).

Their intersection is the point $(\mu^2, \mu\lambda, \lambda^2)$, and this is *a parametric representation* for points on the conic. As we vary (λ, μ) over the complex field, we obtain all the points of the conic. The value of the parameter gives the point, and the point gives the value of the parameter. The point $(1, 0, 0)$ is attached to the value $(\lambda, \mu) = (0, 1)$, and the point $(0, 0, 1)$ is attached to the value $(\lambda, \mu) = (1, 0)$. Both these points are on the conic. The coordinates (λ, μ) are, of course, homogeneous coordinates, so that $(k\lambda, k\mu)$ $(k \neq 0)$ gives the same point on the conic as (λ, μ).

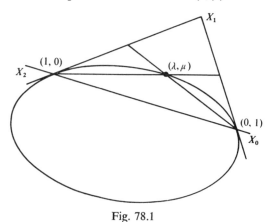

Fig. 78.1

Except for the point $(0, 0, 1)$ on the conic, we may use non-homogeneous parameters, and take any point on the conic as $(1, \lambda, \lambda^2)$. This is easier as far as notation goes. Since the cross-ratio of the four lines

$$\lambda_i X_0 - X_1 = 0 \quad (i = 1, 2, 3, 4)$$

is $[\lambda_1, \lambda_2; \lambda_3, \lambda_4]$, and these are lines joining the points $(1, \lambda_i, \lambda_i^2)$ to a point on the conic, we may say:

The cross-ratio of the four points $(1, \lambda_i, \lambda_i^2)$ $(i = 1, 2, 3, 4)$ on the conic $X_0 X_2 - X_1^2 = 0$ is $[\lambda_1, \lambda_2; \lambda_3, \lambda_4]$.

In order to deal with points such as $(0, 0, 1)$, without returning to homogeneous coordinates, we write

$$(1, \lambda, \lambda^2) = (\lambda^{-2}, \lambda^{-1}, 1)$$

and write $\lambda^{-1} = 0$ for the value of the parameter attached to $(0, 0, 1)$.

The equation of the line joining the points $(1, \lambda, \lambda^2)$ and $(1, \mu, \mu^2)$ on the conic is found in determinantal form as

$$\begin{vmatrix} X_0 & X_1 & X_2 \\ 1 & \lambda & \lambda^2 \\ 1 & \mu & \mu^2 \end{vmatrix} = 0.$$

On dividing by $\lambda - \mu$, which is not zero, we obtain the equation

$$X_0\lambda\mu - X_1(\lambda + \mu) + X_2 = 0.$$

If we put $\mu = \lambda$, we obtain *the equation of the tangent to the point-conic* at the point $(1, \lambda, \lambda^2)$ as

$$X_0\lambda^2 - 2X_1\lambda + X_2 = 0.$$

If a tangent is also written in the form

$$U^{(0)}X_0 + U^{(1)}X_1 + U^{(2)}X_2 = 0,$$

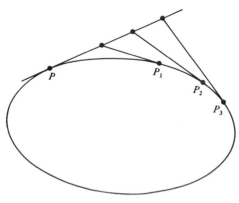

Fig. 78.2

it follows that we may write $U^{(0)} = \lambda^2$, $U^{(1)} = -2\lambda$, $U^{(2)} = 1$, and therefore

$$4U^{(0)}U^{(2)} - (U^{(1)})^2 = 0.$$

If we compare this with the equation of an irreducible line-conic, found in Thm II of the preceding Section, we see that the equations are essentially the same, since an evident change of coordinates changes one equation into the other. We have therefore proved:

Theorem I *The tangents at the points of a point-conic form a line-conic.*

We recall that the dual of Thm I of §77.1 states that the ranges of points in which the lines of a line-conic cut two fixed lines of the line-conic are projective. Now that we know that the tangents to our point-conic form a line-conic, we may obtain a stronger theorem. The tangent at $(1, \lambda, \lambda^2)$ to our point-conic cuts the line $X_0 = 0$, which is a tangent at $\lambda^{-1} = 0$, in the point $(0, 1, 2\lambda)$ (Fig. 78.2). This range of points for varying λ is evidently projective with the pencil of lines $\lambda X_0 - X_1 = 0$, which joins the point $(0, 0, 1)$ on the conic to the point $(1, \lambda, \lambda^2)$ on the conic. We therefore have the theorem:

Theorem II *The tangents at the points P, P_1, P_2, . . . of a point-conic cut any fixed* tangent to the conic in a range which is projective on the line to the set of points on the conic.

78.2   Projective correspondences on a conic

Our development of the projective generation of point-conics has established that we may deal with projective ranges on a given point-conic as if we were considering projective ranges on a line. Returning to homogeneous parameters for a moment, the point (λ, μ) on our point-conic is projective with the point (λ^*, μ^*) if a relation of the form

$$\lambda^* = p\lambda + q\mu,$$
$$\mu^* = r\lambda + s\mu$$

exists between the parameters of corresponding points, where $ps - qr \neq 0$. If we use non-homogeneous parameters, the relation may be written

$$\lambda^* = \frac{p\lambda + q}{r\lambda + s},$$

which is usually written in the form

$$A\lambda\lambda^* + B\lambda + C\lambda^* + D = 0.$$

If $\lambda^* = \lambda$, we obtain the equation

$$A\lambda^2 + (B + C)\lambda + D = 0,$$

which gives the *fixed points* of the projective correspondence. Since the equation is quadratic, there are at most two fixed points.

We are interested in the case when a given projective correspondence is an involution of order two on the conic. We use Thm I of §76.2 which says that if a projectivity between the points of a line interchanges one pair of distinct points, then it interchanges every pair, and is an involution. We are working with points of a conic, but we may use this theorem. We have the two equations

$$A\lambda\lambda^* + B\lambda + C\lambda^* + D = 0,$$

and

$$A\lambda^*\lambda + B\lambda^* + C\lambda + D = 0,$$

assuming that the point λ is mapped on λ^* and the point λ^* is mapped on λ. If we subtract these equations, we obtain

$$(B - C)(\lambda - \lambda^*) = 0,$$

and since we are assuming that $\lambda \neq \lambda^*$, we deduce that $B = C$. Hence we have:

Theorem I   *The projectivity $A\lambda\lambda^* + B\lambda + C\lambda^* + D = 0$ between points of a conic (or a line) is an involution if and only if $B = C$.*

If we look at the equation of a chord of the conic,

$$X_0\lambda\mu - X_1(\lambda + \mu) + X_2 = 0,$$

we observe that if the point (X_0, X_1, X_2) be fixed, the intersections of the resulting pencil of lines traces out the pairs of points of an involution on the conic. Conversely, an involution on the conic is given by the equation

$$A\lambda\lambda^* + B(\lambda + \lambda^*) + D = 0,$$

and if we compare this equation with that of the chord joining the points λ, λ^*, we see

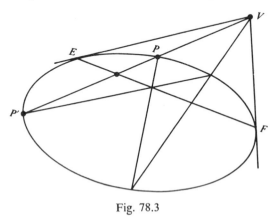

Fig. 78.3

that the chord joining these points passes through the fixed point $(A, -B, D)$. We have therefore proved:

Theorem II   *An involution on a point-conic is necessarily cut out by a pencil of lines of the containing plane S_2.*
   The fixed points of the involution are given by putting $\lambda^* = \lambda$ in

$$A\lambda\lambda^* + B(\lambda + \lambda^*) + D = 0,$$

and this gives us the quadratic equation

$$A\lambda^2 + 2B\lambda + D = 0.$$

Since we are using numbers from the field of complex numbers, we may assume that this equation has roots. These roots would coincide if $B^2 - AD = 0$. But our equation between λ and λ^* is

$$\lambda^* = (-B\lambda - D)/(A\lambda + B),$$

and the determinant of this transformation is $-B^2 + AD \neq 0$. Therefore our involution always has *distinct fixed points*.
   Geometrically, these are the points at which tangents to the point-conic pass through the point $(A, -B, D)$. But this can also be interpreted as the theorem that *two tangents can be drawn from a given point to a given point-conic* (Fig. 78.3).

Of course, another way of looking at this is to assert that *two lines of a given line-conic pass through a given point*, and this is the dual of the theorem that *two points of a given point-conic lie on a given line*.

This is a theorem we have not stressed as yet, since we have been examining the internal structure of the conic. It is clear from the fact that the equation of the conic is of the second degree, and also from the projective generation, and we shall return to this aspect of the point-conic. But before we leave involutions on the conic, we remark that *if, E, F are the two fixed points of an involution on a conic, and P, P' any pair in the involution, then the points E, F, P, P' form a harmonic range on the conic.* This follows from the theorem that cross-ratios are preserved under a projectivity (§69.1), and a harmonic range is determined by its cross-ratio being equal to -1. Our involution is a projectivity, and maps E onto E, F onto F, and P on P'. Hence since it also maps P' on P, we have equality for the cross-ratios:

$$[E, F; P, P'] = [E, F; P', P].$$

But in §69.1 we saw that $[E, F; P', P] = [E, F; P, P']^{-1}$. Hence

$$[E, F; P, P']^2 = 1,$$

and since $P \neq P'$ we cannot have $[E, F; P, P'] = 1$, so we must have

$$[E, F; P, P'] = -1,$$

and our assertion follows.

If the tangents at E and F meet in V, we have $E(E, F, P, P')$ is a harmonic pencil, and we are clearly justified in taking the apparently meaningless join EE to be the tangent at E to our conic. Taking sections of this pencil on the line VPP', we see that the range $\{V, EF \cap VP, P, P'\}$ is harmonic, and so we obtain the theorem:

If a chord be drawn through a point V, cutting a point-conic in the points P, P', the locus of the harmonic conjugate of V with respect to P and P' is the chord of contact of the tangents from V to the conic.

This chord is called *the polar of V with respect to the conic*. The point V is called *the pole* of the chord of contact.

(We encountered the polar of a point with respect to a quadric in S_3 in §35.1. The ideas and definition of the polar of a point with respect to a quadric in any number of dimensions are very similar to each other, and we shall examine these ideas again very shortly. We did not stop to define the polar of a point with regard to a circle in Chapter II, since we did not need to make use of the concept there. But we remark here that a circle in the real Euclidean plane is a conic, and all the projective notions we develop for conics hold for circles, but since circles are specialized conics (see §80.1) in a sense to be specified later, circles have properties not possessed by all conics).

If two points X and Y in S_2 are such that the intersection of XY with a point-conic are points P, P' which are harmonic conjugates with respect to X and Y, then X and Y are said to be *conjugate points* with regard to the conic. We have proved above that if we keep X fixed, the locus of points conjugate to X is a line, the polar of X with regard to the conic.

If the polar of X passes through Y, we have the situation just described, with X and Y as conjugate points, and so the polar of Y passes through X.

If X and Y are two distinct points on a line p (Fig. 78.4), let the polars of X and Y with respect to our point-conic intersect at P. Then since the polar of X contains P, the polar of P contains X. Since the polar of Y contains P, the polar of P contains Y. The polar of P contains X and Y, and is therefore the line p. Hence P is the pole of p, and we have the theorem:

The polars of the range of points on a line p form a pencil of lines, with vertex at the pole P.

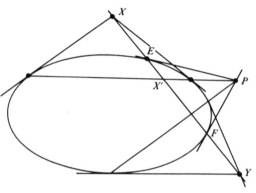

Fig. 78.4

Let X be a point on p, and let the polar of X, which passes through P, intersect p in X'. If the intersections of p with the point-conic be E, F, we know that the pair of points X, X' are harmonic conjugates with respect to the pair of points E, F. Since E, F are fixed, the ranges $\{X\}$ and $\{X'\}$ are projective, the projectivity being an involution. We may write

$$\{X\} \barwedge \{X'\} \barwedge P\{X'\}.$$

Stated in words, this result is:

A set of points on a line is projective with the pencil of polars of this set with respect to a given conic.

We leave some of the consequences of applying the dual notions arising from the concept of pole and polar, to the line-conic formed by the tangents, to be investigated as Exercises. We end this section with a proof of the Cross-Axis Theorem, which involves Pascal's Theorem as a consequence.

Theorem III   *If $(P, Q, R, U, V, \ldots) \barwedge (P', Q', R', U, V, \ldots)$ on a given point-conic, so that the points U, V, assumed distinct, are the fixed points of the projectivity on the*

*conic, then all intersections of cross-joins, such as $PQ' \cap P'Q$ lie on a fixed line, the
line UV (the so-called cross-axis).*

Proof. By the theory of projective ranges on a conic, we have the projective relation
between pencils given by

$$P'(P, Q, R, U, V, \ldots) \,\overline{\wedge}\, P(P', Q', R', U, V, \ldots).$$

In this projective relationship the line $P'P$ of the one pencil corresponds to the line
PP' of the second pencil (Fig. 78.5). The locus of intersection of corresponding rays
of the two pencils is therefore a line (§77.1). Hence the points $PQ' \cap P'Q$, $PR' \cap P'R$,

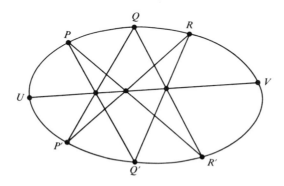

Fig. 78.5

$PU \cap P'U$, $PV \cap P'V$, . . . lie on a line. But $PU \cap P'U = U$, and $PV \cap P'V = V$, so
all the intersections lie on the line UV.

Corollary. Pascal's Theorem. If A, B', C, A', B, C' are six points on a conic, considered
in this order, the opposite sides of the hexagon being AB', $A'B$; $B'C$, BC' and CA',
$C'A$, then the intersections of opposite sides of the hexagon lie on a line.

Proof. If we set up the uniquely defined projectivity in which

$$(A, B, C) \,\overline{\wedge}\, (A', B', C')$$

and suppose that the fixed points of this projectivity are U and V, the Pascal line is the
line UV.

Remark. The Pascal line intersects the conic in the fixed points of the projectivity
determined by the map $(A, B, C) \,\overline{\wedge}\, (A', B', C')$, and this provides us with a simple
and effective method of constructing the fixed points in certain problems, which will
be indicated in the Exercises.

Given six points on a conic, there are 60 distinct hexagons which arise from the set of points, by considering the different orders which may be assigned to them, and therefore 60 Pascal lines. Some properties of these lines will be obtained by stereographic projection from a quadric in S_3 (§86.2).

The Pascal Theorem enables us to draw a conic through five given points in the real Euclidean plane. If we are given A, B, C, A', B', the theorem enables us to find the point C' in which any line through A, say, intersects the conic again. The fact that a conic is determined uniquely by passing through five points follows from the projective generation, since the correspondence

$$A(B, C, B') \mathbin{\overline{\wedge}} A'(B, C, B')$$

determines a unique projectivity, and the resulting conic goes through A, A', B, C and B'.

78.3   The polar of a point

It is very useful to have the equation of the polar of a given point with respect to a given point-conic, when the conic is not referred to any special triad of reference. Let the equation of the conic be

$$X^T A X = 0,$$

in matrix form, where A is a symmetric non-singular 3×3 matrix. The algebra which follows will be exactly the same whatever the order of the projective space we are working in. If P and Q are distinct points in S_2, let $P\lambda + Q\mu$ be on the conic. Then

$$(\lambda P + \mu Q)^T A (\lambda P + \mu Q) = 0,$$

and this, on expansion, is the quadratic equation in λ, μ:

$$\lambda^2 P^T A P + \lambda\mu(P^T A Q + Q^T A P) + \mu^2 Q^T A Q = 0.$$

Each coefficient in this equation is a 1×1 matrix, and can be regarded as an ordinary complex number. Since A is symmetric, and $A^T = A$, so that

$$(P^T A Q)^T = Q^T A P,$$

we may write the coefficient of $\lambda\mu$ as $2(P^T A Q)$, and the equation is

$$\lambda^2(P^T A P) + 2\lambda\mu(P^T A Q) + \mu^2(Q^T A Q) = 0.$$

This equation gives the values of $\lambda : \mu$ corresponding to the points of intersection of the line PQ with the conic. If these points of intersection are harmonic conjugates with respect to P and Q, we know that they will be

$$P + tQ, \quad P - tQ$$

for some value of t (§64.2). Hence the sum of the roots of the quadratic equation above must equal zero, and therefore

$$P^T A Q = 0$$

is the condition that P and Q be separated harmonically by the intersections of PQ with the conic. In this case P and Q are said to be *conjugate points* with regard to the conic.

If we keep P fixed, and let Q vary, the locus of points Q conjugate with respect to P is the line

$$P^TAX = 0,$$

and this is called *the polar of P with respect to the conic.*

If P lies on the conic, $P^TAP = 0$, and one value of $\mu = 0$, as we expect. The other root in $\lambda:\mu$ gives the other intersection of PQ with the conic. If this other intersection also coincides with P, so that

$$P^TAQ = 0,$$

Q lies on the tangent to the conic at P. Hence when the point P lies on the conic, the equation for the polar of P gives *the tangent at P to the conic.*

If $A = [a_{ij}]$ $(i, j = 0, 1, 2)$, the equation for the polar of $P = (p_0, p_1, p_2)$ is:

$$p_0(a_{00}X_0 + a_{01}X_1 + a_{02}X_2) + p_1(a_{10}X_0 + a_{11}X_1 + a_{12}X_2)$$
$$+ p_2(a_{20}X_0 + a_{21}X_1 + a_{22}X_2) = 0.$$

and this may also be written in the form:

$$X_0(a_{00}p_0 + a_{10}p_1 + a_{20}p_2) + X_1(a_{01}p_0 + a_{11}p_1 + a_{21}p_2)$$
$$+ X_2(a_{02}p_0 + a_{12}p_1 + a_{22}p_2) = 0,$$

where, since A is symmetric, $a_{ij} = a_{ji}$.

The polar of $(1, 0, 0)$ is the line

$$a_{00}X_0 + a_{01}X_1 + a_{02}X_2 = 0.$$

If our triad of reference is such that this is the line $X_0 = 0$, we must have $a_{01} = a_{02} = 0$. Similarly, if the polar of $(0, 1, 0)$ is the line $X_1 = 0$, we must have $a_{10} = a_{12} = 0$, and if the polar of $(0, 0, 1)$ is the line $X_2 = 0$, we have $a_{20} = a_{21} = 0$, conditions already implied by the first two equations, since the matrix A is symmetric. The triangle of reference when these conditions are fulfilled is said to be *a self-polar triangle* with regard to the conic, and we have shown that *the equation of a conic with regard to a self-polar triangle as triangle of reference is of the form*

$$a_{00}X_0^2 + a_{11}X_1^2 + a_{22}X_2^2 = 0.$$

If A, B, C, D *are four distinct points on a conic, the diagonal point triangle of the quadrangle $ABCD$ is a self-polar triangle with regard to the conic* (Fig. 78.6). It will be recalled that the diagonal point triangle consists of the points of intersection of opposite sides of the quadrangle $ABCD$. If $U = AC \cap BD$, $V = AD \cap BC$, and $W = AB \cap CD$, we know (Exercise 64.4) that U and V, say, are harmonic conjugates with respect to the points in which UV intersects AB and CD. The pencil $W(V, U; B, C)$ is harmonic, in fact, and this means that B and D are harmonic conjugates with respect to U and the point $DB \cap VW$, and C and A are harmonic conjugates with respect to U and the point $CA \cap VW$. Hence the polar of U with respect to any conic through A, B, C, and D is VW. Similarly the polar of V is UW, and as a consequence W is the pole of UV.

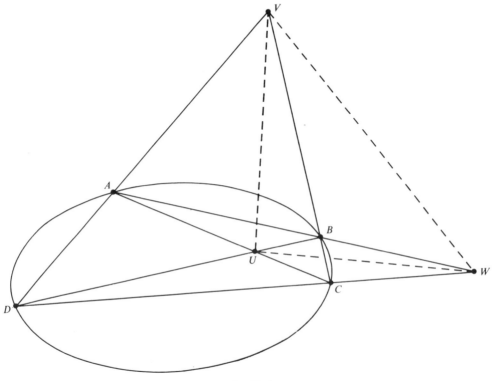

Fig. 78.6

Exercises

78.1   Verify that the tangents to the conic $X_0X_2 - X_1{}^2 = 0$ at $(1, 0, 0)$ and $(0, 0, 1)$ are respectively $X_2 = 0$ and $X_0 = 0$, and that $(0, 1, 0)$ is the pole of $X_1 = 0$.

78.2   Show that if X, Z are any two distinct points of an irreducible point-conic, and Y is the pole of XZ, then if the triangle XYZ is taken as triangle of reference the equation of the conic may be written in the form $X_0X_2 - X_1{}^2 = 0$. (Take the general equation of a conic, and find the conditions that the points $(1, 0, 0)$ and $(0, 0, 1)$ satisfy it. Then find the condition that the polar of $(0, 1, 0)$ be $X_1 = 0$. The equation should then reduce to $X_0X_2 + kX_1{}^2 = 0$, $(k \neq 0)$, and absorption of k produces the desired equation.)

78.3   Show that if a point-conic with equation $X_0X_2 - X_1{}^2 = 0$ has the homogeneous parameters $(\lambda^2, \mu\lambda, \lambda^2)$, any projective correspondence between the points may be written in the form

$$A\lambda\lambda^* + B\lambda\mu^* + C\lambda^*\mu + D\mu\mu^* = 0.$$

Show that if the fixed points of this projectivity are at the points $(1, 0, 0)$ and $(0, 0, 1)$, then $A = D = 0$. Continue by showing that the equation of a chord $U^{(0)}X_0 + U^{(1)}X_1 + U^{(2)}X_2 = 0$ which joins pairs of points on the conic related by this projectivity satisfies an equation of the form $U^{(0)}U^{(2)} - k(U^{(1)})^2 = 0$, determining the value of k. What do you deduce?

78.4 If the line $X_0\lambda^2 - 2X_1\lambda + X_2 = 0$, which is tangent at the point with parameter λ on our given conic $X_0X_2 - X_1^2 = 0$ passes through the point (p_0, p_1, p_2), we have the equation $p_0\lambda^2 - 2p_1\lambda + p_2 = 0$. Deduce that the chord of contact of the tangents from the point to the conic has the equation $p_0X_2 + p_2X_0 - 2p_1X_1 = 0$. Compare this equation with that of the polar of the point given in the preceding Section.

78.5 A variable line through a point V intersects a given conic in the pairs of points (P, P'). Show that the pairs of tangents to the conic at P, P' intersect on a fixed line. Formulate a theorem on the intersection of pairs of tangents to a conic which are in involution.

78.6 The points A, B, C and D lie on a given conic, and the diagonal triangle of the quadrangle A, B, C, D is U, V, W. Show that the quadrilateral formed by the four tangents to the conic at A, B, C, D has UV, VW, WU as its diagonal line triangle.

78.7 Write down the dual of the Cross-Axis Theorem (p. 334), and obtain the dual of Pascal's Theorem, discovered, centuries after, by Brianchon.

78.8 We consider the uniquely defined projectivity on a line l for which

$$(A, B, C) \,\overline{\wedge}\, (A', B', C'),$$

and we wish to find the fixed points of this projectivity on l. Let V be a point not on l, S a conic through V, and let the further intersections of VA, VB, VC and VA', VB', VC' with the conic S be P, Q, R and P', Q', R'. If the cross-axis determined by the points $PQ' \cap P'Q$, $QR' \cap Q'R$ and $RP' \cap R'P$ intersects the conic S in X and Y, and $VX \cap l = E$, and $VY \cap l = F$, show that E and F are the required fixed points of the projectivity on l.

78.9 We require to draw a triangle PQR so that P lies on l, Q lies on m and R on n, where l, m and n are given lines, and PQ passes through D, QR passes through E and RP passes through F, where D, E and F are given points. We proceed by 'the method of false positions', choosing any point P on l. Let $DP \cap m = Q$, $EQ \cap n = R$ and $FR \cap l = P'$. Show that

$$\{P\} \,\overline{\wedge}\, \{Q\} \,\overline{\wedge}\, \{R\} \,\overline{\wedge}\, \{P'\},$$

and that the problem is solved when P is chosen at one of the two fixed points of the projectivity $\{P\} \,\overline{\wedge}\, \{P'\}$. Use the result of the previous Exercise to determine these fixed points, and construct the two possible triangles. (Three false positions determine the true positions.)

78.10 If A' is the pole of BC, B' the pole of CA and C' the pole of AB with respect to a conic S, prove that the triangles ABC, $A'B'C'$ are in perspective. (The triangles ABC, $A'B'C'$ are called *polar* triangles w.r.t. to S.)

78.11 Deduce from the preceding Exercise that if two pairs of opposite sides of a given quadrangle are each conjugate lines with respect to a given conic, then so is the third pair. What is the dual of this theorem?

78.12 If two triangles are inscribed in a conic S, prove that their six sides touch a conic Σ. (Six lines a, b, c, d, e, f touch a conic if the range of four points cut on e by a, b, c, and d is projective with the range cut on f by a, b, c and d; that is, if $e(a, b, c, d) \,\overline{\wedge}\, f(a, b, c, d)$.)

78.13 BC is the chord of contact of tangents from A to a conic S, and EF is the chord of contact of tangents from D. Prove that the sides of the two triangles ABC, DEF touch a conic.

78.14 An involution T_1 on a conic S is cut out by chords which pass through a point U, and another involution T_2 is cut out by chords which pass through a point V. If P is any point on the conic, show that $(P)T_1T_2 = (P)T_2T_1$, so that the composition of the involutions is commutative, if and only if U and V are conjugate points with regard to the conic. (Compare Exercise 76.6.)

Apply this result to the representation of inversion in a plane, given in §39.1, to show that the operation of inversion in two distinct circles is commutative under composition if and only if the circles are orthogonal. (Exercise 25.2. See also Exercise 39.6.)

78.15 The quadratic equation $a_0z^2 - 2a_1z + a_2 = 0$ over the complex field is mapped onto the point (a_0, a_1, a_2) of the complex plane S_2. Show that if the equation has coincident roots equal to t, the representative point is the point $(1, t, t^2)$ on the conic $X_1^2 - X_0X_2 = 0$ of S_2. Find the equation of the chord joining points with parameters t, t' on the conic, and verify that if the quadratic has roots t and t', the representative point in S_2 is the pole with respect to the conic of the chord which connects the points with parameters t and t' on the conic. Using the condition (Exercise 69.7) that two given quadratics have roots which are harmonically separated, show that in the representation the points which represent such quadratics are conjugate with regard to the conic.

79.1 Conics through four points

We saw, at the end of §78.2, using the projective generation of a conic, that a conic through five points is uniquely determined. This is also seen if we notice that the equation of a conic is a homogeneous equation of the second degree and contains six constants. The equation is

$$a_{00}X_0^2 + a_{11}X_1^2 + a_{22}X_2^2 + 2a_{01}X_0X_1 + 2a_{12}X_1X_2 + 2a_{02}X_0X_2 = 0.$$

The conic is determined by the *ratio* of five of these constants to the sixth. If the conic is assigned to pass through a given point, say (p_0, p_1, p_2), we obtain the equation

$$a_{00}p_0^2 + a_{11}p_1^2 + a_{22}p_2^2 + 2a_{01}p_0p_1 + 2a_{12}p_1p_2 + 2a_{02}p_0p_2 = 0,$$

which is a *linear* homogeneous equation connecting the five ratios. If the conditions for the conic to pass through five given points determine five linearly independent equations in the five ratios, we obtain a unique solution for the five ratios, and therefore a unique conic through the five points. It is easy to determine conditions which would *not* give five independent equations. For example, if four of the given points were on a line, the conic would automatically, once it were made to contain three collinear points, contain the line, and therefore the fourth point. The fourth equation would therefore be linearly dependent on the first three equations.

If the five equations are independent, we say that the five points *impose independent conditions* on conics which pass through them, there being only one such conic. If we take four points which impose independent conditions, the solutions of our equations involve one parameter linearly, and the equation of a conic through the four points is

$$S_0 + \lambda S_1 = 0,$$

where $S_0 = 0$ and $S_1 = 0$ are conics through the four points. Conversely, if $S_0 = 0$ and $S_1 = 0$ are two distinct conics, the equation above represents a conic which, for all values of λ passes through the points common to both conics. We can show that two distinct conics intersect in four points (see Exercise 79.6), and therefore we are led to consider conics through four points as being given by the *pencil* of conics

$$S_0 + \lambda S_1 = 0.$$

We have already discussed pencils of circles, under the title *Coaxal Systems* (§27.1). We shall find that pencils of conics have a number of interesting properties. The properties of pencils of circles will be deducible from these, at a later stage.

Theorem I   *The conics which pass through four given points A, B, C and D cut pairs of points in involution on any line l which does not pass through any of the four points.*

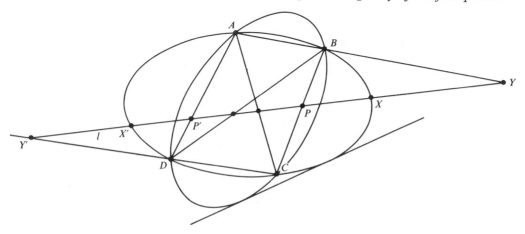

Fig. 79.1

Proof. Let $BC \cap l = P$, $AD \cap l = P'$, $BA \cap l = Y$, $DC \cap l = Y'$, and let a conic through the four given points intersect l in the points X and X' (Fig. 79.1). On this conic we have the projectivity

$$B(C, A, X, X') \barwedge D(C, A, X, X'),$$

and if we take sections of these pencils with the line l, we have:

$$(P, Y, X, X') \barwedge (Y', P', X, X').$$

But

$$(Y', P', X, X') \barwedge (P', Y', X', X) \text{ (§69.1)}.$$

Hence

$$(P, Y, X, X') \barwedge (P', Y', X', X),$$

and therefore X, X' are pairs in an involution which contains the pairs (P, P') and (Y, Y'). (Exercise 76.1). This is an involution determined by the point-pairs (P, P')

and (Y, Y'); and this is determined by the four points A, B, C and D, and independent of the particular conic through the four points considered.

Remark. Since the lines AC and BD constitute a particular conic through the points A, B, C and D, we recover the theorem (Exercise 76.2) that the opposite sides of a quadrangle cut any line not through the vertices in three pairs of points in involution.

Since an involution has two distinct fixed points, we deduce that:

There are two points of a given line l at which conics of a given pencil touch the line.

Since a conic is uniquely determined by the points A, B, C, D and a point of contact with l, we may also say:

Two conics of a given pencil touch a given line l.

Remembering that any pair of an involution forms a harmonic range with the fixed points (§78.2), we deduce that *the point-pairs in which conics of the pencil intersect l are harmonic conjugates with respect to the two points on l at which conics of the pencil touch l.*

Various deductions about conics in the real Euclidean plane will be made later (§80.1).

We note that if the line l passes through a diagonal point of the quadrangle $ABCD$, say the point of intersection of AC and BD, this point is one of the fixed points of the involution cut on l by conics through the points A, B, C and D.

Theorem II *If P is a given point, the polars of P with respect to the various conics through four given points A, B, C and D form a pencil of lines.*

Proof. Let P' be the intersection of the polars of P with respect to two conics of the pencil. If these conics cut the line PP' in the points (X, X') and (Z, Z') respectively, then the points $(X, X'; P, P')$ form a harmonic range, by the definition of the polar of P, and also the points $(Z, Z'; P, P')$ form a harmonic range. We know, from Thm I, that the pairs of points (X, X'), (Z, Z') are pairs of an involution on PP', and that the fixed points of the involution form harmonic ranges with each of the pairs (X, X') and (Z, Z'). Since an involution is uniquely determined by the assignment of two pairs in the involution, (being the uniquely defined projectivity which maps (X, Z, X') on (X', Z', X)), we see that P and P' must be the fixed points of the involution. Hence every conic of the conics through A, B, C and D cuts PP' in a pair of points which form a harmonic range with P and P'. The polar of P therefore passes through P' for every conic through A, B, C and D.

We now investigate the locus of P' as the point P moves on a line l.

Theorem III *The point P' moves on a conic which passes through eleven points with a simply defined geometrical relation to the quadrangle $ABCD$ and the line l.*

Proof. Let U^*, V^* be the poles of l with regard to two distinct conics of the pencil, say

S_0 and S_1. Then the polar of P with regard to S_0 passes through U^*, since the polar of U^*, which is l, passes through P, and similarly the polar of P with regard to S_1 passes through V^*. As P moves on l we determine P' by taking the polars of P with regard to S_0 and S_1 in each case. We know that as P moves on l, the pencil $U^*(P')$ is projective with the set of points $\{P\}$, and the pencil $V^*(P')$ is also projective with the set of points $\{P\}$ (p. 334). Hence

$$U^*\{P'\} \;\overline{\wedge}\; V^*\{P'\}$$

and therefore the point P' moves on a conic, by the definition of a conic.

This conic, by our proof, contains the poles of l with respect to *every* conic through A, B, C, and D. There are two conics of the pencil which touch l. These points are the pole of l for those particular conics, and therefore lie on the locus. The locus passes through the vertices of the diagonal point-triangle of the quadrangle $ABCD$. For if this be UVW, and l intersects VW in T, the polar of T with respect to every conic of the pencil passes through U, since the polar of U with regard to every conic of the pencil is VW and contains T. Finally, let X be the intersection with l of a side of the quadrangle $ABCD$, say the side AB. Then if X' be the harmonic conjugate of X with respect to A and B, the polar of X with regard to all conics through A, B, C and D passes through X'. There are six sides to the quadrangle $ABCD$. We have therefore found eleven points connected geometrically with the line l and the quadrangle $ABCD$ which lie on the locus. The conic-locus is naturally called the *eleven-point* conic associated with the line l and the quadrangle $ABCD$. We shall see (Exercises 80.7–9) that it is connected with the nine-point circle of a triangle. (§12.1.)

Remark. Both Theorems II and III can be proved algebraically with ease, and we indicate proofs in the Exercises. The synthetic proofs given above are attractive, and underline the importance of the fundamental theorems we have proved.

Exercises

79.1 If the conic S_0 has equation $X^T A X = 0$, and the conic S_1 equation $X^T B X = 0$, show that the equation of the polar of $P = (p_0, p_1, p_2)$ with regard to the conic $S_0 + \lambda S_1 = 0$ is

$$P^T(A + \lambda B)X = P^T A X + \lambda(P^T B X) = 0.$$

Deduce Thm II of the preceding section.

79.2 Show that the equation of a conic through the points A, B, C, D, the triangle of reference being the diagonal point triangle of the quadrangle formed by the four points, may be written in the form

$$a_{00} X_0{}^2 + a_{11} X_1{}^2 + a_{22} X_2{}^2 = 0,$$

where $a_{00} + a_{11} + a_{22} = 0$.

79.3 Prove that the locus of the poles of a given line with respect to the conics given by the equation in the preceding Exercise is a conic, and deduce Thm III of the preceding Section. (Note that the equation, with the condition on the coefficients, is that of a pencil of conics.)

79.4 Prove algebraically that if p be the polar of P with regard to a given conic, then as P moves on a given line l,

$$\{P\} \;\overline{\wedge}\; \{p\}.$$

79.5 Consider the results which are dual to those obtained in §79.1, and in particular the changes which are needed notationally in the four Exercises above to obtain the results algebraically.

79.6 Show that substitution of $(1, \lambda, \lambda^2)$ for $(X_0, X_1, \cdot X_2)$ in the equation $X^T A X = 0$ produces a quartic equation in λ. What do you deduce?

80.1   Conics in the real Euclidean plane

We described a model of the projective plane in §61.1. We show here that a real Euclidean plane can be embedded in a real projective plane, and then, following Poncelet, obtain valuable insight into metrical properties by extending our ground field to that

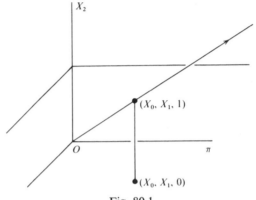

Fig. 80.1

of the complex numbers, and examining the relation of circles to the 'line at infinity'.

Let π be a real Euclidean plane, and $P = (X_0, X_1)$ a point of π (Fig. 80.1). We suppose that π is in real Euclidean space of three dimensions, with coordinates (X_0, X_1, X_2), and we map the points of π onto the points of $X_2 = 1$ by the mapping $(X_0, X_1, 0) \leftrightarrow (X_0, X_1, 1)$. That is, we draw lines through the points P parallel to the OX_2 axis to intersect the plane $X_2 = 1$ in the points $(X_0, X_1, 1)$. We now join the points in $X_2 = 1$ to the origin, and allow the resulting ray, which consists of the points $(X_0 k, X_1 k, k)$ to represent the point (X_0, X_1).

The set of rays appeared as a model for the projective plane in §61.1. Having shown the geometrical content of what we are doing, we can now simply examine the mapping

$$(X_0, X_1) \leftrightarrow (X_0 k, X_1 k, k) \quad (k \neq 0)$$

between the points of a real Euclidean plane and the points of a real projective plane. If we write this in the form

$$(X_0, X_1) \leftrightarrow (T_0, T_1, T_2),$$

the relationship is

$$X_0 = T_0 T_2^{-1}, \quad X_1 = T_1 T_2^{-1},$$

$$(T_0, T_1, T_2) = (X_0, X_1, 1).$$

By means of this relationship, non-homogeneous equations of loci in the real Euclidean plane are mapped onto homogeneous equations of loci in the real projective plane. For example, the equation

$$a_0 X_0 + a_1 X_1 + a_2 = 0,$$

the equation of a line in the real Euclidean plane, is mapped onto the equation

$$a_0 T_0 + a_1 T_1 + a_2 T_2 = 0$$

in the real projective plane, with a one-to-one correspondence between points on the two loci except for the point of intersection of the line in the real projective plane with the line $T_2 = 0$. This is the 'point at infinity' on the line in the real Euclidean

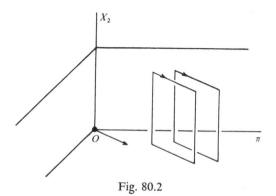

Fig. 80.2

plane (see Fig. 80.2). This point is $(-a_1, a_0, 0)$, and we note that *the set of parallel lines* in the real Euclidean plane given by

$$a_0 X_0 + a_1 X_1 + k = 0,$$

where a_0, a_1 are kept fixed, and k is variable, map onto *the pencil of lines $a_0 T_0 + a_1 T_1 + k(T_2) = 0$* in the real projective plane, which all pass through the point $(-a_1, a_0, 0)$. There is a one-to-one correspondence between *directions* in the real Euclidean plane and *points* on the line $T_2 = 0$. Thus the Euclidean axiom that through a point there passes one line parallel to a given line is merely the equivalent of the projective axiom that there is a unique line joining two points.

The conic in the real Euclidean plane given by the equation

$$a_{00} X_0^2 + a_{11} X_1^2 + 2a_{01} X_0 X_1 + 2a_{12} X_1 + 2a_{20} X_0 + a_{22} = 0$$

is mapped onto the conic

$$a_{00} T_0^2 + a_{11} T_1^2 + 2a_{01} T_0 T_1 + 2a_{12} T_1 T_2 + 2a_{20} T_0 T_2 + a_{22} T_2^2 = 0.$$

There is a one-to-one correspondence between points on the two conics except for possible points of the conic in the projective plane which lie on the line $T_2 = 0$. These are given by the equation

$$a_{00}T_0^2 + 2a_{01}T_0T_1 + a_{11}T_1^2 = 0.$$

These are lines through the point $(0, 0, 1)$ and correspond to the lines

$$a_{00}X_0^2 + 2a_{01}X_0X_1 + a_{11}X_1^2 = 0$$

through the origin in the Euclidean plane. These lines are given by equating the highest degree terms in the equation of the conic to zero, and they give *the directions of the asymptotes*, if any, of the conic.

If the lines are real and distinct, the conic approaches the line at infinity in two

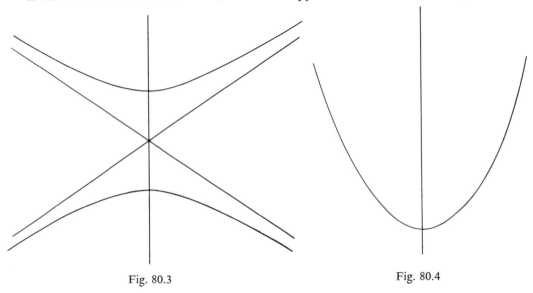

Fig. 80.3   Fig. 80.4

distinct directions. We then say the conic is an *hyperbola* (Fig. 80.3). If the lines are coincident, and then necessarily real, we say the conic is a parabola (Fig. 80.4). In both cases we are assuming that the conic is irreducible. If the lines are not real, that is, the homogeneous equation does not factorize into real factors, we say the conic is an *ellipse*. It is then confined entirely to a finite part of the plane.

The theory of conics in the real Euclidean plane is greatly simplified by considerations similar to those initiated above.

If A, B, C, D be four collinear points, the cross-ratio $[A, B; C, D]$ is the same number, whether we are in the Euclidean plane or the projective plane, the points corresponding in the mapping defined above. The cross-ratio in the Euclidean plane may be written in terms of signed distances, thus:

$$[A, B; C, D] = (\overline{AC} : \overline{CB})/(\overline{AD} : \overline{DB}) = (\overline{AC} . \overline{DB})/(\overline{AD} . \overline{CB}).$$

As the point D moves towards infinity in the Euclidean plane, the corresponding point D in the projective plane moves towards the line $T_2 = 0$, and the ratio $\overline{AD}:\overline{DB}$ tends towards the limit -1. In the projective plane we have a perfectly proper cross-ratio when D is on the line $T_2 = 0$, but in the Euclidean plane we have three finite points A, B, C, where

$$\overline{AC}:\overline{CB} = -[A, B; C, D],$$

the cross-ratio on the right being evaluated in the projective plane.

This device enables us to introduce the ratio of the lengths of segments on the same line via projective geometry. In particular, if C is the midpoint of AB, we can say the range A, B, C, D is harmonic, where D is *the point at infinity*.

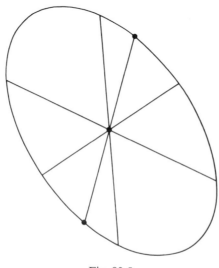

Fig. 80.5

To define *the center* of a conic in the real Euclidean plane, we imagine it mapped into the real projective plane, as above, and say *the center is the pole of the line at infinity*. The center is thus uniquely defined, and only lies on the conic when the conic touches the line at infinity, which means it is a parabola. In any other case, since any line through the center is divided harmonically by the center, the intersection with its polar, and its intersections with the conic, we deduce, one point being on the line at infinity, that *in the Euclidean plane the center of a conic bisects every chord of the conic which passes through it* (Fig. 80.5).

The *asymptotes* of a conic are defined to be the tangents to the conic at the points where it intersects the line at infinity. Again, this definition refers to the image of the conic in the projective plane, but it enables us to find the equation of the asymptotes with ease for the Euclidean case (Exercise 80.18).

We now come to the method of regarding *angles* from the projective point of view. First, we introduce *the circular points at infinity*.

The equation of any circle in the Euclidean plane is of the form

$$X_0{}^2 + X_1{}^2 + 2gX_0 + 2fX_1 + c = 0,$$

if we return to the notation of §18.1. In the projective plane, this becomes

$$X_0{}^2 + X_1{}^2 + 2gX_0X_2 + 2fX_1X_2 + cX_2{}^2 = 0.$$

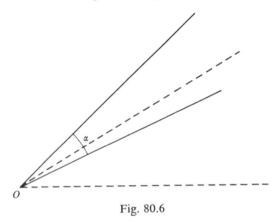

Fig. 80.6

If we consider the intersections with the line $X_2 = 0$, these are given by the equation

$$X_0{}^2 + X_1{}^2 = 0,$$

and *this equation is the same for all circles*.

Of course, this equation does not factorize into real factors. Over the field of complex numbers we may factorize the equation and obtain

$$(X_0 + iX_1)(X_0 - iX_1) = 0,$$

and we may say that *all circles intersect the line at infinity in the same pair of conjugate complex points*. These are *the circular points at infinity*.

The lines joining a given point to the circular points at infinity are called *the isotropic lines through the point*. If the point be (h, k) in the real Euclidean plane, the equation of the isotropic lines is:

$$(X_0 - h)^2 + (X_1 - k)^2 = 0.$$

We note that this is also *the equation of a point-circle*, a circle, center (h, k) and of radius zero. We now see that point-circles are reducible conics, the isotropic lines through the center, over the field of complex numbers.

By means of the isotropic lines through a point, we can define an angle with vertex at the point, using a projective concept, that of cross-ratio. For convenience take the point at $(0, 0)$, and let the lines $X_1 - X_0m_1 = 0$, $X_1 - X_0m_2 = 0$ enclose an angle α (Fig. 80.6). Then $m_1 = \tan A_1$, say, and $m_2 = \tan A_2$, and $A_1 - A_2 = \alpha$. Now the lines $X_1 - iX_0 = 0$, and $X_1 + iX_0 = 0$ are the isotropic lines through the origin, and the cross-ratio of the pencil formed by the four lines is that of the four points $(m_1, m_2, i, -i)$. We evaluate the cross-ratio

$$[m_1, m_2; i, -i] = \frac{(i - m_1)(m_2 + i)}{(m_2 - i)(-i - m_1)} = \frac{(1 + im_1)(1 - im_2)}{(1 + im_2)(1 - im_1)}$$

$$= \frac{(\cos A_1 + i \sin A_1)(\cos A_2 - i \sin A_2)}{(\cos A_2 + i \sin A_2)(\cos A_1 - i \sin A_1)}$$

$$= (\cos A_1 + i \sin A_1)^2 / (\cos A_2 + i \sin A_2)^2$$

$$= \cos 2(A_1 - A_2) + i \sin 2(A_1 - A_2),$$

$$= \cos 2\alpha + i \sin 2\alpha.$$

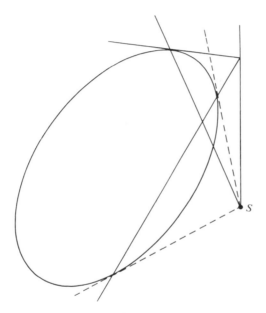

Fig. 80.7

This *formula of Laguerre* enables us to define angle via a projective concept. The chief use we shall make of this formula is in the case $\alpha = \pi/2$, when the cross-ratio is -1, and so the pencil is a harmonic pencil. Hence *a pair of orthogonal lines are harmonically conjugate to the isotropic lines through their intersection*.

We give one final set of definitions and an application before we close this section. A *focus* of a given conic is a point S such that the tangents to the conic from S are the isotropic lines through S (Fig. 80.7). The polar of S with respect to the conic is then called *the corresponding directrix*. One of the metrical results obtained with some difficulty from the equation of a given point-conic is *the focus-directrix property*:

If S be a focus of a given conic, and l the corresponding directrix, the point P any point on the conic, PN the perpendicular from P onto the directrix, then the ratio $|SP| : |PN|$ is constant.

We obtain this property of a conic from our enlarged conception of a conic, using our projective theorems, and then some simple metrical theorems.

If a conic intersects a given line in the points U, V, and P, P' are points on the line which are conjugate, we have $[U, V; P, P'] = -1$, and the points P, P' are in an involution with fixed points U, V. The dual of this theorem says that if through a fixed point S we draw pairs of lines which are conjugate with respect to a given conic,

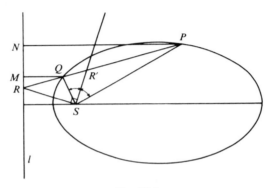

Fig. 80.8

each containing the pole of the other, then they are in an involution of lines through S of which the fixed rays are the tangents from S to the given conic.

If the tangents from S to the given conic are the isotropic lines through S, any pair of conjugate lines through S will be orthogonal, since they form a harmonic pencil with the fixed lines of the involution. We therefore have a defining property of a focus of a given conic:

the point S is a focus of a given conic if and only if pairs of conjugate lines through S are orthogonal.

Let S be a focus of a given conic (Fig. 80.8), l the corresponding directrix, and P, Q two points on the conic, the line PQ intersecting the directrix in R. Let R' be the harmonic conjugate of R with respect to P and Q. Then the polar of R with respect to the conic passes through R', and it also passes through S because the polar of S, which is l, passes through R. Hence the lines SR and SR' are conjugate with respect to the conic. They are therefore orthogonal.

We now use the metrical theorem that if in a harmonic pencil, $S(P, Q, R, R')$, a pair of conjugate rays SR, SR' are orthogonal, then they are necessarily the internal and external bisectors of the angle PSQ. (p. 319.) We now have $|RQ| : |RP| = |SQ| : |SP|$, and if N is the foot of the perpendicular from P onto l, M the foot of the perpendicular from Q onto l,

$$|RQ|:|RP| = |QM|:|PN|,$$

so that

$$|SQ|:|SP| = |QM|:|PN|,$$

or

$$|SP|:|PN| = |SQ|:|QM|,$$

and this establishes the focus-directrix property of a conic.

The manifold applications of the ideas discussed above are indicated in the following Exercises.

80.2　Conics in the real affine plane

There is a geometry which appears in the real Euclidean plane if we forget the idea of distance between two points and angle between two lines and use coordinate transformations of the type

$$X'_0 = a_{00}X_0 + a_{01}X_1 + a_{02},$$
$$X'_1 = a_{10}X_0 + a_{11}X_1 + a_{12},$$

where $a_{00}a_{11} - a_{01}a_{10} \neq 0$. This last condition ensures that we can solve for (X_0, X_1) uniquely in terms of (X'_0, X'_1). Such transformations are called 'general coordinate transformations' in the older books on analytic geometry.

We see that if we write these transformations in terms of the homogeneous coordinates (T_0, T_1, T_2), introduced in the preceding Section, where

$$X_0 = T_0 T_2^{-1}, \quad X_1 = T_1 T_2^{-1},$$

they are

$$T'_0 = a_{00}T_0 + a_{01}T_1 + a_{02}T_2,$$
$$T'_1 = a_{10}T_0 + a_{11}T_1 + a_{12}T_2,$$
$$T'_2 = \qquad\qquad\qquad a_{23}T_2,$$

where $a_{23}(a_{00}a_{11} - a_{01}a_{10}) \neq 0$. This projective transformation *maps the line* $T_2 = 0$ *onto itself.*

Since we regarded $T_2 = 0$ as *the line at infinity* in the preceding Section, we see that so-called 'general coordinate transformations' preserve the line at infinity, if we regard them as one-to-one mappings of the real Euclidean plane. It is easily verified that the mapping above is the most general non-singular mapping of the projective plane which maps $T_2 = 0$ onto itself. Such a mapping is called *an affine mapping*, and the affine geometry of the plane is the investigation of the invariants which are unchanged under affine mappings. If we restrict ourselves to the field of real numbers for our coefficients a_{ij}, the geometry is called *real affine geometry*.

If we regard our real Euclidean plane as immersed in a projective plane, the affine transformations clearly form a group under composition, and this group is a subgroup of $PGL(n, K)$, where $n = 2$ and K is the field of real numbers. We shall investigate the invariants under affine mappings, and clearly we must have a name for points on the line $T_2 = 0$, the line at infinity, which do not appear in the real plane. We call these points 'improper' points. We could call them 'ideal' points. Neither description is really proper.

In affine geometry we have the concept of *parallel lines*. Lines which intersect on the line we have singled out for special treatment and called $T_2 = 0$ can be called *parallel*. Through a point P which does not lie on a given line l there is a unique parallel to l. The concept of *parallelogram* is an affine concept. But, as we shall see, any two triangles are equivalent under affine transformation, so the shape of a triangle is not an affine concept.

We may choose a point O as origin in the plane, and set up a coordinate system (X_0, X_1). A two-dimensional vector space may be introduced, where the vectors are point-pairs (x_0, x_1), that is, localized at the origin, and the addition of vectors (x_0, x_1) and (y_0, y_1) is defined in the usual manner as

$$(x_0, x_1) + (y_0, y_1) = (x_0 + y_0, x_1 + y_1).$$

Scalar multiples $r(x_0, x_1) = (rx_0, rx_1)$ are also defined as usual, and the usual axioms for a vector space are easily verified. *Translations*, given by

$$X'_0 = X_0 + p,$$

$$X'_1 = X_1 + q$$

are evidently affine transformations, and, in the one given, the point (p, q) is the image of the point $(0, 0)$. The point (x_0, x_1) is mapped onto the point $(x_0 + p, x_1 + q)$, and the vector defined by the points $(0, 0)$ and (x_0, x_1) is mapped onto the vector defined by the points (p, q) and $(x_0 + p, x_1 + q)$. Hence we have the concept of free vectors in affine space also, as in §1.1

If $A = (a_0, a_1)$ and $B = (b_0, b_1)$ are two points, the points

$$P = \frac{kA + k'B}{k + k'} = \left(\frac{ka_0 + k'b_0}{k + k'}, \frac{ka_1 + k'b_1}{k + k'} \right)$$

describe the line joining A to B, as k and k' vary. If we are in the field of real numbers, we may say that the point P divides the segment AB in the ratio $k':k$, and write $\overline{AP}:\overline{PB} = k':k$; if $k = k' = 1$, we may talk of the *midpoint* of a segment. These concepts are invariant under affine mappings, the mappings being real, of course. The equation of a line is, of course, a linear equation.

If V, A, A', B, B' are such that A' divides VA in the same ratio in which B' divides VB, we can show that $A'B'$ is parallel to AB, and we may therefore compare the ratios of segments on parallel lines, by saying that $\overline{AB}:\overline{A'B'} = \overline{VA}:\overline{VA'}$. Since an affine mappings maps parallel lines onto parallel lines, and the ratio $\overline{VA}:\overline{VA'}$ is unchanged by the mapping, the ratio of segments on parallel lines is an affine concept.

It is also possible to introduce the notion of *area*, and we indicate this in an Exercise (Exercise 80.24).

Affine mappings are a matter of common experience, when we witness a *parallel projection*. If a light-source is so distant that the rays are parallel, the shadows of objects cast upon a screen are parallel projections. If the rays are perpendicular to the screen, we obtain *orthogonal* projection, and the properties of affine transformations mentioned above are perhaps most easily verified in this case.

The affine transformations which map the origin onto itself are given by

$$X'_0 = a_{00}X_0 + a_{01}X_1,$$
$$X'_0 = a_{10}X_0 + a_{11}X_1,$$

where $\triangle = a_{00}a_{11} - a_{01}a_{10} \neq 0$, and this transformation can be interpreted as mapping any two given independent vectors localized at the origin onto any two given independent vectors localized at the origin. Since a translation is also an affine mapping, we see how any triangle ABC can be mapped by an affine transformation onto any triangle $A'B'C'$. Let T be the translation which maps A onto A', and let $A'B''C''$ be the map of ABC. Then if T^* is the affine transformation which maps $A'B''$ onto $A'B'$ and $A'C''$ onto $A'C'$, the affine transformation TT^* maps ABC onto $A'B'C'$.

With these preliminaries about affine transformations, let us consider conics in the real affine plane. They can be distinguished by their behaviour relative to the line at infinity. That is, we immerse the plane in a real projective plane, and then consider projective transformations which preserve the line $T_2 = 0$. Conics which do not possess improper points, that is, do not intersect the line at infinity, preserve this property, and are all of the type $X_0^2 + X_1^2 = 1$, where, of course, we no longer distinguish between circles and ellipses, since they all belong to one class under affine transformations. Conics which intersect the line at infinity are in the class $X_0^2 - X_1^2 = 1$, and conics which *touch* the line at infinity are in a third class, and can be written in the form $X_0^2 = X_1$. Our three types of conic in the affine plane are therefore ellipses, hyperbolas and parabolas. The first two have centers at proper points.

80.3   The Euclidean group

Our investigations in §80.1 show us that Euclidean geometry appears when our affine transformations of the plane are real, and map not only the line at infinity onto itself, but also two special points on that line, *the circular points at infinity* onto themselves. The isotropic lines through $(0, 0)$, which join the origin to the circular points, are given by the equation $X_0^2 + X_1^2 = 0$. Let us consider affine transformations

$$X_0 = a_{00}Y_0 + a_{01}Y_1$$
$$X_1 = a_{10}Y_0 + a_{11}Y_1$$

which leave the quadratic form

$$X_0{}^2 + X_1{}^2 = X^T I X$$

invariant, and therefore leave the isotropic lines through $(0, 0)$ unchanged. The transformation $X = AY$ changes $X^T I X$, where $\mathbf{I}$ is the 2×2 unity matrix, into

$$(AY)^T I(AY) = Y^T(A^T I A)Y = Y^T(A^T A)Y.$$

We therefore require that $A^T A = I$. Matrices which have this property are called *orthogonal* matrices. Since A^T is the inverse of A, they are non-singular, and we can prove that $AA^T = I$ (see Exercise 80.28).

In the case of a 2×2 orthogonal matrix, which we have here, we may also show that over the real field they are of two types:

$$\begin{pmatrix} \cos \alpha & -\sin \alpha \\ \sin \alpha & \cos \alpha \end{pmatrix} \quad \begin{pmatrix} \cos \alpha & \sin \alpha \\ \sin \alpha & -\cos \alpha \end{pmatrix},$$

the first with determinant $+1$, and the second with determinant -1 (see Exercise 80.29).

Transformations of the type

$$X_0 = Y_0 \cos \alpha - Y_1 \sin \alpha + p,$$
$$X_1 = Y_0 \sin \alpha + Y_1 \cos \alpha + q$$

are of the type already encountered as *direct isometries* in §43.1, arising from the mapping $z = X_0 + iX_1 = (\cos \alpha + i \sin \alpha)(Y_0 + iY_1) + (p + iq)$, that is, $z' = az + b$, where $|a| = 1$. The transformation

$$X_0 = Y_0 \cos \alpha + Y_1 \sin \alpha + p,$$
$$X_1 = Y_0 \sin \alpha - Y_1 \cos \alpha + q$$

corresponds to an *indirect isometry*, $z' = a\bar{z} + b$, $|a| = 1$. It will be remembered that both arise from a change of rectangular cartesian coordinate axes. Under these transformations distances are preserved, and thus angles.

These isometries, as we saw, form a group, and this is a subgroup of the group of real affine transformations. Euclidean geometry is a study of the invariants of the group of isometries. It should be noted that an affine transformation which merely leaves the *equation* $X_0{}^2 + X_1{}^2 = 0$ unchanged leads to a geometry in which distances are stretched by the same amount, corresponding to the direct and indirect *similitudes* of §47.1 (see Exercise 80.25).

It can be shown that if I, J are the circular points at infinity, the direct isometries map I onto I and J onto J, whereas the indirect isometries interchange the points. In both cases the line at infinity is mapped onto itself as a consequence. When it was recognized that the circular points at infinity may be regarded as a degenerate line-conic, the line-equation being $(U^0)^2 + (U^1)^2 = 0$, and the line-equation of the circular points $U^0 + iU^1 = 0$, $U^0 - iU^1 = 0$, the idea arose of considering the invariants of a group of projective mappings of the plane which map a given *irreducible* conic onto itself. This geometry is also a non-Euclidean geometry (see Veblen and Young, Vol. II (23), and Appendix III).

Our Poincaré model of a hyperbolic non-Euclidean geometry, which involves the subgroup of the group of Moebius transformations which map a given circle onto itself, and the interior into the interior is, in many ways, a more complicated example than that suggested above, but we explored the world of the Poincaré model in §56.1 because it was a more familiar one when it was introduced at that stage of this book.

Exercises

80.1 Show that if an involution of lines through a point P in the real Euclidean plane contains two pairs of lines which are orthogonal, the fixed lines must be the isotropic lines through P, and every line-pair in the involution consists of pairs of orthogonal lines.

80.2 The point P lies on a given conic, and points Q, Q' on the conic are such that the lines PQ, PQ' are orthogonal. Show that the points Q, Q' are pairs in an involution on the conic, and deduce that the join QQ' passes through a fixed point. By taking a special case, show that the fixed point lies on the *normal* to the conic at P (the perpendicular through P to the tangent at P).

80.3 Write down the dual of the theorem that the opposite sides of the quadrangle determined by four points A, B, C, D intersect any line in pairs of points in involution.

80.4 Show, by giving the dual of a known theorem, that the pairs of tangents from a given point P to the conics which touch four given lines a, b, c, d are pairs in involution, and that the pairs of lines joining P to pairs of opposite vertices of the quadrilateral formed by the four lines are in the same involution.

80.5 Let P be an intersection of two circles described on EE' and FF' respectively as diameters, where E, E' are one pair of opposite vertices, and F, F' another pair of opposite vertices of a quadrilateral formed by lines a, b, c and d. Using Exercise 80.1, show that if G, G' be the third pair of opposite vertices, the circle on GG' as diameter also passes through P. From the deduction that the three constructed circles pass through the same two points, obtain again the theorem that the midpoints of the diagonals of a complete quadrilateral are collinear (§9.4).

80.6 Show that all conics which contain the vertices A, B, C of a triangle and also the orthocenter H must be rectangular hyperbolas (which are defined as conics with perpendicular asymptotes). Deduce that if ABC is a triangle, and H is the intersection of the perpendicular from A onto BC with the perpendicular from B onto CA, then CH is the perpendicular from C onto AB.

80.7 From the discussion on eleven-point conics in §79.1, show that the locus of the centers of conics through four points A, B, C, D is a conic through the vertices of the diagonal point triangle of the quadrangle, the midpoints of the six sides of the quadrangle, and with asymptotes in the directions of the axes of the two parabolas which pass through A, B, C, and D. (The axis of a parabola is in the direction of its contact with the line at infinity.)

80.8 If A, B, C, D are the vertices of a triangle and its orthocenter, show that the locus of the centers of the rectangular hyperbolas which pass through the four points is a conic which passes through the circular points at infinity, that is, a circle.

80.9 Show that this circle bisects the midpoints of the six sides of the quadrangle, and passes through the vertices of the diagonal point triangle of A, B, C, H, that is through the points $AB \cap CH$, $BC \cap AH$ and $CA \cap BH$, which means it is the nine-point circle of triangle ABC (§12.1).

80.10 The asymptotes of a conic have been defined as the tangents to the conic at the points where the conic intersects the line at infinity. Deduce that the asymptotes are the tangents to the conic from the center of the conic.

80.11 The tangents to a given conic from a point C touch at E and F, and a tangent to the conic at a point P of the conic intersects CE at X, CF at Y, and EF at P'. Prove that $(X, Y; P, P')$ is a harmonic range. Deduce that if a tangent to a hyperbola at a point P cuts the asymptotes at X and Y, then P is the midpoint of XY.

80.12 In the following set of Exercises, **80.12–20**, the conic $S = 0$ lies in the real affine plane, and has the equation:

$$S(X, Y) \equiv aX^2 + 2hXY + bY^2 + 2gX + 2fY + c = 0.$$

We shall write

$$\Delta = \begin{vmatrix} a & h & g \\ h & b & f \\ g & f & c \end{vmatrix},$$

and $A = bc - f^2$, $H = fg - ch$, $G = hf - bg$, $B = ac - g^2$, $F = hg - ac$, $C = ab - h^2$, the cofactors of the determinant.

Show that the pole of the line $lX + mY + n = 0$ is the point $(X'/Z', Y'/Z')$, where $X' : Y' : Z' = Al + Hm + Gn : Hl + Bm + Fn : Gl + Fm + Cn$.

80.13 Show that the line-equation of the conic, the condition that the line $[l, m, n]$ touch the conic, is

$$Al^2 + 2Hlm + Bm^2 + 2Gnl + 2Fmn + Cn^2 = 0.$$

80.14 Show that the equation to the polar of a point (X', Y') may be written

$$X'\partial S/\partial X + Y'\partial S/\partial Y + Z'\partial S/\partial Z = 0,$$

where we write X/Z for X and Y/Z for Y and make the equation of the conic homogeneous, eventually writing $Z = Z' = 1$.

80.15 If $U = 0$ and $V = 0$ are given lines, show that the equation

$$S(X, Y) + \lambda UV = 0$$

is that of a conic which passes through the points of intersection of the conic with each of the lines $U = 0$ and $V = 0$.

80.16 Show that the equation

$$S(X, Y) + \lambda U^2 = 0$$

is that of a conic which has the same tangents at the points P, Q, where $U = 0$ intersects the conic $S = 0$, as the conic S itself. (We can derive this result as a limiting argument from the preceding Exercise, or use Exercise 80.14 to find the tangent at a point where $S = 0$ and $U = 0$.)

80.17 If $U(X, Y) \equiv pX + qY + r$, the equation

$$S(X, Y) + \lambda(pX + qY + r)^2 = 0$$

represents a pencil of conics with two fixed tangents, the tangents to the conic $S(X, Y) = 0$ at its intersections with the line $U(X, Y) = 0$. The condition that a conic of the pencil should be reducible, and so consist of a pair of lines, is

$$\begin{vmatrix} a + \lambda p^2 & h + \lambda pq & g + \lambda pr \\ h + \lambda pq & b + \lambda q^2 & f + \lambda qr \\ g + \lambda pr & f + \lambda qr & c + \lambda r^2 \end{vmatrix} = 0.$$

Verify that this determinant is linear in λ. What pair of lines do we obtain if we substitute the unique value of λ given by the determinantal equation in the equation of conics of the pencil? (If you are uncertain, experiment with the pencil $X^2 - YZ + \lambda X^2 = 0$.)

80.18   If we introduce the line at infinity as $0 \cdot X + 0 \cdot Y + r = 0$, the equation

$$S(X, Y) + \lambda(r)^2 = 0$$

is that of a pencil of conics which all touch the tangents to $S = 0$ at its intersections with the line at infinity. That is, the pencil of conics consists of conics which all have the same asymptotes. Show that by finding the reducible conics of this pencil, we obtain the equation of the asymptotes of the conic $S(X, Y) = 0$ in the form

$$S(X, Y) - \triangle/C = 0.$$

Find the asymptotes of the conic

$$2X^2 + 5XY + 2Y^2 + 5X + 4Y = 0.$$

80.19   Show that if $S(X, Y) = 0$ is not a parabola, its equation after the coordinate axes have been moved parallel to themselves so as to pass through the center of the conic is

$$aX'^2 + 2hX'Y' + bY'^2 + \triangle/C = 0$$

in the new coordinates. Does this suggest another method for obtaining the result of the preceding Exercise?

80.20   Show that the equation of the pair of tangents which can be drawn from a point (X', Y') to the conic $S(X, Y) = 0$ is

$$S(X, Y)S(X', Y') - P^2 = 0,$$

where

$$P \equiv X(aX' + hY' + g) + Y(hX' + bY' + f) + (gX' + fY' + c).$$

80.21   From the condition that if (X', Y') be a focus of $S(X, Y) = 0$ the equation of the tangents from (X', Y') to the conic is that of a point-circle, center (X', Y'), show that the equations satisfied by the foci of S are:

$$\frac{(aX' + hY' + g)^2 - (hX' + bY' + f)^2}{a - b} = S(X', Y') = \frac{(aX' + hY' + g)(hX' + bY' + f)}{h}.$$

(In this Exercise we are in the real Euclidean plane.)

80.22   Show that the theorems of Menelaus and Ceva (Sections 4.1 and 4.3) hold in an affine plane.

80.23   Does the Desargues Theorem (§5.1, §5.2) hold in an affine plane?

80.24   If $\triangle$ denotes the expression

$$\frac{1}{2} \begin{vmatrix} X_1 & Y_1 & 1 \\ X_2 & Y_2 & 1 \\ X_3 & Y_3 & 1 \end{vmatrix}$$

prove that $\triangle$ is a *relative* affine invariant, that is that any affine transformation of the coordinates (X, Y) multiplies it by a constant which depends only on the transformation.

80.25   Show that the transformations of the form

$$X_0 = \quad c(Y_0 \cos \alpha + Y_1 \sin \alpha) + p,$$
$$X_1 = \pm c(- Y_0 \sin \alpha + Y_1 \cos \alpha) + q,$$

are the only affine transformations which leave the distance function $(Y_0 - Y_0')^2 + (Y_1 - Y_1')^2$ relatively invariant, that is which multiply it by the same constant factor for all pairs of points (Y_0, Y_1), (Y_0', Y_1').

80.26   Obtain the equations for an orthogonal projection in the form:

$$X_0 = X_0', \qquad X_1 = kX_1'.$$

Show that, by the choice of a suitable projection, a given ellipse may be transformed into a circle. If the ellipse touches the sides of a triangle ABC at the midpoints of the sides, what can one say about the projected triangle?

80.27   A given triangle ABC of sides a_0, b_0 and c_0 and area $\triangle$ is projected orthogonally into an equilateral triangle. Show that the side x of the equilateral triangle satisfies the equation

$$3x^4 - 2(a_0{}^2 + b_0{}^2 + c_0{}^2)x^2 + 16\triangle^2 = 0.$$

Prove that the roots of this equation are always real, and deduce that for any triangle ABC, $a_0{}^2 + b_0{}^2 + c_0{}^2 \geqslant 4\sqrt{3}\,\triangle$, with equality if and only if the triangle is equilateral.

80.28   If A is a matrix which satisfies the equation $A^TA = I$, where I is a unity matrix, show that $AA^T = I$, and that $det\ A = \pm 1$.

80.29   If A is a 2×2 orthogonal matrix with $A^TA = I$, write down the equations which connect the elements a_{ij} of the matrix, and show that A is either of the type

$$\begin{bmatrix} \cos \alpha & -\sin \alpha \\ \sin \alpha & \cos \alpha \end{bmatrix} \quad \text{or} \quad \begin{bmatrix} \cos \alpha & \sin \alpha \\ \sin \alpha & -\cos \alpha \end{bmatrix}$$

80.30   If A and B are both $n \times n$ matrices which satisfy the equations $A^TA = I = B^TB$, show that $C = AB$ satisfies the equation $C^TC = I$.

80.31   The four points A_i ($i = 1, 2, 3, 4$) on the circle $X^2 + Y^2 = 1$ have angular coordinates α_i, so that, regarded as complex numbers they are $z_i = \cos \alpha_i + i \sin \alpha_i$. Show that, regarded as points on the circle, their cross-ratio is equal to their cross-ratio regarded as complex numbers. (For the first cross-ratio we may find the cross-ratio of the points in which the tangents to the circle at the points A_i intersect the fixed tangent at the point $(1, 0)$ on the circle. Show that this is the same as the value for $(z_1, z_2; z_3, z_4)$ given in Exercise 53.11.)

80.32   If four distinct points are orthocyclic, and also lie on a circle, show that they form a harmonic set on the circle. (By Exercise 53.12, the cross-ratio of the four points, regarded as complex numbers, has modulus 1. By Thm 53.5, since they lie on a circle the cross-ratio is a real number. It cannot equal 1, since the points are distinct, and it must therefore be equal to -1. By the preceding Exercise the cross-ratio of the four points, regarded as points on the circle, is therefore equal to -1. The points therefore form a harmonic set on the circle.)

80.33   If the points A_i ($i = 1, 2, 3, 4$) form a harmonic set on a circle, the tangents to the circle at A_1 and A_2 meet on the chord A_3A_4 (§78.2, Thm II). Deduce that if the points A_1, A_2 and A_3, A_4 form a harmonic set on a circle, there is a circle through A_1 and A_2 in which A_3, A_4 are inverse points, so that the points form an orthocyclic set.

81.1   The quadric

Let a and b be skew lines in S_3, and suppose that a projectivity has been established between the planes of the pencils, axes a and b. Then corresponding planes in the projectivity intersect in a line. This set of lines is called a *regulus*.

We note that since a and b are skew, corresponding planes never coincide, so the lines of the regulus are well-determined. We note also that the regulus can also be defined as the locus of lines joining corresponding points of projective ranges on a and on b, since planes through b intersect a, and planes through a intersect b, and the points cut on a and on b are projective with the pencils, axes b and a, respectively (Fig. 81.1).

This remark shows that a regulus is *a self-dual construct*, since in S_3 the dual of the lines of intersection of corresponding planes of two projective pencils is the set of lines joining corresponding points of projective ranges on two lines.

We now treat the concept of a regulus algebraically, the algebra being very similar to that given in the case of a conic, but the interpretation in S_3 is different. Let the line a be given by the planes $U_0 = U_1 = 0$, the line b by the planes $V_0 = V_1 = 0$, and let corresponding planes in the projectivity between the planes of each pencil be given by the equations:

$$\alpha_\lambda \equiv \lambda_0 U_0 + \lambda_1 U_1 = 0,$$
$$\beta_\lambda \equiv \lambda_0 V_0 + \lambda_1 V_1 = 0.$$

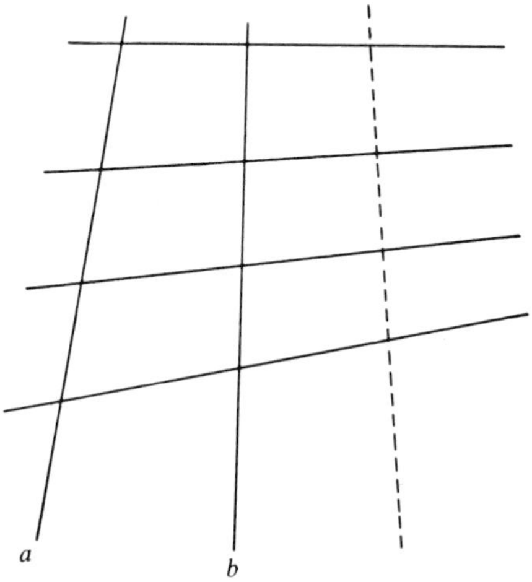

Fig. 81.1

Since a and b are skew, the four planes $U_0 = 0$, $U_1 = 0$, $V_0 = 0$, $V_1 = 0$ have no point in common, and must therefore be *linearly independent*. For any three planes have at least one point in common, and linear dependence would involve the four planes having at least one point in common. The planes α_λ, β_λ intersect in a line l_λ, which describes a regulus $\{l\}$ as $\lambda_0 : \lambda_1$ varies. Each line of $\{l\}$ arises from and corresponds to a unique value of $\lambda_0 : \lambda_1$. For suppose that the values (λ_0, λ_1) and (λ_0', λ_1'), where $\lambda_0 : \lambda_1 \neq \lambda_0' : \lambda_1'$, give the same line l_λ. Then the planes

$$\lambda_0 U_0 + \lambda_1 U_1 = 0,$$

and

$$\lambda_0' U_0 + \lambda_1' U_1 = 0,$$

both of which contain a, must intersect b in the same point. That is, they must represent the same plane. Hence there is a value k for which $U_0 \equiv k U_1$, whereas we assumed that $U_0 = 0$ and $U_1 = 0$ were distinct planes through a.

The line l_λ lies in the plane

$$\mu_0\alpha_\lambda + \mu_1\beta_\lambda \equiv \lambda_0(\mu_0 U_0 + \mu_1 V_0) + \lambda_1(\mu_0 U_1 + \mu_1 V_1) = 0,$$

for any μ_0, μ_1 which are not both zero. For fixed $\mu_0:\mu_1$, this plane describes, as $\lambda_0:\lambda_1$ varies, a pencil with axis g_μ given by the equations

$$\mu_0 U_0 + \mu_1 V_0 = 0 = \mu_0 U_1 + \mu_1 V_1.$$

This pencil, axis g_μ (Fig. 81.2), is projective with the pencils axes a and b respectively with which we began, and corresponding planes contain l_λ, that is they intersect in a line of $\{l\}$. The lines a, b with which we began are also particular cases of the lines g_μ, corresponding to $(\mu_0, \mu_1) = (1, 0)$ and $(\mu_0, \mu_1) = (0, 1)$ respectively.

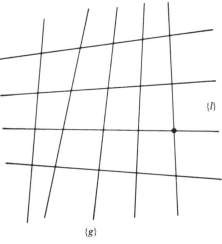

{l}

{g}

Fig. 81.2

As $\mu_0:\mu_1$ varies, the two planes $\mu_0 U_0 + \mu_1 V_0 = 0$, $\mu_0 U_1 + \mu_1 V_1 = 0$ describe projective pencils, corresponding planes intersecting in the lines g_μ. The axes of these projective pencils of planes are the lines given by the equations $U_0 = V_0 = 0$, and $U_1 = V_1 = 0$, and these lines do not intersect, since we know that the four planes involved are linearly independent, whereas four planes through a point are always dependent. Moreover the line g_μ lies in the plane $\mu_0\alpha_\lambda + \mu_1\beta_\lambda = 0$, which, as $\mu_0:\mu_1$ varies describes a pencil with axis l_λ, and all the pencils so obtained, corresponding to different values of $\lambda_0:\lambda_1$, are projective.

Thus *the lines g_μ form a second regulus $\{g\}$, and the relation between $\{l\}$ and $\{g\}$ is symmetrical.*

The lines of either regulus are the axes of projective pencils of planes, corresponding planes of which intersect in the lines of the other.

Two such reguli are said to be *complementary* reguli.

Both sets of lines lie on the surface

$$U_0V_1 - U_1V_0 = 0,$$

as we see immediately by eliminating $\lambda_0:\lambda_1$ from $\alpha_\lambda = \beta_\lambda = 0$, or $\mu_0:\mu_1$ from

$$\mu_0U_0 + \mu_1V_0 = 0 = \mu_0U_1 + \mu_1V_1.$$

This surface, which is of the second degree in X_0, X_1, X_2, X_3, is called *a quadric surface*.

We shall examine quadric surfaces from the algebraic point of view later (§84.1). Here we pursue some consequences of the projective generation of a quadric surface.

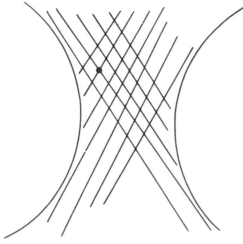

Fig. 81.3

We saw, in the projective generation of a conic, that the vertices of the two projective pencils whose intersections gave the conic can be replaced by any two points of the conic. In the projective generation of a regulus, we began with two skew lines a and b, and considered the intersection of corresponding planes of two pencils of planes, axes a and b respectively, which are in projective correspondence. These lines of intersection cut a and b in projective ranges. Our work shows that the lines a and b can be replaced by any two lines of the set $\{g\}$, and therefore the lines of intersection of corresponding planes, which generate the regulus $\{l\}$, intersect all lines of the regulus $\{g\}$.

Hence, *a line of a regulus intersects every line of the complementary regulus*. But *the lines of a regulus are mutually skew*. If two lines of $\{l\}$ intersected, the lines a and b would be coplanar, and they are skew, and if two lines of $\{g\}$ intersected, the lines of $\{l\}$ would be coplanar, which they are not.

The same quadric surface is generated by the set of lines $\{l\}$ and $\{g\}$ (Fig. 81.3). The surface is the locus of points on the set of lines $\{l\}$ and $\{g\}$. Hence, through every point of the surface there passes at least one line of $\{l\}$, and one line of $\{g\}$. There cannot be more, since the lines of $\{l\}$ are mutually skew to each other, as are the lines of $\{g\}$ to each other. Hence

Through every point of the quadric surface there passes one line of each of the complementary reguli $\{l\}$ and $\{g\}$.

81.2   Intersection of a plane and a quadric surface

We prove that any plane which does not contain a pair of intersecting lines l_λ, g_μ of the two complementary reguli meets the quadric surface in an irreducible conic.

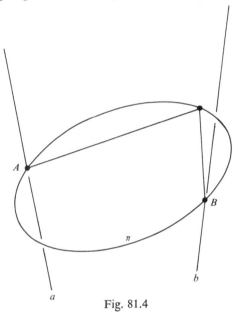

Fig. 81.4

Let π be any plane in S_3. If π contains a line g_0 of the regulus $\{g\}$, it also contains a line l_0 of the regulus $\{l\}$, since the plane π through g_0 is one of the planes in a projective pencil, axis g_0, by which the regulus $\{l\}$ can be generated, and the corresponding plane, axis any other g, intersects π in l_0.

Such a plane contains no other point of the quadric surface. For if P were such a point, the generator g_1 of $\{g\}$ which passes through P meets l_0, and lies in π, since it passes through two distinct points of π. It would then intersect g_0, and we know that distinct generators of the same regulus do not intersect.

If our plane π contained a line l_0 of the regulus $\{l\}$, the reasoning is similar to that already given, and π would contain a generator of each regulus, and no other point of the quadric surface. We now assume that π does not contain a generator of either regulus. Let it intersect a in A, and b in B (Fig. 81.4).

Since π contains neither a nor b, the projective pencils of planes, axes a and b, which generate the quadric, intersect π in projective pencils of lines, vertices A and B. Since AB is not a line of l, the plane π not containing such a line, the line AB does not correspond to the line BA. (If there were this correspondence, the plane of the pencil, axis a, through B would correspond to the plane of the pencil, axis b, through A, and the line AB would be the intersection of corresponding planes of the two pencils, that is, a line of $\{l\}$.) The intersection of corresponding rays of the pencils vertices A and B is therefore an irreducible conic, which passes through A and B. This constitutes the complete intersection of π with the quadric surface.

All this, of course, can be proved algebraically, but again, the fundamental ideas of

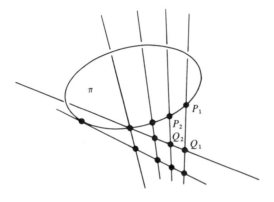

Fig. 81.5

projective generation enable us to prove these results very simply. We end this section with a remark on projectivities on the conic intersection of π and the quadric.

A pencil of planes intersects a given plane π in a pencil of lines which is projective with the pencil of planes (Fig. 81.5). The set of points $\{P\}$ on the conic intersection of the quadric and the plane π is projective, by definition, with the pencil of lines joining the points P to the point A on the conic. This pencil of lines is projective with the pencil of planes, axis a, and this pencil of planes is projective with the set of points $\{Q\}$ in which the generators l of the regulus $\{l\}$ through P intersect any given generator g of the set $\{g\}$.

We may interchange l and g in this discussion. It is now clear that we may talk of *a projectivity between generators of a regulus*, say $\{l\}$, on a quadric, and this is equivalent to a projectivity between the points of a conic in which any given plane π intersects the quadric, corresponding generators intersecting the conic in corresponding points; or to a projectivity between the points of any given generator of the complementary regulus $\{g\}$, the corresponding points being where the corresponding generators of $\{l\}$ intersect the given generator g.

The concept of a projectivity between generators of complementary reguli will be discussed in §83.1.

Finally, we note that if we take two sections of the same quadric, by planes π and π', the generators of either system of the quadric cut the conics in π and π' in projective point-sets (Fig. 81.6). The planes π and π' intersect in a line, and this will intersect the quadric in two points. The conic-intersections will therefore have two points in common, and these points are fixed in the projective correspondence between the points of the two conics, since only one generator of each regulus passes through a given point of the quadric.

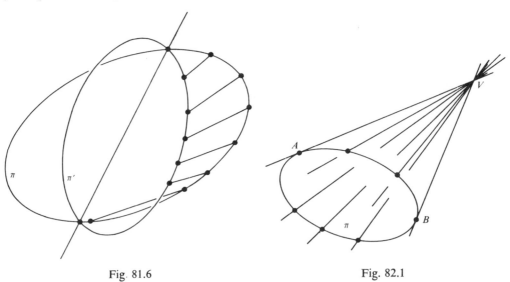

Fig. 81.6   Fig. 82.1

Exercises

81.1   The lines l and m are skew in S_3, and A, B, C are three points such that the planes lA, lB and lC are distinct, as are the planes mA, mB and mC. Prove that a unique quadric surface can be drawn to contain the lines l and m, and the points A, B and C.

81.2   Two conics S and S' lie in distinct planes π and π' in S_3, and meet the line of intersection of π and π' in the same two distinct points A and B. A projectivity is established between the points of S and S' which is such that the points A and B are fixed. Prove that the lines joining corresponding points of S and S' in this projectivity belong to a regulus.

81.3 S is a conic and l a line which does not lie in the plane of S, but meets S in a point P. A projectivity is established between the points of l and the points of S which is such that P is fixed. Prove that the joins of corresponding points of l and S in this projectivity form a regulus.

82.1   Quadric cones

We assumed that the lines a and b, used as the axes of projective pencils of planes to define a regulus, were skew lines in S_3. We now suppose that they intersect in a point V (Fig. 82.1). Then the line of intersection of corresponding planes of the projective pencils also passes through V. If π is any plane which does not contain V, and $\pi \cap a = A$, $\pi \cap b = B$, then the plane π intersects the projective pencils of planes through a and b in projective pencils of lines through A and B, and corresponding lines of the pencils intersect in a point which traces out a conic through A and B. The lines through V which are the intersections of corresponding planes are merely the joins of V to the points of this conic, and therefore describe *a quadric cone*, with V as vertex.

In this case there is only one regulus of lines on the quadric, the joins of V to the points of the conic section of the cone by the plane π. If the conic section is itself

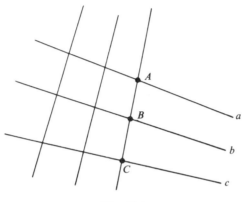

Fig. 83.1

reducible, splitting into two lines in π, the quadric cone becomes two planes through V, in which case every line in either plane lies on the quadric. In the most degenerate case the two lines may coincide, and then the two planes coincide.

The algebraic treatment which follows in §84.1 makes this very clear.

Conics, historically, were first considered as sections of a right circular cone, which is the reason for their name. We observe that the section of a quadric cone by any plane gives us a conic. If the plane passes through the point V itself, our conic is two lines, two generators of the quadric cone, and if the plane is a *tangent* plane, the two lines coincide. The concept of *tangent plane to a quadric* will be examined in §85.1.

83.1   Transversals of three skew lines

We began our consideration of the projective generation of a quadric surface by considering two skew lines a, b in S_3, and obtained a regulus of lines on the quadric, all skew to each other and to a and b. The lines a and b belong to this regulus. We now show that a regulus is completely determined by any three of its lines. Suppose that a, b and c are three mutually skew lines in S_3 (Fig. 83.1). From any point C on c there is a unique transversal CAB to the lines a and b, intersecting a in A and b in B. This line is the intersection of the planes $[Ca]$ and $[Cb]$. As the point C varies on c, the pencils of planes $[Ca]$, $[Cb]$ are projective, and therefore the set of transversals

ABC form a regulus. Hence, *the set of transversals of three skew lines in S_3 form a regulus.*

The complementary regulus contains the lines *a*, *b* and *c*. Since the set of transversals of *a*, *b* and *c* is uniquely defined, and form a regulus, and therefore the regulus complementary to this set of transversals is uniquely defined, and contains *a*, *b* and *c*, we can say:

A regulus is uniquely defined by any three lines of the regulus.

We know that a projectivity between two lines is uniquely determined by the assignment of three pairs of corresponding points on the lines. If the lines be skew in S_3

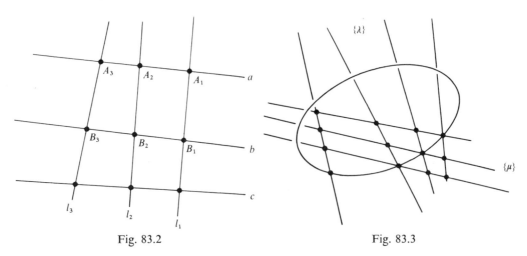

Fig. 83.2   Fig. 83.3

we now have a simple geometrical construction for this projectivity (Fig. 83.2). Let the skew lines be *a* and *b*, and suppose that A_1, A_2, A_3 on *a* correspond to B_1, B_2, B_3 respectively on *b*. The lines l_i ($i = 1, 2, 3$) joining A_i to B_i are necessarily skew to each other. If *c* is any transversal of the lines l_i distinct from *a* and *b*, then *c* is also skew to *a* and to *b*. The transversals drawn from points of *c* to *a* and *b* form a regulus, as we have seen above, and the lines l_i are members of this regulus. This regulus of lines traces out projective ranges on *a* and *b* in which A_i, B_i ($i = 1, 2, 3$) are corresponding points.

We first considered this kind of construct in §65.1, (*b*). We recall that a transversal from any point of *c* to *a* and *b* intersects a transversal from any point of l_3, say, to l_1 and l_2 (Fig. 83.3), and that with a suitable coordinate system the *equation* of the quadric surface may be taken as

$$X_1X_3 - X_0X_2 = 0.$$

This anticipated the result of §81.1. Any point on the surface has the coordinates $(1, \mu, \mu\lambda, \lambda)$. If we take $\lambda = \lambda_0$, a constant, the point $(1, \mu, \mu\lambda_0, \lambda_0)$ on the surface describes a generator, and as we vary λ_0 we obtain all the lines of one regulus. These are in one-to-one correspondence with the values of λ. If we take $\mu = \mu_0$, a constant, the point $(1, \mu_0, \mu_0\lambda, \lambda)$ describes a generator of the complementary regulus, and as we vary μ_0 we obtain all the generators of this complementary regulus. The point $(1, \mu_0, \mu_0\lambda_0, \lambda_0)$ of the surface is the intersection of the λ_0-generator with the μ_0-generator. We know that through every point of the surface there passes one and only one generator of each system.

Our statements about projective correspondences between lines of the same regulus, made at the end of §81.2, are very evident geometrically. But a projective correspondence between lines of complementary reguli is handled more easily, perhaps, algebraically. We know that a projective correspondence between the parameters λ, μ is of the form (§78.2)

$$A\lambda\mu + B\lambda + C\mu + D = 0.$$

The points of intersection of corresponding lines of the two reguli satisfy the equation

$$AX_2 + BX_3 + CX_1 + DX_0 = 0,$$

and we have therefore proved:

Theorem I *Corresponding lines in a projective correspondence between the two reguli on a quadric surface intersect in the points of a plane section, that is, a conic.*

The converse theorem is evidently true, and we have:

Theorem II *There is a projective correspondence between the members of the two reguli on a quadric which intersect at the points of a given plane section of the surface.*

We shall obtain even another geometrical insight into the behavior of generators on a quadric surface when we consider *stereographic projection* (§87.1).

A line in S_3 intersects a quadric surface in two points. This is an immediate deduction from the equation of a quadric surface being of the second degree. Or we can argue that the projective pencils of planes, axes a and b, which generate the surface, cut projective ranges on the line, and it is at fixed points of this projectivity that the quadric itself cuts the line. But a projectivity on a line has two fixed points at most, if it is not the identity. If it is the identity, the line lies on (or in) the surface. We can say, then, that if a line intersects a quadric surface in three or more points, it is a generator of that surface.

We conclude this section with a theorem we assumed in Exercise 36.8.

Theorem III *Given four mutually skew lines in S_3, there are in general two lines which intersect all four lines.*

Proof. If the lines are a, b, c and d, we consider the regulus of lines which intersect a, b and c. This regulus fills a quadric surface, and d intersects this surface in two points. Through each of these points there passes a generator of the regulus of lines which intersect a, b and c.

Exercises

83.1 l and m are given skew lines in S_3, and the transversal to l and m from a variable point P meets them in R and S. P' is the point on this transversal which is the harmonic conjugate of P with respect to R and S. Show that as P moves on a line n which is skew to l and m, the point P' also moves on a line skew to the three lines l, m and n.

83.2 $\{P_r\}$ and $\{Q_r\}$ are projective ranges on the skew lines l and m in S_3, and A is a given point on the line P_1Q_1. Prove that the planes AP_rQ_r all pass through a given line.

83.3   The lines l, m and n in S_3 all pass through a point O, but are not coplanar. The ranges $\{P\}$, $\{Q\}$ and $\{R\}$ on l, m and n are projective in pairs, and O is a fixed point for the projectivity between each pair of lines. Prove that the plane PQR contains a fixed line.

83.4   The mutually skew lines a, b, c and d in S_3 meet the lines l_i ($i = 1, 2, 3, 4$) in the points A_i, B_i, C_i, D_i. Prove that if the points A_1, B_2, C_3 and D_4 are coplanar, then the points A_2, B_1, C_4 and D_3 are coplanar, and so are the points A_3, B_4, C_1, D_2 and the points A_4, B_3, C_2, D_1; and the ranges (A_1, B_1, C_1, D_1) and (A_1, A_2, A_3, A_4) are projective.

84.1   The quadratic form in four variables

We obtained the equation of a quadric surface in §81.1 in the form

$$U_0V_1 - U_1V_0 = 0,$$

and this can be written in the form

$$(U_0 + V_1)^2 - (U_0 - V_1)^2 + (U_1 - V_0)^2 - (U_1 + V_0)^2 = 0.$$

Since U_0, U_1, V_0, V_1 are assumed to be linearly independent, so are the linear forms $U_0 + V_1$, $U_0 - V_1$, $U_1 - V_0$, $U_1 + V_0$. By a change of coordinate system we may therefore write the equation of the quadric as

$$X_0{}^2 - X_1{}^2 + X_2{}^2 - X_3{}^2 = 0,$$

or we may work with the form $U_0V_1 - U_1V_0 = 0$ and take $X_0 = U_0$, $X_1 = V_0$, $X_2 = V_1$, $X_3 = U_1$, and obtain

$$X_0X_2 - X_1X_3 = 0,$$

the form obtained in §65.1, (b).

Any quadratic form in four variables may be written as X^TAX, where $\mathbf{A}$ is a symmetric 4×4 matrix. Under coordinate transformations $X = \mathbf{P}Y$, this quadratic form becomes

$$Y^T(\mathbf{P}^T\mathbf{AP})Y,$$

and the symmetric matrix of the derived quadratic form is $\mathbf{P}^T\mathbf{AP}$. Under these transformations the set of matrices $\mathbf{P}^T\mathbf{AP}$ are said to be *congruent* to $\mathbf{A}$, and this relation is an equivalence relation. The proof is independent of the order of the square symmetric matrix $\mathbf{A}$. Over the complex field the rank of $\mathbf{A}$ is the one invariant of the matrices $\mathbf{P}^T\mathbf{AP}$, and $\mathbf{P}$ can be chosen so that this matrix is in diagonal form. Hence, since the rank of any matrix which does not consist entirely of zero entries can be 1, 2, 3, 4, the matrix being 4×4, the form $\mathbf{X}^T \mathbf{A} \mathbf{X}$ may be transformed into one of the following:

$$Y_0{}^2, \quad Y_0{}^2 + Y_1{}^2, \qquad Y_0{}^2 + Y_1{}^2 + Y_2{}^2, \qquad Y_0{}^2 + Y_1{}^2 + Y_2{}^2 + Y_3{}^2.$$

The equation $Y_0{}^2 = 0$ represents a plane, taken twice, that is, a pair of identical planes. The equation $Y_0{}^2 + Y_1{}^2 = 0$ can be factorized as $(Y_0 + iY_1)(Y_0 - iY_1) = 0$, and therefore represents *a pair of distinct planes*. The equation

$$Y_0{}^2 + Y_1{}^2 + Y_2{}^2 = 0$$

represents the *cone* joining the point $(0, 0, 0, 1)$ to the points of the *conic* given by

$$Y_0{}^2 + Y_1{}^2 + Y_2{}^2 = 0 = Y_3.$$

This is *an irreducible conic* (§77.2), and the quadric cone is *an irreducible quadric cone*. Finally,

$$Y_0{}^2 + Y_1{}^2 + Y_2{}^2 + Y_3{}^2 = 0$$

may be written as

$$(Y_0 + iY_1)(Y_0 - iY_1) - (iY_2 + Y_3)(iY_2 - Y_3) = 0,$$

and this is the original form for an irreducible quadric surface.

If we restricted ourselves to *real* coordinate transformations of a real quadratic form, the various diagonal forms obtained would contain both positive and negative terms. We should obtain $Y_0{}^2 + Y_1{}^2 = 0$, which contains no real points, whereas $Y_0{}^2 - Y_1{}^2 = 0$ represents two real distinct planes, if our original quadratic form were of rank two. Geometrically, the first case is distinct from the second, and both cases cannot arise from the same form under real transformations. In fact there is a theorem known as *Sylvester's Law of Inertia* which states that the difference between the number of positive signs and negative signs, when the form is expressed as a sum or difference of squares, is, besides the rank, another invariant of a real quadratic form under real coordinate transformations.

Thus, for $r = 3$, there are two types of real quadric cone, given by

$$Y_0{}^2 + Y_1{}^2 + Y_2{}^2 = 0, \quad \text{and} \quad -Y_0{}^2 + Y_1{}^2 + Y_2{}^2 = 0,$$

the former containing no real points. The distinct types of real irreducible quadric surface are of three types, given by the equations:

$$Y_0{}^2 + Y_1{}^2 + Y_2{}^2 + Y_3{}^2 = 0, \qquad -Y_0{}^2 + Y_1{}^2 + Y_2{}^2 + Y_3{}^2 = 0,$$

and

$$-Y_0{}^2 - Y_1{}^2 + Y_2{}^2 + Y_3{}^2 = 0.$$

The first of these types contains no real points, the second contains no real generators, and the third is of the same form as the quadric we have been investigating, and contains real generators of two systems, given by

$$\left.\begin{array}{rl} Y_2 - Y_0 &= \lambda(Y_1 - Y_3) \\ \lambda(Y_2 + Y_0) &= \quad Y_1 + Y_3, \end{array}\right\}$$

and

$$\left.\begin{array}{rl} Y_2 - Y_0 &= \mu(Y_1 + Y_3), \\ \mu(Y_2 + Y_0) &= \quad Y_1 - Y_3. \end{array}\right\}$$

Some Exercises on real quadrics will be given following §88.1.

85.1   The polar of a point

As we remarked in §78.3, the algebra which leads to the idea of the polar of a point with respect to a conic is the same as that which leads to the idea of the polar of a point with respect to a quadric, but, of course, we now have a surface, not a curve to deal with. Let P and Q be distinct points in S_3, and let $P\lambda + Q\mu$ lie on the quadric $X^T AX = 0$. Then

$$(\lambda P + \mu Q)^T A(\lambda P + \mu Q) = 0,$$

and on expansion, we obtain the quadratic equation in λ, μ:

$$\lambda^2 P^T AP + \lambda\mu(P^T AQ + Q^T AP) + \mu^2 Q^T AQ = 0, \tag{1}$$

which, since A is symmetric, may be written:

$$\lambda^2(P^T AP) + 2\lambda\mu(P^T AQ) + \mu^2(Q^T AQ) = 0.$$

Once again, the condition

$$P^T AQ = 0$$

expresses the condition that *the intersections of the line PQ with the quadric are harmonic conjugates with respect to P and Q.*

In this case the points P and Q are said to be *conjugate points* with regard to the quadric.

If we keep P fixed, the locus of points $Q = X$ which are conjugate with regard to P is given by the equation

$$P^T AX = 0.$$

This is the equation of a plane, *the polar plane* of P with regard to the quadric.

If we suppose that this plane is given, and that A is non-singular, we can find a unique point P such that the given plane is its polar. This point is called *the pole* of the given plane with regard to the quadric. If the given plane be given by the equation

$UX = 0$, then if this is to be identical with the polar plane of a point P, we must have

$$P^TA = U,$$
$$P^T = UA^{-1},$$

or
$$P = A^{-1}U^T.$$

We verify immediately that

$$(UA^{-1})A(X) = 0$$

gives the equation $UX = 0$, of the given plane.

Returning to equation (1) above, we first note that *if P lies on the quadric, then $P^TAP = 0$, so that P lies in its own polar plane*. That the converse is also true is equally evident.

Assuming now that P lies on the quadric, equation (1) becomes

$$2\lambda\mu(P^TAQ) + \mu^2Q^TAQ = 0.$$

If Q is any point, distinct from P, in the polar plane of P, then $P^TAQ = 0$, and the equation reduces further to

$$\mu^2(Q^TAQ) = 0.$$

If Q does *not* lie on the quadric, the coefficient $Q^TAQ \neq 0$, and therefore $\mu^2 = 0$. Hence

If P lies on the quadric, and Q lies in the polar plane of P, but not on the quadric, the line PQ has no further intersections with the quadric other than P.

Conversely, let P lie on the quadric, and suppose that Q is such that the equation (1) reduces to the form $\mu^2(Q^TAQ) = 0$, with $Q^TAQ \neq 0$, so that PQ only intersects the quadric at P. Then we must have $P^TAQ = 0$, which means that Q lies in the polar plane of P.

Such a line PQ, where P is on the quadric, and PQ only intersects the quadric at P, is called *a tangent line to the quadric at P*.

These tangent lines lie in the polar plane of P, and this plane is called *the tangent plane to the quadric at P*.

It should be noted that the polar plane always exists if and only if the rank of A is the maximum, that is, A is non-singular. If A is of rank $r < 4$, there are solutions to the equations in $P = (p_0, p_1, p_2, p_3)$:

$$P^TA = 0.$$

If P' is a solution, the equation of the polar of P' is $0 . X = 0$, which can only be interpreted as *the whole of S_3*. If the rank of A is equal to three, the equations

$$p_0a_{0i} + p_1a_{1i} + p_2a_{2i} + p_3a_{3i} = 0 \qquad (i = 0, 1, 2, 3)$$

have a unique solution, that is, there is a unique point which has as its polar plane the equation $0 . X = 0$. This point is *the vertex of the cone represented by the equation*

$X^T A X = 0$. The equation of the polar of a point is, of course, invariant under co-ordinate transformations, and if we take the equation of a cone as $X_0{}^2 + X_1{}^2 + X_2{}^2 = 0$, the polar plane of $P = (p_0, p_1, p_2, p_3)$ with regard to this quadric is

$$p_0 X_0 + p_1 X_1 + p_2 X_2 = 0,$$

and if, and only if $P = (0, 0, 0, 1)$, this plane is of the form

$$0 \cdot X_0 + 0 \cdot X_1 + 0 \cdot X_2 + 0 \cdot X_3 = 0.$$

It will be noticed that the polar plane of P with regard to the cone passes through the vertex, which is $(0, 0, 0, 1)$. For quadrics we have the theorem: *if the polar of P contains Q, the polar of Q contains P.* Since the polar planes of all points P of S_3 pass through the vertex of the cone, the polar plane of the vertex of the cone must contain all points of S_3, which it can hardly do if it is a proper plane! Hence $0 \cdot X = 0$ is the only possibility open to it!

Some other possibilities, when the rank of A is less than four, are mentioned in the Exercises.

Once more returning to Equation (1), we assume once more that P lies on the quadric, so that $P^T A P = 0$, but now we also assume that Q both lies in the polar plane of P and on the quadric. Then $P^T A Q = 0$, and $Q^T A Q = 0$, and equation (1) becomes of the form

$$0 \cdot \lambda^2 + 2\lambda\mu(0) + 0 \cdot \mu^2 = 0,$$

so that every value of $\lambda:\mu$ satisfies the equation. But the equation gives the values of $\lambda:\mu$ corresponding to intersections of the line PQ with the quadric. We deduce that every point of the line PQ lies on the quadric, so that PQ is a generator of the quadric. We know that two generators of the quadric pass through P. We deduce that:

The tangent plane at P cuts the quadric in the two generators of the quadric which pass through P.

85.2   Polar lines with regard to a quadric

In the theory of conics, we saw that as a point P moves along a line, the polar of P with regard to a given conic always passes through a point, the pole of the line with regard to the conic. In S_3, let us see what happens as a point P moves along the line joining the points R, S. Any point on the line can be taken as $P = \lambda R + \mu S$, and the polar plane of this point with regard to the quadric $X^T A X = 0$ is

$$(\lambda R + \mu S)^T A X = \lambda R^T A X + \mu S^T A X = 0.$$

This is a pencil of planes, the axis of the pencil being given by the planes

$$R^T A X = 0 = S^T A X.$$

Hence *the polar planes of the points on a line l always pass through a line l', and the polar planes of points on l' always pass through the line l.*

These lines are called *polar lines* with regard to the quadric. We note that if P, P' are any pair of points on l, l' respectively, the intersection of the line PP' with the quadric gives a pair of points harmonic to P and P'. We came across polar lines in §37.2, where we saw that in our mapping of the circles of E_2 onto points of E_3, polar lines, with regard to the quadric which represents zero circles, arose from orthogonal systems of coaxal circles. We prove:

Theorem I *A necessary and sufficient condition that l intersect its polar line l' with regard to a quadric is that l be a tangent to the quadric.*

Proof. Let l intersect l', its polar line, in the point X. Then the polar plane of X, which lies on both l and l', contains both lines l' and l. Since the plane determined by the intersecting lines l and l' contains the intersection of l and l', which is X, we have a point X which lies in its polar plane. As we saw, X then lies on the quadric, and its polar plane is the tangent plane to the quadric at X. The polar lines l and l' lie in this plane, and pass through X. They are therefore both tangent lines at X to the quadric.

If we are given that l is a tangent to the quadric, at a point X, say, the tangent plane at X is the polar of X, and contains l'. But it also contains l. Hence l intersects l'. By the above, the point of intersection lies on the quadric, and its polar plane is the tangent plane at the point, and this contains both l and l'. Since the pole of a plane is uniquely defined, this point must be the point X.

We leave the property that l and l', at their intersection, form a harmonic pencil with the generators through their point of intersection, as Exercise 85.2.

85.3 Conjugate lines

If l and l' are polar lines with regard to a quadric, and m is any line which intersects l', then the polar line m' of m lies in the polar plane of the point $l' \cap m$, and this plane contains l. Hence m' intersects l. The two lines l and m are such that *each intersects the polar line of the other*. Two such lines are said to be *conjugate lines* with regard to the quadric. A line which intersects its own polar line is called *a self-conjugate line*. We have seen that such lines are tangent lines to the quadric.

85.4 Correlations

We have discussed at some length projective transformations in which a point of S_n is mapped onto a point (§70.1), and briefly considered the map of a point onto a hyperplane, when considering the Principle of Duality in §66.2. We now investigate correlations, in which a point of S_n is mapped onto a hyperplane, in more detail.

A projective *correlation* is a mapping of points of S_n onto hyperplanes of S_n, given by the equations

$$V^T = \mathbf{A} Y, \tag{1}$$

where Y is the column-vector which represents a point of S_n, $\mathbf{A}$ is a non-singular $(n + 1) \times (n + 1)$ matrix, and V is the row-vector which represents the coordinates of a hyperplane in S_n. It will be remembered that we write the coordinates $[V^0, V^1, \ldots, V^n]$ of the hyperplane

$$V^0 X_0 + V^1 X_1 + \ldots + V^n X_n = 0$$

as a row-vector, so that the equation of the hyperplane is simply

$$VX = 0,$$

X being a column-vector. Since the equation (1) produces a column-vector on the right-hand side of the equation, it gives the transpose of the row-vector V. When we considered the Principle of Duality, in §66.2, we took the matrix $\mathbf{A}$ to be the unity matrix.

The inverse mapping to (1) is given by

$$Y = \mathbf{A}^{-1} V^T. \tag{2}$$

The mapping is therefore one-to-one.

Now suppose that the point Y lies in a hyperplane U, so that $UY = 0$. Then

$$U\mathbf{A}^{-1} V^T = 0.$$

If we take the transpose of this equation, which is $V(\mathbf{A}^{-1})^T U^T = 0$, and compare it with the equation $VX = 0$ of a hyperplane, we deduce that the hyperplane V always contains the point

$$X = (\mathbf{A}^{-1})^T U^T. \tag{3}$$

Conversely, if the hyperplane V passes through a fixed point X, so that $VX = 0$, then we have $(\mathbf{A}Y)^T X = Y^T \mathbf{A}^T X = 0$, which we may also write

$$X^T \mathbf{A} Y = 0, \tag{4}$$

so that the point Y always lies in the hyperplane U, where

$$U = X^T \mathbf{A}. \tag{5}$$

The equations (3) and (5), which are also those of a correlation, are said to be the correlation *associated with* the original correlation. The relation is a symmetrical one, and the connecting link is:

If the point Y lies in the hyperplane U, then the hyperplane V contains the point X, and conversely.

We may call a correlation and its associated correlation a *complete correlation*, and can then say:

A complete correlation maps every point Y of S_n bijectively onto a hyperplane V, and every hyperplane U bijectively onto a point X, so that incidence relations between points and hyperplanes are preserved.

We may show, as we did in §66.2, where our matrix $\mathbf{A}$ was the unity matrix, that a correlation maps a subspace S_m of S_n onto a subspace S_{n-m-1}, and that the relation $S_p \subseteq S_q$ become the relation $S_{n-p-1} \supseteq S_{n-q-1}$.

A correlation, like a projective transformation, is uniquely determined as soon as the maps of $n + 2$ points are assigned, where no $n + 1$ points lie in a hyperplane, and no $n + 1$ of the image hyperplanes pass through a point. The proof is the same as that in §71.1.

It is clear that the composition of two correlations is a projective transformation, and that the product of a correlation and a projective transformation is a correlation. The totality of projective collineations and correlations in S_n form a group.

85.5 Polarities and null-polarities

Two correlations are identical if and only if their matrices differ by a non-zero multiplicative factor. That is, the correlation $V^T = \mathbf{A}Y$ is identical with the correlation $V^T = \mathbf{B}Y$ if and only if $\mathbf{B} = k\mathbf{A}$, where $k \neq 0$.

We consider *the involutory correlations*, that is those correlations for which a correlation and its associated correlation are identical. Since, if a correlation is given by $V^T = \mathbf{A}Y$, the associated correlation is given by $U = X^T\mathbf{A}$, or $U^T = (X^T\mathbf{A})^T = \mathbf{A}^TX$, a necessary and sufficient condition for a correlation to be identical with its associated correlation is

$$\mathbf{A}^T = k\mathbf{A} \qquad (k \neq 0).$$

But since $(\mathbf{A}^T)^T = k\mathbf{A}^T = k^2\mathbf{A} = \mathbf{A}$, and $\mathbf{A}$ has at least one non-zero element, this leads to $k^2 = 1$, and $k = +1$, or $k = -1$. In the case $k = 1$, $\mathbf{A}^T = \mathbf{A}$, and the matrix $\mathbf{A}$ is *symmetric*, and in the case $\mathbf{A} = -1$, the matrix $\mathbf{A}$ is *anti-symmetric*, or *skew-symmetric*.

When $\mathbf{A}$ is symmetric the correlation is called *a polarity*, and the equation $V^T = \mathbf{A}Y$ produces *the polar hyperplane* of the point Y with respect to the quadric $X^T\mathbf{A}X = 0$ (§85.1). We note that this quadric can be called *the incidence locus* of the correlation, *the locus of points Y which lie in the hyperplanes associated with them in the polarity*. For if Y lies in the hyperplane $V = (\mathbf{A}Y)^T$, then $VY = Y^T\mathbf{A}^TY = Y^T\mathbf{A}Y = 0$, since $\mathbf{A}^T = \mathbf{A}$.

In the case when $\mathbf{A}^T = -\mathbf{A}$, so that $\mathbf{A}$ is skew-symmetric, the correlation is called *a Null-System*. These share certain properties, as we shall see, with polarities, but also have some distinctive features.

In the first instance, $\mathbf{A}$ is only non-singular, which was our initial assumption, when n is an odd integer. For, from $\mathbf{A}^T = -\mathbf{A}$, we have

$$\det(\mathbf{A}^T) = \det(\mathbf{A}) = \det(-\mathbf{A}) = (-1)^{n+1}\det(\mathbf{A}),$$

since $\mathbf{A}$ is an $(n + 1) \times (n + 1)$ matrix, and if $n + 1$ is odd, since our field is that of the complex numbers, we have $2\det(\mathbf{A}) = 0$, and so $\det(\mathbf{A}) = 0$.

Hence, by our definition, there are no non-singular null-systems in S_2, and the first non-singular null-system which arises geometrically lies in S_3.

For a polarity the hyperplane V given by $V^T = \mathbf{A}Y$ contains the point Y if and only if Y lies on a certain quadric locus given by $X^T \mathbf{A} X = 0$. In a null-system *the null-hyperplane V given by $V^T = \mathbf{A}Y$ always contains the point Y.* In fact $Y^T \mathbf{A} Y$ is a (1×1) matrix, and is unaltered if we take the transpose. But the transpose is $(Y^T \mathbf{A} Y)^T = Y^T \mathbf{A}^T Y = Y^T(-\mathbf{A})Y = -Y^T \mathbf{A}Y$, and so

$$Y^T \mathbf{A} Y = -Y^T \mathbf{A} Y,$$

from which we deduce that $Y^T \mathbf{A} Y = 0$.

Hence, also, there is no incidence-locus for a null-system, every point of S_n lying on the incidence-locus.

For a polarity we have had the fundamental result: if the polar hyperplane V of Y contains X, then the polar hyperplane of X contains Y. In fact, if $VX = 0$, then $Y^T \mathbf{A} X = 0$, which leads to $X^T \mathbf{A} Y = 0$, and $X^T \mathbf{A}$ is the polar hyperplane of X.

This result also holds for null-polarities, the argument being the same, except that $(Y^T \mathbf{A} X)^T = X^T(-\mathbf{A})Y = -X^T \mathbf{A} Y = 0$, from which $X^T \mathbf{A} Y = 0$. Let us write this relation between what we may call *conjugate points*, X, Y, in terms of the elements a_{ik} of the matrix $\mathbf{A}$.

It is

$$\sum_i \sum_k a_{ik} x_i y_k = 0 \qquad (i, k = 0, 1, \ldots, n).$$

But since $\mathbf{A}^T = -\mathbf{A}$,

$$a_{ii} = 0 \qquad (i = 0, 1, \ldots, n),$$

and

$$a_{ik} = -a_{ki} \qquad (i \neq k).$$

Hence we may write the relation between conjugate points X, Y in the form:

$$\sum_{i<k} a_{ik}(x_i y_k - x_k y_i) = 0 \qquad (i, k = 0, 1, \ldots, n). \qquad (1)$$

If we regard X as a given point, the points Y which satisfy this relation lie in the null-hyperplane of X, which passes through X.

Let us see what happens in projective space of three dimensions, where $n = 3$. If the point X moves on a line l, its null-plane passes through a line l', *the polar line of l.* If, however, we consider the line l which joins X to any point Y in the null-plane of X, *this line is self-polar in the null-polarity we are considering.* This is because the null-plane of X contains Y, and therefore the null-plane of Y contains X, and so the polar line of the line joining X to Y is the line joining Y to X, the null-plane of Y containing Y also, as well as X.

Hence, through any point X, there is a pencil of lines, lying in the null-plane of X, which are self-polar in the null-polarity. Any given plane has a uniquely defined pole, and all self-polar lines which lie in the plane pass through the pole.

The relation (1) above which holds for two points X, Y on a self-polar line also holds for any two points $X' = \lambda X + \mu Y$, $Y' = \lambda' X + \mu' Y$. In fact, if we evaluate $x'_i y'_k - x'_k y'_i$, we find that it is equal to $(\lambda \mu' - \lambda' \mu)(x_i y_k - x_k y_i)$. If we write $p_{ik} = x_i y_k - x_k y_i$, we may write (1) in the form

$$\sum_{i<k} a_{ik} p_{ik} = 0 \qquad (i, k = 0, 1, 2, 3). \tag{2}$$

The p_{ik} are called the *Pluecker* or *Grassmann coordinates* of the line joining X to Y (see Appendix II, p. 436). Lines satisfying equation (2) are said to form *a linear complex of lines*, and properties of a linear complex are indicated in the following Exercises.

Finally, a word about the origin of the term *null-polarity*. In investigating the equilibrium of a set of forces acting on a rigid body in real Euclidean space of three dimensions, it was found that there were lines about which the forces had zero moment. These are the *null-lines*, as they were called, of the system. The null-lines through any point P form a pencil through P, lying in a plane through P, and this plane is the null-plane of P (see §88.3, p. 397).

A null-polarity provides a very ready method for producing Moebius tetrahedra, that is tetrahedra which are inscribed and escribed to each other. If A, B, C and D be the vertices of a tetrad of independent points in S_3, and π_A, π_B, π_C and π_D be the respective null-planes of the four points with regard to any given null-polarity, the four planes pass, respectively, through A, B, C and D, and the intersection D' of the three planes π_A, π_B and π_C is the pole of the plane ABC, and therefore lies in the plane ABC, and so on. All that is needed to set up a null-polarity is a skew-symmetric matrix $\mathbf{M}$ (compare §65.2), and $\pi_A = (\mathbf{M}A)^T$, and so on.

85.6   The dual of a quadric

We have discussed the quadric surface in S_3 produced by the transversals of three skew lines in §83.1, and the quadratic form in four variables in §84.1. We saw that the quadratic form of rank 4 gives the equation of the quadric surface obtained geometrically, and that the two concepts are equivalent under non-singular transformations of coordinates. We may write the quadratic form in matrix notation as $X^T A X$, where A is a non-singular 4×4 matrix, and we may consider quadrics in S_n defined by the equation $X^T A X = 0$, where A is now a square matrix of order $n + 1$. This is an evident generalization of the algebraic approach. We shall not generalize the geometric approach here.

We remarked that the definition of a quadric surface in S_3 which considers the locus of lines which join the corresponding points of projective ranges on two skew lines in S_3 shows that a quadric surface in S_3 is a self-dual construct. The dual of the construction is the intersection of corresponding planes of two projective pencils of planes. This intersection is a set of lines which intersect the axes of the pencils of planes in projective ranges, and so we are back at the original construct. The dual of a line in

S_3 is a line, of course. If we begin with points on the quadric surface satisfying the equation $X^T A X = 0$, the dual of a point in S_3 is a plane, and we must consider the equation $U^T B U = 0$ satisfied by planes, where $\mathbf{B}$ is some 4×4 non-singular matrix. We show that these planes touch a quadric surface, as we now expect.

For a quadric in S_n the relation $P^T A Q = 0$ is that satisfied by conjugate points P and Q, the quadric being given by the equation $X^T A X = 0$. The algebra given in §85.1 is independent of the value of n. If the matrix $\mathbf{A}$ is non-singular, and the problem is to find the point P such that the locus of harmonic conjugates of P with respect to the intersections with the quadric of lines through P is a given hyperplane $U X = 0$, then P, the *pole* of the hyperplane, is uniquely determined. In fact, as in §85.1, we have $P = \mathbf{A}^{-1} U^T$.

If P lies on the quadric, the polar hyperplane is the *tangent* hyperplane to the quadric at the point P, and so the tangent hyperplanes U satisfy the condition

$$U(\mathbf{A}^{-1})^T U^T = 0.$$

We may write this

$$U^T \mathbf{B} U = 0,$$

where $\mathbf{B} = \mathbf{A}^{-1}$, and since if $\mathbf{A}$ is non-singular, so is $\mathbf{A}^{-1}$, and if $\mathbf{A}$ is symmetric, so is $\mathbf{A}^{-1}$, the construct formed by tangent hyperplanes to a non-singular quadric in S_n is essentially the same, algebraically, as the dual construct which arises from a given non-singular quadric, considered as a point-locus, in S_n. We may clearly go from the locus of tangent hyperplanes back to the point-locus, and so also derive, from $U^T \mathbf{B} U = 0$, a point-locus which is a quadric, such that the hyperplanes U which satisfy $U^T \mathbf{B} U = 0$ are simply the *tangent* hyperplanes at points of the quadric point-locus.

Since this reasoning depends on the matrix $\mathbf{A}$ being non-singular, we cannot expect matters to be so simple when $\mathbf{A}$ is singular. We note, in the first instance, that the polar hyperplanes of all points P in S_n contain a fixed subspace of S_n. The equation of the polar hyperplane of P is $P^T A X = 0$, and if $\mathbf{A}$ has rank r less than $n + 1$, the set of linear homogeneous equations $\mathbf{A} X = 0$ has solutions X which fill a linear space of $n - r$ dimensions. The polar hyperplanes of all points P in S_n therefore contain this S_{n-r}. Looked at from another point of view, the polar hyperplane of a point P in this S_{n-r} can be written

$$X^T(AP) = 0,$$

and *every coefficient of X_i is zero*. For such points *there is no polar hyperplane*. Since also such points P satisfy the equation $X^T A X = 0$, *the points of S_{n-r} lie on the quadric*. This phenomenon arises most evidently in S_3, when $r = 3$ instead of 4, and our quadric surface is a *cone*, therefore. We take its equation in the form $X_0{}^2 + X_1{}^2 + X_2{}^2 = 0$, and note that the polar plane of any point (Y_0, Y_1, Y_2, Y_3) is $Y_0 X_0 + Y_1 X_1 + Y_2 X_2 = 0$, which plane always contains the point $(0, 0, 0, 1)$, which point lies on the cone. This point is the vertex of the cone, and has no tangent plane.

Returning to polar hyperplanes of points in S_n, if $\mathbf{A}$ is singular, we cannot expect to take any hyperplane, and find its pole, since polar hyperplanes all pass through a certain S_{n-r}, and an arbitrary hyperplane will not do this.

But we can, of course, interpret geometrically the equation $U^T B U = 0$, where $\mathbf{B}$ is singular, and U is a hyperplane. A case of special interest, which arises in §94.2, is when $n = 3$ and $\mathbf{B}$ is of rank 3. We may take the equation as $(U^{(0)})^2 + (U^{(1)})^2 + (U^{(2)})^2 = 0$, satisfied by a plane $U^{(0)} X_0 + U^{(1)} X_1 + U^{(2)} X_2 + U^{(3)} X_3 = 0$ in S_3. This plane cuts the plane $X_3 = 0$ in the line $U^{(0)} X_0 + U^{(1)} X_1 + U^{(2)} X_2 = 0$, and the algebraic condition we are investigating says that this line is a tangent to the conic $X_0^2 + X_1^2 + X_2^2 = 0$ in the plane $X_3 = 0$. Hence, *any plane in S_3 which intersects $X_3 = 0$ in a tangent to the conic* satisfies the algebraic condition. We may regard the conic as a degenerate quadric, if we wish. Through any line of S_3 there pass two tangent planes to this degenerate quadric, so it shares some of the properties of a respectable quadric surface. (Exercise 85.15.)

Exercises

85.1 Show that if the line l is not a tangent line to a given quadric, two tangent planes to the quadric pass through l, and l', the polar line of l, is the line joining their points of contact.

85.2 Prove that if the line l is tangent to a quadric, and l' is its polar line, then the lines l and l' are harmonic conjugates with respect to the two generators of the quadric which pass through the point $l \cap l'$.

85.3 The lines l and l' are polar lines with regard to a quadric, and l intersects the quadric in A and B, and l' intersects the quadric in A' and B'. Prove that the lines AA', BB', AB' and $A'B$ are generators of the quadric. (The ground field is the complex field.)

85.4 If a line m intersects a pair of lines l, l' which are polar lines with regard to a quadric, prove that the generators of either regulus on the quadric which intersect m are harmonic conjugates with respect to the generators of the same regulus which intersect l.

85.5 In S_3 suppose that the tetrad of reference is such that the vertex D has the plane DAB as its null-plane, and the vertex C has the plane CAB as its null-plane, so that DC and AB are polar lines. Show that the null-plane of (p, q, r, s) with respect to this null-polarity has the form

$$c'(sZ - rT) + c(pY - qX) = 0.$$

85.6 Prove that two distinct lines in S_3 which are polar lines with respect to a given null-polarity do not intersect.

85.7 Prove that any line which meets two polar lines in a given null-polarity belongs to the linear complex defined by the null-polarity (is a null-line).

85.8 If l and m are two null-lines of a given null-polarity, and the polar plane of a point P on l (which contains l) intersects m in P', prove that the ranges described by P and P' are projective. If the point Q on l is mapped on the point Q' on m in this projectivity, show that the lines PQ' and $P'Q$ are polar lines.

85.9 Deduce from the results of the two preceding Exercises that if the range $(A, B, C, D, \ldots)$ on a line l is projective with the range $(A', B', C', D', \ldots)$ on a skew line m in S_3, and we draw transversals from a given point O to the lines AB' and $A'B$, to AC' and $A'C$, to BC' and $B'C$, to AD' and $A'D, \ldots$, then all these transversals lie in a plane, and are the lines of a linear complex in which AB is the polar line of $A'B'$.

85.10   Prove the first part of the preceding Exercise by projecting the lines *l* and *m* from *O* onto a plane, and using the Pappus Theorem (Thm III, §63.1).

85.11   'A null-polarity, by its definition, is a self-dual construct.' Justify this statement, and give the dual of Exercise 85.9 above.

85.12   If a line intersects two null lines of a given null-polarity, show that the polar line also intersects the two null-lines. If we have four mutually skew lines, which have only two common transversals, and the four mutually skew lines are all null-lines in a given null-polarity, show that the two transversals must be polar lines in the null-polarity, and that any line meeting the two transversals is a null-line of the null-polarity.

85.13   Assuming that we can find a linear complex which contains five mutually skew lines in S_3, prove the following theorem: five mutually skew lines in S_3, omitting one of them in turn, produce five sets of four mutually skew lines. Each set of four lines has two transversals. From a given point *O* the transversals are drawn to these sets of two lines. Prove that the five single lines thus produced lie in a plane.

85.14   Show that a given linear complex of lines in S_3, as defined by Equation (2) of the preceding Section, defines a null-polarity. (If the join of the points *X* and *Y* gives a null-line, consider the locus of points *Y* which arise from a given point *X*.)

85.15   Show both algebraically and geometrically that through a given line in S_3 there pass two planes $UX = 0$ which satisfy the equation

$$(U^{(0)})^2 + (U^{(1)})^2 + (U^{(2)})^2 = 0.$$

(See the end of §85.6.)

86.1   Another look at Pascal's theorem

There is a considerable corpus of theorems in the geometry of two dimensions which may be deduced from the geometry of three dimensions. To cite one extreme instance, Desargues' Theorem in the plane is true if the plane can be immersed in a three-dimensional projective space (Exercise 86.1). But there are also many theorems on conics, and in particular circles, which may be deduced from constructs in three dimensions. Sometimes a theorem becomes much clearer when it is regarded as a section, or a projection from a construct in a space of higher dimension. (Chapter IV, §72.1 and elsewhere.)

Here we consider once more the Pascal Theorem for six points on a conic (Corollary, Thm III, §78.2), and show how the theorem itself, and also a theorem due to Steiner, on the concurrence of Pascal lines, may be deduced from the theory of quadrics already developed.

We suppose that we have an irreducible conic in S_2, which we shall call *S*, with six distinct points on it, which we denote by the letters *I, J, K, L, M, N* (Fig. 86.1). We wish to show that the three points of intersection of the lines

$$IM, JL; JN, KM; IN, LK;$$

that is

$$(IM) \cap (JL), \quad (JN) \cap (KM), \quad (IN) \cap (LK)$$

are collinear. These points can either be considered as the intersections of the *cross-joins* of the points I, J, K paired with L, M, N, or as the intersections of *opposite sides* of the hexagon formed by the six points, the successive vertices running in the order: I, M, K, L, J, N. Different permutations of the vertices may produce different hexagons. (§86.2.) In fact, there are sixty permutations which produce different hexagons, and thus sixty Pascal lines associated with the six given points. We shall discover some properties of these sixty Pascal lines.

We first construct an irreducible quadric S^* which contains the conic S as a plane

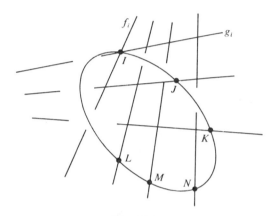

Fig. 86.1

section. This may be done in a variety of ways, one being indicated in Exercise 81.3. We give here a direct algebraic proof. We choose a coordinate system in the plane so that the equation of the conic S is

$$X_0 X_2 - (X_1)^2 = 0,$$

this being the equation of an irreducible conic (Thm I, §77.2), and then take any point not in the plane of the conic as the fourth vertex of our tetrad of reference for S_3. We now consider the quadric S^* given by the equation:

$$X_0 X_2 - X_1{}^2 + X_3{}^2 = 0.$$

The section of this quadric by the plane $X_3 = 0$ is the conic S, and we now show that our quadric S^* is irreducible, and of maximum rank, so that it is not a cone.

If we write the equation of S^* in the form:

$$(X_0 + X_2)^2/4 - (X_0 - X_2)^2/4 - X_1{}^2 + X_3{}^2 = 0,$$

an evident transformation of coordinates produces the equation

$$Y_0{}^2 - Y_1{}^2 - Y_2{}^2 + Y_3{}^2 = 0,$$

which is of maximum rank four. The quadric S^* contains real generators, since we may write the original equation in the form:

$$X_0X_2 = X_1{}^2 - X_3{}^2 = (X_1 - X_3)(X_1 + X_3),$$

and the two systems of generators are given by the equations:

$$\begin{aligned} X_0\lambda &= X_1 - X_3, \\ X_2 &= (X_1 + X_3)\lambda \end{aligned} \Big\}$$

and

$$\begin{aligned} X_0\mu &= X_1 + X_3, \\ X_2 &= (X_1 - X_3)\mu \end{aligned} \Big\}$$

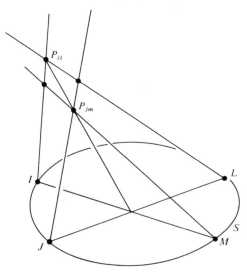

Fig. 86.2

Let the two reguli on S^* be $\{f\}$ and $\{g\}$, where $f_a, f_b \ldots$ and $g_a, g_b, \ldots$ are individual generators of each system. Let P_{mn} be the point of intersection of f_m and g_n.

We name the generators by their intersections with the conic S. Thus, f_i and g_i are the generators which pass through the point I on S, and so P_{ij} is the intersection of the generator f_i through I and the generator g_j through J on S.

Pascal's Theorem for the conic S is now very evident. The point $(IM) \cap (JL)$ is the intersection with the plane of S of the line of intersection of the planes $[f_ig_m]$ and $[f_jg_l]$. (Note that we are constantly making use of the theorem that generators of complementary reguli intersect each other, and are thus coplanar.) The point $P_{il} = f_i \cap g_l$ lies in both planes, $[f_ig_m]$ and $[f_jg_l]$, and the point $P_{jm} = f_j \cap g_m$ also lies in both planes. Hence *the point $(IM) \cap (JL)$ is the point where the line $P_{il}P_{jm}$ meets the plane of S.*

Similarly, $(JN) \cap (KM)$ is the point where the line $P_{jm}P_{kn}$ meets the plane of S, and finally $(IN) \cap (KL)$ is the point where the line $P_{il}P_{kn}$ meets the plane of S (Fig. 86.2).

The three points of the Pascal Theorem therefore lie on the intersection of the plane $P_{il}P_{jm}P_{kn}$ with the plane of S, and thus we have proved the Pascal Theorem, since planes in S_3 intersect in a line.

This method of viewing the Pascal Theorem also leads to a theorem on the various Pascal lines associated with six given points on a conic S.

86.2   Steiner's theorem on Pascal lines

If we are given six points in S_2, we may choose one of them and then imagine a connected path which begins and ends at this point and traverses each of the six points, each point being joined to only two other of the six points. We have then described a hexagon, whose vertices are at the six points, and this hexagon can be designated by giving the order in which the points occur. A side of the hexagon is obtained by picking any two consecutive points, and the *opposite* side of the hexagon is obtained by omitting the next vertex, if we are proceeding in serial order, and choosing the next two vertices, which give the opposite side. Thus, if our vertices be

$$I, M, K, L, J, N,$$

in this order, the side opposite to the line (IM) is (LJ), the side opposite to (MK) is (JN), and the side opposite to (KL) is (NI). The Pascal line of this hexagon, if the six points are on a conic, is the line joining the points

$$(IM) \cap (LJ), (MK) \cap (JN), (KL) \cap (NI).$$

We obtain the same hexagon, and therefore the same Pascal line, from the points

$$N, J, L, K, M, I$$

taken in reverse order, or if we shift the points along, as if they were beads on a circular necklace, so that the sequences:

$$(M, K, L, J, N, I), (K, L, J, N, I, M), (L, J, N, I, M, K),$$

$$(J, N, I, M, K, L), (N, I, M, K, L, J)$$

give the same hexagon as the original one. Since there are $6! = 720$ permutations of six points in serial order, and each hexagon arises from $6 \times 2 = 12$ permutations, the number of distinct hexagons is $720/12 = 60$. We therefore expect not more than 60 distinct Pascal lines from six unordered points on a conic. To show that if the six points are "general" points on a conic, the 60 lines are distinct, this merits discussion, and is left as an Exercise.

We take the Pascal line of six points as being given by cross-joins, and so a Pascal line is given by a symbol, (IJK, LMN), where the I, J, K are respectively paired with the L, M and N, and the Pascal line contains the points:

$$(IM) \cap (JL), (JN) \cap (KM) \text{ and } (IN) \cap (LK).$$

Theorem I   *The lines associated with the symbols*

$$(IJK, LMN), (IJK, MNL), (IJK, NLM)$$

pass through a point, and the lines associated with the symbols

$$(IJK, LNM), (IJK, NML), (IJK, MLN)$$

pass through a conjugate point with respect to the conic S.

Proof. We prove this theorem on the basis of two lemmas. Using the notation of §86.1, we have:

Lemma I. The lines $P_{jn}P_{km}$, $P_{kl}P_{in}$ and $P_{im}P_{jl}$ are concurrent. (See Fig. 86.3.)
    We indicate the *f*-generators through the points *I, J, K* and the *g*-generators through

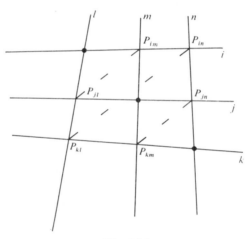

Fig. 86.3

the points *L, M* and *N*. The three lines we wish to prove concurrent are marked with dashes. If these three lines are *concurrent*, their respective polar lines, with regard to *S**, are *coplanar*, lying in the polar plane of the point of concurrency. Conversely, if three lines are coplanar, their respective polar lines all pass through the pole of the plane. We prove that the polar lines of the three designated lines are coplanar.
    We recall that the polar plane of a point on the quadric *S** is the tangent plane at the point, and this is the plane which contains the two generators, one from each regulus, which pass through the point. The polar line of $P_{jn}P_{km}$ is the intersection of the polar plane of P_{jn} with the polar plane of P_{km}. The polar plane of P_{jn} is the plane $[f_j g_n]$, by the remark above, and the polar plane of P_{km} is the plane $[f_k g_m]$, so that the polar line of $P_{jn}P_{km}$ is the line joining the points $f_k \cap g_n$ and $f_j \cap g_m$, that is the line $P_{kn}P_{jm}$.

Similarly, the polar line of $P_{kl}P_{in}$ is $P_{il}P_{kn}$, and that of $P_{im}P_{jl}$ is $P_{jm}P_{il}$, and so our three polar lines are the joins, in pairs, of the three points

$$P_{il}, \; P_{jm}, \; P_{kn}$$

and therefore lie in a plane (see also Exercise 86.2).

Corollary. Interchanging j and k, the lines $P_{kn}P_{jm}, \; P_{jl}P_{in}, \; P_{im}P_{kl}$ are concurrent. Interchanging l and n, the lines $P_{jl}P_{km}, \; P_{kn}P_{il}, \; P_{im}P_{jn}$ are concurrent.

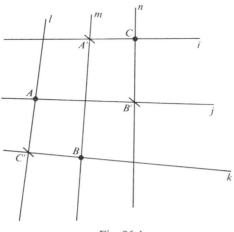

Fig. 86.4

We now prove:

Lemma II. The planes $P_{il}P_{jm}P_{kn}, \; P_{im}P_{jn}P_{kl}$ and $P_{in}P_{jl}P_{km}$ meet in a line p, the planes $P_{il}P_{jn}P_{km}, \; P_{in}P_{jm}P_{kl}$ and $P_{im}P_{jl}P_{kn}$ meet in a line p', and $p, \; p'$ are polar lines with regard to the quadric S^*.

In Fig. 86.4, the plane $P_{im}P_{jn}P_{kl}$ is the plane $A'B'C'$, and the plane $P_{in}P_{jl}P_{km}$ is the plane CAB. We proved that AA', BB' and CC' are concurrent, in Lemma I. Therefore AB intersects $A'B'$, and AC intersects $A'C'$, and these two points of intersection define the line of intersection of the planes ABC, $A'B'C'$.

Now $AB = P_{jl}P_{km}$, and $A'B' = P_{im}P_{jn}$, and by the Corollary to Lemma I the line $P_{kn}P_{il}$ passes through the point $AB \cap A'B'$. Again,

$$CA = P_{in}P_{jl}, \text{ and } C'A' = P_{kl}P_{im},$$

and, again by the Corollary, the line $P_{kn}P_{jm}$ passes through $CA \cap C'A'$. Hence the plane $P_{il}P_{jm}P_{kn}$ contains the line of intersection of the planes $P_{im}P_{jn}P_{kl}$ and $P_{in}P_{jl}P_{km}$. Call this line p.

By the proof of Lemma I, the plane $P_{il}P_{jm}P_{kn}$ is the polar plane of the point of concurrence of the lines $P_{jl}P_{im}$, $P_{km}P_{jn}$ and $P_{in}P_{kl}$. If we now interchange j and k, we have the following results:

The planes $P_{il}P_{km}P_{jn}$, $P_{im}P_{kn}P_{lj}$ and $P_{in}P_{kl}P_{jm}$ meet in a line. Call this line p'. Also the plane $P_{il}P_{km}P_{jn}$ is the polar plane of the point of concurrence of the lines $P_{kl}P_{im}$, $P_{jm}P_{kn}$ and $P_{in}P_{jl}$.

The line $P_{kl}P_{im}$ lies in the plane $P_{im}P_{jn}P_{kl}$, the line $P_{jm}P_{kn}$ lies in the plane $P_{il}P_{jm}P_{kn}$ and the line $P_{in}P_{jl}$ lies in the plane $P_{in}P_{jl}P_{km}$. These three planes all pass through the line p. Therefore the three lines designated intersect in a point of p. But the point of concurrence is the pole of the plane $P_{il}P_{km}P_{jn}$. This plane contains p'. Hence we have found that the polar plane of a point of p contains p'. Similarly, the pole of the plane $P_{im}P_{kn}P_{jl}$, which contains p', lies on p, and the pole of the plane $P_{in}P_{kl}P_{jm}$, which contains p', lies on p, and therefore p and p' are polar lines with regard to the quadric S.

Thm I now follows immediately, since the section of polar lines by a plane, that of S, gives points conjugate with regard to S, and we recall that the Pascal line (IJK, LMN) is the intersection of the plane of S with the plane $P_{il}P_{jm}P_{kn}$, and so on.

Exercises

86.1 ABC, $A'B'C'$ are two triangles in a plane π, and AA', BB', CC' pass through a point O. Let l be a line through O which does not lie in the plane π (so that π lies in an S_3), and choose two distinct points V, V' on l, distinct from O. Show that VA, $V'A'$ intersect, in A^*, say, that VB, $V'B'$ intersect, in B^*, say, and that VC, $V'C'$ intersect, in C^*, say. Show that A^*, B^*, C^* do not lie in π, and are distinct from V and V'. Show also that A^*, B^* and C^* are not collinear. Finally, show that the points $BC \cap B^*C^*$, $CA \cap C^*A^*$ and $AB \cap A^*B^*$ are collinear, lying on the line of intersection of the planes $A^*B^*C^*$ and π. Applying a similar argument to the triangles $A'B'C'$ and $A^*B^*C^*$, deduce that the points $BC \cap B'C'$, $CA \cap C'A'$ and $AB \cap A'B'$ are collinear, which is the Desargues Theorem.

86.2 Give another proof of Lemma I of §86.2, that the lines $P_{jn}P_{km}$, $P_{kl}P_{in}$ and $P_{im}P_{jl}$ are concurrent by observing that these lines are the lines of intersection of the planes $[f_i g_l]$, $[f_j g_m]$ and $[f_k g_n]$, taken in pairs, and three planes have a common point of intersection. (You should investigate what happens if they intersect in a line, just as you should check, in the proof of the Lemma, that the points P_{il}, P_{jm} and P_{kn} do not lie on a line.)

86.3 A point V which does not lie on a given quadric is joined to the f-generators and the g-generators. Show that all these planes touch a quadric cone, and therefore intersect any plane which does not contain V in the tangents to a conic, a line-conic.

86.4 From the preceding Exercise we deduce that the join of V to the six generators $f_i, f_j, f_k, g_l, g_m, g_n$ on the quadric S^* produces six planes which intersect a plane in six tangents to a conic S. Show that the result of Exercise 86.2 indicates Brianchon's Theorem, that if we have a hexagon escribed to a conic, the joins of pairs of opposite vertices are concurrent.

86.5 In §86.1 we constructed an irreducible quadric S^* to contain a given conic S as a plane section, with a view to proving Pascal's Theorem. Show what dual procedures are necessary to prove Brianchon's Theorem for a given line-conic, generated by tangents.

86.6 Can you prove that six distinct points on a given conic S determine 60 distinct Pascal lines? (This is not easy, and drawing is hardly the answer. For some investigations into the subject, see Pedoe, **13**.)

87.1   Stereographic projection

We have already seen that the coordinates of a point on a quadric can be expressed as rational functions of two independent parameters, namely: (§83.1)

$$(X_0, X_1, X_2, X_3) = (1, \mu, \mu\lambda, \lambda),$$

the curves $\lambda = $ constant and $\mu = $ constant being generators of the complementary reguli on the surface. For convenience we rename the coordinates, so that this becomes:

$$(Y_0, Y_1, Y_2, Y_3) = (1, \lambda, \mu, \lambda\mu).$$

We may now consider $(1, \lambda, \mu)$ as coordinates (X_0, X_1, X_2) in a plane π (Fig. 87.1),

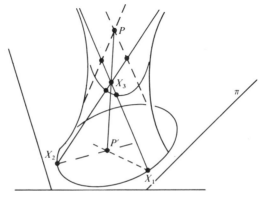

Fig. 87.1

and see that the point $(1, \lambda, \mu, 0)$ of the plane $X_3 = 0$ is mapped on the point $(1, \lambda, \mu, \lambda\mu)$ of the quadric, and conversely, the geometric derivation of the points, one from the other, being effected thus:

The join of $P = (1, \lambda, \mu, \lambda\mu)$ on the quadric to $P' = (1, \lambda, \mu, 0)$ in the plane $X_0 X_1 X_2$ passes through the point $X_3 = (0, 0, 0, 1)$ of the tetrad of reference.

The point X_3 is a point of the quadric, as we see at once by going into homogeneous parameters, and taking the point on the quadric as

$$(\lambda_0\mu_0, \lambda\mu_0, \mu\lambda_0, \lambda\mu)$$

and taking $\lambda_0 = \mu_0 = 0$. Or, as a short-cut, we can take $\lambda^{-1} = \mu^{-1} = 0$ in the non-homogeneous representation.

This mapping of the points of a quadric onto the points of a plane, by projecting the points of the quadric by means of lines through a fixed point on the quadric, is called *stereographic projection*. It is familiar when the quadric is a sphere, as we shall see in §88.1.

We note that the generator $\lambda^{-1} = 0$ is the locus

$$(0, 1, 0, \mu) = (0, 1, 0, 0) + \mu(0, 0, 0, 1),$$

which is the line $X_3 X_1$, and the generator $\mu^{-1} = 0$ is the locus

$$(0, 0, 1, \lambda) = (0, 0, 1, 0) + \lambda(0, 0, 0, 1),$$

which is the line $X_3 X_2$.

This is evident if we find where the plane $X_0 = 0$ intersects the quadric, which has the equation $X_0 X_3 - X_1 X_2 = 0$.

Any plane section of the quadric is a conic, and therefore projects from X_3 onto the plane $X_0 X_1 X_2$ as a conic. Since the plane section intersects $X_3 X_1$ and $X_3 X_2$, and these points of intersection are points of the quadric, the projected conic passes through the points X_1 and X_2.

To see this algebraically, let the plane be

$$U^{(0)} Y_0 + U^{(1)} Y_1 + U^{(2)} Y_2 + U^{(3)} Y_3 = 0.$$

At points on the quadric we have:

$$U^{(0)} + U^{(1)}\lambda + U^{(2)}\mu + U^{(3)}\lambda\mu = 0,$$

and since $\lambda = X_1/X_0$, $\mu = X_2/X_0$, in the plane $X_3 = 0$, the equation of the projected conic is

$$U^{(0)} X_0{}^2 + U^{(1)} X_1 X_0 + U^{(2)} X_2 X_0 + U^{(3)} X_1 X_2 = 0.$$

This is the equation of a conic which passes through the points $(0, 1, 0)$ and $(0, 0, 1)$, but not through the point $(1, 0, 0)$, unless $U^{(0)} = 0$.

The generators $\lambda = $ constant of the quadric project into $X_1/X_0 = $ constant, that is, into lines $X_1 - k X_0 = 0$, which are lines through X_2, and the generators $\mu = $ constant of the quadric project into $X_2/X_0 = $ constant, that is into lines $X_2 - k X_0 = 0$, which are lines in the plane through X_1. Hence:

The representation of the reguli on the quadric is by means of pencils of lines in the plane.

This is evident geometrically. The vertices of these pencils are special points in the representation. We see from the geometric interpretation of the projection that the whole generator $X_3 X_1$ is mapped onto the point X_1, and the whole generator $X_3 X_2$ is mapped onto the point X_2. Here the general point-point mapping breaks down, and we have two line-point special cases.

Since only two generators of the quadric pass through the vertex of projection X_3, we do not expect any other line on the quadric to map onto a point of the plane $X_3 = 0$. But what is the map of the vertex X_3 itself?

We cannot expect to find this directly, since the geometric mapping breaks down if we try to join the vertex X_3 to the point X_3 on the quadric. We look again at the conic projection of a plane section:

$$U^{(0)}X_0{}^2 + U^{(1)}X_1X_0 + U^{(2)}X_2X_0 + U^{(3)}X_1X_2 = 0.$$

The matrix of this quadratic form is:

$$\begin{bmatrix} U^{(0)} & U^{(1)}/2 & U^{(2)}/2 \\ U^{(1)}/2 & 0 & U^{(3)}/2 \\ U^{(2)}/2 & U^{(3)}/2 & 0 \end{bmatrix},$$

and the determinant is $U^{(3)}(U^{(1)}U^{(2)} - U^{(0)}U^{(3)})/4$. A necessary and sufficient condition that the projected conic be *reducible* is that this determinant be zero, for then the rank of the matrix is less than three. If $U^{(3)} = 0$, the plane section of the quadric passes through X_3, and the projected conic becomes

$$X_0(U^{(0)}X_0 + U^{(1)}X_1 + U^{(2)}X_2) = 0,$$

which is a reducible conic, consisting of the line $X_0 = 0$, that is the line X_1X_2, and another line, which is the intersection of the plane through X_3 with the plane $X_0X_1X_2$. This last intersection is, of course, the projection from X_3 of the points of the conic section cut on the quadric by the plane through X_3.

It seems to be indicated that we should accept that the map of the vertex of projection X_3 is the line $X_0 = 0$, which is the line X_1X_2. We can make this suggestion even more plausible by noting, if we are thinking metrically, that points on the quadric *near* to X_3 lie in the tangent plane to the quadric at X_3, and this is the plane $X_3X_1X_2$, so that these near points project from X_3 into the points of the line X_1X_2. More specifically, and less metrically, we can show that all curves on the quadric which pass through X_3 and touch a given line through X_3, which naturally must lie in the tangent plane at X_3 to the quadric, project into curves in the plane $X_0X_1X_2$ which pass through a given point on the line X_1X_2. Hence *directions* in the tangent plane at X_3 and through the point X_3 correspond in the mapping to *points* on the line X_1X_2, and we can therefore say that *the neighbourhood of the point X_3 is mapped onto the line X_1X_2*.

We can attack the problem by algebraic methods also. We write down the equations of the mapping as a set of homogeneous equations, and accept all solutions of the equations which arise when the coordinates of special points of the object variety (the quadric) are substituted. This is the modern algebraic viewpoint (Hodge and Pedoe, Vol. II, p. 246, (8)).

Among the equations of the mapping are those which arise by comparison of the equation of a plane and that of the projection of its intersection with the quadric. These are

$$Y_0: Y_1: Y_2: Y_3 = X_0{}^2: X_0X_1: X_0X_2: X_1X_2,$$

and we write these as the set of equations homogeneous in both sets of variables:

$$Y_0 X_0 X_1 - Y_1 X_0{}^2 = 0,$$
$$\cdots \qquad \cdots$$
$$Y_2 X_1 X_2 - Y_3 X_0 X_2 = 0,$$
$$\cdots \qquad \cdots$$

without any attempt to find a basis for all the equations of the mapping (as above, pp. 247–8). If we now substitute $(Y_0, Y_1, Y_2, Y_3) = (0, 0, 0, 1)$ in these equations, we obtain, as the equations for the map of X_3,

$$X_0{}^2 = X_0 X_1 = X_0 X_2 = 0,$$

which give the solution $X_0 = 0$, which is the equation of the line $X_1 X_2$.

In any investigation of curves on a quadric by means of stereographic projection, the irregularities of the mapping are naturally of the utmost importance, and must be taken into account, if we are not to be led into erroneous conclusions. By *irregularities* we mean the points where the one-to-one relationship between object and image-point breaks down.

Exercises

87.1   In the Section above we found that the projection from X_3 of a plane section of the quadric is reducible if and only if $U^{(3)}(U^{(1)}U^{(2)} - U^{(0)}U^{(3)}) = 0$, and we investigated the meaning of $U^{(3)} = 0$. Show that $U^{(1)}U^{(2)} - U^{(0)}U^{(3)} = 0$ if and only if the plane $(UX) = 0$ is a tangent plane to the quadric, in which case the intersection with the quadric is a pair of generators through the point of contact, and the projection from X_3 consists of a pair of lines, one through X_1, and the other through X_2.

87.2   Investigate *the quadratic transformation* between two planes given by the equations

$$Y_0 : Y_1 : Y_2 = X_1 X_2 : X_0 X_2 : X_0 X_1,$$

showing that a line in the one plane corresponds to a conic through the vertices of the triad of reference in the other plane. Show that the conics with a given tangent at the vertex X_0 in the one plane are mapped onto a pencil of lines in the other plane, with vertex at an assigned point on the line $Y_1 Y_2$, the relation between the tangents through X_0 and the points on $Y_1 Y_2$ being projective. What is the map of the point X_0?

88.1   The sphere as a quadric

There is an evident extension of the methods used, in §80.1, for investigating conics in the real Euclidean plane, to quadrics in real Euclidean space of three dimensions, E_3. We map the point (T_0, T_1, T_2, T_3) of S_3 onto the point (X_0, X_1, X_2) of E_3, with the relationship

$$X_0 = T_0 T_3{}^{-1}, \; X_1 = T_1 T_3{}^{-1}, \; X_2 = T_2 T_3{}^{-1},$$

which has meaning as long as $T_3 \neq 0$. This time, moving off *to infinity* in E_3 corresponds to moving towards the *plane* $T_3 = 0$ in S_3, and it is easy to show that parallel lines in

E_3 correspond to sets of lines in S_3, which all pass through a definite point in the plane $T_3 = 0$.

When spheres in E_3 are mapped onto quadrics in S_3 by the above transformation, which corresponds to making the equation

$$X_0{}^2 + X_1{}^2 + X_2{}^2 + 2gX_0 + 2fX_1 + 2hX_2 + c = 0$$

of a sphere in E_3 *homogeneous*, in the form:

$$T_0{}^2 + T_1{}^2 + T_2{}^2 + 2gT_0T_3 + 2fT_1T_3 + 2hT_2T_3 + cT_3{}^2 = 0,$$

in S_3, we see that *all spheres* intersect $T_3 = 0$ in *the same conic*,

$$T_0{}^2 + T_1{}^2 + T_2{}^2 = 0 = T_3.$$

By abuse of language, this is often called *the imaginary circle at infinity*, since a sphere intersects any plane in a circle.

We may show that metrical properties in E_3 may be interpreted as projective properties in S_3 with regard to this conic in the plane $T_3 = 0$. Let us call this special conic Ω. *Lines in E_3 which are perpendicular intersect the plane at infinity in points which are conjugate with regard to Ω.* Once again, we are using a verbal shorthand to cover the transformations of coordinates which occur before we can make such a statement. Let the lines in E_3 be

$$\frac{X_0 - a_0}{l} = \frac{X_1 - a_1}{m} = \frac{X_2 - a_2}{n}, \quad \frac{X_0 - b_0}{p} = \frac{X_1 - b_1}{q} = \frac{X_2 - b_2}{r},$$

so that one line has direction-ratios $l:m:n$, and the other $p:q:r$, and since they are perpendicular,

$$lp + mq + nr = 0.$$

After transformation into lines in S_3, these lines intersect the plane $T_3 = 0$ in the points $(l, m, n, 0)$ and $(p, q, r, 0)$ respectively, and the condition that these points be conjugate with regard to Ω is precisely $lp + mq + nr = 0$.

The theorem that *any two skew lines in E_3 have a unique common perpendicular*, that is, a unique transversal which intersects each line at right angles, is an immediate deduction from the projective notion of orthogonality. Let the lines intersect the plane at infinity in the points L and M respectively, and let N be the pole of the line LM with regard to Ω. Then N is conjugate to L, and also to M. Through N there is a unique transversal to the two given skew lines (§65.1, (a)), and this transversal is perpendicular to each of the given lines.

Some further applications will be given in the Exercises.

We note that a plane section of a quadric which contains Ω intersects Ω in the circular points at infinity in that plane. Hence the plane section of a sphere is a circle. We now examine the process of *stereographic projection from a point V on a sphere*. We project onto the tangent plane at the point on the sphere which is diametrically opposite to V. We call this point O.

A *small circle* on a sphere is a section by a plane which does not pass through the center of the sphere (Fig. 88.1). We show that *the projection from V of small circles on the sphere are circles in the plane onto which they are projected.* This theorem, surprising at first impact, is an immediate consequence of our discussion on stereographic projection in §87.1.

The sphere, over the field of the complex numbers, is, of course, a quadric of maximum rank, and possesses two reguli of generators. The generators of the sphere which pass through V lie in the tangent plane at V, which is parallel to the tangent plane at O, this being diametrically opposite to V. The generators at V intersect the generators of opposite systems at O, and these intersections lie at infinity, and being on the sphere,

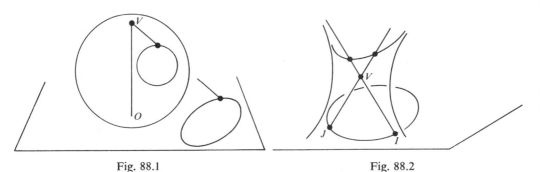

Fig. 88.1   Fig. 88.2

they also lie on Ω (Fig. 88.2). The intersections are therefore the circular points at infinity in each of the tangent planes, that at V, and that at O. Since any plane section of the sphere intersects both VI and VJ, where I and J are the circular points, the plane section projects from V into a conic in the plane OIJ which passes through I and J. Such a conic is a circle.

Hence, any plane section of the sphere which does not pass through V (this was tacitly assumed) projects into a circle in the tangent plane at O, or any parallel plane.

This theorem can be proved by using inversion with regard to spheres, and the method is indicated in Exercise 88.1.

Stereographic projection has already appeared, although we did not call it this, in §37.1, in connection with the representation of circles in the OXY-plane by points in E_3. We found that the circle

$$X^2 + Y^2 - 2pX - 2qY + r = 0,$$

which is represented by the point (p, q, r) of E_3, can also be represented by the plane section of the paraboloid of revolution

$$\Omega \equiv X^2 + Y^2 - Z = 0 \tag{1}$$

cut by the plane

$$Z - 2pX - 2qY + r = 0.$$

This plane section is obtained by erecting parallels to the OZ-axis at all points of the given circle in the OXY-plane, and finding their points of intersection with Ω. These points lie in the designated plane. Conversely, a plane $aX + bY + cZ + d = 0$ cuts Ω in a conic which projects down onto the OXY-plane into the circle

$$c(X^2 + Y^2) + aX + bY + d = 0.$$

This process is stereographic projection, the vertex of projection being at infinity on the OZ-axis.

We may easily find a projective transformation of the paraboloid which maps it onto a sphere, and converts the one stereographic mapping into the other. Writing the equation of the paraboloid as

$$X^2 + Y^2 - ZT = 0,$$

introducing homogeneous coordinates, we can write the equation

$$X^2 + Y^2 - \tfrac{1}{4}(Z + T)^2 + \tfrac{1}{4}(Z - T)^2 = 0.$$

If we now put $x = X, y = Y, z = (Z - T)/2, t = (Z + T)/2$, the equation becomes

$$x^2 + y^2 + z^2 - t^2 = 0, \tag{2}$$

which, on returning to non-homogeneous coordinates, is the equation of a sphere. The point at infinity on the OZ-axis is $(0, 0, 1, 0)$ in homogeneous coordinates, and maps onto the point $(0, 0, 1, 1)$, which is on the sphere, and the plane $Z = 0$ maps onto the plane $z + t = 0$, which is the tangent plane to the sphere at the point $(0, 0, -1, 1)$,

88.2 Tetracyclic coordinates

We return to the mapping of circles onto points of a space of three dimensions (§36.1). If we write the equation of a circle in the form:

$$a_0(X^2 + Y^2) + 2a_1X + 2a_2Y + a_3 = 0,$$

and map this circle onto the point (a_0, a_1, a_2, a_3) of projective space S_3, the circles of zero radius, the point-circles of the plane, correspond to the points of S_3 which satisfy the equation

$$X_0X_3 - X_1{}^2 - X_2{}^2 = 0. \tag{1}$$

There is a one-to-one correspondence between the points of this quadric and the points of the plane, if we exclude the points of the quadric which satisfy the equation $X_0 = 0$. Such a system of homogeneous coordinates for points of the plane, (a_0, a_1, a_2, a_3), where the relation (1) is satisfied, is called *a tetracyclic system* of coordinates. The same name is used for any non-singular transformation of the coordinates. We investigate such a transformation.

We now write circles of the plane in the form:

$$b_0(X^2 + Y^2 + 1) + b_1(X^2 + Y^2 - 1) + 2b_2 X + 2b_3 Y = 0,$$

using the four linearly independent circles

$$X^2 + Y^2 + 1 = 0, \ X^2 + Y^2 - 1 = 0, \ X = 0, \text{ and } Y = 0.$$

These circles, in terms of which any given circle may be expressed, have already appeared in §40.1, in connexion with the development of an algebra of circles. We note that the first is a virtual circle, and the third and fourth lines, while any two of the four are orthogonal. In terms of the coordinates (b_0, b_1, b_2, b_3), which also give a one-to-one mapping of the circles of the plane onto an S_3, point-circles are mapped onto the quadric

$$- X_0{}^2 + X_1{}^2 + X_2{}^2 + X_3{}^2 = 0 \tag{2}$$

and the relation between the quadrics given by (1) and (2) is exactly the same as that between the quadrics (1) and (2) in the preceding Section. As we expect, if two of the circles in the second representation are orthogonal, their representative points are conjugate with respect to the quadric given by (2). The four fundamental circles of the mapping, given by the points $(1, 0, 0, 0)$, $(0, 1, 0, 0)$, $(0, 0, 1, 0)$ and $(0, 0, 0, 1)$, form a self-polar simplex of reference for the quadric given by (2), being mutually orthogonal circles. This is also shown by the form of the equation (2).

88.3   The Euclidean group in three dimensions

We have already seen how Euclidean geometry manifests itself when our Euclidean space E_3 is imbedded in a real projective space S_3. (§88.1). Once again, a special quadric appears, this time the 'imaginary circle at infinity', with respect to which metrical properties may be interpreted in a projective fashion. If we regard Euclidean geometry as the study of the invariants of a certain group of transformations of a space, we begin with a real projective space, and consider in the first place the affine group of transformations which map a given plane, say $T_3 = 0$, onto itself. If we change to non-homogeneous coordinates, that is to affine coordinates, given by

$$X_0 = T_0 T_3^{-1}, \ X_1 = T_1 T_3^{-1}, \ X_2 = T_2 T_3^{-1},$$

our affine transformations are of the form

$$Y_i = a_{i0} X_0 + a_{i1} X_1 + a_{i2} X_2 + p_i \quad (i = 0, 1, 2),$$

which we may write in matrix form as

$$Y = \mathbf{A} X + P.$$

We note that the transformation $Y = X + P$ is a *translation*, and can be interpreted as simply moving the coordinate axes, parallel to themselves, to a new origin. The transformation $Y = \mathbf{A} X$, on the other hand, is a transformation of axes in which the origin of coordinates is kept fixed. The general transformation $Y = AX + P$ is a composition of both. We also see that if the transformation $Y_1 = AX + P$ is followed by the transformation $Y_2 = BY_1 + Q$, the composition is the transformation

$$Y_2 = B(AX + P) + Q = (BA)X + (BP + Q),$$

and if the matrices A and B are both non-singular, which we assume in affine transformations, the matrix BA is also non-singular, and the composition is also an affine transformation. We shall use the fact that

$$\det(BA) = \det(B) \det(A),$$

and so the determinant of the composition, which is $\det(BA)$, is the product of the determinants of each transformation.

In Euclidean transformations we are interested in keeping the special quadric, in this case the conic at infinity given by

$$T_0{}^2 + T_1{}^2 + T_2{}^2 = 0 = T_3$$

invariant, returning to homogeneous coordinates to define the conic. If we merely aim at doing this, we shall obtain *similarity* transformations (compare Exercise 80.25), in which lengths are transformed by a fixed multiplier, not necessarily equal to 1. For Euclidean transformations, which preserve lengths, we require that the square of the distance between two points be unchanged by the transformation. We can do this by considering what transformations $Y = AX$, which leave the origin of coordinates invariant, transform the quadratic form

$$X_0{}^2 + X_1{}^2 + X_2{}^2$$

into itself. This represents the square of the distance of the point X from the origin, which maps into itself. If we write this form as $X^T X$, and write the transformation as $X = BY$, where $B = A^{-1}$,

$$X^T X = (BY)^T(BY) = Y^T(B^TB)Y,$$

and our requirement is that $B^TB = I$, where I is the unity matrix. Once again (see §80.3), we call the matrix B an *orthogonal* matrix, and it is proved very simply that if B is orthogonal, so is its inverse, A.

If A is orthogonal, the general transformation

$$Y = AX + P$$

leaves the square of the distance *between any two points* unchanged. If the points are X_1 and X_2, then $Y_1 - Y_2 = A(X_1 - X_2)$, and the square of the distance between two points involves only the difference of their coordinates, and the matrix A being orthogonal ensures that the square of the distance is preserved. This set of transformations constitute the transformations of the Euclidean group, which preserve lengths, and therefore also angles.

If A is orthogonal, we have $\det(A^T A) = \det(I) = 1 = \det(A^T) \det(A)$, and since $\det(A^T) = \det(A)$, we end up with $\det(A) = +1$ or -1. If $\det A = +1$, we call the isometry arising from the transformation $Y = AX + P$ a *direct* isometry, and if $\det A = -1$, we call the isometry an *opposite* or *indirect* isometry.

The direct isometries form a subgroup of the Euclidean group, since $\det(AB) = \det(A) \det(B) = +1$ if both $\det(A) = +1$ and $\det(B) = +1$.

In Chapter V we investigated the isometries of the real Euclidean plane with the help of a natural and effective tool, the complex numbers. The natural tool for the investigation of the isometries of E_3, the *quaternions*, were invented by Hamilton as a natural generalization of the complex numbers. They form a non-commutative field. We shall not discuss their properties here (see Veblen and Young, Vol. II, (23)), and we shall content ourselves with an enumeration of the properties of isometries in E_3 which may be discovered by using quaternions, or other methods.

Translations, given by $Y = X + P$, are evidently direct isometries in E_3. Reflexions in a given plane are indirect isometries. Rotations about a given point would be meaningless in E_3, but we have rotations about a given line. These are direct isometries.

It can be shown that direct isometries are always given by a rotation about a line followed by a translation. The translation, of course, may be the identity translation, which leaves a rotation about a line only. Since the composition of direct isometries is a direct isometry, we deduce that the composition of two rotations about lines through the same point, which is necessarily left invariant, is itself a rotation about a line through the point.

In E_3 we also have *rotatory reflexions*, obtained by reflexion in a plane with a simultaneous rotation about a line perpendicular to the plane. If the rotation is through 180 degrees, the total effect is that of reflexion in the point where the line meets the plane. (Exercise 88.11.) This special case of rotatory reflexion is called *point-reflexion*. It can be shown that *opposite* isometries are either reflexions in a plane followed by a translation, or a rotatory reflexion followed by a translation.

It will be remembered that a point-reflexion in E_2 is a *direct* isometry. In E_3 it is an opposite isometry. The point-reflexion in E_2 is a special case of a rotation about a point, the angle of rotation being 180 degrees. The point-reflexion in E_3 cannot be considered as a rotation of any sort.

We remarked that successive rotations about two lines through a point O in E_3 can be replaced by a rotation about a line through O. Conversely, if in E_3 two distinct lines l and m through a point O are given, then every rotation about a line through O can be generated (in the group sense) by rotations about l and m.

A rotation about a given line in E_3, accompanied by a translation along the line is called *a screw motion*, for obvious reasons. In our section on null polarities (85.4) we mentioned the connexion with the equilibrium of a set of forces acting on a rigid body in E_3. Such a set can always be reduced to a *screw*, that is to a single force together with *a couple*, the plane of the couple being perpendicular to the direction of the force. A *couple* consists of two equal, parallel but opposite forces. The force will tend to move the rigid body along the axis of the screw, and the couple will tend to rotate the body about the axis of the screw. The lines through any given point P about which the set of forces has zero moment (turning force) is the pencil of lines through P which lie in a plane through P which contains the perpendicular to the axis of the screw (Fig. 88.3). These are the *null lines*, mentioned in §85.4.

It is not surprising to learn that a direct isometry in E_3, without fixed points, is either a translation or a screw motion. A screw motion with zero rotation is, of course, a translation, and so we may say that any direct isometry in E_3 without fixed points is a screw motion.

Finally, for opposite isometries, we can show that any opposite isometry of E_3 which

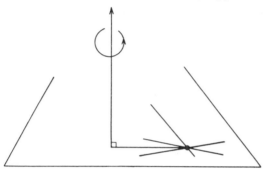

Fig. 88.3

has a fixed point is a rotatory reflexion, and, defining a *glide reflexion* as a reflexion in a plane followed by a translation along a line lying in the plane of reflexion, we can show that every opposite isometry of E_3 without invariant points is a glide reflexion.

88.4   The affine group in three dimensions

As we saw in the preceding section, this is the group of transformations given in matrix form as $Y = \mathbf{A}X + P$, where the matrix $\mathbf{A}$ is non-singular, and the coordinates are non-homogeneous coordinates (X_0, X_1, X_2). In homogeneous coordinates (T_0, T_1, T_2, T_3), where

$$X_0 = T_0 T_3^{-1}, \quad X_1 = T_1 T_3^{-1}, \quad X_2 = T_2 T_3^{-1},$$

an affine transformation is a transformation of the projective space which maps the plane $T_3 = 0$ onto itself. If we call this plane *the plane at infinity*, we have *an affine geometry of three dimensions*. We can see what concepts form a part of this geometry by keeping the projective picture in the background, but using the terminology suitable for 'improper' points at infinity, as we called them in §80.2, that is points in the plane $T_3 = 0$.

We have parallel lines in affine geometry, since lines which intersect at a point in the plane $T_3 = 0$ continue to do so after a transformation. We have the concept of parallel planes, these being planes which intersect in the same line in the plane $T_3 = 0$. We also have the concept of the ratio of signed segments for points on a line, or on parallel lines, just as in the real affine plane, if we restrict ourselves to the field of the real numbers.

If we consider the different types of quadric which arise in real affine geometry, we have, for quadrics of maximum rank, five types, given by the equations:

(1) $X_0^2 - X_1^2 - X_2^2 = 1$ (2) $X_0^2 + X_1^2 - X_2^2 = 1$ (3) $X_0^2 + X_1^2 + X_2^2 = 1$

(4) $X_0^2 - X_1^2 = 2X_2$ (5) $X_0^2 + X_1^2 = 2X_2$.

Quadric (1) is a hyperboloid of two sheets, contains no real generators, and meets the plane at infinity in an irreducible conic. Quadric (2) is a hyperboloid of one sheet, contains real generators, and also meets $T_3 = 0$ in a real irreducible conic. Quadric (3) is an ellipsoid (there are no spheres), and does not meet $T_3 = 0$ in real points. Quadric (4) is called a hyperbolic paraboloid, contains real generators and meets $T_3 = 0$ in a pair of real lines: that is it *touches* the plane at infinity. Quadric (5) is an elliptic paraboloid. It contains no real generators, and meets the plane at infinity in only one real point. It also must be regarded as touching the plane at infinity, since in the field of complex numbers the intersection with this plane is a pair of lines passing through the real point.

In three dimensions our quadrics of rank 3 are cones. Cones with vertex at a proper point, which contain real points, can always be transformed into

$$X_0^2 + X_1^2 - X_2^2 = 0.$$

If the vertex of the cone lies in the plane at infinity, we have cylinders, which contain a family of parallel generators. The plane at infinity may intersect the cylinder in a real pair of lines through the vertex, or it may touch the cylinder, so that the intersection is a pair of coincident lines, or it may only intersect the cylinder in one real point, the vertex of the cone. We therefore have three types of cylinder, a hyperbolic cylinder, a parabolic cylinder and an elliptic cylinder. Their forms of equation are:

$$X_0^2 - X_1^2 = 1, \qquad X_0^2 = 2X_1, \qquad X_0^2 + X_1^2 = 1.$$

Although we have not said so explicitly, our classification of quadrics has been restricted to those which contain real points, so that we have not mentioned such types as $X_0^2 + X_1^2 + X_2^2 = -1$. If we had been classifying real *quadratic forms* under the affine group, such types would have entered into the classification.

Exercises

88.1 In the process of stereographic projection from a point V on a sphere onto the tangent plane π at a point O on the sphere diametrically opposite to V, let P be a point on the sphere, and let VP intersect the plane π in P'. Show that $VP \cdot VP' = VO^2$. Deduce that if the sphere be inverted with respect to a sphere center V, and radius $|VO|$, it will invert into the plane π. Since a small circle on the sphere is cut out by some other sphere, and the inverse of a sphere which does not pass through V in a sphere, deduce again the result of the preceding section, that the stereographic projection of a small circle on the sphere is a circle in π.

88.2 Deduce from the theorem (Exercise 78.11, Hesse's Theorem) that if two pairs of opposite vertices of a quadrilateral are conjugate points with respect to a conic S, then the third pair are also conjugate points, that if two pairs of opposite edges of a tetrahedron in E_3 are perpendicular, then so are the third pair.

88.3 Prove that if a line l in E_3 is perpendicular to each of two lines m and n which lie in a plane π, then l is orthogonal to every line in π.

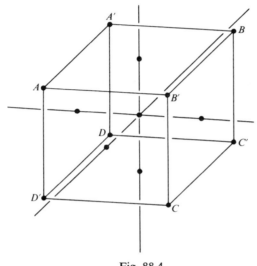

Fig. 88.4

88.4 $ABCD$ is a tetrahedron in E_3, and l is a line perpendicular to the plane ABC which passes through the orthocenter of ABC (the point of intersection of the perpendiculars from A, B, C onto the opposite sides BC, CA and AB.) Prove that the perpendiculars from A onto the plane BCD, from B onto the plane CDA and from C onto the plane DAB intersect l.

88.5 $ABCD$ is a tetrahedron in which pairs of opposite sides are perpendicular. Show that the plane through the line AC and the perpendicular from A onto the plane BCD intersects the plane BCD in the line through C perpendicular to BD. Deduce that the perpendicular from A onto the plane BCD passes through the orthocenter of triangle BCD. Prove that the perpendiculars from the vertices A, B, C, D of the tetrahedron onto the opposite faces all pass through a point.

88.6 Show that a set of three *desmic* tetrahedra (described in Exercise 65.16) can be realized by taking two of them at the alternate vertices of a cube (that is $ABCD$, $A'B'C'D'$ in Fig. 88.4) and the third at the center of the cube and the three points at infinity in the directions of the sides of the cube.

88.7 Show that the real solutions of the equation $X_0{}^2 - X_1{}^2 - X_2{}^2 = 1$ fall into two disjoint sets, justifying the description of the surface (§88.4) as a hyperboloid of two sheets.

88.8 Show that the parabolic cylinder $X_0{}^2 = 2X_1$ can be regarded as a cone with its vertex in the plane at infinity, intersecting this plane in two coincident lines.

88.9 Show that in real three-dimensional affine space the equations of any two skew lines may be taken to be

$$X_0 = 0 = X_2 - c, \qquad X_1 = 0 = X_2 + c,$$

by a suitable choice of the coordinate system.

   Investigate the locus of the midpoint of a variable transversal of two given skew lines which is parallel to a given plane.

88.10 Solve the preceding Exercise without using algebra, as a problem in projective space of three dimensions, in the first instance. (Remember that if P' is the improper point on a line AB, and P the midpoint of the segment AB, the points P, P' form a harmonic range with A and B. Look back at §83.1, also.)

88.11 Show that in E_3 a rotatory reflexion becomes a point reflexion when the angle of rotation is 180 degrees.

88.12 π and π' are two intersecting and distinct planes in E_3, and V is a point which does not lie on either plane. The points P in π are projected from V onto the points P' in π'. The planes are then rotated about their line of intersection until they coincide, and in this one plane, which we still call π, we consider the mapping $P \rightarrow P'$. Show that it is a collineation, and that if the line l is mapped onto the line l', then $l \cap l'$ always lies on a fixed line v^*. Deduce that the line PP' always passes through a fixed point V^*. If v^* and V^* are given, show that the collineation we are studying is uniquely determined if we are also given one pair of corresponding points, P, P'. Show that the points in π which have no map lie on a line v (the vanishing line) which is parallel to v^*. Would you expect this? Show that if Q is any point on v, any line through Q is mapped onto a line parallel to V^*Q. Show that if V^* is given, a pair of corresponding points and also v, then v^* is uniquely determined. Choose V^*, v and a pair of corresponding points so that a given quadrangle in π is mapped (i) onto a parallelogram, (ii) onto a square. Investigate the maps of circles which have (i) no intersections with v (ii) touch v and (iii) intersect v in distinct points. (See Pedoe, (12), Chapter I.)

X

PRELUDE TO ALGEBRAIC GEOMETRY

In this final chapter of our book we discuss curves and surfaces of geometrical interest, as before, but the stress is less on the algebra, and more on the geometry. There is a good reason for this. A complete algebraic treatment is possible, but involves a fair amount of algebra, and, so far, we have assumed a minimum of knowledge of this very important mathematical discipline. There is no space to give all the algebraic theory in this book, nor would we wish to do so. There are books which do this, and here we have a modest aim: to arouse the interest of the student in algebraic geometry, and to prepare him for some of the ideas he may encounter when he begins a serious study of algebraic curves, with which the curtain rises on any study of algebraic geometry.

Of course, the whole of this volume may be subsumed under the title of this final chapter. But since there may be many readers who have no intention of proceeding to the study of algebraic geometry proper, we have stressed that this book deals with the methods in use in the study of elementary geometry.

89.1   Curves on a quadric

We have encountered conic sections of a quadric, in §87.1, and since any such section is a plane curve, and a plane intersects every member of a regulus of the quadric in a point, conic sections of a quadric intersect every member of each regulus in one point. Stereographic projection of a conic section from a point X_3 on the quadric produces a conic in the plane $X_0 X_1 X_2$ which passes through two points, the points X_1 and X_2. Two conic sections of the quadric intersect in two points, where the line of intersection of the planes of each conic intersects the quadric. In the stereographic projection, these two points of intersection correspond to the two points of intersection of two conics, each through X_1 and X_2, *not counting the intersections at X_1 and X_2.*

There are other curves on the quadric, besides lines and conics. Let us define *a space cubic curve* (sometimes called a *twisted* cubic curve, to distinguish it from *a plane cubic curve*) as a curve, defined by algebraic equations, not contained entirely in a plane, and such that *any plane in S_3 intersects it in three points.*

We shall see that there are such curves. Of course, we are already using terms without defining them completely, but, as we said above, this cannot be helped.

We now suppose that a space cubic curve, which we call $\mathscr{C}^3$, lies on our quadric.

Take two intersecting generators of the quadric. These define a plane. This plane intersects the curve $\mathscr{C}^3$, by hypothesis, in three points, and these points must lie on the two generators, since the curve lies on the quadric.

Hence one of the generators contains two points of the curve, and the other generator contains one point of the curve. If one generator contained *three* points of the curve, and we joined this generator to any other point of the curve, we should have a plane which intersects the curve in at least four points, contrary to hypothesis. The generator which contains two points of $\mathscr{C}^3$ is said to be *a chord* of the cubic curve. If we consider the pencil of planes through this chord as axis, each plane of the pencil intersects the quadric in a generator of the complementary regulus, and each of these generators contains one point of $\mathscr{C}^3$.

Similarly, if we take a pencil of planes through a generator which intersects $\mathscr{C}^3$ in one point, we find that all generators of the complementary regulus are chords of $\mathscr{C}^3$, meeting it in two points.

Hence we may think of $\mathscr{C}^3$ as being a $(2, 1)$ curve on the quadric, meeting generators of the regulus $\{f\}$, say, in two points, and generators of the regulus $\{g\}$ in one point, or we may think of the curve $\mathscr{C}^3$ as being a $(1, 2)$ curve on the quadric, meeting generators of $\{f\}$ in one point, and generators of $\{g\}$ in two points. We saw that conics on the quadric are $(1, 1)$ curves, and generators themselves are $(1, 0)$ and $(0, 1)$ curves.

If we project our curve $\mathscr{C}^3$ from X_3 onto the plane $X_0X_1X_2$, assuming that $\mathscr{C}^3$ does not pass through the point X_3 on the quadric, we obtain a cubic curve in the plane, in the sense that *every line in the plane intersects the plane curve in three points*. This arises from the fact that the join of any line in the plane $X_0X_1X_2$ to the point X_3 gives a plane which meets $\mathscr{C}^3$ in three points, and these points, projected from X_3, lie on the line in the plane $X_0X_1X_2$ and on the projected curve.

If $\mathscr{C}^3$ intersects the generator X_3X_1 in one point (Fig. 89.1), and therefore intersects the generator X_3X_2 in two points, the projected curve passes through the point X_1, and has a unique tangent there. We say it passes *simply* through X_1. On the other hand, the projected cubic passes *doubly* through X_2, with *two tangents*, corresponding to the fact that $\mathscr{C}^3$ intersects the generator X_3X_2 in two distinct points, and the part of $\mathscr{C}^3$ in the neighbourhood of each of these points projects into a part of the projected curve in the neighbourhood of X_2. Each part of the projected curve is called a *branch*, and we say that the projected cubic curve has a *node*, or *double point* with distinct tangents at the point X_2.

Any line in the plane $X_0X_1X_2$ through X_2 intersects the plane cubic curve in one further point, which varies with the line, corresponding to the fact that any plane through the generator X_3X_2 intersects the space cubic in just one point which varies with the plane. Both space curve and projected curve are said to be *rational* curves. We explain this description in the next paragraph.

If we take the line $X_0 + \lambda X_1 = 0$ in the plane of the projected cubic, it passes through X_2, where the cubic has a node, and intersects the plane cubic in *one variable point*.

From the equation of the plane cubic, we must be able to determine the coordinates of this point, and they must be of the form:

$$X_0 : X_1 : X_2 = p(\lambda) : q(\lambda) : r(\lambda),$$

where the p, q, r are polynomials in λ, at most of degree three. The non-homogeneous coordinates of a point on the curve are therefore *rational functions* of the parameter λ. This is why the curve is called a *rational* curve.

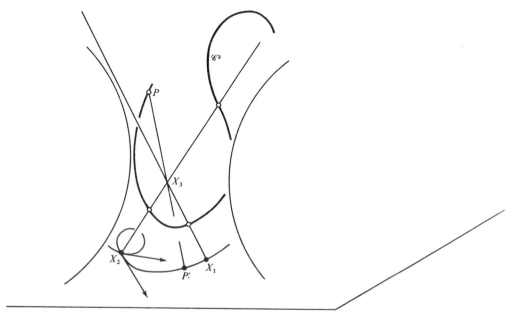

Fig. 89.1

Since the pencil of planes in S_3 through the generator $X_3 X_2$ also meet the space cubic in only one variable point, the non-homogeneous coordinates of a point on the space cubic are also rational functions of a parameter. It is a theorem of the theory of algebraic curves that if a curve is a rational curve, so are its projections, but in our case we have seen directly that both $\mathscr{C}^3$ and its projection are rational curves.

89.2 Generation of $\mathscr{C}^3$

We wish to show that the space cubic can be generated as the locus of the point of intersection of corresponding planes of three projectively related pencils of planes. This is a natural generalization of the generation of a conic, which is, of course, also a rational curve, as the locus of intersection of the points of intersection of corresponding lines of two projective pencils of lines.

Suppose that we take three generators f_1, f_2, f_3 of the regulus of generators on the quadric which are chords of $\mathscr{C}^3$. If we join each of these generators to a generator g of the opposite regulus, we obtain three planes which are projectively generated as g varies in $\{g\}$. These planes intersect in a line g, and although each line g contains a point of $\mathscr{C}^3$, we wish to show that we can find three projective pencils of planes which are such that corresponding planes meet in a point only, and this point describes $\mathscr{C}^3$. To do this we must not choose pencils of planes whose axes are all generators belonging to the regulus $\{f\}$.

We first show that if V is a point on $\mathscr{C}^3$, *the projection of $\mathscr{C}^3$ from V onto a plane which does not contain V is a conic.* The argument is as follows: the join of any line in

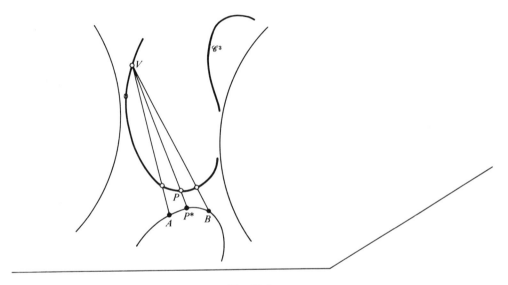

Fig. 89.2

the plane to V gives a plane, which intersects $\mathscr{C}^3$ in three points. One of these points is at V. Therefore any line in the plane intersects the projected curve in two points. We assume that an algebraic curve which intersects any line in its plane in two points is a conic.

The cone which has vertex V is therefore a *quadric cone* (Fig. 89.2). It now follows that if VA, VB are fixed chords of $\mathscr{C}^3$ through V, then the pencils of planes $VA(P)$, $VB(P)$ are projectively related, where P is a variable point on $\mathscr{C}^3$. To see this we choose a fixed plane section of the quadric cone which does not contain V, and choose A and B to lie on this section. If VP, which is a generator of the quadric cone, intersects the conic section in P^*, we know that the pencils $A(P^*)$ and $B(P^*)$ are projectively related. Therefore this is true for the pencils of planes $VA(P)$ and $VB(P)$.

From this we move on to proving that if VA and UB are any two chords of $\mathscr{C}^3$, then the pencils of planes $VA(P)$ and $UB(P)$ are projectively related, where P is any point on $\mathscr{C}^3$. This follows from the relations

$$VA(P) \barwedge VU(P),$$

and

$$VU(P) \barwedge BU(P),$$

where $\barwedge$ is the symbol for projective relationship.

Finally, to show that $\mathscr{C}^3$ can be generated as the locus of intersection of three projectively related pencils of planes, all that we have to do is to choose as axes of these pencils three chords of $\mathscr{C}^3$ of which two are in the regulus of chords $\{f\}$, but the third is not. Let these chords be f_1, f_2 and h. Then if P is any point on $\mathscr{C}^3$, we have

$$f_1(P) \barwedge f_2(P) \barwedge h(P),$$

by what we have just proved, and $f_1(P)$ always intersects $f_2(P)$ in a generator g of the quadric, and this generator contains P. The plane $h(P)$ cannot contain g, for if it did, the axis h would have to be a member of the regulus $\{f\}$, and it is not.

Hence the corresponding planes in the projective pencils always intersect in a point of $\mathscr{C}^3$, and never in a line, and we have proved:

Theorem I *The space cubic curve* $\mathscr{C}^3$ *can be generated as the locus of the points of intersection of the corresponding planes of three projectively related pencils of planes.*

As we have already pointed out, this curve $\mathscr{C}^3$ is the natural generalization of the conic.

89.3 The space cubic as a residual curve

Since the projection of $\mathscr{C}^3$ from a point V on the curve is a quadric cone, if we project the curve from another point on it, $U \neq V$, we obtain another quadric cone which contains $\mathscr{C}^3$. We now have two quadric cones containing $\mathscr{C}^3$, and we can therefore say that $\mathscr{C}^3$ is the *residual intersection* of two quadric cones. The intersection is termed residual because the two cones have a common generator, the line UV, and *the complete intersection* of the two cones consists of the line UV together with the curve $\mathscr{C}^3$.

But the surfaces cutting out $\mathscr{C}^3$ need not be cones. If $\mathscr{C}^3$ is projectively generated by pencils with axes a, b and c respectively, the pencils axes a and b generate a regulus, which is complementary to the regulus which contains a and b. The pencils with axes b and c also generate a regulus, and this is complementary to the regulus which contains b and c. Hence $\mathscr{C}^3$ is the residual intersection of two quadric surfaces with a common generator, in this case the line b. If a and b are skew, and b and c are skew, neither surface is a quadric cone.

The space cubic $\mathscr{C}^3$ may be introduced in this way, of course, and its generation by projective pencils of planes then follows immediately.

Let Q and Q' be two quadrics which have $\mathscr{C}^3$ as a residual intersection. The pencil of quadric surfaces $Q + \lambda Q' = 0$, by a familiar argument, also contains $\mathscr{C}^3$, since every point of $\mathscr{C}^3$ satisfies both the equation $Q = 0$ and the equation $Q' = 0$. If P is any given point in S_3 which does not lie on both quadrics, we can find a unique quadric of the pencil which passes through P. All we have to do is to determine the value of λ from the equation

$$Q(P) + \lambda Q'(P) = 0.$$

Now this quadric, which we shall call Q_0, contains $\mathscr{C}^3$, and therefore the two systems of generators on Q_0 intersect $\mathscr{C}^3$, one system, say the $\{f\}$ system, being chords of $\mathscr{C}_3$, and the other system intersecting $\mathscr{C}^3$ in just one point. Through P, which lies on Q_0, there is a generator of the $\{f\}$ system. This is a chord of $\mathscr{C}^3$. Hence we have proved:

Theorem I *Through any given point of S_3 there passes a unique chord of $\mathscr{C}^3$.*

The uniqueness of the chord through P is clear, since two chords through P would determine a plane which has four intersections with $\mathscr{C}^3$.

We have already seen that conics on a quadric are represented, after stereographic projection from a point X_3 on the quadric, by conics through two fixed points of the plane $X_0X_1X_2$, the points X_1 and X_2 (§87.1). We have also seen that the projection of a $(2, 1)$ cubic curve on the quadric is a plane cubic curve, with a double point at X_2, if the cubic curve cuts the generator X_3X_2 in two points, and a simple point at X_1, the cubic curve cutting the generator X_3X_1 in one point.

If the cubic curve passes through the point X_3 on the quadric, and is of the same $(2, 1)$ type, it will cut the generator X_3X_2 in one further point, and will not cut the generator X_3X_1 again. Hence the projection of this $\mathscr{C}^3$ will be a *conic* in the plane $X_0X_1X_2$, passing simply through the point X_2, and *not* passing through the point X_1. We note that a conic which passes simply through X_2 and through X_1 will intersect a conic which goes simply through X_2 and does not pass through X_1 in $4 - 1 = 3$ points which are not at X_2 or X_1. (We are assuming that two conics intersect in $2 \times 2 = 4$ points.) This number *three* corresponds to the intersection of a plane section of the quadric with the cubic curve $\mathscr{C}^3$.

Let us suppose that there pass through the point X_3 a space cubic of the $(2, 1)$ type, and a space cubic of the $(1, 2)$ type. We can find the number of points in which these two space cubics intersect, by counting the number of points in which their stereographic projections intersect *outside the two fundamental points X_1 and X_2*. The reason for this is that the two cubics do not intersect on the generators X_3X_1 and X_3X_2, since we assume that the two cubics are such that their only point of intersection on X_3X_2 is at X_3, and their only point of intersection on X_3X_1 is at X_3 also.

The stereographic projections from X_3 are both conics, one passing through X_2 and not through X_1, and the other passing through X_1 and not through X_2. The number of intersections of these two conics is four, none being at X_1 or X_2, so that there are four intersections of the two cubics besides the point X_3. Hence: *a $(2, 1)$ cubic on a quadric intersects a $(1, 2)$ cubic on the quadric in five points.*

Exercises

89.1 a, b and c are three mutually skew lines in S_3, and T is a projectivity on c with two distinct fixed points C_1 and C_2. If C and C' are related in this projectivity, show that the projective pencils $a(C)$ and $b(C')$ generate a regulus $\{l\}$ of which two and only two lines intersect c. Show that these lines are distinct, and also show how to construct a pencil of planes, axis c, projectively related to $a(C)$ and $b(C')$, such that no three corresponding planes can meet in a line.

89.2 Six points are given in S_3. Show that unless they are in a special relationship to each other, there is a unique twisted cubic curve $\mathscr{C}^3$ which passes through them.

89.3 Five general points and a general line are given in S_3. Prove that there is a unique space cubic curve through the five points which has the given line as a chord.

89.4 Given three points and three lines in S_3, in general position, show that there is a unique space cubic curve which passes through the three points and has the three given lines as chords.

89.5 Prove that two (2, 1) space cubics on a quadric have four points of intersection.

90.1　Normal equations for a space cubic

We suppose that we are generating a space cubic curve $\mathscr{C}^3$ by means of three projectively related pencils of planes, with respective axes a, b and c, and that no three corresponding planes of the pencils ever intersect in a line. We may take the equations of the pencils to be:

$$\lambda_0\alpha_0 + \lambda_1\alpha_1 = 0,$$
$$\lambda_0\beta_0 + \lambda_1\beta_1 = 0,$$
$$\lambda_0\gamma_0 + \lambda_1\gamma_1 = 0,$$

where $\alpha_0 = \alpha_1 = 0$ are suitably chosen planes through the line a, $\beta_0 = \beta_1 = 0$ are suitably chosen planes through the line b, and $\gamma_0 = \gamma_1 = 0$ are suitably chosen planes through the line c.

We may now solve these equations for their point of intersection, using Cramer's rule, if we so desire, since we know in advance that the rank of the system of equations is always three. We then obtain a solution in the form:

$$X_0 : X_1 : X_2 : X_3 = f_0(\lambda_0, \lambda_1) : f_1(\lambda_0, \lambda_1) : f_2(\lambda_0, \lambda_1) : f_3(\lambda_0, \lambda_1),$$

where each $f_i(\lambda_0, \lambda_1)$ is equal to a three-rowed determinant formed from the matrix of the above equations, and is therefore a cubic form (homogeneous polynomial) in λ_0, λ_1.

The four homogeneous polynomials are linearly independent, since if they were not, we should have a relation of the form

$$U^{(0)}f_0 + U^{(1)}f_1 + U^{(2)}f_2 + U^{(3)}f_3 \equiv 0,$$

where the $U^{(i)}$ are elements of the ground field. This would involve all points of $\mathscr{C}^3$ lying in the plane with equation

$$U^{(0)}X_0 + U^{(1)}X_1 + U^{(2)}X_2 + U^{(3)}X_3 = 0,$$

and we have assumed that $\mathscr{C}^3$ is not a plane curve.

Since the f_i are linearly independent, we may solve the equations

$$kX_i - f_i(\lambda_0, \lambda_1) = 0 \qquad (k \neq 0, i = 0, 1, 2, 3)$$

for λ_0^3, $\lambda_0^2\lambda_1$, $\lambda_0\lambda_1^2$, λ_1^3, and we obtain linearly independent forms Y_0, Y_1, Y_2 and Y_3 in the coordinates (X_0, X_1, X_2, X_3) such that

$$Y_0:Y_1:Y_2:Y_3 = \lambda_0^3:\lambda_0^2\lambda_1:\lambda_0\lambda_1^2:\lambda_1^3.$$

We may now change the coordinate system, write $\lambda = \lambda_0/\lambda_1$, and have the desired parametric representation for the space cubic in normal form:

$$Y_0:Y_1:Y_2:Y_3 = \lambda^3:\lambda^2:\lambda:1.$$

In this form it is clear that our curve is not a plane curve, and that it is intersected by a general (that is, non-special) plane in three distinct points. This is because any plane $U^{(0)}Y_0 + U^{(1)}Y_1 + U^{(2)}Y_2 + U^{(3)}Y_3 = 0$ meets the curve where

$$U^{(0)}\lambda^3 + U^{(1)}\lambda^2 + U^{(2)}\lambda + U^{(3)} = 0, \tag{1}$$

and this equation has distinct roots unless there is a relation between the coefficients of the plane.

Similar curves arise in space of any dimension n, and can be defined by the equations:

$$Y_0:Y_1:Y_2:\ldots:Y_n = \lambda^n:\lambda^{n-1}:\lambda^{n-2}:\ldots:1.$$

They are called *normal rational curves*, and play an important part in the theory of algebraic curves. We note that $n = 2$ gives the conic, and $n = 1$ the line.

The space cubic $\mathscr{C}^3$ has no double points, whereas its projection from any point V of S_3 onto a plane is a plane cubic with a double point. We have already seen this, and we prove it again, by proving, again, Thm I of §89.3, but with a different proof.

Theorem   *Through any given point of S_3 there passes a unique chord of $\mathscr{C}^3$.*

Proof. In equation (1) above, suppose that the roots are λ_1, λ_2 and λ_3. Then since $\Sigma\lambda_i = -U^{(1)}/U^{(0)}$, $\Sigma\lambda_i\lambda_j = U^{(2)}/U^{(0)}$, and $\lambda_1\lambda_2\lambda_3 = -U^{(3)}/U^{(0)}$, by the elementary theory of equations, we have the equation of the plane through the three points of $\mathscr{C}^3$ with parameters λ_1, λ_2 and λ_3, in the form:

$$Y_0 - Y_1\Sigma\lambda_i + Y_2\Sigma\lambda_i\lambda_j - Y_3\lambda_1\lambda_2\lambda_3 = 0.$$

Suppose now that we keep λ_1 and λ_2 fixed, and let $\lambda_3 = \lambda$ be variable. We may write this equation in the form:

$$Y_0 - Y_1(\lambda_1 + \lambda_2) + Y_2\lambda_1\lambda_2 - \lambda[Y_1 - Y_2(\lambda_1 + \lambda_2) + Y_3\lambda_1\lambda_2] = 0.$$

This is the equation of a pencil of planes, with axis

$$Y_0 - Y_1(\lambda_1 + \lambda_2) + Y_2\lambda_1\lambda_2 = 0 = Y_1 - Y_2(\lambda_1 + \lambda_2) + Y_3\lambda_1\lambda_2.$$

This, then, is the equation of *the chord* joining the points with parameters λ_1 and λ_2 on $\mathscr{C}^3$. To show that there is a unique chord through a given point (Y_0', Y_1', Y_2', Y_3') of S_3, we solve the equations

$$Y_0' - Y_1'(\lambda_1 + \lambda_2) + Y_2'\lambda_1\lambda_2 = 0,$$

$$Y_1' - Y_2'(\lambda_1 + \lambda_2) + Y_3'\lambda_1\lambda_2 = 0,$$

for $\lambda_1 + \lambda_2$ and $\lambda_1\lambda_2$. We find that λ_1 and λ_2 are the roots of the quadratic equation:

$$(Y_1'Y_3' - (Y_2')^2)T^2 - (Y_0'Y_3' - Y_1'Y_2')T + (Y_0'Y_2' - (Y_1')^2) = 0.$$

Hence the theorem is proved, unless the point Y' is such that all the coefficients of this quadratic equation are zero. We can prove that in this case the point Y' must lie on $\mathscr{C}^3$, and we leave this as an Exercise. (Exercise 90.1.)

The projection of $\mathscr{C}^3$ from a point V which does not lie on it is a plane cubic curve with a double point, corresponding to the chord of $\mathscr{C}^3$ which passes through V. If $\mathscr{C}^3$ had a double point, any plane of S^3 would intersect $\mathscr{C}^3$ in two points at the double point, and since a plane can be drawn through any three points, a plane through the double point and any other two points of $\mathscr{C}^3$ would intersect $\mathscr{C}^3$ in four points at least, which is impossible.

There are plane cubic curves without a double point, but these can never arise as the projections of a space $\mathscr{C}^3$, and the theory of algebraic curves shows that they are not rational curves.

Exercises

90.1 Show that the surfaces $Y_1Y_3 - Y_2^2 = Y_0Y_3 - Y_1Y_2 = Y_0Y_2 - Y_1^2 = 0$ intersect in the space curve $\mathscr{C}^3$, and have no other point in common. (The three equations are equivalent to the equations

$$\frac{Y_0}{Y_1} = \frac{Y_1}{Y_2} = \frac{Y_2}{Y_3}.$$

Put these fractions equal to λ.)

90.2 Show that the equation of the plane which intersects $\mathscr{C}^3$ in only one point, with parameter λ, is

$$Y_0 - 3\lambda Y_1 + 3\lambda^2 Y_2 - \lambda^3 Y_3 = 0.$$

Calling this plane *the osculating plane* at the point $(\lambda^3, \lambda^2, \lambda, 1)$, show that three osculating planes to $\mathscr{C}^3$ pass through a point (a_0, a_1, a_2, a_3) of S_3.

90.3 If the three osculating planes in the preceding Exercise which pass through the point (a_0, a_1, a_2, a_3) osculate $\mathscr{C}^3$ at the points P, Q and R, prove that the equation of the plane PQR is

$$a_3Y_0 - 3a_2Y_1 + 3a_1Y_2 - a_0Y_3 = 0.$$

90.4 If we call the plane associated with the point (a_0, a_1, a_2, a_3) in the preceding Exercise the *polar plane* of the point, show that we have set up *a null-polarity* (§85.5) with regard to $\mathscr{C}^3$.

91.1 Rational surfaces

In §87.1 we studied the stereographic projection of a quadric surface, and saw that plane sections of the quadric by the planes

$$U^{(0)}X_0 + U^{(1)}X_1 + U^{(2)}X_2 + U^{(3)}X_3 = 0 \tag{1}$$

are represented in the plane of projection by the conics

$$U^{(0)}X_0{}^2 + U^{(1)}X_1X_0 + U^{(2)}X_2X_0 + U^{(3)}X_1X_2 = 0. \tag{2}$$

These conics all pass through the fixed points $(0, 1, 0)$ and $(0, 0, 1)$, which we call *the fundamental points* in the plane $X_0X_1X_2$.

We note that two such conics intersect in two points other than the fundamental points, corresponding to the fact that two plane sections of the quadric surface intersect in a line which meets the quadric surface in two points. There is, of course, a one-to-one correspondence between points of the quadric surface and points of the plane, but this breaks down at the fundamental points, and at the point of projection $(0, 0, 0, 1)$ of the surface. We saw that the generators X_3X_1 and X_3X_2 of the surface map into the points X_1 and X_2 respectively of the plane, and that the point X_3 itself may be considered to map onto the line X_1X_2 of the plane.

Now, we could have started our investigation with a system of conics in a plane through two fixed points, X_1 and X_2, noted that the equation of any conic of the system can be written in the form of equation (2) above, which involves four homogeneous constants $U^{(0)} : U^{(1)} : U^{(2)} : U^{(3)}$, and considered the algebraic variety in S_3 given by the equations:

$$Y_0 : Y_1 : Y_2 : Y_3 = X_0{}^2 : X_1X_0 : X_2X_0 : X_1X_2. \tag{3}$$

A plane section of this algebraic variety, given by the intersection with the plane

$$U^{(0)}Y_0 + U^{(1)}Y_1 + U^{(2)}Y_2 + U^{(3)}Y_3 = 0,$$

is represented in the plane by the conic we started with, given by equation (2) above.

The equations (3) give *a map of points of the plane* onto points of an algebraic construct (we use the word *variety*) in S_3, and since the equations involve *rational functions* of the ratios $X_0 : X_1 : X_2$, we call the *variety a rational algebraic variety*.

In this case we have *a surface*, since two independent parameters are involved, say the ratios $X_1 : X_0$ and $X_2 : X_0$, and it can be verified that a point on the variety arises, in general, from at most *a finite number of points in the plane* (see Exercise 91.3).

Since two conics of the plane representation of our surface, given by equations (2), intersect in *two variable points*, we deduce that *a general line of S_3 intersects the surface in two points*. If we investigate lines in the plane which pass through X_1 or X_2, we rapidly rediscover the two systems of generators on the quadric.

In fact, a line through X_1, taken together with a line through X_2, constitute a special plane section representation of the surface, being a special case of equation (2). Since each of these lines meets a fixed and irreducible conic through X_1 and X_2 in *one variable point, each line represents a line on the surface.* The curve on the surface represented by the line through X_1, say, intersects a variable plane in one point, and we assume that such a curve must be a line (see Exercise 91.4).

Two lines through X_1 have no variable point of intersection, and therefore each represents a line on the surface, *but these two lines have no intersection on the surface.* Hence we rediscover the two reguli of lines on the surface, each consisting of a set of mutually skew lines, and, since any line through X_1 intersects any line through X_2, each line of one regulus meets each line of the other regulus on the surface.

What do the points X_1 and X_2 represent on the surface? What does the line X_1X_2 represent on the surface? We know the answer, of course. To obtain the results, that

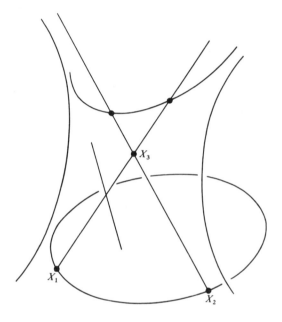

Fig. 91.1

X_1 represents the line X_3X_1, that X_2 represents the line X_3X_2, and *the line X_1X_2 represents the point X_3,* this would need more *algebraic* theory than we have at our disposal in this book. But we can feel confident *geometrically* about certain matters, our confidence being based on the study of how things work in the kind of representation we are studying.

We observe that X_1 and X_2 are simple points on the conics which pass through them and represent plane sections of our surface (Fig. 91.1). The variable tangents at X_1 and X_2 are related to the variable points in which a plane section of the quadric cuts the loci corresponding to the points X_1 and X_2. These loci must therefore be *lines* on the surface.

The line X_1X_2 does not meet conics through X_1 and X_2 in variable points, and so can only correspond to a point on the quadric. Particular conics of the system through X_1 and X_2 are given by taking the line X_1X_2 together with any line of the plane $X_0X_1X_2$, and these particular conics correspond to the plane sections of the quadric which *pass through* the point on the quadric which corresponds to the line X_1X_2.

Exercises

91.1 If the point (p, q, r) be a given point in the $X_0X_1X_2$-plane, show that the point $(kp, 1 + kq, kr)$ is a variable point on the join of the given point to $X_1 = (0, 1, 0)$ which tends to X_1 as k tends to zero. Show that the map of this variable point, as given by equation (3) of the preceding Section, gives a point on the quadric which moves towards the point $(0, p, 0, r)$ on the generator X_3X_1 as k tends to zero, so that there is a projective relation between the *directions* in the plane through X_1 and the *points* on the generator X_3X_1.

91.2 Show that the map of any point $(0, q, r)$ on the line X_1X_2, under the mapping given by the equations (3) of the preceding Section, is the point X_3.

91.3 We asserted, in the preceding Section, that a point on the quadric arises, in general, from at most a finite number of points in the plane, under the mapping given by equations (3). Show that if p, q, r are not all zero, the equations $X_0{}^2 : X_1X_0 : X_2X_0 : X_1X_2 = p:q:r:s$, if solved for X_0, X_1, X_2, give the unique solution, $X_0 : X_1 : X_2 = p:q:r$. Why should we have anticipated this result?

91.4 Prove directly, using the equations of the mapping given by (3), that the map of a line through X_1, given by the locus $(kp, 1 + kq, kr)$, where k is variable, and p, q, r are fixed, is a line in S_3.

92.1 The cubic surface

We can obtain some information about an interesting rational surface in S_3, the cubic surface, intersected by a general line in three points, if we consider the plane cubic curves which pass through six points $A_1, A_2, \ldots, A_6$ in a plane, where these six points are non-special, and, in particular, do not lie on a conic.

The equation of a plane cubic curve has the form:

$$a_{000}X_0{}^3 + a_{111}X_1{}^3 + a_{222}X_2{}^3 + 3a_{001}X_0{}^2X_1 + 3a_{002}X_0{}^2X_2 + 3a_{011}X_0X_1{}^2$$
$$+ 3a_{022}X_0X_2{}^2 + 3a_{122}X_1X_2{}^2 + 3a_{211}X_2X_1{}^2 + 6a_{012}X_0X_1X_2 = 0.$$

This equation involves ten homogeneous parameters, the coefficients a_{ijk} in the equation. The condition that the cubic pass through a given point is a linear condition, or equation, involving the coefficients. If the cubic be made to pass through six points A_i, we obtain six linear equations connecting the a_{ijk}, and we assume that the points A_i are sufficiently general for these equations to have solutions which depend on $10 - 6 = 4$ homogeneous parameters. If we then substitute the values of the a_{ijk} we have found in the equation of the cubic curve, and gather together the terms which multiply each of the four parameters, we find that the equation of a cubic curve through the six points A_i is of the form:

$$U^{(0)}f_0 + U^{(1)}f_1 + U^{(2)}f_2 + U^{(3)}f_3 = 0,$$

where the $U^{(i)}$ are the four homogeneous parameters, and the f_i are cubic forms in X_0, X_1, X_2, and $f_i(X_0, X_1, X_2) = 0$ is the equation of a particular cubic curve which passes through the six points A_i.

It is important to show that the four cubic forms f_i are *linearly independent*, that is, there are no values $p^{(0)}$, $p^{(1)}$, $p^{(2)}$, $p^{(3)}$ of the $U^{(0)}$, $U^{(1)}$, $U^{(2)}$ and $U^{(3)}$ such that

$$p^{(0)}f_0 + p^{(1)}f_1 + p^{(2)}f_2 + p^{(3)}f_3 \equiv 0$$

identically in the variables X_0, X_1, X_2, X_3, unless all the $p^{(i)} = 0$.

We do this by showing that we can find four forms f_i which are cubic, such that $f_i = 0$ is a cubic curve through the six points A_i, and which are linearly independent. Let $S = 0$ be the equation of an irreducible conic through the points A_1, A_2, A_3, A_4 and A_5, and let $S' = 0$ be the equation of an irreducible conic through the points A_2, A_3, A_4, A_5 and A_6. The points A_i are such that these conics are irreducible, and the conic $S = 0$ does *not* pass through A_6, nor does the conic $S' = 0$ pass through A_1. Let the equation of the line A_1A_6 be $u = 0$, the equation of a distinct line through A_6 be $v = 0$, and the equation of a line through A_1 distinct from A_1A_6 be $w = 0$. Then the four composite cubic curves

$$Su = 0, \ Sv = 0, \ S'u = 0, \ S'w = 0$$

each pass through the *six* points A_i. The system of cubic curves

$$U^{(0)}(Su) + U^{(1)}(Sv) + U^{(2)}(S'u) + U^{(3)}(S'w) = 0$$

consists of cubic curves through the six points A_i, and we cannot have

$$p^{(0)}(Su) + p^{(1)}(Sv) + p^{(2)}(S'u) + p^{(3)}(S'w) \equiv 0$$

identically, for then we should have

$$S(p^{(0)}u + p^{(1)}v) \equiv -S'(p^{(2)}u + p^{(3)}w),$$

and since S is irreducible, it cannot have the factor $p^{(2)}u + p^{(3)}w$, nor, even if the linear factors on both sides of the identical relation were the same, could S and S' be the same form, except for a constant factor, since $S = 0$ does *not* pass through A_6, and $S' = 0$ *does* pass through A_6.

We have now seen that if the six points A_i through which our cubic curves are to pass are chosen so that A_2, A_3, A_4 and A_5 are the distinct intersections of two irreducible conics $S = 0$ and $S' = 0$, and A_1 is a further point on $S = 0$, and A_6 a final point on $S' = 0$, then the six points A_i are sufficiently non-special for the cubic curves through them to be of the form

$$U^{(0)}f_0 + U^{(1)}f_1 + U^{(2)}f_2 + U^{(3)}f_3 = 0,$$

where the f_i are linearly independent.

We now consider the locus in S_3 given by the equations

$$Y_0 : Y_1 : Y_2 : Y_3 = f_0(X_0, X_1, X_2) : f_1(X_0, X_1, X_2) : f_2(X_0, X_1, X_2) : f_3(X_0, X_1, X_2).$$

Since the f_i are linearly independent, this locus *does not lie in a plane*. It is given as rational functions of two parameters, and is *a surface* if the mapping does *not* map an infinity of points of the plane $X_0X_1X_2$ onto each point of the surface.

If we assume that two cubic curves in the plane intersect in $3 \times 3 = 9$ points, so that any two cubic curves through the six points A_i intersect in $9 - 6 = 3$ further points, we see that any two plane sections of our surface have three points in common. A line of S_3 therefore meets the surface in three points, and it is *a cubic surface*.

We shall not stop to investigate any further details of the mapping, but consider the curves on the cubic surface suggested by the mapping of plane sections of the surface onto the cubic curves which pass through six given points A_i.

Each of the six points A_i represent *a line on the surface*, the variable tangents to the cubic curves at each point A_i representing the variable intersections of the corresponding plane section with the corresponding line. We call these lines $a_1, a_2, \ldots, a_6$. They do not intersect each other.

Now consider the conics which pass through sets of five of the points A_i. The conic through the points A_2, A_3, A_4, A_5, A_6 represents a curve on the cubic surface which intersects the lines $a_2, a_3, a_4, a_5,$ and a_6, but not the line a_1. Since the conic is met by a cubic curve through the six points A_i in $2 \times 3 - 5 = 1$ variable point, the curve on the cubic surface represented by the conic in the plane is met by a variable plane section in one point. It is therefore *a line*. We call this line b_1, and therefore have six lines b_i on the cubic surface, each intersecting five lines of the six lines a_i. Each line of the six lines a_i intersects five lines of the six lines b_i.

Such a system of twelve lines is called *a double six*.

It will have been noted that we are assuming a theorem in the theory of algebraic plane curves called BEZOUT'S THEOREM:

Theorem   *A plane algebraic curve of degree m intersects a plane algebraic curve of degree n in mn points.*

So far we have merely used this theorem for plane cubic curves and conics. The proof, and further explanation, belong to the theory of algebraic curves, and will not be given here, but the theorem can be understood and used in our present context.

There are more lines on our cubic surface! Consider the line in the plane which joins A_1 to A_2. This intersects the cubic curves through the six points A_i in one variable point, and therefore represents a line on the surface. We call this line c_{12}. There are fifteen such lines c_{ij}. The line c_{12} intersects a_1 and a_2, and also the lines b_1 and b_2. The first statement follows from the fact that the line A_1A_2 passes both through A_1 and A_2, and has a definite direction at each of these points. The second statement follows from the fact that the line A_1A_2 intersects the conic through the points A_2, A_3, A_4, A_5, A_6 in a point distinct from A_1 and A_2. This point represents the intersection of the lines c_{12} and b_1. In a similar way we show that c_{12} intersects b_2.

Since c_{ij} intersects the lines a_i, a_j and b_i, b_j, it is the intersection of the planes $[a_ib_j]$ and $[a_jb_i]$.

We have now found *twenty-seven* lines on our cubic surface, and there are many properties of this configuration of lines. The question arises: Does a general cubic surface in S_3, one whose equation has non-specialized coefficients, contain such a configuration of twenty-seven lines? The answer is that it does, and such a surface contains no further lines. All this, of course, is over the field of complex numbers. But models of real cubic surfaces containing the full number of twenty-seven lines have been constructed. For a picture of a double-six, called a Schlaefli double-six, see Hilbert and Cohn-Vossen, p. 164, (7). Ludwig Schlaefli was the Swiss mathematician who first investigated the double six. The notation a_i, b_j, c_{ij} for the twenty-seven lines is due to him.

We conclude this short discussion of cubic surfaces by studying a special cubic surface, one with a node. This arises when the six points A_i in our plane representation lie on a conic. Let this conic have equation $S = 0$.

Among the plane cubic curves through the six points A_i there is a subsystem of curves with the equation

$$S(U^{(0)}X_0 + U^{(1)}X_1 + U^{(2)}X_2) = 0,$$

which consists of the fixed conic through the six points A_i, together with a general line of the plane. We note that if C^3 is any irreducible cubic curve through the six points, that is, with equation $C^3 = 0$, the system

$$U^{(3)}C^3 + U^{(0)}SX_0 + U^{(1)}SX_1 + U^{(2)}SX_2 = 0$$

consists of cubic curves through the six points A_i, and the component curves are linearly independent, since the identical equation

$$p^{(3)}C^3 + p^{(0)}SX_0 + p^{(1)}SX_1 + p^{(2)}SX_2 = 0$$

would involve the form C^3 being divisible by the form S, and C^3 does not factorize. If we are uncertain whether there is an irreducible cubic through the six points A_i, we may begin by taking an irreducible cubic curve, and *choosing* the six points A_i as given by the intersections with the cubic curve of an irreducible conic $S = 0$.

Hence, if we begin with cubic curves through six points A_i which lie on a conic, we can be certain that the map does give a cubic surface.

Our map of points of the plane onto points of the cubic surface is:

$$Y_0: Y_1: Y_2: Y_3 = SX_0: SX_1: SX_2: C^3,$$

and therefore *all points of the conic $S = 0$ are mapped onto the point $(0, 0, 0, 1)$ of the surface.* Any two plane sections of the cubic surface which pass through this point, which we shall call V, arise from the plane curves

$$S(p^{(0)}X_0 + p^{(1)}X_1 + p^{(2)}X_2) = 0 = S(q^{(0)}X_0 + q^{(1)}X_1 + q^{(2)}X_2).$$

The number of *variable* intersections of the line through V which is the intersection of the planes through V with the surface corresponds to the number of variable intersections of the two reducible curves shown which do not lie on $S = 0$. This is *one*, given by the intersection of the two lines which are components, and we can therefore say:

The point V on the cubic surface which corresponds to the conic S in the plane is a double point, or node, since any line through it meets the cubic surface in only one further point.

Sections of the surface by planes through the node are cubic curves with a double point. These correspond to the lines of the plane representation.

The points A_i correspond to lines on the cubic surface, and since they lie on the conic $S = 0$, *they all pass through the node V*. In fact, the cone of tangent lines to the surface at V is a quadric cone, and this intersects the cubic surface in $2 \times 3 = 6$ lines through V.

The lines $A_i A_j$ correspond to lines c_{ij} on the surface. In fact, a plane through two lines of the surface which meet at V must intersect the surface again in another line, so c_{ij} intersects a_i and a_j. The six b_i lines have disappeared.

Exercises

92.1 Show that on the cubic surface with a node the points

$$c_{12} \cap c_{45}, \quad c_{23} \cap c_{56}, \quad c_{34} \cap c_{61}$$

lie on a plane section of the surface through the node.

92.2 Write down the equation of a cubic surface with a node at the point $(0, 0, 0, 1)$, and show that the plane representation discussed above can be obtained by projection of points of the surface from the node.

92.3 The lines a, b, c and d are coplanar and form a quadrilateral. Show that cubic curves through the six vertices of this quadrilateral represent a cubic surface with four nodes, each side of the quadrilateral representing a node. Show also that the sides of the diagonal line triangle of the quadrilateral represent three lines on the cubic surface which are coplanar.

92.4 Show that the equation

$$Y_0 Y_3{}^2 - Y_1 Y_2{}^2 = 0$$

is that of a ruled (scrollar) cubic surface, with an infinity of lines on it given by the equations

$$Y_3 - \lambda Y_2 = 0 = Y_0 \lambda^2 - Y_1,$$

these lines intersecting the lines $Y_0 Y_1$ and $Y_2 Y_3$, both of which lie on the surface. (These lines are called *directrices* of the surface.)

What is the difference between the residual intersection with the surface of a plane through the line $Y_0 Y_1$ and a plane through the line $Y_2 Y_3$?

Show that two of the above system of generators pass through any given point of $Y_0 Y_1$, and that these lines coincide at the points Y_0 and Y_1 (called *pinch-points* of the surface).

92.5 Show that the system of conics which pass through $(1, 0, 0)$ and for which the points $(0, 1, 0)$ and $(0, 0, 1)$ are always a conjugate pair may be defined by the equation:

$$a_{11} X_1{}^2 + a_{22} X_2{}^2 + 2a_{01} X_0 X_1 + 2a_{02} X_0 X_2 = 0.$$

Show that this system of conics represents the plane sections of the cubic surface which has the parametric equations:

$$Y_0 : Y_1 : Y_2 : Y_3 = \lambda_1{}^2 : \lambda_2{}^2 : 2\lambda_0\lambda_1 : 2\lambda_0\lambda_2,$$

whose equation may also be written in the form

$$Y_0 Y_3{}^2 - Y_1 Y_2{}^2 = 0.$$

Show that the point $(1, 0, 0)$ of the plane represents the simple directrix of the surface, through the points of which *one* generator of the surface passes, and show also that the line joining $(0, 1, 0)$ to $(0, 0, 1)$ represents the double directrix, through the points of which *two* generators of the surface pass.

Which pairs of points in the plane impose only one condition on conics of the system which pass through them, and thus represent a single point on the double directrix? Which curves in the plane represent the lines on the cubic surface?

93.1   The intersection of two quadrics

We need to know something about the curve of intersection of two quadrics. We denote these by F_2 and $F_2{}^*$. A general plane section of a quadric is a conic, and two conics intersect in four points. Hence the intersection of F_2 and $F_2{}^*$ is a curve which is met by a plane in four points. Hence we describe the intersection as *a space quartic $\mathscr{C}^4$*.

Consider the plane representation of F_2, the stereographic projection of curves on F_2 from a point X_3 on F_2. Since $F_2{}^*$ intersects each generator through X_3 in two points, the curve of intersection $\mathscr{C}^4$ of F_2 and $F_2{}^*$ is mapped on a plane quartic curve which has nodes at X_1 and X_2.

Suppose that $F_2{}^*$ contains a plane conic section of F_2. Then the curve of intersection is a conic, together with another curve. To find this other curve, we note that the plane representation of a conic section of F_2 is a conic which passes simply through X_1 and X_2. A quartic curve which has nodes at X_1 and X_2 now reduces to a conic through X_1 and X_2, and another curve. This other curve must also be a conic through X_1 and X_2, since the composite curve has to continue to be a quartic curve with nodes at X_1 and X_2. Since the space representation of a conic through X_1 and X_2 is a conic on F_2, we have the theorem:

Theorem   *If part of the intersection of two quadrics is a conic section, the remaining part is also a conic section.*

This theorem arose in connexion with the inverse of a circle being a circle, in a more special form (§39.1).

The two conics in which two quadrics may intersect can be special conics. Suppose that $F_2{}^*$ contains a section of F_2 by a tangent plane. This section will consist of two generators of reguli of complementary systems on F_2. Hence, if $F_2{}^*$ contains two intersecting generators of F_2, its further intersection with F_2 is a plane section. This may also be a tangent plane to F_2, and so we conclude that two quadrics may intersect in the sides of a skew quadrilateral, consisting of two pairs of generators of the two reguli on each quadric.

In the plane representation of F_2, the quartic curve of intersection with $F_2{}^*$ is represented by a pair of lines through X_1, and a pair of lines through X_2, when the intersection consists of the sides of a skew quadrilateral.

93.2   The intersection of three quadrics

The intersection of three quadrics, F_2, $F_2{}^*$ and F_2', say, is represented in the plane representation of F_2 by the intersections, outside the points X_1 and X_2, of two quartic curves, each of which has nodes at X_1 and X_2. There are good reasons for asserting that curves with a node at a given point use up $2 \times 2 = 4$ points of their points of intersection at this point. Hence the number of intersections of the two quartics outside the points X_1 and X_2 is

$$4 \times 4 - 2 \times 2 \times 2 = 8.$$

We can assert that three quadrics in general position intersect in eight points.

These points have a remarkable property, in that *quadrics which are made to contain seven of them will automatically also contain the eighth point.*

To investigate this important phenomenon, let us consider the equation of a quadric surface in S_3, $\Sigma a_{ij} X_i X_j = 0$ $(i, j = 0, 1, 2, 3)$. This contains ten terms. If we wish the quadric to pass through a given point Y' of S_3, this imposes one linear condition on the coefficients a_{ij}, of the equation. Hence, if we wish to make the quadric pass through r points, we expect the equation of the resulting quadric surface to depend linearly on $10 - r$ homogeneous parameters, *provided that the r prescribed conditions are linearly independent.*

This may not be the case. For example, if we make the quadric contain three collinear points, it will automatically contain the line through the three points, and so automatically pass through any other point on the line. Hence, making the quadric pass through four collinear points imposes only three independent conditions on the coefficients, not four.

We illustrate this by working with a conic, imposing the conditions that the conic $\Sigma a_{ij} X_i X_j = 0$, $(i, j = 0, 1, 2)$ pass through the collinear points $(0, 0, 1)$, $(0, 1, 0)$ and $(0, 1, 1)$. The first condition leads to $a_{22} = 0$, the second condition to $a_{11} = 0$, so that at this stage the equation is:

$$a_{00} X_0{}^2 + 2a_{01} X_0 X_1 + 2a_{02} X_0 X_2 + 2a_{12} X_1 X_2 = 0,$$

involving four homogeneous coefficients instead of six. If we now make this conic pass through the point $(0, 1, 1)$, we find that $a_{12} = 0$, and the equation may be written:

$$X_0(a_{00} X_0 + 2a_{01} X_1 + 2a_{02} X_2) = 0.$$

This conic contains the line $X_0 = 0$ as a component, and now the equation contains three independent parameters. The condition that the conic contain a fourth point $(0, q, r)$ on $X_0 = 0$ is automatically satisfied at this stage, and does not diminish the number of independent homogeneous parameters on which the equation of the conic depends. If we had written the linear conditions for the conic to contain the four collinear points, in terms of the coefficients a_{ij}, these conditions would have been linearly dependent. In fact, they are

$$a_{22} = 0,\ a_{11} = 0,\ a_{12} = 0,\ \text{and}\ a_{11}q^2 + a_{22}r^2 + 2a_{12}qr = 0.$$

We may not arbitrarily assume that the linear conditions imposed by making curves or surfaces pass through a given set of points are linearly independent. Every case should be examined. In particular, when the points lie on some curve or surface, they are already algebraically connected, and not *general* points of the space they lie in.

To take another example, if we can find nine points in S_3 which furnish independent conditions to a general quadric surface, then there is a unique quadric which passes through them. We saw that such a set of points is furnished by taking them in sets of three, on three mutually skew lines in S_3 (§83.1). But we cannot find nine independent points, in this sense, on the curve of intersection of two quadric surfaces $F_2 = 0$, $F_2{}^* = 0$. For the pencil of quadrics $F_2 + \lambda F_2{}^* = 0$ contains the curve of intersection given by $F_2 = 0 = F_2{}^*$. If the nine points on the curve of intersection furnished independent conditions, there would be a *unique* quadric through the curve, not an *infinity* of quadrics.

However, we must be able to find sets of *eight* points which impose independent conditions. For if there were only sets of *seven* points which impose independent conditions, on the curve of intersection of $F_2 = 0$ and $F_2{}^* = 0$, this would mean that all quadrics through the seven points would necessarily contain any other point of the curve, and therefore would contain the whole curve. Remembering that a quadric surface can be found to satisfy *nine* independent conditions, we should then be able to find a quadric through *two* arbitrary points of S_3, and also through the curve of intersection of the two quadrics $F_2 = 0$ and $F_2{}^* = 0$. This leads to a contradiction. A plane through the two points would cut the quadric in a conic through the two points, and this conic would also contain the four points of intersection of the plane with the curve $F_2 = 0 = F_2{}^*$. Expressed differently, it would be possible to find a conic which passes through four *given* coplanar points and two *arbitrary* points in the plane. We can ensure that no three of these six points are collinear. We know that only one conic can be drawn through *five* points, no three of which are collinear, and this does not contain every other point of the plane. This contradiction proves that *there are sets of eight points on the curve of intersection of two quadric surfaces which present independent conditions to quadrics passing through them.*

93.3   Sets of eight associated points

We now consider a set of eight points on the curve of intersection of two quadrics, $F_2 = 0$ and $F_2{}^* = 0$ which impose independent conditions on quadrics passing through them. If we choose from this set of eight points a subset of seven points, these seven points must also impose independent conditions on the quadrics which pass through them. Among these quadrics are certainly the quadrics $F_2 = 0$ and $F_2{}^* = 0$, which define, by their intersection, the curve on which the seven points lie. If we now consider the linear system of quadrics which pass through the seven points, it can be written in the form:

$$\lambda_0 F_2 + \lambda_1 F_2{}^* + \lambda_2 F_2' = 0,$$

where $F_2 = 0$ and $F_2{}^* = 0$ are the quadrics already mentioned, and $F_2' = 0$ is a third quadric surface, *not contained in the pencil* $\lambda_0 F_2 + \lambda_1 F_2{}^* = 0$. This *net* of quadrics all pass through the seven points on the curve of intersection $F_2 = 0 = F_2{}^*$.

But the three quadric surfaces

$$F_2 = 0, \; F_2{}^* = 0, \; F_2' = 0$$

intersect in *eight* points, and these points necessarily lie on the curve $F_2 = 0 = F_2{}^*$, and seven of the eight are the subset we started with. *The eighth point is uniquely determined by the given set of seven.* It does not, of course, coincide with the eighth point of the set of eight points imposing independent conditions with which we began.

The set of eight points we have found, which are such that all quadrics which pass through a certain subset of seven of them also pass through the eighth, is called *an associated set of eight points*, and the very title implies a theorem, which we now prove.

Theorem *Given eight distinct points which are the set of intersections of three quadric surfaces, all quadrics through any subset of seven of the points must pass through the eighth point.*

Proof. Let the points be $P_1, P_2, \ldots, P_8$, and the quadrics $F_2 = 0$, $F_2{}^* = 0$ and $F_2' = 0$. These three surfaces are assumed to have the eight points P_i in common, and no other points. Let $G_2 = 0$ be a quadric which passes through the seven points $P_1, P_2, \ldots, P_7$. By renumbering the points we can ensure that this subset is an arbitrary subset of seven points. If the quadric $G_2 = 0$ is linearly dependent on $F_2 = 0$, $F_2{}^* = 0$ and $F_2' = 0$, so that, identically,

$$G_2 \equiv \lambda F_2 + \mu F_2{}^* + \nu F_2',$$

then $G_2 = 0$ contains the point P_8 also, since this lies on the three quadrics $F_2 = 0$, $F_2{}^* = 0$ and $F_2' = 0$.

To prove our theorem by contradiction, we therefore assume that $G_2 = 0$ is *not* linearly dependent on the three quadrics we are considering. The system of quadrics

$$\lambda F_2 + \mu F_2{}^* + \nu F_2' + \rho G_2 = 0 \tag{1}$$

can now be made to satisfy *three* linear conditions, and can therefore be made to contain any three given points of S_3. We show that this assumption leads to a contradiction.

We first note that of the eight points P_i no three can be collinear, since the three quadrics which defined the set of points would contain the line of collinear points, and therefore intersect in an infinity of points, not just eight. Again, no four of the seven points $P_1, \ldots, P_7$ can be coplanar, with the hypothesis that $G_2 = 0$ is not linearly dependent on the three quadrics whose intersection defines the eight points. For suppose that P_1, P_2, P_3 and P_4 are coplanar. We can find a quadric of the system given by equation (1) above to contain two points P, Q in the plane of the four points, where P and Q do not lie on any conic through the four points P_1, P_2, P_3 and P_4, and also to contain an arbitrary point R of S_3. This quadric must be reducible, one component being the plane through the four points P_1, P_2, P_3, P_4. For if it were not reducible, it would cut this plane in a conic through the points P_1, P_2, P_3, P_4, P and Q. The other component of the quadric must be a plane, and this plane has to contain the points P_5, P_6, P_7 and R. We have already ruled out the possibility that P_5, P_6 and P_7 are collinear. There is consequently a unique plane through these three points, and this cannot contain R, which is an arbitrarily chosen point of S_3. This contradiction shows that with our hypothesis, the points P_1, P_2, P_3 and P_4 cannot be coplanar. If they are coplanar, we have a contradiction, and our theorem is proved. Let us assume then that they are not coplanar.

We now have no four of the seven points $P_1, P_2, \ldots, P_7$ coplanar, and choose three points P, Q and R in the plane $P_1P_2P_3$ such that the six points do not lie on a conic. We then find a quadric of the system given by equation (1) above to contain P, Q and R. This quadric will contain the plane $P_1P_2P_3$, and is therefore reducible into two planes. The other plane must contain the points P_4, P_5, P_6 and P_7. But these points are not coplanar. We therefore have a contradiction.

Our initial assumption, that quadrics through the seven points $P_1, P_2, \ldots, P_7$ contain a set of *four* linearly independent quadrics, and not three, is false. We have therefore proved the theorem of the eight associated points.

Remark. We shall give some of the many applications of this theorem in §94.1, but we note in the meantime that if four of the eight points in which three quadrics intersect are coplanar, so are the other four. The set of eight points is an associated set, and if P_1, P_2, P_3 and P_4 lie in a plane π, and the plane through P_5, P_6 and P_7 is called π', the quadric which consists of the two planes π and π' passes through seven of the eight associated points, and must therefore pass through the eighth point P_8. If this point is not coplanar with P_1, P_2, P_3 and P_4, it must lie in the plane π'.

The theorem that, of a set of eight associated points five cannot lie in a plane, is left as an Exercise (Exercise 93.1).

93.4   A theorem for plane cubic curves

We show that the theorem we have just proved on eight associated points in S_3 is equivalent to the following plane theorem:

Theorem   *If two plane cubic curves intersect in nine distinct points, then every cubic curve which contains eight of these points automatically contains the ninth point.*

Proof. Let $F_2 = 0$ be a quadric surface, X_3 a point on it, X_3X_1 and X_3X_2 the two generators through the point X_3, where the plane $X_0X_1X_2$ intersects the generators in X_1 and X_2, so that we have the situation we discussed in the section on stereographic projection (§87.1). Let C^3 be a cubic curve in the plane $X_0X_1X_2$ which passes simply through the points X_1 and X_2. *The projection of this curve onto the quadric gives a quartic curve which passes simply through the point X_3, and is* (2, 2) *on the quadric, so that it meets X_3X_1 in one further point besides X_3, and X_3X_2 in one further point besides X_3.*

To prove this, we remark that the cubic cone which arises when we join X_3 to the plane cubic curve intersects the quadric in a curve of order $3 \times 2 = 6$, and part of the intersection is a curve of order two, namely the generators X_3X_1 and X_3X_2. The residual intersection is therefore a curve intersected by a plane in four points, that is, a space quartic curve. The projection of this curve from X_3 back onto the plane from X_3 must give the plane cubic curve. Therefore the space quartic curve intersects a plane which passes through X_3 in only three points distinct from X_3. The space quartic therefore passes simply through the point X_3. As a quartic curve on the quadric surface, it is either (2, 2), or (3, 1). If we draw a variable plane through the generator X_3X_1, it intersects the plane $X_0X_1X_2$ in a variable line through the point X_1. This line intersects the plane cubic curve in two variable points, and these points correspond to the points in which the further intersection, besides X_3X_1, of the plane through X_3X_1 with the quadric intersects the space quartic. But this further intersection is a generator of the same system as X_3X_2. Hence generators of this system meet the space quartic in two points, and so generators of the complementary system also meet the space quartic in two points.

The converse is more evident. If the quadric surface F_2^* passes through X_3, the quartic curve of intersection of $F_2 = 0 = F_2^*$ is a (2, 2) curve on $F_2 = 0$, and projects from X_3 into a cubic, in the plane $X_0X_1X_2$, which passes simply through the points X_1 and X_2.

If we now intersect $F_2 = 0$ with two quadric surfaces which pass simply through the point X_3, and call the surfaces $F_2^* = 0$ and $F_2' = 0$, the space quartic curves given by $F_2 = F_2^* = 0$ and $F_2 = F_2' = 0$ intersect in a set of eight associated points which lie on $F_2 = 0$, one of these points being the point X_3. The projection of this set of eight points from X_3 gives a set of seven points in the plane $X_0X_1X_2$, and the projection of the quartic curves $F_2 = F_2^* = 0$ and $F_2 = F_2' = 0$ from the point X_3 gives two cubic curves in the plane $X_0X_1X_2$ through these seven points and the two points X_1 and X_2. Since any quadric through X_3 and six of the set of the remaining seven of the eight associated points automatically passes through the last point of the set, this is true for (2, 2) curves on the quadric $F_2 = 0$ which pass through X_3, and since these project into cubic curves through X_1, X_2, we can say that the cubic curves in the plane $X_0X_1X_2$ which arise in this way are such that any one of them which passes through eight of the nine points of intersection of two of them, two of these points being X_1 and X_2, automatically passes through the ninth point.

Remark. It will be noted that we have not proved that *all* cubic curves in the plane $X_0X_1X_2$ through X_1 and X_2 arise as projections from the curves of intersection of $F_2 = 0$ and another quadric surface. We have proved that such a cubic curve projects from a (2, 2) curve on the quadric which passes through X_3, but we have not proved that such a curve can always be cut out by a quadric surface. We shall not try to do this. Our aim is to show the connection between a theorem in S_3 and in S_2. A satisfactory proof of the theorem on plane cubics is indicated in Exercise 93.4.

Exercises

93.1 Show that if three quadric surfaces pass through five given coplanar points, no three of which are collinear, they necessarily intersect in the unique conic which passes through the five points, and possibly other points.

93.2 If three quadric surfaces intersect in the points of a given conic, show that they also have in common two points which do not lie on the conic.

93.3 If three quadric surfaces intersect in the points of a given line, show that they also have in common four points which do not lie on the line.

93.4 Prove the theorem for plane cubic curves of §93.4, using a method similar to that used in the proof of the theorem on associated points in §93.3. (Suppose that the cubic curves $C = 0$ and $C^* = 0$ intersect in the nine distinct points $P_1, P_2, \ldots, P_9$, and that there are three linearly independent cubic curves through $P_1, P_2, \ldots, P_8$, the curves $C = 0$, $C^* = 0$ and $C' = 0$. Show that this leads to contradiction.)

93.5 Show that the theorem on plane cubic curves of §93.4 leads to a proof of Pascal's Theorem, which we state in the form: if $l_1, l_2, l_3, l_4, l_5, l_6$ are the successive sides of a hexagon inscribed in a conic S, then the three points $l_1 \cap l_4$, $l_2 \cap l_5$ and $l_3 \cap l_6$ of intersection of opposite sides are collinear. (Thm III, Corollary, §78.2.) (Consider the nine points of intersection of the two (reducible) cubic curves $l_1l_3l_5 = 0$, $l_2l_4l_6 = 0$. Let L be the line joining two of the points mentioned in the theorem, and consider the cubic curve $SL = 0$.)

93.6 If $P_1, P_2, \ldots, P_5$ are points in S_2 which are such that when $1 < k \leqslant 5$ we can find *a* conic which passes through $P_1, P_2, \ldots, P_{k-1}$, but not through P_k, then the five points offer independent conditions to conics which pass through them. Show that as long as no four of the points are collinear, there is a unique conic through them. (The trial conics can be line-pairs.)

94.1   Applications of the associated points theorem

Our first application is to the proof of the existence of Moebius tetrads. It will be recalled that Moebius showed that it is always possible to find tetrahedra which are both inscribed and escribed to each other (§65.2). We prove again

Theorem I *If two tetrahedra $ABCD$ and $A'B'C'D'$ are such that $A' \in BCD$, $B' \in CDA$, $C' \in DAB$ and $D' \in ABC$, and also $A \in B'C'D'$, $B \in C'D'A'$, $C \in D'A'B'$ then necessarily $D \in A'B'C'$.*

Proof. We observe that a plane contains the points A', B, C, D and another the points A, B', C', D'. Again, a plane contains the points B', C, D, A and another the points

B, C', D', A'. Finally, a plane contains the points C', D, A, B and another the points C, D', A', B'.

Hence, the points A, B, C, D, A', B', C' and D' are the complete intersection of three (reducible) quadric surfaces. They form a set of associated points. The quadric surface which consists of the planes ABC, $A'B'C'$ contains the points A, B, C, A', B', C' and D'. It must therefore also contain the eighth associated point D, and this must lie in the plane $A'B'C'$.

We now prove a theorem which deals with the intersections of a given quadric surface F_2 in S_3 and the four sides of a skew quadrilateral.

Theorem II *Let A, B, C and D be four distinct points in S_3, and F_2 a given quadric surface. Suppose that AB cuts F_2 in the points P_1, Q_1, that BC cuts F_2 in the points P_2, Q_2, that CD cuts F_2 in the points P_3, Q_3, and finally that DA cuts F_2 in the points P_4, Q_4. Then if the four points P_1, P_2, P_3, P_4 are coplanar, so are the four points Q_1, Q_2, Q_3 and Q_4.*

Proof. The planes ABC, ACD contain the eight points P_i, Q_i ($i = 1, 2, 3, 4$) and so do the planes ABD, BCD. These eight points therefore lie on F_2 and on two (reducible) quadrics. These three quadrics are distinct, and since $(ABC) \cap (ABD) = AB$, $(ABC) \cap (BCD) = BC$, $(ACD) \cap (ABD) = AD$, and $(ACD) \cap (BCD) = CD$, the eight points P_i, Q_i are the only intersections of the three quadrics, and the eight points are therefore associated points. If the four points P_i lie in a plane π, and we consider the plane $\pi' = Q_1Q_2Q_3$, the quadric $\pi\pi'$ must contain Q_4 also, and so Q_4 lies in the plane $Q_1Q_2Q_3$. If Q_4 were in the plane π, the eight points P_i, Q_i would be coplanar, and the theorem trivial.

We shall use this theorem to obtain a theorem on the intersections of four circles, and its dual theorem to obtain a theorem on *oriented* circles. We state the dual theorem as a Corollary.

Corollary. If a, b, c and d are four distinct planes in S_3, and if the tangent planes to a given quadric through $a \cap b$ are p_1, q_1, through $b \cap c$ are p_2, q_2, through $c \cap d$ are p_3, q_3 and finally through $d \cap a$ are p_4, q_4, then if the four planes p_i all pass through a point, so do the four planes q_i.

We now use Thm II to obtain a theorem called the *Six-Circles Theorem*.

Theorem III *If $\mathscr{C}_1$, $\mathscr{C}_2$, $\mathscr{C}_3$, and $\mathscr{C}_4$ are four circles in the real Euclidean plane, and if the intersections of $\mathscr{C}_1$ and $\mathscr{C}_2$ are P_1 and Q_1, the intersections of $\mathscr{C}_2$ and $\mathscr{C}_3$ are P_2 and Q_2, the intersections of $\mathscr{C}_3$ and $\mathscr{C}_4$ are P_3 and Q_3, and finally, the intersections of $\mathscr{C}_4$ and $\mathscr{C}_1$ are P_4 and Q_4, then if the four points P_i lie on a circle, the four points Q_i lie on a circle (Fig. 94.1).*

Proof. We use our representation of circles in E_2 by points in E_3 (Chapter IV). The circles $\mathscr{C}_i$ are represented by the points A, B, C and D respectively, but the lines AB, BC, CD and DA do *not* cut the quadric Ω which represents the zero (point)-circles in E_2. If we find the *polar lines* of AB, BC, CD and DA, the intersections with Ω will give us what we require. More specifically, the polar line of AB represents the

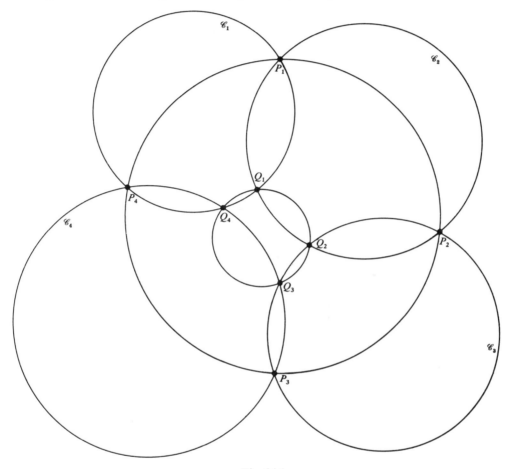

Fig. 94.1

coaxal system in E_2 which consists of circles orthogonal to all the circles of the coaxal system defined by the circles $\mathscr{C}_1$ and $\mathscr{C}_2$. This system contains, as point-circles, the points of intersection, P_1 and Q_1 of $\mathscr{C}_1$ and $\mathscr{C}_2$. Hence, the polar line of AB intersects Ω in the points which are the representatives of point-circles situated at P_1 and Q_1. These points are *vertically above* P_1 and Q_1.

Now, if we find the polar lines of AB, BC, CD and DA respectively, we find four points A', B', C' and D' such that $A'B'$ is the polar line of AB, $B'C'$ the polar line of

BC, *C'D'* the polar line of *CD* and *D'A'* the polar line of *DA*. The reason for this is that the polar lines of, say, *AB* and *DA*, lie in the polar plane with regard to Ω of the point *A*, and therefore meet in the point *A'*, say. The lines *A'B'*, *B'C'*, *C'D'* and *D'A'* intersect Ω in the points which represent the points P_1, Q_1; P_2, Q_2; P_3, Q_3 and P_4, Q_4. We can now apply Thm II above, and say that if the representative points of the P_i on Ω lie in a plane, so do the representative points of the Q_i. But the points on Ω representing the P_i lie in a plane if and only if the actual points P_i in E_2 lie on a circle (see the end of §37.1). Similarly for the points Q_i. We therefore obtain the Six Circles Theorem.

Remark. Applications of the Six-Circles Theorem are usually made to the inversive plane, and then we use the fact that all lines pass through the special point ∞, the inverse of the center of a circle of inversion. Thus if two of our circles are lines, their intersections are two points, one being the finite point of intersection of the lines, and the other being the point ∞. An alternative proof of the theorem, using the cross-ratio of complex numbers, has already appeared, in Exercise 53.9. In the Exercises which follow another proof by inversion is also suggested. The Six-Circles Theorem leads to some interesting *chains of theorems*, and two such are indicated in the Exercises.

The dual of Thm II, stated as a Corollary, leads to an interesting theorem on *oriented circles*, sometimes called *cycles*, and we discuss this in the next Section. We state the dual theorem here again, as:

Theorem IV *If a, b, c, d are four distinct planes in S_3, and if the tangent planes to a given quadric which pass through the line $a \cap b$ are p_1, q_1, the tangent planes to the quadric which pass through the line $b \cap c$ are p_2, q_2, the tangent planes to the quadric which pass through the line $c \cap d$ are p_3, q_3, and finally, the tangent planes to the quadric which pass through the line $d \cap a$ are p_4, q_4, then if the four planes p_i pass through one point, so do the four planes q_i.*

94.2 Oriented circles

We have already discussed one method of representing circles by points in space of three dimensions (Chapter IV). A method which appears, at first sight, to be simpler, is to erect a perpendicular, at the center of a given circle, to the plane of the circle, and to mark off on this perpendicular, from the center, a length equal to the radius of the circle. The endpoint of this segment is to represent the circle (see Fig. 94.2).

This representation is not as useful as the one we have given for the representation of coaxal systems of circles which were represented by lines in our representation. But the present method lends itself to the representation of *oriented* circles, as we shall see.

An *oriented* circle is simply a circle which is traversed in a given sense, so that it can be suggested in a figure by a circle with an arrow to show the sense. Any given circle is the matrix for two oriented circles, the directions of the arrows being different.

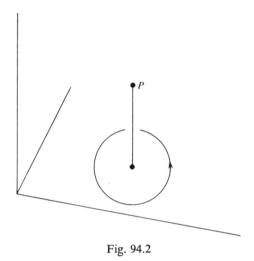

Fig. 94.2

We regard a line as being a special kind of circle, and can therefore talk of *oriented lines* also. An oriented line is said to *touch* an oriented circle if and only if the unoriented line touches the unoriented circle, and the orientations of line and circle are the same at the point of contact.

It follows immediately that two given oriented circles, external to each other, have

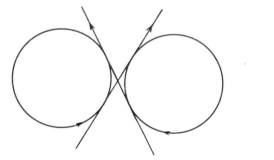 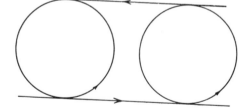

Fig. 94.3

only *two* oriented common tangents. The two possibilities are shown in Fig. 94.3.

Returning to the representation of oriented circles in the plane E_2 by points in E_3, we represent the circle

$$X^2 + Y^2 - 2pX - 2qY + r = 0$$

by the point $(p, q, \sqrt{(p^2 + q^2 - r)})$ in E_3 if the circle is regarded as being traversed in an anti-clockwise sense, and we represent the circle by the point $(p, q, -\sqrt{(p^2 + q^2 - r)})$ if the circle is regarded as being traversed in a clockwise sense. Let P be the point which represents a given circle $\mathscr{C}$. Then P is the vertex of a right circular cone, of semi-vertical angle 45 degrees, formed by joining P to the points of the circle $\mathscr{C}$. This cone intersects the plane at infinity in the points of a conic C_∞, which is the same for every circle $\mathscr{C}$ in E_2. This is clear geometrically, and easily proved algebraically by finding the equation of the cone (see Exercise 94.8).

If circles $\mathscr{C}$ and $\mathscr{D}$ are represented by points P and Q respectively (Fig. 94.4), we note that there are just two tangent planes through the line PQ to the right circular

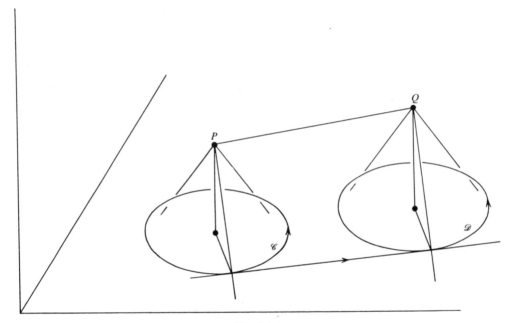

Fig. 94.4

cones with vertices at P and Q, and these tangent planes intersect the plane E_2 in the two tangents common to $\mathscr{C}$ and $\mathscr{D}$, assumed to be external to each other. These two tangent planes through the line PQ touch the conic at infinity, C_∞, and can be obtained by drawing the two tangent planes to C_∞ which pass through PQ.

We remark that the conic C_∞ in the plane at infinity may be regarded as a degenerate quadric (see §85.6), and two tangent planes to C_∞ pass through any given line, as for any quadric. Of course these are obtained by finding where the line intersects the plane at infinity, and drawing the two tangents to C_∞ from the point of intersection.

Returning to the two circles $\mathscr{C}$ and $\mathscr{D}$ represented by points P and Q, we also see that *all circles in E_2 represented by points on the line PQ have the same two common tangents.*

We leave as an Exercise (Exercise 94.9) the theorem that *the points of a given plane in E_3 represent circles which cut the line of intersection of the plane and E_2 at a fixed angle.*

If the plane touches C_∞, it represents all circles which touch the line of intersection of the plane and E_2.

We regain the theorem that *circles represented by the points on a line PQ, where P and Q represent circles $\mathscr{C}$ and $\mathscr{D}$ respectively, have the same two common tangents, these being the intersections with E_2 of the two planes through the line PQ which touch C_∞.*

We now consider the application of Thm IV of the preceding Section to the representation of oriented circles, the quadric of the theorem being C_∞. Since three planes in E_3 meet in a point, the theorem states that if A, B, C, D are four points in E_3, and the tangent planes to C_∞ through AB are p_1, q_1, the tangent planes to C_∞ through BC are

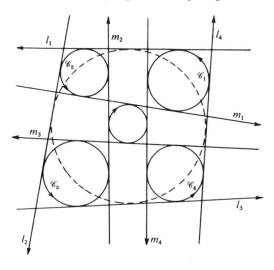

Fig. 94.5

p_2, q_2, the tangent planes to C_∞ through CD are p_3, q_3, and finally, the tangent planes to C_∞ through DA are p_4, q_4, then if the planes p_i are concurrent, so are the planes q_i.

The four planes p_i passing through a point, since they all touch C_∞, give us a circle, represented by the point, which touches four given lines, given by the intersections of the planes p_i with the plane E_2. The theorem for oriented circles which we derive is therefore the following:

Theorem I $\mathscr{C}_1$, $\mathscr{C}_2$, $\mathscr{C}_3$, $\mathscr{C}_4$ *are four oriented circles in the plane E_2, and the tangents to $\mathscr{C}_1$ and $\mathscr{C}_2$ are l_1, m_1, the tangents to $\mathscr{C}_2$ and $\mathscr{C}_3$ are l_2, m_2, the tangents to $\mathscr{C}_3$ and $\mathscr{C}_4$ are l_3, m_3 and finally, the tangents to $\mathscr{C}_4$ and $\mathscr{C}_1$ are l_4, m_4. Then if there is an oriented circle which touches the lines l_i, there is also an oriented circle which touches the lines m_i. (See Fig. 94.5.)*

Remark. The theorem is not necessarily true for circles which are not oriented. In

Fig. 94.6 there is a circle which touches the sides of the square, but not one which touches the sides of the inner rectangle.

Exercises

94.1 Look at Exercise 53.9 again, where another proof of the Six-Circles Theorem is suggested, using the theorem that the cross-ratio of four complex numbers is real if and only if the four points representing the numbers lie on a circle or a line.

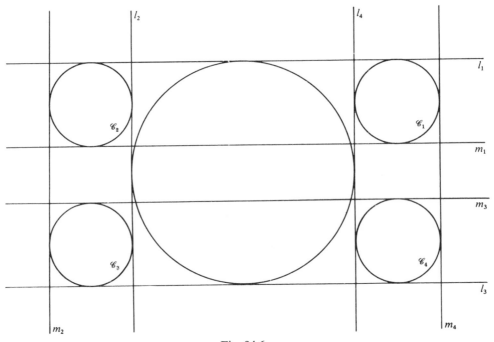

Fig. 94.6

94.2 The Pivot Theorem states that if C' is a point on AB, B' a point on AC, and A' a point on BC, and the circles $BC'A'$, $CB'A'$ meet again in P, then there is a circle through the points A, C', P and B'. Prove this theorem, and then invert it with regard to any point O as center, showing that it becomes the Six-Circles Theorem. (In this case the four points P_i are already concyclic.)

94.3 Show that the Pivot Theorem described in the preceding Exercise can itself be derived, as a theorem in the inversive plane, from the Six-Circles Theorem. (We use the fact that all lines in the inversive plane pass through the point ∞. There is a circle through the points A, C', P and B' because the points B, A', C are collinear, and therefore the points B, A', C, ∞ are on a line. The four circles whose intersections give the points A, C', P, B', and B, A', C, ∞ are the two circles $BC'A'$, $CB'A'$ and two sides of the triangle ABC.)

94.4 Verify that your proof of the Pivot Theorem (Exercise 94.2) still holds if the points A', B' and C' are collinear.

94.5 The circumcircle of a triangle, is, as we know, the circle which passes through the vertices of the triangle formed by three lines which are not concurrent. Suppose that we take four lines in a plane, which form four triangles, if we take them in sets of three. Show, using the Pivot Theorem, that the four circumcircles of the four triangles pass through a point. (Let the fourth line intersect the sides of triangle ABC in the points A', B' and C' respectively. The Pivot Theorem shows that circles $BC'A'$, $CA'B'$ and $AB'C'$ pass through a point P. Now consider $B'A'C$ as the given triangle, with points C' on $B'A'$, B on $A'C$ and A on CB', and apply the Pivot Theorem again.)

94.6 If we take five lines in a plane, no three of which are concurrent, every four of them define a point, as shown in the preceding Exercise. Show that the five points obtained by taking the lines in sets of four all lie on a circle. (It suffices to show that any four of the points are concyclic. Use the Six-Circles Theorem.) Can you generalize the results of the preceding Exercise and this one?

94.7 We take two intersecting lines, and on each line we choose a point distinct from the point of intersection. Let us call the circle through the two points and the point of intersection of the two lines the *pivot circle* of the two lines and the two given points. If we now take three lines which form a triangle, and points on the sides, the Pivot Theorem (Exercise 94.2) states that the three pivot circles pass through a point, which we call *the pivot point* of the given triangle and the set of three points. Now take four lines in the plane, no three of which are concurrent, and on each of them choose a point, the four points chosen lying on a circle or a line. Each triad of lines which can be selected from the four lines determines a pivot point. Show that the four pivot points obtained from the four lines all lie on a circle. (This leads to another chain of theorems.)

94.8 In E_3 show that the equation of the cone which joins the point (p, q, R) to the points of the circle $X^2 + Y^2 - 2pX - 2qY + r = 0 = Z$, where $R^2 = p^2 + q^2 - r$, is $(X - p)^2 + (Y - q)^2 - (Z - R)^2 = 0$, so that the intersection of this cone with the plane at infinity is $X^2 + Y^2 - Z^2 = 0$, the same conic for all circles.

94.9 If $aX + bY + d = 0 = Z$ is the equation in E_2 of the line of intersection of a given plane with E_2, which is taken as $Z = 0$, show that the equation of the plane itself is of the form $aX + bY + d = kZ$, for some value of k. Since the length of the perpendicular from the point $(p, q, 0)$ onto the line of intersection is $(ap + bq + d)/(a^2 + b^2)^{1/2}$, show that the plane whose equation we have just found represents circles which cut the line of intersection of the plane with $Z = 0$ at a fixed angle.

94.10 Show that the plane $aX + bY + d = kZ$ represents circles which touch the line in $Z = 0$ given by $aX + bY + d = 0$ if and only if the plane makes an angle of 45 degrees with the plane $Z = 0$. Show that in this case the intersection of the plane with the plane at infinity is a tangent to the conic $X^2 + Y^2 - Z^2 = 0$ in that plane.

Appendix I

EUCLID'S DEFINITIONS, POSTULATES AND FIRST THIRTY PROPOSITIONS FROM BOOK I

Definitions

1 A *point* is that which has no part.

2 A *line* is breadthless length.

3 The extremities of a line are points.

4 A *straight line* is a line which lies evenly with the points on itself.

5 A *surface* is that which has length and breadth only.

6 The *extremities* of a surface are lines.

7 A *plane surface* is a surface which lies evenly with the straight lines on itself.

8 A *plane angle* is the inclination to one another of two lines in a plane which meet one another and do not lie in a straight line;

9 And when the lines containing the angle are straight, the angle is called *rectilineal*.

10 When a straight line set up on a straight line makes the adjacent angles equal to one another, each of the equal angles is *right*, and the straight line standing on the other is called a *perpendicular* to that on which it stands.

11 An *obtuse angle* is an angle greater than a right angle.

12 An *acute angle* is an angle less than a right angle.

13 A *boundary* is that which is an extremity of anything.

14 A *figure* is that which is contained by any boundary or boundaries.

15 A *circle* is a plane figure contained by one line such that all the straight lines falling upon it from one point among those lying within the figure are equal to one another;

16 And the point is called the *centre* of the circle.

17 A *diameter* of the circle is any straight line drawn through the centre and terminated in both directions by the circumference of the circle, and such a straight line also bisects the circle.

18   A *semicircle* is the figure contained by the diameter and the circumference cut off by it. And the centre of the semicircle is the same as that of the circle.

19   *Rectilineal figures* are those which are contained by straight lines, *trilateral* figures being those contained by three, *quadrilateral* those contained by four, and *multilateral* those contained by more than four straight lines.

20   Of trilateral figures, an *equilateral triangle* is that which has its three sides equal, an *isosceles triangle* that which has two of its sides alone equal, and a *scalene triangle* that which has its three sides unequal.

21   Further, of trilateral figures, a *right-angled triangle* is that which has a right angle, an *obtuse-angled triangle* that which has an obtuse angle, and an *acute-angled triangle* that which has its three angles acute.

22   Of quadrilateral figures, *a square* is that which is both equilateral and right-angled; an *oblong* that which is right-angled but not equilateral; a *rhombus* that which is equilateral but not right-angled; and a *rhomboid* that which has its opposite sides and angles equal to one another but is neither equilateral nor right-angled; and let quadrilaterals other than these be called *trapezia*.

23   *Parallel* straight lines are straight lines which, being in the same plane and being produced indefinitely in both directions, do not meet one another in either direction.

Postulates

Let the following be postulated:

1   To draw a straight line from any point to any point.

2   To produce a finite straight line continuously in a straight line.

3   To describe a circle with any centre and distance.

4   That all right angles are equal to one another.

5   That, if a straight line falling on two straight lines make the interior angles on the same side less than two right angles, the two straight lines, if produced indefinitely, meet on that side on which are the angles less than the two right angles.

Propositions

1   On a given finite straight line to construct an equilateral triangle.

2   To place at a given point (as an extremity) a straight line equal to a given straight line.

3   Given two unequal straight lines, to cut off from the greater a straight line equal to the less.

4   If two triangles have the two sides equal to two sides respectively, and have the angles contained by the equal straight lines equal, they will also have the base equal to the base, the triangle will be equal to the triangle, and the remaining angles will be equal to the remaining angles respectively, namely those which the equal sides subtend.

5   In isosceles triangles the angles at the base are equal to one another, and, if the equal straight lines be produced further, the angles under the base will be equal to one another.

6   If in a triangle two angles be equal to one another, the sides which subtend the equal angles will also be equal to one another.

7   Given two straight lines constructed on a straight line (from its extremities) and meeting in a point, there cannot be constructed on the same straight line (from its extremities), and on the same side of it, two other straight lines meeting in another point and equal to the former two respectively, namely each to that which has the same extremity with it.

8   If two triangles have the two sides equal to two sides respectively, and have also the base equal to the base, they will also have the angles equal which are contained by the equal straight lines.

9   To bisect a given rectilineal angle.

10   To bisect a given finite straight line.

11   To draw a straight line at right angles to a given straight line from a given point on it.

12   To a given infinite straight line, from a given point which is not on it, to draw a perpendicular straight line.

13   If a straight line set up on a straight line make angles, it will make either two right angles or angles equal to two right angles.

14   If with any straight line, and at a point on it, two straight lines not lying on the same side make the adjacent angles equal to two right angles, the two straight lines will be in a straight line with one another.

15   If two straight lines cut one another, they make the vertical angles equal to one another.

16   In any triangle, if one of the sides be produced, the exterior angle is greater than either of the interior and opposite angles.

17 In any triangle two angles taken together in any manner are less than two right angles.

18 In any triangle the greater side subtends the greater angle.

19 In any triangle the greater angle is subtended by the greater side.

20 In any triangle two sides taken together in any manner are greater than the remaining one.

21 If on one of the sides of a triangle, from its extremities, there be constructed two straight lines meeting within the triangle, the straight lines so constructed will be less than the remaining two sides of the triangle, but will contain a greater angle.

22 Out of three straight lines, which are equal to three given straight lines, to construct a triangle: thus it is necessary that two of the straight lines taken together in any manner should be greater than the remaining one.

23 On a given straight line and at a point on it to construct a rectilineal angle equal to a given rectilineal angle.

24 If two triangles have the two sides equal to two sides respectively, but have the one of the angles contained by the equal straight lines greater than the other, they will also have the base greater than the base.

25 If two triangles have the two sides equal to two sides respectively, but have the base greater than the base, they will also have the one of the angles contained by the equal straight lines greater than the other.

26 If two triangles have the two angles equal to two angles respectively, and one side equal to one side, namely, either the side adjoining the equal angles, or that subtending one of the equal angles, they will also have the remaining sides equal to the remaining sides and the remaining angle to the remaining angle.

27 If a straight line falling on two straight lines make the alternate angles equal to one another, the straight lines will be parallel to one another.

28 If a straight line falling on two straight lines make the exterior angle equal to the interior and opposite angle on the same side, or the interior angles on the same side equal to two right angles, the straight lines will be parallel to one another.

29 A straight line falling on parallel straight lines makes the alternate angles equal to one another, the exterior angle equal to the interior and opposite angle, and the interior angles on the same side equal to two right angles.

30 Straight lines parallel to the same straight line are also parallel to one another.

Appendix II

THE GRASSMANN–PLUECKER COORDINATES OF LINES IN S_3

It might be thought that a convenient way to represent lines in S_3 is the following: take two fixed planes, and represent a line by the coordinates of the two points in which it meets the two planes. This representation fails, however, for all the lines which lie in each of the two planes. A representation which turns out to be one-to-one without exception is obtained by introducing redundant coordinates, six homogeneous coordinates for a line which are connected by a homogeneous quadratic relation. The coordinates arise naturally in the science of statics, when we consider forces in E_3, and their components parallel to a fixed set of orthogonal cartesian axes, and their moments (turning force) about these axes.

Let $x = (x_0, x_1, x_2, x_3)$ and $y = (y_0, y_1, y_2, y_3)$ be any two distinct points on a given line l in S_3, and consider the set of coordinates defined thus:

$$p_{ij} = x_i y_j - x_j y_i \qquad (i, j = 0, 1, 2, 3).$$

The p_{ii} are zero, but since the points x, y are distinct, not all the p_{ij}, $i \neq j$ can be zero. Of course, $p_{ij} = -p_{ji}$. We take the six numbers:

$$p_{01}, p_{02}, p_{03}, p_{12}, p_{13}, p_{23}$$

as a set of homogeneous coordinates for the line l.

If $z = \lambda_{11}x + \lambda_{12}y$, $t = \lambda_{21}x + \lambda_{22}y$ are any two distinct points on the line l, which involves the condition $\lambda_{11}\lambda_{22} - \lambda_{12}\lambda_{21} \neq 0$, a simple calculation shows that

$$z_i t_j - z_j t_i = (\lambda_{11}\lambda_{22} - \lambda_{21}\lambda_{12})p_{ij},$$

so that if the two points x, y on l are replaced by any other two distinct points, the coordinates of the line, evaluated for the two new points, are merely all multiplied by the same non-zero factor. The coordinates are therefore homogeneous coordinates for a given line.

If we consider the determinant with columns given by the vectors x, y, x, y, its value is zero, and evaluation of this 4×4 determinant, by a method called after Laplace, gives the following identity between the six coordinates of any line in S_3:

$$p_{01}p_{23} + p_{02}p_{31} + p_{03}p_{12} = 0.$$

It can be shown that given six numbers p_{ij}, not all zero, which satisfy this equation,

[436]

there is a unique line in S_3 which has the given p_{ij} as its coordinates. If we consider the six coordinates as the coordinates of a point in S_5, the fundamental relation we have just obtained shows that the point lies on a quadric. There is a one-to-one correspondence without exception between the points of this quadric and the lines of S_3. The *linear complexes* with equation

$$\Sigma a_{ik} p_{ik} = 0 \qquad (i, k = 0, 1, 2, 3)$$

which arose in §85.5, when we considered the self-polar lines (also called null lines) which arise in a null-polarity, can be considered as the lines in S_3 represented by hyperplane sections of the quadric in S_5.

There is an extensive theory, generalized to coordinates of S_p in S_n, and it is clear that the methods of the exterior calculus, encountered in §9.1, can be profitably utilized in this theory. For further details on the Pluecker coordinates of lines in S_3 we refer to Todd (21) and for the Grassmann coordinates of S_p in S_n we refer to Hodge and Pedoe (8), Vol. II.

Appendix III

THE GROUP OF CIRCULAR TRANSFORMATIONS

There are a number of ideas which it is appropriate to discuss here, since they link together concepts which have arisen in different parts of this book. In Chapter VI we discussed Moebius transformations of the inversive plane. These are given by equations of the form $z' = (az + b)/(cz + d)$, where we are in the field of complex numbers, and $ad - bc \neq 0$. There is an entity we have not mentioned yet, called *the complex line*. This is merely the set of all complex numbers, augmented by the special symbol ∞. The inversive plane is the real Euclidean plane augmented by the special symbol ∞, after we have agreed to use the points (a, b) of the plane as the representative of the complex number $z = a + ib$. Our Moebius transformations of the inversive plane are also projective transformations of the complex line onto itself. If we recall that the Fundamental Theorem of Projective Geometry for the case $n = 1$ says that a projective mapping is uniquely determined on a line when three distinct points and their distinct images under the mapping are given, we are not surprised that an exactly similar theorem holds for Moebius transformations of the inversive plane.

If, in the inversive plane, we consider the set of all inversions, including the set of all reflexions in lines, and understand that when we talk of circles we are including lines as well, it is clear that under composition of transformations these mappings generate a group, and that under mappings of this group circles always map into circles. It is therefore appropriately called *the group of circular transformations*.

Let us look at the dissection of a Moebius transformation given in §52.4. We found that it is a composition of a translation, a line reflexion, an inversion, a dilative rotation and a translation. The dilative rotation is the composition of a dilatation and a rotation. Each one of these mappings, if it is not already an inversion or line reflexion, can be regarded as the composition of inversions or line reflexions. In particular a translation can always be regarded as the composition of two line reflexions (Exercise 14.2), a rotation as the composition of two line reflexions (Exercise 14.1), a dilatation as the composition of inversions in two concentric circles (Exercise 21.10), so that if we add up the total number of inversions or line reflexions which can enter a Moebius transformation we find that it is 10. We also note that if $c = 0$ the number reduces to 6, that if $b = 0$ also, the number reduces further to 4, that if a/d is real, the number reduces to 2, and finally, if we have the identity mapping, this may be regarded as the composition of the same inversion, so that the number is still 2, or we can take the number as zero. In every case the number of inversions or line reflexions is *an even number*.

The converse of this theorem is also true. If we look at the formula for inversion in terms of complex numbers given in Exercise 57.11, which is:

$$z' = (p\bar{z} + b)/(c\bar{z} - \bar{p}) \quad (p\bar{p} + bc > 0),$$

the inversion being in the circle

$$cz\bar{z} - z\bar{p} - \bar{z}p - b = 0,$$

we note that inversion is what we called a *conjugate* Moebius transformation, a particular case being, when $c = 0$, line reflexion, and that the composition of two inversions is a Moebius transformation. Hence the composition of an even number of inversions always produces a Moebius transformation.

If we call the elements of the group of circular transformations by the name *circular transformations*, and distinguish those which are the product of an even number of inversions as *direct* circular transformations, the others being *opposite* circular transformations, we have the theorem that the direct circular transformations are the Moebius transformations, and the opposite or indirect circular transformations are the conjugate Moebius transformations.

We recall that in our discussion of the Poincaré model for a hyperbolic non-Euclidean geometry (§56.1) we considered the subgroup of the Moebius transformations which map the unit circle onto itself, and the interior onto the interior. We showed that this subgroup could be considered to be that of the *direct* isometries of the non-Euclidean plane. We found that this subgroup preserved both the measure and the orientation of angles. We know that inversion and line reflexion preserve the measure but change the orientation of angles. The composition of an *even* number of inversions preserves the measure and the orientation of angles. We know that a circle is unchanged by inversion in an orthogonal circle, and that its interior is mapped into itself. We are not surprised to learn that the subgroup which maps ω onto itself and Ω onto Ω is made up of elements each of which is the composition of an even number of inversions in circles orthogonal to ω, or lines which are diameters of ω.

If we turn to the representation of circles by means of points in E_3 (§39.1), we see that inversion in a given circle $\mathscr{C}$ is effected in E_3 by a harmonic homology. The vertex (or center) of the homology is the point V which represents the circle $\mathscr{C}$, and the plane of fixed points of the homology is the polar plane of V with respect to the fundamental quadric Ω. This kind of homology is called a *point-plane reflexion*, and is denoted by $\{V, \pi\}$, where π is the plane of fixed points. Under this harmonic homology the quadric is mapped onto itself, points of Ω lying on lines through V being interchanged. If we consider harmonic homologies in projective space (§76.2), we see that with a suitable simplex of reference the equation of the mapping we call a point-plane reflexion is $y' = Ay$, where the matrix A has zeros everywhere except on the principal diagonal, where it has three -1's and one $+1$. Since $\det A = -1$ in this case, we know that the determinant of the corresponding mapping in E_3 is also negative, and can be called *an indirect collineation* of E_3. An inversion in the OXY-plane therefore induces an indirect collineation of the quadric Ω into itself. The product of two harmonic homologies of the point-plane reflexion type is a harmonic homology in which there are two lines of fixed points (Exercise 76.8). This type is called *skew involution*, or *line reflexion in the two lines*, and also appears in §76.2. The determinant of a skew involutory mapping in E_3 is positive, and if it is the product of two homologies which arise from inversions, we have a *direct* collineation of Ω into itself. We see that a study of the geometry of the inversive plane under the group of circular transformations leads to a study of the geometry of a real quadric without real generators under the group of collineations which map the quadric onto itself. If we examine the effect of a point-plane reflexion which maps a quadric with real generators onto itself, we find that the pair of generators through any fixed point of the quadric are interchanged by the mapping, and we may change our definition of direct and indirect collineation of a quadric into itself so that it is based on such a property, the direct collineations *not* interchanging generators at a fixed point, the indirect collineations interchanging them. For all this see Veblen and Young, Vol. II, (**23**).

In §80.3 we mentioned a method of constructing non-Euclidean geometries by considering the collineations of a real projective plane which map a given irreducible conic S onto itself. The method is due to Arthur Cayley. We know that such collineations exist. A harmonic homology, with vertex at any point V not on S, and line of fixed points the polar of V with respect to S, will map S onto itself. We can define the *inside* of S as being the set of points from which real tangents *cannot* be drawn to S. A collineation of the ambient projective plane which maps S onto itself preserves such a property for points, and therefore such a transformation maps the inside of S onto itself. If we regard the conic S as being the absolute of the plane, and consider only points inside S, and portions of projective lines lying inside S, we have a hyperbolic non-Euclidean geometry. Lines which intersect on S are said to be parallel, lines meeting outside S are said to be ultra-parallel. Since a line of this geometry intersects S in two points, we see that through a point V not lying on a line l there are two parallels to l. The hyperbolic distance between two points P and Q is defined by introducing the points α, β in which the line PQ intersects S, and considering the cross-ratio $[P, Q; \alpha, \beta]$. A collineation of the plane which maps S onto itself maps the line PQ onto a line $P'Q'$, and the points α, β onto points on PQ which also lie on S. Since a projective collineation preserves cross-ratios, we have $[P, Q; \alpha, \beta] = [P', Q'; \alpha', \beta']$, and so the group of collineations is the group of isometries for the space we are considering. Of course, details have to be filled in. Besides Veblen and Young, Vol. II, (**23**), the reader may consult Artzy (**1**). The notion of orthogonal lines in the space can be based on that of *conjugate* lines with respect to the conic S. Lines are said to be orthogonal when each contains the pole of the other line with regard to S. There is a relation between the Poincaré model we investigated in detail in §56.1 and this model, and this is discussed in Artzy (**1**). Finally, if the conic S be taken as an imaginary conic in the real projective plane, and we consider the *polarity* defined by such a conic, we can construct an elliptic non-Euclidean geometry, in which there are no parallels through a point V to a given line l.

Similar notions to those mentioned above apply if we wish to construct a three-dimensional hyperbolic geometry. We can use the inside of a real quadric which has no real generators as our space, the quadric being taken as the absolute. In some treatments the circles in a plane are mapped onto the planes of a hyperbolic space of three dimensions. The reader who has followed our treatment of the representation of circles both by points and planes in E_3 will have no difficulty in understanding this development, and may even be encouraged to work it out for himself. But at this point we must stop, and wish the reader a happy journey in his further explorations of geometry.

BIBLIOGRAPHY AND REFERENCES

1. Artzy, R. *Linear Geometry* (Addison Wesley, New York, 1965).
2. Baker, H. F. *Principles of Geometry*, Vols. 1, 2, 3, 4 (Cambridge, 1925).
3. Coxeter, H. S. M. *Introduction to Geometry* (John Wiley, New York, 1961).
4.   *Non-Euclidean Geometry* (Toronto, 1942).
5. Eves, H. *A Survey of Geometry* (Allyn and Bacon, London, 1965).
6. Forder, H. G. *Calculus of Extension* (Chelsea, New York, 1960).
7. Hilbert, D., and Cohn-Vossen, S. *Geometry and the Imagination* (Chelsea, New York, 1952.)
8. Hodge, W. V. D., and Pedoe, D. *Methods of Algebraic Geometry*, Vol. I, Vol. II, (Cambridge, Paperback edition, 1968).
9. Klein, F. *Hoehere Geometrie* (Chelsea, New York, 1957).
10. Pedoe, D. *Circles: A Mathematical View* (Dover, New York, 1957, 1979).
11.   *A Geometric Introduction to Linear Algebra* (Chelsea, New York, 1978).
12.   *Introduction to Projective Geometry* (Pergamon, Oxford, 1963).
13.   *How Many Pascal Lines has a Sixpoint?* Math. Gazette, **25,** No. 264, May, 1941, p. 110–111.
14.   *A Geometric Proof of the Equivalence of Fermat's Principle and Snell's Law. Am. Math. Monthly*, **5,** May, 1964, p. 543–544.
15.   *On a Theorem in Geometry. Am. Math. Monthly*, **74,** June, 1967, p. 627–640.
16.   *On (what should be) a well-known Theorem in Geometry. Am. Math. Monthly*, **74,** August, 1967, p. 839–841.
17.   *On a Geometrical Theorem in Exterior Algebra. Canadian Jnl of Math.*, **19,** 1967, p. 1187–1191.
18.   *The Gentle Art of Mathematics* (Dover, New York, 1973).
19. Schwerdtfeger, H. *The Geometry of Complex Numbers* (Dover, New York, 1979).
20. Sommerville, D. M. Y. *The Elements of non-Euclidean Geometry* (Bell, London, 1914).
21. Todd, J. A. *Projective and Analytical Geometry* (Pitman, London, 1947).
22. Van der Waerden, B. L. *Einfuehrung in die Algebraische Geometrie* (Julius Springer, Berlin, 1939).
23. Veblen, O. and Young, J. W. *Projective Geometry*, Vols I, II (Ginn, New York, 1938).
24. Walker, R. J. *Algebraic Curves* (Princeton, 1950).

INDEX

(The numbers refer to the pages, except those in parentheses, which are Exercise numbers)